FUNDAMENTAL INTERACTIONS AND TWISTOR-LIKE METHODS

Related Titles from AIP Conference Proceedings

FUNDAMENTAL INTERACTIONS AND TWISTOR-LIKE METHODS

XIX Max Born Symposium

Wroclaw, Poland 28 September – 1 October 2004

EDITORS
Jerzy Lukierski
University of Wroclaw
Wroclaw, Poland
Dmitri Sorokin
INFN, Sezione di Padova
Padova, Italy

SPONSORING ORGANIZATIONS
University of Wroclaw, Poland
University of Padua, Italy
Polish Academy of Sciences, Committee of Physics
Bogolyubov-Infeld Foundation, Russia

AMERICAN INSTITUTE OF PHYSICS

Melville, New York, 2005

AIP CONFERENCE PROCEEDINGS ■ VOLUME 767

Editors:

Jerzy Lukierski
Institute of Theoretical Physics
University of Wroclaw
Pl. Maxa Borna 9
50-204 Wroclaw
POLAND
E-mail: lukier@ift.uni.wroc.pl

Dmitri Sorokin
INFN, Sezione di Padova
& Dipartimento di Fisica "Galileo Galilei"
via F. Marzolo 8
35131 Padova
ITALY
E-mail: sorokin@pd.infn.it

Technical Editor:

Anna Jadczyk, M.Sc.
Institute of Theoretical Physics
University of Wroclaw
Pl. Maxa Borna 9
50-204 Wroclaw
POLAND
E-mail: poff@ift.uni.wroc.pl

L.C. Catalog Card No. 2005924403
ISBN 0-7354-0252-3
ISSN 0094-243X
Printed in the United States of America

CONTENTS

PREFACE

Max Born Symposia started in 1991 as the continuation of series of Leipzig-Wroclaw seminars in theoretical physics, which were held in seventies and eighties. Since 1991 we have already held eighteen of these scientific meetings in different fields of theoretical physics. The XIX-th Max Born Symposium is devoted to the application of methods based on the notion of twistors and supertwistors to problems of the theory of fundamental interactions. During the Symposium we were given talks dealing directly with twistorial geo- metric construction (Bette, Cederwall, Fedoruk, Hughston, Odzijewicz), ones strongly linked to twistorial methods and supersymmetric techniques (Bandos, Bars, de Azcarraga, Perjes, Sorokin, Zheltukhin) as well as those on other aspects of fundamental interactions theory (Bergshoeff, Ivanov, Gunayadin, Toppan). This volume contains the last, post mortem, contribution of Zoltan Perjes. As we recently learned from our colleagues in Budapest, Zoltan died last October after a long fight with acute diabetics.

This Symposium was jointly organized by the Institute for Theoretical Physics of Wroclaw University and by the Department of Physics, University of Padova. It was financially supported by several sources, by Wroclaw and Padova University, Polish Ministry of Education and Sciences as well as by Bogoliubov-Infeld Foundation. The last source provided grants for the lecturers from the former Soviet Union. The members of the Organizing Committee A. Borowiec, A. Frydryszak, M. Mozrzymas and P. Pasti were of a great help to us in preparation of this meeting. We are also grateful to Anna Jadczyk for having taken care of the Symposium Web page and for having prepared the Proceedings in the camera-ready form.

Finally the Organizers would like to thank the American Institute of Physics for publishing this volume of Proceedings.

Jerzy Lukierski
Dmitri Sorokin

TWISTORS—GEOMETRICAL FORMALISM AND TWISTORAL DYNAMICS

Twistors and 2T–physics

I. Bars

Department of Physics and Astronomy
University of Southern California, Los Angeles, CA 90089-0484

Abstract. Two-Time physics applies broadly to the formulation of physics and correctly describes the physical world as we know it. Recently it was applied to a 2T re-formulation of the $d = 4$ twistor superstring, which was suggested by Witten as an efficient approach for computations of physical processes in the maximally supersymmetric $N = 4$ Yang-Mills field theory in four dimensions. The 2T formalism provides a six dimensional view of this theory and suggests the existence of other $d = 4$ dual forms of the same theory. Furthermore the 2T approach led to the first formulation of a twistor superstring in $d = 10$ appropriate for $AdS_5 \times S^5$ backgrounds, and a twistor superstring in $d = 6$ related to the little understood superconformal theory in $d = 6$. The proper generalization of twistors to higher dimensions is an essential ingredient which is provided naturally by 2T-physics. These developments are summarized in this lecture.

1. INTRODUCTION

Two-Time Physics (2T-physics) is a natural framework in higher spacetime with 2T signature that encodes and unifies many aspects of physics, from simple quantum mechanics to strings [1]. The 2T-physics formalism is free from any problems with unitarity or causality, thanks to appropriate gauge symmetries. One-time Physics (1T-physics) is correctly embedded in the 2T-physics framework.

The 1T interpretation of a 2T-physics system depends on the *perspective* of embedding phase space in $(d-1,1)$ dimensions into phase space in $(d,2)$ dimensions. This is done by making a gauge choice, which yields a holographic image of the physical subspace of the $(d,2)$ phase space. Consequently, the same system in $(d,2)$ is viewed as various holographic dynamical images in various $(d-1,1)$ embeddings. From the point of view of 2T-physics, many aspects of 1T-physics, such as the Hamiltonian with interactions, *time and space, are all emergent concepts* that depend on the embedding. In particular *twistors* provide a particular holographic image of the system in $(d,2)$. The 2T-physics formalism leads to the proper generalization of twistors to any dimension [2] as will be outlined in this lecture and presented in more detail in [3].

The $d = 4$ twistor superstring developed by Witten [4, 5] and Berkovits [6, 7, 8] coincide with a 2T superstring [9] in $(4,2)$ dimensions with $SU(2,2|4)$ supersymmetry, when the 2T superstring is discussed from the perspective of twistors. In addition, the 2T superstring approach in $(6,2)$ dimensions with $OSp(8|2)$ supersymmetry, and $(10,2)$ dimensions with $SU(2,2|4)$ supersymmetry, yielded new twistor superstrings that were conceived for the first time, thus demonstrating the usefulness of the 2T-physics formalism. The $(10,2)$ case yields a holographic twistor description of the space $AdS_5 \times S^5$, with a twistor superstring whose particle limit spectrum is the full Kaluza-Klein towers

CP767, *Fundamental Interactions and Twistor-Like Methods,* edited by J. Lukierski and D. Sorokin

of type IIB supergravity compactified on $AdS_5 \times S^5$. This spectrum contains information about hidden dimensions with $(10,2)$ signature as discussed earlier [10][11], and the new superstring extends the $(10,2)$ view of $AdS_5 \times S^5$ to the realm of strings. Similarly, the $(6,2)$ case yields a new twistor superstring whose particle limit describes a supermultiplet of a peculiar self-dual superconformal theory in $(5,1)$ dimensions whose physical space (in a lightcone gauge) consists of 8 bosons and 8 fermions. Another description [9] of this supermultiplet is the unitary representation of $OSp(8|4)$ in the oscillator formalism [12] which coincides with the $OSp(8|4)$ doubleton given in [13]. This six dimensional conformal theory is expected to exist as an interacting theory, but it cannot be described in the form of a field theory [14]. The twistor superstring may be a possible description of this interacting theory.

Perhaps I should give some of the history that motivated the development of 2T-physics. It is often stated that 32 supersymmetries is the maximum possible number of supersymmetries, and therefore 11 dimensions, which has a spinor of 32 real components, is the maximum number of dimensions in a supersymmetric theory of fundamental physics. However, the Weyl spinor in 12 dimensions, with signature (10,2), also has a real spinor with 32 components. Furthermore, the maximally extended supersymmetry algebra, called the M-algebra, has a symmetry of isomorphisms that include $SO(10,2)$, which can be interpreted as acting on a 12-dimensional spacetime with signature $(10,2)$. This point of view was expressed for the first time in 1995 and related to dualities in one of my talks [15] and later further developed in [16]. The possibility of hidden timelike dimensions in M-theory was strengthened further by the hidden symmetry structures in the web of dualities involving D-branes, as in F-theory [17] and S-theory [16].

Of course hints coming from symmetries, although suggestive, are not enough to infer extra spacetime dimensions. However, a dynamical theory involving the higher spacetime, which describes the recognizable world, would go a long way toward understanding the higher spacetime. This was the motivation behind the development of 2T-physics, which after some attempts [18] finally took the correct physical form starting in 1998 as described in [1]. The backbone of the 2T structure is an $Sp(2,R)$ gauge symmetry that acts in phase space. By generalizing this symmetry in several appropriate ways 2T-physics makes contact with the real world. By now it is abundantly clear that the 2T framework describes correctly simple everyday physics as well as complicated structures in string theory.

We don't have to wait until we discover the correct formulation of M-theory to know that 2T-physics is correct, and that it teaches us that there is a sense in which $(d,2)$ dimensions provide a higher unifying framework. This view is already born out in simple classical and quantum mechanical systems, and there are useful nontrivial consequences that follow from it. We have now come back full circle to using 2T-physics techniques to try to construct corners of M-theory, such as the twistor superstrings given in [9] and described briefly in this lecture. 2T-physics suggests that we should look for a formulation of M-theory in $(11,2)$ dimensions with global symmetry $OSp(1|64)$ [16][19][20].

In this lecture I will first give a brief description of the concepts in 2T-physics, and then describe the twistors and the twistor superstrings in $d = 3,4,5,6,10$ that were constructed by using the formalism. The twistors that emerge in the new twistor superstrings can be discussed without the full 2T formalism, as will be done in part of this lecture and in [3], so this aspect can be carried away and applied usefully elsewhere without the need for the full 2T package. However, the full 2T formalism is what provides the easy proof that the new twistors in the higher dimensions describe $AdS_5 \times S^5$ (for $d = 10$), the six dimensional conformal theory (for $d = 6$) respectively, and of course the Super Yang Mills (SYM) theory (for $d = 4$). The 2T version of the theory is far richer because it relates the twistors to other dual forms of the same theory, and this aspect may be crucial ultimately for deeper understanding and for practical progress.

2. 2T-PHYSICS

2.1. Gauge symmetry is the origin of spacetime signature

First I suggest the point of view that gauge symmetry is at the origin of the 1T spacetime signature $(-,+,\dots,+)$, and then show that the same point of view leads to the 2T signature $(-,-,+,\cdots,+)$.

Consider the action of a particle on the worldline which is invariant under τ reparametrizations

$$S = \int_1^2 d\tau \left(\partial_\tau x^\mu p_\mu - e(\tau) Q(x,p)\right), \tag{1}$$

where $e(\tau)$ is the gauge field that transforms as $\delta_\varepsilon e(\tau) = \partial_\tau \varepsilon(\tau)$, while the infinitesimal transformations of $x(\tau), p(\tau)$ are given by the Poisson brackets $\delta_\varepsilon x^\mu = \varepsilon(\tau)\{Q, x^\mu\} = -\varepsilon(\tau)\partial Q/\partial p_\mu$ and $\delta p_\mu = \varepsilon(\tau)\{Q, p_\mu\} = \varepsilon(\tau)\partial Q/\partial x^\mu$. The Lagrangian transforms into a total derivative so that $\delta_\varepsilon S = \int_1^2 d\tau \partial_\tau(\varepsilon Q) = 0$, with boundary conditions $\varepsilon(\tau_1) = \varepsilon(\tau_2) = 0$. The well known free massless relativistic particle action corresponds to $Q(x,p) = \frac{1}{2}\eta^{\mu\nu} p_\mu p_\nu$, while the general $Q(x,p)$ can describe all possible interactions of the particle in any background. For example for an electromagnetic background we have $Q = \frac{1}{2}\eta^{\mu\nu}(p_\mu - qA_\mu(x))(p_\nu - qA_\nu(x))$. Evidently Q is the generator of the local gauge symmetry. The equation of motion for e requires $Q(x,p) = 0$. The space in which the gauge symmetry generator vanishes is evidently gauge invariant (a singlet under gauge transformations). Therefore, this equation is interpreted to mean that the physical space (either classical or quantum), defined to be the solution space of $Q(x,p) = 0$, is gauge invariant.

Consider at first the simplest case $Q(x,p) = p^2 = 0$. We notice that if the signature in $p^2 = \eta^{\mu\nu} p_\mu p_\nu$ is Euclidean the only solution of $p^2 = 0$ is $p_\mu = 0$, so that no non-trivial solution exists for physical space for Euclidean signature. To describe non-trivial motion, target space-time must have 1 time

$$p \cdot p = -p_0^2 + \vec{p}^{\,2} = 0.$$

There are nontrivial solutions also with more timelike dimensions, however τ reparametrization is insufficient to remove the ghosts of more than 1 timelike dimension. Therefore unitarity of the theory requires that spacetime cannot have more than 1 time. Thus, τ reparametrization requires just one time coordinate no more and no less. Causality corresponds to admitting only nonwinding maps $\tau \to x^\mu(\tau)$.

From the simplest case $Q(x,p) = p^2$ we have learned that the signature of the parameter $\varepsilon(\tau)$ in τ reparametrization is timelike. Thus, a timelike (or lightlike, but not spacelike) degree of freedom can be removed from $x^\mu(\tau)$ and similarly a timelike degree of freedom can be removed from p^μ by solving the constraint $p^2 = 0$. For the more general $Q(x,p)$ the signature of $\varepsilon(\tau)$ is the same as before, therefore the gauge symmetry will remove a timelike degree of freedom, not a spacelike one, and the constraint $Q(x,p) = 0$ can have a solution provided target spacetime has signature $(-,+,\ldots,+)$ that includes a timelike degree of freedom. Therefore we deduce that the gauge symmetry requires that there has to be one timelike degree of freedom in any target spacetime (relativistic, nonrelativistic, curved, etc.).

This reasoning is broadened by starting with a worldline action that is invariant under Sp$(2,R)$ gauge symmetry introduced in [1]. For the simplest case Sp$(2,R)$ acts on phase space as a doublet, and an invariant action is written as follows

$$
\begin{aligned}
\text{Sp(2,R) doublet: } & \begin{pmatrix} X^M(\tau) \\ P^M(\tau) \end{pmatrix} \equiv X_i^M \; i = 1,2\,, \\
A_i^{\;j} \text{ gauge field: } & D_\tau X_i^M = \partial_\tau X_i^M - A_i^{\;j} X_j^M\,, \\
S = & \frac{\eta_{MN}}{2} \int d\tau (\varepsilon^{ij} \partial_\tau X_i^M X_j^N - A^{ij} X_i^M X_j^N)\,, \\
\text{Sp(2,R) generators } & : X\cdot X, X\cdot P, P\cdot P, \rightarrow \; X_i \cdot X_j = 0\,.
\end{aligned}
\tag{2}
$$

We deduce that physical space must be Sp(2,R) singlet $X_i \cdot X_j = 0$, and then ask for which signature η^{MN} can we find a nontrivial physical space? We quickly learn that there is no non-trivial content for zero times or one time, and therefore we must admit that target space-time *must* have 2 times

$$-X_{0'}^2 - X_0^2 + X_I^2 = 0,\; etc.\, X\cdot P = P\cdot P = 0\,.$$

Compared to τ reparametrization, Sp(2,R) has 2 more gauge symmetries and 2 more constraints. These eliminate 2 more degrees of freedom from both X^M and P^M. Thus by starting from a space with signature $(d,2)$ we end up with an emergent spacetime with signature $(d-1,1)$ by making various gauge choices

$$(d,2) - (1,1)_{\substack{\text{signature of extra} \\ \text{gauge parameters}}} = (d-1,1)_{\substack{\text{emergent} \\ \text{space-time}}}\,.$$

We conclude that physical spacetime has $(d-1)$ space and 1 time, just like before, but these *must be embedded in a higher spacetime* with signature $(d,2)$.

This would not be very deep if there were a single solution to this embedding. The non-trivial aspect is that there are many ways in which phase space in $(d-1,1)$ is embedded in phase space in $(d,2)$, and this provides many ways in which time (or Hamiltonian) is defined in the emergent spacetime. The embedding provides a holographic image of the events and motion in the $(d,2)$ space, which can be interpreted very differently from the perspective of each of the emergent spaces in $(d-1,1)$ since each such space defines time (and Hamiltonian) differently than one another. Even though we start from a single well defined 2T-physics system in $(d,2)$, we end up with many holographic pictures that are interpreted differently, with different Hamiltonians, in 1T-physics [21].

Considering also unitarity we find that no more than 2 times are possible since the Sp$(2,R)$ gauge symmetry cannot remove the ghosts from a spacetime with more timelike dimensions. In this case causality is satisfied since the situation in the emergent $(d-1,1)$ is no different than the one time situation.

The simple model above has been generalized to include arbitrary interactions with all possible background fields [22]. The generalized action is $S=\int d\tau(\partial_\tau X^M P_M - \frac{1}{2}A^{ij}Q_{ij}(X,P)$ and the gauge symmetry is still Sp$(2,R)$, with $\delta A_i^{\ j} = \partial_\tau \omega_i^{\ j} + [A,\omega]_i^{\ j}$ and Q_{ij} the generator for infinitesimal transformations $\delta X^M = \omega^{ij}(\tau)\{Q_{ij},X^M\} = \omega^{ij}(\tau)\partial Q_{ij}/\partial P_M$, and $\delta P^M = \omega^{ij}(\tau)\{Q_{ij},P^M\} = -\omega^{ij}(\tau)\partial Q_{ij}/\partial X^M$. The generalized $Q_{ij}(X,P)$ depend on background fields in $(d,2)$ dimensions, such as $A_M(X), G_{MN}(X)$, etc., and those are constrained by the requirement that $Q_{ij}(X,P)$ must satisfy the Sp$(2,R)$ algebra. This leads to differential equations for the background fields. All possible solutions are obtained in [22], and it is shown that this covers all possible interactions with backgrounds in 1T-physics, including the Maxwell field $A_\mu(x)$, the gravitational field $g_{\mu\nu}(x)$, etc. as described in Eq. (1). A similar but less complete analysis has been done also for spinning particles [26].

In this way we can argue that possibly all of 1T-physics can be embedded in 2T-physics. It is evident that from the same 2T-physics model, with fixed backgrounds, one can obtain in principle many 1T-physics systems in the form of various holographic images, thus showing that 2T-physics unifies the various dynamics in 1T into a parent theory in 2T that reveals the hidden relationships and symmetries that are not evident at all in the 1T approach.

2.2. Some examples of emergent dynamics and spacetimes

Consider the simplest case of a 2T-physics action for a particle in flat $(d,2)$ spacetime as given in Eq. (2). In this lecture I will illustrate two solutions of the constraints that are inequivalent from the point of view of 1T-physics, but which are evidently equivalent under the Sp$(2,R)$ gauge transformations of the 2T theory, and therefore dual to one another. The first case is the relativistic massless particle and the second case is the hydrogen atom. Sometimes it will be convenient to express the flat $(d,2)$ metric in lightcone type basis. By using the extra dimensions $X^{0'}, X^{1'}$ we define $X^{\pm'} = (X^{0'} \pm X^{1'})/\sqrt{2}$ so that the metric is $ds^2 = -\left(dX^{0'}\right)^2 + \left(dX^{1'}\right)^2 + +dX^\mu dX^\nu \eta_{\mu\nu} = -2dX^{+'}dX^{-'} + dX^\mu dX^\nu \eta_{\mu\nu}$ with $\eta_{\mu\nu}$ the $(d-1,1)$ Minkowski metric.

2.2.1. *Relativistic spacetime gauge*

In the lightcone type basis we choose two gauges by fixing $X^{+'}(\tau)=1,\ \ P^{+'}(\tau)=0$ for all τ, and then solve two of the constraints $X^2 = X\cdot P = 0$. This solution is given by

$$X^M = \begin{pmatrix} +' & -' & \mu \\ 1, & x^2/2, & x^\mu \end{pmatrix} \quad X\cdot X = -2X^{-'}X^{+'} + X^\mu X^\nu \eta_{\mu\nu} = 0, \tag{3}$$

$$P^M = \begin{pmatrix} +' & -' & \mu \\ 0, & x\cdot p, & p^\mu \end{pmatrix} \quad X\cdot P = 0\,,\quad P\cdot P = p^2\,.$$

There remains one more gauge choice to be made and one more constraint $p^2 = 0$ to be solved, but we refrain from doing those steps for now. To interpret our choice of independent variables x^μ, p^μ we investigate the form of the gauge fixed action

$$\text{gauge fixed } S = \int d\tau \left(\dot{x}\cdot p - \frac{1}{2}A^{22}p^2 \right)$$

and note that x^μ, p^μ are canonical variables which describe the massless relativistic particle in $(d-1,1)$ dimensions. To be sure of this fact, we can also investigate that the original equations of motion $\dot{X}^M = A^{22}P^M + A^{12}X^M$ and $\dot{P}^M = -A^{11}X^M - A^{12}P^M$ are fully consistent with the equations of motion that follow from the gauge fixed action.

The original gauge invariant action was symmetric under the SO$(d,2)$ global symmetry. The conserved gauge invariant generators of that symmetry are $L^{MN} = \varepsilon^{ij}X_i^M X_j^N = X^M P^N - X^N P^M$. Since both the action and the generators are gauge invariant, the gauge fixed action must have a hidden SO$(d,2)$ symmetry, with generators given by the gauge fixed form of L^{MN}. Indeed, the gauge fixed generators become the conformal symmetry of the massless particle

$$\text{gauge fixed } L^{MN} \text{ become conformal SO}(d,2)$$

$$L^{\mu\nu} = x^{[\mu}p^{\nu]},\quad L^{+'-'} = x\cdot p,$$

$$L^{+'\mu} = p^\mu\,,\; L^{-'\mu} = \frac{x^2}{2}p^\mu - x\cdot p\, x^\mu\,.$$

When the system is quantized in terms of the relativistic variables x^μ, p^μ the L^{MN} must be carefully quantum ordered. The ordering must insure Hermitian L^{MN} relative to a relativistic norm for the quantum states. When this is done [1] one can compute the Casimir eigenvalue of the SO$(d,2)$ representation that describes the massless spinless particle. The result is in complete agreement with covariant quantization of the 2T system, and is given by

$$\text{After quantum ordering: } C_2 = \frac{1}{2}L^{MN}L_{MN} = 1 - \frac{d^2}{4}$$

same as SO$(d$,2) *covariant* quantization in Sp(2,R) invariant space

This representation is known as the singleton in $d = 3,4$ and thus we will call it the singleton for any d. Note that at the classical level (not watching orders of operators) one obtains zero for the Casimir since $L^{MN}L_{MN}$ is constructed from $X^2, P^2, X\cdot P$ which vanish in the physical sector.

2.2.2. H-atom gauge

Another solution of the constraints $X^2 = P^2 = X \cdot P = 0$, is [21]

$$X^M = [\overset{0'}{r\cos u}, \overset{0}{r\sin u}, \overset{1'}{\frac{\vec{r}\cdot\vec{p}}{-\alpha} r\sqrt{-2H}}, (\overset{i}{\vec{r}^i - \frac{\vec{r}\cdot\vec{p}}{\alpha/r}\vec{p}^i})]\ r \equiv |\vec{r}|\,, \tag{4}$$

$$P^M = [\frac{-\alpha\sin u}{r},\ \alpha\cos u,\ \left(\frac{\alpha}{r} - \vec{p}^2\right),\ \ \sqrt{-2H}\,\vec{p}^i\ ](-2H)^{-1/2},$$

where $H = \frac{\vec{\mathbf{p}}^2}{2} - \frac{\alpha}{r}$, and $u = (\vec{r}\cdot\vec{p} - 2\tau H)\frac{\sqrt{-2H}}{\alpha}$. The interpretation of the emergent dynamics is found by examining the gauge fixed action

$$\text{gauge fixed } S = \int d\tau\,(\partial_\tau\vec{r}\cdot\vec{p} - H) \leftrightarrow \int d\tau(\frac{1}{2}(\partial_\tau\vec{r})^2 + \frac{\alpha}{r}).$$

Evidently, this is the spinless Hydrogen atom (or a planetary system, etc.). The original $\mathrm{SO}(d,2)$ global symmetry $L^{MN} = X^M P^N - X^N P^M$ must be a hidden symmetry of this action. The gauge fixed generators are (at the classical level)

$$
\begin{aligned}
L^{0'0} &= \frac{\alpha}{\sqrt{-2H}}, \\
L^{ij} &= r^i p^j - r^j p^i, \\
L^{1'i} &= \frac{1}{\sqrt{-2H}}\left(r\cdot p\ p^i - r^i p^2 - \alpha\frac{\mathbf{r}^i}{r}\right), \\
L^{0'1'} &= -r\cdot p\sin u + \frac{\alpha}{\sqrt{-2H}}(1 - \frac{r\mathbf{p}^2}{\alpha})\cos u, \\
L^{01'} &= r\cdot p\cos u + \frac{\alpha}{\sqrt{-2H}}(1 - \frac{r\mathbf{p}^2}{\alpha})\sin u, \\
L^{0'i} &= rp^i\cos u + \frac{\alpha}{\sqrt{-2H}}(\frac{\mathbf{r}^i}{r} - \frac{\mathbf{r}\cdot\mathbf{p}}{\alpha}p^i)\sin u, \\
L^{0i} &= rp^i\sin u - \frac{\alpha}{\sqrt{-2H}}(\frac{\mathbf{r}^i}{r} - \frac{\mathbf{r}\cdot\mathbf{p}}{\alpha}p^i)\cos u.
\end{aligned}
$$

In the first line one can recognize the angular momentum and the Runge-Lenz vector that are long known to be conserved quantities of the H-atom system (i.e. commute with H) and that they correspond to a hidden $\mathrm{SO}(d)$ symmetry (better known as $\mathrm{SU}(2)\times\mathrm{SU}(2)=\mathrm{SO}(4)$ in 3 space dimensions). However, 2T-physics gives a stronger symmetry, not of the Hamiltonian, but of the action. According to 2T-physics the H-atom action is invariant under $\mathrm{SO}(d,2)$. Indeed this symmetry can be verified directly[1]. Before this was understood in 2T-physics no-one seems to have been aware of the symmetry of the action, although there has been discussions of a dynamical $\mathrm{SO}(4,2)$ algebra of the H-atom system.

[1] See a homework problem and its solution at http://physics.usc.edu/~bars/papers/Hatom.pdf.

We can go further by quantizing the system, ordering properly the operators and computing the Casimir operator. We find again that C_2 reduces to a pure number which corresponds to the singleton representation [21]

$$\text{After quantum ordering: } C_2 = \frac{1}{2}L^{MN}L_{MN} = 1 - \frac{d^2}{4}.$$

This is what it should be according to the general properties of 2T-physics. Indeed L_{MN} are gauge invariant and they should give the same Casimir in any gauge. This also fits the idea of a duality between the free relativistic particle and the H-atom, since these are derived from the same 2T action by gauge fixing, and therefore they can be transformed to each other by the $\mathrm{Sp}(2,R)$ gauge transformations. Such $\mathrm{Sp}(2,R)$ transformations are easily constructed classically between the gauges (3) and (4). This is expected to succeed also at the quantum level in the form of unitary transformations among dual bases, since the quantum states in either gauge belong to the same representation of $\mathrm{SO}(d,2)$ with the same Casimir operators.

2.2.3. More examples of emergent dynamics/spacetimes

Many more 1T dynamical systems emerge holographically from the same 2T theory given in Eq. (2). The diagram below illustrates some of the cases that have been investigated [21].

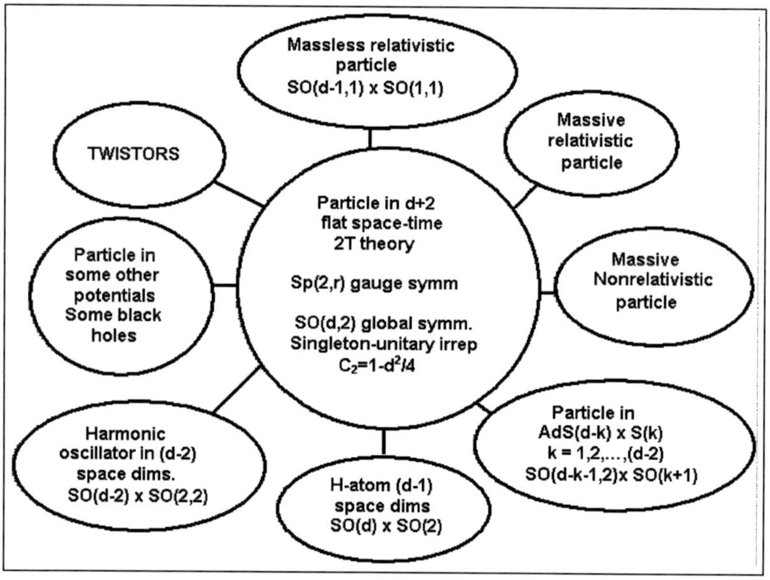

FIGURE 1. d+2 to d holography gives many 1T systems. Each one is a basis for $C_2 = 1 - d^2/4$ irrep of $SO(d,2)$.

These include interacting as well as free systems. The quantum theory has been investigated, and the quantum ordering of the L^{MN} operators has been obtained for the cases of the massless relativistic particle, H-atom, harmonic oscillator, particle on $AdS_{d-k} \times S^k$ background, particle in the $SL(3,R)$ black hole, and twistors. In each case it is shown that $C_2 = 1 - d^2/4$.

The 1T systems are derived from the same 2T system by making an $Sp(2,R)$ gauge choice. In each case some combination of $X^M(\tau), P^M(\tau)$ is gauge fixed to be τ. The canonical conjugate to that choice is always the Hamiltonian written as a function of the remaining phase space degrees of freedom. Although the Hamiltonian (time) looks very different in each case, it still represents a holographic image of the original 2T particle. Thus, each one of these systems represents the same 2T theory although they each have a different interpretation in 1T-physics. In the 1T context they must be interpreted as being dual to each other. In the quantized version each one provides a basis for the same singleton representation of $SO(d,2)$. Within the same representation they must correspond to different bases (which diagonalize the respective Hamiltonian) related by unitary transformations. The existence of such relationships are not at all evident in the 1T approach.

Further generalizations include 2T formulations of spinning particles [23], spacetime supersymmetry [24], twistors [2, 9, 3], and some study of 2T-physics in the context of field theory [26] and string theory [25, 9]. A lot more basic research is awaiting to be developed in 2T-physics.

With what we know so far about 2T-physics, it is evident that it applies broadly to physics and correctly describes the physical world as we know it. The advantage of 2T-physics over 1T-physics is its unification of various 1T systems into a single 2T system, thus providing a more unified perspective. This aspect could be illuminating for the physics that we already understand in the 1T formalism by taking advantage of the revealed hidden symmetries and by exploring the unsuspected duality type relationships among various 1T dynamical systems. Knowing these facts should shed light on the solution and interpretation of physical systems. Furthermore, the 2T approach could be used as a tool for formulating the physics we don't fully understand yet, such as M-theory. The mystery of M-theory has been my main motivation so far in pursuing and developing this approach.

3. TWISTORS AS A GAUGE CHOICE IN 2T-PHYSICS

Twistors were obtained as one of the possible gauge choices in 2T-physics in [2], and this has been further explored in [9] and [3]. For this purpose we consider a group or supergroup G that contains $SO(d,2)$ as a subgroup ($G = SO(d,2)$ is the smallest choice). A group element $g(\tau) \in G$ is introduced as a degree of freedom in addition to $X^M(\tau), P^M(\tau)$, and it is taken in the smallest representation of G such that $SO(d,2) \in G$ is the spinor representation. When G is a supergroup it contains spacetime fermionic spinors that will be useful for spacetime supersymmetric theories. The group element g contains also more bosons beyond (X,P) which, in most cases, can be gauged away by additional gauge symmetries.

g is taken as a singlet under the $Sp(2,R)$ gauge group, while $\left(X^M, P^M\right)$ form a doublet. Next we introduce a further gauge symmetry embedded in G that acts on the *left side of*

g, as well as on X^M, P^M as specified below, to have the *correct number of physical degrees of freedom* for a (spinning) particle or superparticle after gauge degrees of freedom are eliminated. On the right side of g we maintain a full global symmetry G, therefore rows of g transform as spinors under the $\mathrm{SO}(d,2) \in G$. This is where twistors in the spinor representation of $\mathrm{SO}(d,2)$ will come from.

In most applications G is a supergroup, and the fermions in g are used to supersymmetrize the 2T system. However, it is also possible to discuss purely bosonic cases and even specialize to $G = \mathrm{Spin}(d,2)$. In this setting it is possible to choose gauges to eliminate degrees of freedom from g and/or from X, P. If all of g is eliminated, as in the purely bosonic $G = \mathrm{Spin}(d,2)$ case, we remain only with X, P which give the 2T system discussed in the previous sections. However, if all of X, P and some of the g is eliminated we remain with some of the degrees of freedom of g which describe the same 2T system in terms of twistors. If G is an appropriate supergroup that yields the superparticle in one gauge, then in the twistor gauge (with X, P completely eliminated) we obtain the *supertwistor* description of the superparticle. In this way we derive the correct supertwistor representation of several systems of interest in several dimensions as given in [2, 9, 3] and briefly described here.

3.1. Supersymmetric 2T-physics

The most general case studied corresponds to the supersymmetrization of the spinning particle of spin $n/2$. For the spin=$n/2$ generalization we introduce n fermions $\psi_1^M, \cdots, \psi_n^M$ in addition to X^M, P^M, g. Although in this lecture we will mainly discuss the $n = 0$ case (i.e. only X, P, g), we first give the more general Lagrangian

$$L = \frac{\eta_{MN}}{2}\left(q^{ab}\partial_\tau Y_a^M Y_b^N - A^{ab} Y_{\{a}^M Y_{b]}^N\right) \quad \underset{q^{ab}\,=\text{metric}}{\text{local OSp(n|2)}}$$
$$- \frac{1}{2^{[d/2]}}\left(L^{MN} + S^{MN}\right) Str\left(\Gamma^{MN}\partial_\tau g g^{-1}\right)\ ,\quad g \in G_d \text{ supergroup}. \tag{5}$$

For $n = 0$ the first line is the same as Eq. (2) which describes the scalar 2T particle. The second line generalizes it with spacetime supersymmetry. In a specific gauge this system yields the standard superparticle as one of the holographic images [24, 2]. The nonzero spin case with supersymmetry has not appeared yet in the literature, it will be discussed in [3], here we provide a brief description.

The first line in Eq. (5) by itself describes the spinning particle with spin $n/2$. This generalizes the scalar 2T-particle in Eq. (2) by replacing the Sp(2) doublet (X^M, P^M) by the $\mathrm{OSp}(n|2)$ fundamental representation with n fermions, and introducing the corresponding gauge fields A_a^b (which include the $\mathrm{Sp}(2,R)$ gauge fields along with $\mathrm{SO}(d)$ and fermionic counterparts)

$$\underset{\text{fundamental of OSp(n|2)}}{Y_a = \left(X^M, P^M, \psi_1^M, \cdots, \psi_n^M\right)},\ \underset{\text{OSp(n|2) gauge field}}{D_\tau Y_a^M = \partial_\tau Y_a^M - A_a^b Y_b^M},\ J^{MN} = q^{ab} Y_a^M Y_b^N = L^{MN} + S^{MN},\ \underset{S^{MN} = \psi_i^{[M}\psi_i^{N]}\quad \text{spin}=\frac{n}{2}}{} \tag{6}$$

This system (without the supersymmetrization of the second line) was discussed in detail in [23]. Since we don't have time to discuss it here, we will specialize to only $n=0$ in this lecture.

The second line in Eq. (5) corresponds to supersymmetrizing the system in the first line with or without spin. Here the supergroup G_d with N supersymmetries is taken for various dimensions d as one of the following supergroups

$$G_d = \mathrm{OSp}(N|4)_3,\ \mathrm{SU}(2,2|N)_4,\ \mathrm{F}(4)_5,\ \mathrm{OSp}(8|N)_6,\ \mathrm{PSU}(2,2|4)_{10}$$

When $d=3,4,5,6$ the spinor representation of $\mathrm{SO}(d,2)$ corresponds to the first block in the fundamental matrix representation of G_d as shown below

$$\begin{array}{c} G_d{=}\mathrm{OSp(N|4)_3,SU(2,2|N)_4,F(4)_5,OSp(8|N)_6,\ etc.} \\ g(\tau)=\exp\left(\begin{array}{c|c} \frac{1}{4}\Gamma^{MN}\omega_{MN} & \Theta^{i=1\cdots N}_{spinor} \\ \mathrm{Spin}(d,2)\ \mathrm{subgroup} & \mathrm{fermi} \\ \hline \bar{\Theta} & R^a\omega_a \\ \mathrm{fermi} & \mathrm{R\text{-}symmetry} \end{array}\right) \\ d{>}6\ \mathrm{supergroups\ contain\ more\ than\ Spin}(d,2)\rightarrow\Gamma^{M_1\cdots M_p}\mathrm{D\text{-}branes}^2 \end{array} \qquad (7)$$

Therefore, in these cases the coupling $(L^{MN}+S^{MN})\,\Gamma^{MN}$ shown in Eq. (5) takes the matrix form $2^{-[d/2]}\,(L^{MN}+S^{MN})\begin{pmatrix}\Gamma^{MN}&0\\0&0\end{pmatrix}$. This coupling scheme [24, 2] applies to all the cases listed in the first column of the table Eq. (8) and to the $\mathrm{SO}(11,2)$ covariant $d=11$ $\mathrm{OSp}(1|64)$ toy M-model [20] in the second column.

For the $d=10$ $\mathrm{AdS}_5\times\mathrm{S}^5$ case listed in the second column the coupling scheme is slightly different. Namely, we do not keep full covariance $\mathrm{SO}(10,2)$ in 12 dimensions, but rather only under its subgroup $\mathrm{SO}(4,2)\times\mathrm{SO}(6)=\mathrm{SU}(2,2)\times\mathrm{SU}(4)$. The supergroup that contains this subgroup is $\mathrm{PSU}(2,2|4)$. To take this into account we split the 12 coordinates into two groups of six each $X^M=(X^m,X^I)$, and similarly for the momenta $P^M=(P^m,P^I)$, and associate the first six dimensions with the upper block of the $\mathrm{PSU}(2,2|4)$ matrix and the last six dimensions with the lower block (in place of the R-symmetry). Therefore, for $d=10$ the coupling $(L^{MN}+S^{MN})\,\Gamma^{MN}$ shown in Eq. (5) takes the matrix form $\frac{1}{4}(L^{mn}+S^{mn})\begin{pmatrix}\Gamma^{mn}&0\\0&0\end{pmatrix}+\frac{1}{4}(L^{IJ}+S^{IJ})\begin{pmatrix}0&0\\0&-\Gamma^{IJ}\end{pmatrix}$. This $d=10$ 2T model, for $n=0$ (i.e. no spin S^{MN} or S^{IJ}), produces the supersymmetric particle moving on $\mathrm{AdS}_5\times\mathrm{S}^5$ in one of its holographic images, and its physical quantum spectrum is identical to the Kaluza-Klein towers of type IIB supergravity compactified on $\mathrm{AdS}_5\times\mathrm{S}^5$ [10]. This is a model of interest in the context of the $d=10$ twistor superstring described below, and the AdS-CFT correspondence. A similar model for the d=11 $\mathrm{AdS}_4\times\mathrm{S}^7$ or $\mathrm{AdS}_7\times\mathrm{S}^4$ listed in the table does not lead to the corresponding compactification of $d=11$ supergravity, but rather to the first massive level of the d=11 supermembrane.

d=3:	Spin(3,2)=Sp(4)⊂OSp(N|4)	d=10 : $AdS_5\times S^5$	Spin(4,2)×Spin(6)⊂PSU(2,2|4)
d=4:	Spin(4,2)=SU(2,2)⊂PSU(2,2|N)	d=11 : $AdS_4\times S^7$	Spin(3,2)×Spin(8)⊂OSp(8|4)
d=5	Spin(5,2)⊂ $F(4)$ it contains also SU(2)	d=11 : $AdS_7\times S^4$	Spin(6,2)×Spin(5)⊂OSp(8|4)
d=6	Spin(6,2)⊂OSp(8|N)	d=11:	Spin(11,2)⊂OSp(1|64) toy M-model with D branes[2]
any d	Spin(d,2)=G purely bosonic twistors in any d	etc.	generalizations of above

(8)

For $d>6$ supergroups contain bosonic subgroups that are larger than $SO(d,2)$ as long as we insist that $SO(d,2)$ appears in the spinor representation[2]. The extra bosons contained in g correspond to D brane-like degrees of freedom. If we require full covariance for $SO(d,2)$ we are forced to admit these as additional degrees of freedom beyond those of the superparticle. The toy M-model in $d=11$ based on $OSp(1|64)$ is one of the most interesting cases of this type[20]. By breaking the covariance to a subgroup of $SO(d,2)$ we can build 2T models such as the $AdS_{d-k}\times S^k$ cases based only on particle degrees of freedom similar to what was described above.

[2] Here we give a list of the smallest bosonic subgroups in a supergroup G that contain $\mathrm{Spin}(d,2)$ for $3\leq d\leq 12$. We also list the generators, and their numbers in parentheses, as represented by antisymmetrized products of gamma matrices $\Gamma^{M_1\cdots M_n}\equiv\frac{1}{n!}\Gamma^{[M_1}\ldots\Gamma^{M_n]}$ in dimension $d+2$ labelled my M

d	Spin(d,2)	spinor	$\subseteq G_{bose}$	generators of G_{bose} in Spin(d,2) basis	contained in
3	Spin(3,2)	4	Sp(4,R)	Γ^{MN} (10)	$(4\times4)_s$
4	Spin(4,2)	$4_\pm$	SU(2,2)	Γ^{MN} (15)	$(4\times4^*)$
5	Spin(5,2)	8_+	SO*(8)	Γ^{MN} (21) + Γ^{M} (7)	$(8\times8)_a$
6	Spin(6,2)	8_+	SO*(8)	Γ^{MN} (28)	$(8\times8)_a$
7	Spin(7,2)	16	SO*(16)	Γ^{MN} (36) + Γ^{MNK} (84)	$(16\times16)_a$
8	Spin(8,2)	$16_\pm$	SU*(16)	Γ^{MN} (45) + Γ^{MNKL} (210)	$(16\times16^*)$
9	Spin(9,2)	32	Sp*(32)	Γ^{MN} (55) + Γ^{M} (11) + $\Gamma^{M_1\cdots M_5}$ (462)	$(32\times32)_s$
10	Spin(10,2)	32_+	Sp*(32)	Γ^{MN} (66) + $\Gamma_+^{M_1\cdots M_6}$ (462)	$(32\times32)_s$
11	Spin(11,2)	64	Sp*(64)	Γ^{MN} (78) + Γ^{MNK} (286) + $\Gamma^{M_1\cdots M_6}$ (1716)	$(64\times64)_s$
12	Spin(12,2)	$64_\pm$	SU*(64)	Γ^{MN} (91) + Γ^{MNKL} (1001) + $\Gamma^{M_1\cdots M_6}$ (3003)	$(64\times64^*)$

The antisymmetric $\Gamma^{M_1\cdots M_n}$ are associated with group parameters $\omega_{M_1\cdots M_n}(\tau)$ that cannot be eliminated by the gauge symmetries, and therefore they are additional degrees of freedom analogous to D-brane collective coordinates.

3.2. Twistor gauge

The Lagrangian in Eq. (5) has an evident global symmetry G_d which corresponds to group transformations on the right side of g. These leave the Cartan form $(\partial g)\, g^{-1}$ invariant. The conserved Noether charge for this symmetry is the supermatrix $J_A^{\ B}$ in the Lie algebra of G_d

$$\text{Gauge invariant global symm}\quad J_A^{\ B} \sim \frac{i}{2}\left(L_{MN}+S_{MN}\right)\left(g^{-1}\Gamma^{MN}g\right)_A^{\ \ B}.$$

The Lagrangian is also invariant under a number of local symmetries. To begin there is the built in $\mathrm{OSp}(n|2)$ local supersymmetry on the worldline as in Eq. (6). In addition, there are local spacetime supersymmetries. Some of these become easier to notice by rewriting the Lagrangian (5) in the form

$$L=\frac{1}{2^{[d/2]}}q^{ab}Tr\left[\partial_\tau\left(g^{-1}Y_a\cdot\Gamma g\right)\left(g^{-1}Y_b\cdot\Gamma g\right)\right]-\frac{1}{2}A^{ab}Y_a\cdot Y_b.$$

Then it is easy to see that there is an invariance under local transformations [$\mathrm{Spin}(d,2)\times$ R-symmetry]$\in G_d$ that are simultaneously applied on the *left side* of g as well as on the M index of Y_a^M. There is also local fermionic kappa supersymmetry that is also applied on the left side of G_d as well as on A^{ab} as explained in [24, 2, 9]. Let us review what physical degrees of freedom remain after gauge degrees of freedom are removed. We consider $n=0$ in what follows (i.e. supersymmetrizing the scalar particle), the general n is similar.

The local $\mathrm{Spin}(d,2)\,\times$ R-symmetry has enough gauge parameters to remove all of the bosons from g in all the cases listed in the table Eq. (8), except for the toy M-model which has D-branes[2]. Therefore for those cases the bosonic degrees of freedom are just the particle phase space $\left(X^M,P^M\right)$ or their gauge equivalent.

The local kappa supersymmetry has enough fermionic gauge parameters to remove $3/4$ of the fermions Θ shown in Eq. (7) for $d=3,4,5,6$ and the $d=11$ toy M-model. On the other hand for the $d=10$ $AdS_5\times S^5$ case only $1/2$ of the 32 fermions can be removed by the kappa supersymmetry, while for the $d=11$ $AdS_4\times S^7$ or $AdS_7\times S^4$ there is no kappa supersymmety at all.

This summarizes then the physical degrees of freedom up to gauge equivalences. The interesting aspect of 2T-physics is that the gauge equivalence within the 2T-system does not necessarily imply that the 1T interpretation is the same, but rather that there are dualities between various holographic 1T images, as in the figure above.

It was shown in [24, 2, 9] that, for $n=0$ and $d=3,4,5,6$, one can choose a gauge that reduces the 2T system to the standard superparticle in the corresponding number of dimensions. Also it was shown in [10] that the $d=10$ $\mathrm{SU}(2,2|4)$ case can be gauge fixed to the superparticle moving on the $AdS_5\times S^5$ background.

In this lecture we will concentrate on the twistor gauge [2, 9] for the $n=0$ case. By using the local symmetry $\mathrm{Sp}(2,R)\times\mathrm{Spin}(d,2)$, and the constraints $X\cdot P=X^2=P^2=0$, we can gauge fix $X^M(\tau),P^M(\tau)$ for all τ to the following trivial configuration

$$X^M=(\overset{+'}{X^{+'}},\overset{-'}{0},\overset{+}{0},\overset{-}{0},\overset{\nu}{0}),\; P^M=(\overset{+'}{0},\overset{-'}{0},\overset{+}{P^+},\overset{-}{0},\overset{i}{0}),\; i=1,\cdots,(d-2)\,. \tag{9}$$

In this gauge the purely bosonic system for any d, the supersymmetric systems for $d=3,4,6$, and the $d=11$ toy M-model listed in Eq. (8) collapse to the form[3]

$$L=-\frac{1}{2^{[d/2]-1}}Tr\left(\partial_\tau g g^{-1}\begin{pmatrix}\Gamma^{-'-} & 0\\ 0 & 0\end{pmatrix}\right)L^{+'+}=-i\partial_\tau \bar{Z}^{aA}Z_{Aa}\ ,\ a=1,\cdots,k \tag{10}$$

$$J_A^{\ B}\sim L^{+'+}\left(g^{-1}\begin{pmatrix}\Gamma^{-'-} & 0\\ 0 & 0\end{pmatrix}g\right)_A^{\ B}=-2Z_{Aa}\bar{Z}^{aB}\ ,\ \bar{Z}^{aA}Z_{Ab}=0. \tag{11}$$

In an appropriate basis for gamma matrices[4] we find $\Gamma^{-'-}=2\tau^{-}\otimes\sigma^{-}\otimes 1_k$ which is a $4k\times 4k$ matrix with lots of zeroes and k nonzeroes off the diagonal. Therefore only certain off-diagonal rows of g denoted by $\bar{Z}^{aA}$ and certain off-diagonal columns of g^{-1} denoted by Z_{Aa} contribute in the trace in L or to $J_A^{\ B}$. Also the relation $gg^{-1}=1$ implies the constraint $\bar{Z}^{aA}Z_{Ab}=0$ as an off diagonal entry in the matrix 1. The A,B indices label the fundamental representation of G_d and therefore the Z_{Ai} denote k supertwistors with $i=1,\cdots,k$. Thus the theory has now been written in terms of twistors.

Note that for $d=4$ the group is $PSU(2,2|4)$, the gamma matrices are 4×4, and $k=1$. Therefore for $d=4$ there is a single supertwistor Z_A in the fundamental representation of $PSU(2,2|4)$ and it is constrained by $\bar{Z}^A Z_A=0$. These constrained twistors describe $CP^{3|4}$. Thus the 2T formalism for supertwistors is in full agreement with the expectation about twistors in four dimensions. The 2T formalism gives the appropriate generalization to all the other dimensions mentioned earlier. These will be described below case by case for a few dimensions of special interest.

The Lagrangian in (10) suggests that $\bar{Z}^{aA}$ is the canonical conjugate to Z_{Aa} and therefore the twistors can be expressed in terms of oscillators. The current $J_A^{\ B}$ for the global symmetry in Eq. (11) is constructed from these oscillators, and the quantum states are obtained in the Fock space of these oscillators. The physical states are the subset of the Fock space that satisfies the constraint $\bar{Z}^{aA}Z_{Ab}=0$, and form a unitary representation of the global symmetry G_d. This setup precisely coincides with the Bars-Günaydin (BG) oscillator approach to unitary representations of supergroups developed in 1983 [12]. The additional constraint is a gauge invariance condition and is implemented by following the discussion about "color" in the improved oscillator formalism given in [11]. Therefore, we can easily obtain the quantum spectrum and compare to the quantum spectrum in another gauge, such as the superparticle gauge. The agreement is perfect as

---

[3] For the $d=10$ $AdS_5\times S^5$ case the gauge fixed forms of X,P are different and are given in Eq. (17).

[4] $d+2$ gamma matrices that satisfy $\Gamma^M\bar{\Gamma}^N+\Gamma^N\bar{\Gamma}^M=2\eta^{MN}$ are chosen in a Weyl basis as follows

$$\Gamma^{\pm'}=\left(\pm\sqrt{2}\tau^{\pm}\right)\otimes 1\otimes 1_k,\ \ \Gamma^{\pm}=\tau_3\otimes\left(\pm\sqrt{2}\sigma^{\pm}\right)\otimes 1_k,\ \ \Gamma_i=\tau_3\otimes\sigma_3\otimes\gamma_i,\ \ \begin{matrix} i=1,\cdots,(d-2)\\ \gamma_i \text{ is k}\times\text{k matrix}\end{matrix}$$

where $\bar{\Gamma}^M=\Gamma^M$ in odd dimensions, while $\bar{\Gamma}^M$ differs from Γ^M only by inserting an overall sign for the very last γ_i in even dimensions, i.e. $\pm\gamma_{d-2}$. For example for $d=4$ or $SO(4,2)=SU(2,2)$, we have $k=1$, with $\gamma_1=1$, $\gamma_2=i$, while $\bar{\gamma}_1=1$, $\bar{\gamma}_2=-i$, which satisfy $\gamma_i\bar{\gamma}_j+\gamma_j\bar{\gamma}_i=2\delta_{ij}$. Then we construct $\Gamma^{MN}=\frac{1}{2}\left(\Gamma^M\bar{\Gamma}^N-\Gamma^N\bar{\Gamma}^M\right)$.

expected from the 2T approach, since each gauge corresponds to a holographic image of the same 2T system.

The supertwistors are constrained as shown above. The full solution of these constraints in terms of unconstrained degrees of freedom is given as a coset $T_\Gamma \in G_d/H_\Gamma$, where H_Γ is a gauged subgroup H_Γ of G_d, that is a remainder of the original gauge symmetries mentioned before. H_Γ is identified as the subgroup that contains all the generators of G_d that commute with the generator represented by $\begin{pmatrix} \Gamma^{-\prime -} & 0 \\ 0 & 0 \end{pmatrix}$. We can then show [9, 3] that the Lie algebras of h_Γ and of the coset t_Γ form triangular sub-supergroups, and they satisfy (anti)commutation rules of the type [9, 3]

$$[h_\Gamma, h_\Gamma\} \sim h_\Gamma,\ \ [t_\Gamma, t_\Gamma\} \sim t_\Gamma,\ \ [h_\Gamma, t_\Gamma\} \sim h_\Gamma + t_\Gamma\,. \tag{12}$$

Furthermore the system can be written in terms of the unconstrained degrees of freedom $t \in G_d/H_\Gamma$ in the form

$$L = -\frac{1}{2^{[d/2]-1}} Tr\left(\partial_\tau t t^{-1}\begin{pmatrix} \Gamma^{-\prime -} & 0 \\ 0 & 0 \end{pmatrix}\right) L^{+\prime +},\ J_A^{\ B} \sim L^{+\prime +}\left(t^{-1}\begin{pmatrix} \Gamma^{-\prime -} & 0 \\ 0 & 0 \end{pmatrix} t\right)_A^{\ B}. \tag{13}$$

This is like a sigma model based on a coset but the Lagrangian is linear in the Cartan connection $\partial_\tau t t^{-1}$ (as opposed to quadratic form for the sigma model) and there is an unusual insertion $\begin{pmatrix} \Gamma^{-\prime -} & 0 \\ 0 & 0 \end{pmatrix}$.

3.3. Supertwistors for d=4 and SYM spectrum

The twistor must reproduce the physical degrees of freedom and quantum states of the corresponding $d = 4$, $N = 4$ superparticle, as expected from the 2T formalism. Let's see how this is obtained explicitly.

To begin the superparticle has $4x$,$4p$ and 16θ real degrees of freedom in super phase space. We remove $1x$ and $1p$, due to τ reparametrization and the corresponding $p^2 = 0$ constraint. We also remove 8 fermionic degrees of freedom due to kappa supersymmetry. We are left over with $3x, 3p, 8\theta$ physical degrees of freedom. With these we construct the physical quantum states as an arbitrary linear combination of the basis states in momentum space $|\vec{p}, \alpha\rangle$, where α is the basis for the Clifford algebra satisfied by the 8θ. This basis has 8 bosonic states and 8 fermionic states. Viewed as probability amplitudes in position space $\langle x, \alpha|\psi\rangle$ these are equivalent to fields $\psi(x)_{8_B+8_F}$ which correspond to the independent *solutions* of all the constraints. One finds that these are the same as the 8 bose and 8 fermi fields of the Super Yang Mills (SYM) theory which are the solutions of the linearized equations of motion in the lightcone gauge. They consist of two helicities of the gauge field $A_{\pm 1}(x)$, two helicities for the gauginos $\psi^a_{+\frac{1}{2}}(x)$, $\bar{\psi}_{-\frac{1}{2},a}(x)$ in the $\mathbf{4}, \bar{\mathbf{4}}$ of $SU(4)$, and six scalars $\phi^{[ab]}(x)$ in the $\mathbf{6}$ of $SU(4)$.

Now we count the physical degrees of freedom for the twistors. We have already explained following Eq. (10) that for $d = 4$ we have one complex twistor Z_A in the fundamental representation of $PSU(2,2|4)$, with a Lagrangian and a conserved current given by

$$L = i\bar{Z}^A \partial_\tau Z_A,\ J_A^{\ B} = -2Z_A\bar{Z}^B,\ \text{and}\ \bar{Z}^A Z_A = 0 \tag{14}$$

Z_A is in fundamental representation of PSU(2,2|N) $\leftrightarrow CP^{3|N}$

To start Z_A has 4 complex bosons and 4 complex fermions, i.e. $8_B + 8_F$ real degrees of freedom. However, there is one constraint $\bar{Z}^A Z_A = 0$ and a corresponding $U(1)$ gauge symmetry[5], which remove 2 bosonic degrees of freedom. The result is $6_B + 8_F$ physical degrees of freedom whic is equivalent to $CP^{3|4}$, and the same number as $3x, 3p, 8\theta$ for the superparticle, as expected. Instead of constrained twistors we can also express the $CP^{3|4}$ theory in terms of unconstrained coset parameters in the form (12,13) where h_Γ was given in [9], and with more details in [3].

To construct the spectrum we could resort to well known twistor techniques by working with fields $\phi(Z)$ that are holomorphic in Z_A on which $\bar{Z}^A$ acts as a derivative $\bar{Z}^A\phi(Z) = \partial\phi(Z)/\partial Z_A$, as dictated by the canonical structure that follows from the Lagrangian (14). Imposing the constraint amounts to requiring $\phi(Z)$ to be homogeneous with a given degree h, namely $Z_A\bar{Z}^A\phi(Z) = Z_A\partial\phi(Z)/\partial Z_A = h\phi(Z)$. Only one value of h is permitted. Naively h is zero at the classical level, but at the quantum level we have to determine the correct value of h that may arise due to quantum ordering. In the case of the $d = 4$ $N = 4$ superparticle described by the $PSU(2,2|4)$ twistor indeed we find $h = 0$, and the resulting spectrum is again the SYM fields. This is the degree zero wavefunction $\phi(Z)$ described in [8]. Recall that in [8] there are also twistor wavefunctions $f(Z), g(Z)$ that describe the spectrum of conformal gravity; those arise in the same twistor formalism, but at a different value of h. However, since only one value of h is permitted in the current superparticle model, only SYM states $\phi(Z)$ are present. Of course, this is no surprise in the 2T setting. We have simply compared two gauges, and we must agree.

It is also worth analyzing the quantum system in terms of oscillators related to twistors and understand the unitarity of the physical space. We emphasize that $\bar{Z}^A$ is obtained from Z_A by hermitian conjugation and multiplying by the $PSU(2,2|4)$ metric. To see the oscillator formalism clearly we work in a basis of $SU(2,2|4)$ in which the metric is $diag(1_2, -1_2, 1_4)$. This is the $SU(2)\times SU(2)$ basis for $SU(2,2)$, to be contrasted with the $SL(2,C)$ basis in which the metric is off-diagonal and usually used to discuss Lorentz covariant twistors. The two bases are simply related by a linear transformation that diagonalizes the metric. In this diagonal basis we identify the oscillators as (a bar over the symbol means hermitian conjugation)

$$L = i\bar{Z}^A \partial_\tau Z_A = i\bar{a}^i \partial_\tau a_i - ib_I \partial_\tau \bar{b}^I + i\bar{\psi}^r \partial_\tau \psi_r \qquad \begin{array}{c} \overset{SU(2)}{i = 1,2},\ \overset{SU(2)}{I = 1,2} \\ r = 1,\cdots,4\ SU(4) \end{array}$$

$$Z_A = \begin{pmatrix} a_i \\ \bar{b}^I \\ \psi_r \end{pmatrix},\ \bar{Z}^A = \left(\bar{a}^j, -b_J, \bar{\psi}^s\right),\ J_A^B = -2Z_A\bar{Z}^A = -2\begin{pmatrix} a_i\bar{a}^j & -a_i b_J & a_i\bar{\psi}^s \\ \bar{b}^I\bar{a}^j & -\bar{b}^I b_J & \bar{b}^I\bar{\psi}^s \\ \psi_r\bar{a}^j & -\psi_r b_J & \psi_r\bar{\psi}^s \end{pmatrix}.$$

---

[5] This can be restated by reformulating the above system by rewriting $L = \bar{Z}^A(\partial_\tau + A)Z_A$ with an extra $U(1)$ gauge field A, and deriving the constraint by varying A.

It is significant to note that, after taking care of the metric in $\bar{Z}$ as above, the usual canonical rules applied to this Lagrangian identifies the oscillators as being all *positive norm* oscillators $[a_i, \bar{a}^j] = \delta_i^j$, $[b_I, \bar{b}^J] = \delta_I^J$ and $\{\psi_r, \bar{\psi}^s\} = \delta_r^s$. Therefore all Fock states have positive norm. However, among them we must choose only those that satisfy the constraints

$$0 = \bar{Z}^A Z_A = \bar{a}^i a_i - b_I \bar{b}^I + \bar{\psi}^r \psi_r = \bar{a}^i a_i - \left(\bar{b}^I b_I + 2\right) + \bar{\psi}^r \psi_r$$
$$\Leftrightarrow 2 = N_a + N_\psi - N_b \equiv \Delta, \quad N_a, N_\psi, N_b \text{ number operators .}$$

This is precisely the BG oscillator formalism for unitary representations of noncompact groups [12] for a single "color". The constraint $\Delta = 2$ is discussed in [11]. These physical states are characterised by identifying the following lowest supermultiplet

$$\Delta = 2: \left(\underset{(1,0,1)}{\overset{A_{+1}}{\bar{a}^i \bar{a}^j}},\ \underset{(\frac{1}{2},0,4)}{\overset{\psi^r_{+1/2}}{\bar{a}^i \bar{\psi}^r}},\ \underset{(0,0,6)}{\overset{\phi^{[rs]}}{\bar{\psi}^r \bar{\psi}^s}},\ \underset{(0,\frac{1}{2},\bar{4})}{\overset{\psi_{-1/2,a}}{\bar{b}^I \bar{\psi}^r \bar{\psi}^s \bar{\psi}^m}},\ \underset{(0,1,1)}{\overset{A_{-1}}{\bar{b}^I \bar{b}^J \bar{\psi}^r \bar{\psi}^s \bar{\psi}^m \bar{\psi}^n}} \right) |0\rangle \ ,$$

which is annihilated by the double annihilation generators $a_i b_J$ which is part of J_A^B in the conformal subgroup $SU(2,2)$. All other states with $\Delta = 2$ are descendants obtained by applying arbitrary powers of the double creation generator $\bar{a}^j \bar{b}^I$ in $SU(2,2)$. The lowest states correspond to the SYM fields, the descendants are analogous to applying multiple derivatives on a field. The classification of the lowest states under $SU(2) \times SU(2) \times SU(4) \subset PSU(2,2|4)$ is given under each combination of oscillators in the form $(j_1, j_2, \dim(SU(4)))$ where j_1, j_2 correspond to the spin quantum numbers in each $SU(2)$. In arriving at these quantum numbers we took into account that $\bar{a}^i \bar{a}^j$ is symmetric while $\bar{\psi}^r \bar{\psi}^s$ is antisymmetric, etc. Above each of the oscillator combination we indicated one of the physical helicity components of the SYM fields associated with that state. This is because in comparing the compact $SU(2) \times SU(2) \subset SU(2,2)$ to the *helicity* embedded in the noncompact Lorentz group $SL(2,C) \subset SU(2,2)$ we must identify as helicity only the spin up part from the first $SU(2)$ and the spin down part from the second $SU(2)$.

Although we gave a whole supermultiplet of lowest states above, there really is only one lowest oscillator state for the entire unitary representation of $PSU(2,2|4)$. That one is simply $\bar{\psi}^r \bar{\psi}^s |0\rangle$, which satisfies $\Delta = 2$. All other states with $\Delta = 2$ are obtained by applying all powers of J_A^B on this state (note $[\Delta, J_A{}^B] = 0$). This is called the doubleton representation of $PSU(2,2|4)$ (sometimes also called the singleton). If we watch carefully the orders of the oscillators we can show that the generators of $PSU(2,2|4)$ *in this representation* satisfy [11] the following nonlinear constraints

$$(JJ)_A{}^B = 4(J)_A{}^B + 0 \,. \tag{15}$$

The linear J follows from the commutation rules among the generators, the coefficient 4 is related to overall normalization of J, while the coefficient 0 is the $PSU(2,2|4)$ quadratic Casimir eigenvalue $C_2 = 0$. This equation should be viewed as a set of constraints that are satisfied by the generators in this particular representation, and as such this relation identifies uniquely the representation (there is a unique $C_2 = 0$ representation if we also specify the $SU(2,2)$ conformal dimension=1). If the theory is expressed in any other form (such as particle description, or field theory) the doubleton representation can be identified in terms of the global symmetry as one that must satisfy the constraints (15), automatically requiring the 6 scalars $\phi^{[ab]}$ as the lowest SU(4) multiplet. This is a completely PSU$(2,2|4)$ covariant and gauge invariant way of identifying the physical and unitary states of the theory. Of course the $d=4$, $N=4$ SYM fields satisfy this criterion.

3.4. Supertwistors for d=6 and self dual supermultiplet

Now that the concepts have been illustrated clearly for $d=4$, we summarize quickly the equivalent statements for $d=6$. The superparticle in $d=6$ and $N=4$ starts out with $6x, 6p, 16\theta$ real degrees of freedom. Fixing τ, and kappa local gauges and solving constraints, reduces the physical degrees of freedom down to $5x$, $5p$, $16\ \theta$. The superparticle action has a hidden global superconformal symmetry OSp$(8^*|4)$ [24, 2], therefore the physical states should be classified as a unitary representation under this group.

If we quantize in the lightcone gauge we find 8_B+8_F states which should be compared to the fields of a six dimensional field theory taken in the lightcone gauge. There are two possible candidates; (1) the linearized six dimensional SYM theory with $N=4$ SUSY in the lightcone gauge, and (2) the self dual supermultiplet classified (covariantly) as

$$\underset{\text{self dual } F^+_{[\mu\nu\lambda]}=\partial_{[\mu_1}A_{\mu_2\mu_3]}=\varepsilon_{\mu_1\mu_2\mu_3\mu_4\mu_5\mu_6}\partial^{[\mu_4}A^{\mu_5\mu_6]}}{\mathrm{SO}(5,1)\times\mathrm{Sp}(4)\text{: } F^+_{[\mu\nu\lambda]},\ \psi^a_\alpha,\ \phi^{[ab]}} \tag{16}$$

The SYM lightcone degrees of freedom consists of 8_B+8_F, with the 8 bosons being the SO(4) vector A_i and four real scalars in an internal compact SO(4). On the other hand for the self dual multiplet, we have the lightcone fields a self dual antisymmetric $A_{ij}=i\varepsilon_{ijkl}A^{kl}$ (i.e. 3 fields) and the Sp(4) traceless antisymmetric $\phi^{[ab]}$ (5 scalars). These are clearly different. Only the self dual supermultiplet is consistent with the compact USp(4) classification classification (which is not real but pseudo-real for the fundamental representation, and real for $\phi^{[ab]}$). Therefore the initial hidden superconformal symmetry OSp$(8^*|4)$ of the superparticle (and of the 2T superparticle) is consistent only with the field theory for the $d=6$ self dual supermultiplet. An *interacting* quantum conformal field theory with these degrees of freedom is expected but cannot be written down covariantly in the form of a field theory [14].

Let us now examine the twistors that emerged in Eq. (10) for this case. We have (in Table 1.) Z_{Aa} is pseudo real as it is defined through the group element $g\in$OSp$(8^*|4)$ with the correct signature. Then Z_{Aa} takes the form above when we choose an appropriate and natural basis. Thus the second column is related to the first one, but still consistent with a local SU(2) applied on $i=1,2$. We count the degrees of freedom before the constraints, and find that Z_{Aa} has (8_B+4_F)x$2_{\text{(complex)}}=16_B+8_F$ (namely $a_{1i}, a_{2i}, \psi_{1\alpha}, \psi_{2\alpha}$). The constraints are due to a SU(2) gauge symmetry acting on the index $i=1,2$ (although it seems like SU$(2)\times$U(1), the U(1) part is automatically satisfied because of the pseudoreal form of Z_{Aa}). The 3 gauge parameters and 3 constraints remove 6 bosonic degrees of freedom, and we remain with 10_B+8_F physical degrees of freedom. This is the same as the count for the superparticle $(5x, 5p, 8\theta)$. It is obvious we have the same number of degrees of freedom and the same symmetries $\mathrm{OSp}(8^*|4)$, with the symmetry being much more transparent in the twistor basis.

$Z_{Aa} = \begin{pmatrix} 8bose \\ 4fermi \end{pmatrix}$	12x2 rectangular matrix, $A=1,\cdots,12$; a=1,2 2 twistors in fundamental rep of $OSp(8^*|4$	
$Z_{Aa} = (12{,}2)$ of $OSp(8^*|4)_{\mathrm{global}} \times \mathrm{SU}(2)_{\mathrm{local}}$		
$L = \bar{Z}^{Aa}\left((\partial + A)Z\right)_{Aa} \to \bar{Z}^{aA}Z_{Ab} = 0,$	SU(2) gauge invariants in Fock space	
Pseudo-real 1st & 2nd related	$Z_{Aa} = \begin{pmatrix} a_{1i} & a_{2i} \\ \bar{a}^i_2 & -\bar{a}^i_1 \\ \psi_{1\alpha} & \psi_{2\alpha} \\ \bar{\psi}^\alpha_2 & -\bar{\psi}^\alpha_1 \end{pmatrix}$	i: 4 *of* SU(4)⊂SO(6,2) α: 2 *of* SU(2)⊂Sp(4)

Table 1.

Instead of constrained twistors we can also express this theory in terms of unconstrained coset parameters in the form (12,13) where h_Γ was given in [9], with more details in [3].

The quantum theory can proceed in terms of twistors or in terms of constrained oscillators. The resulting representation, after satisfying the $\mathrm{SU}(2)$ constraints in the Fock space of the oscillators defined above, is precisely the doubleton of $\mathrm{OSp}(8^*|4)$, and this is precisely equivalent to the fields in Eq. (16). This oscillator representation was worked out a long time ago in [13] using again the BG method [12]. The selection of the doubleton among many other Fock space states discussed in [13] is the analog of choosing the $\mathrm{SU}(2)$ "color" singlet in analogous discussion to the one in [11]. More details will be given in [3].

3.5. Supertwistors for d=10 and $\mathrm{AdS}_5\times\mathrm{S}^5$ KK towers

This was explained in detail in [10]. Here we will quickly count degrees of freedom for the $\mathrm{AdS}_5\times\mathrm{S}^5$ superparticle. This superparticle starts out with $10x, 10p, 16\theta$ real degrees of freedom. Fixing τ gauges and solving constraints, reduces the physical degrees of freedom down to $9x$, $9p$, $16\ \theta$. With 16θ's we construct the Clifford algebra that is realized on states with $128_B + 128_F$. These correspond to the supergravity multiplet. Hence this case is related to gravity.

The superparticle action has a hidden global superconformal symmetry $\mathrm{PSU}(2,2|4)$ whose generators are given in [10], therefore the physical states should be classified as a unitary representation under this group. The resulting spectrum coincides with all the infinite Kaluza-Klein towers of type IIB supergravity compactified on $\mathrm{AdS}_5\times\mathrm{S}^5$.

Now we turn to the twistor gauge. This is a different gauge choice compared to (9). We have split phase space into two groups of six each, as $X^M = \left(X^m, X^I\right)$, $P^M = \left(P^m, P^I\right)$, but the Sp(2) constraints is overall the 12 dimensions $X^2 = P^2 = X \cdot P$. Using the $\mathrm{Sp}(2)\times\mathrm{SO}(4,2)\times\mathrm{SO}(6)$ gauge symmetries we choose gauges and solve all the Sp(2) constraints with the following configuration

$$\begin{aligned} M &= \left(0' \ 0 \ 1 \cdots 4 \ , I = 1 \ 2 \ 3 \ \cdots 6\right) , \\ X^M &\sim \left(a \ 0 \ 0 \ \cdots 0 \ , \quad a \ 0 \ 0 \ \cdots 0\right) , \\ P^M &\sim \left(0 \ b \ 0 \ \cdots 0 \ , \quad 0 \ b \ 0 \ \cdots 0\right) . \end{aligned} \tag{17}$$

In this configuration the only nonzero angular momenta that couple according to the scheme given above $L^{0'0} = ab \equiv l$ and $L^{12} = ab \equiv l$. Therefore, instead of Eqs. (11,10) we now obtain, with $g \in$PSU(2,2|4),

$$L = -\frac{l}{2} Str\left(\partial_\tau g g^{-1}\hat{\Gamma}\right), \ J_A^{\ B} \sim \left(g^{-1}\hat{\Gamma}g\right)_A^{\ B} \ , \ \hat{\Gamma} \equiv \begin{pmatrix} \Gamma^{0'0} & 0 \\ 0 & -\Gamma^{12} \end{pmatrix} , \tag{18}$$
$$g \in \mathrm{PSU}(2,2|4)\ /\ [\mathrm{PSU}(2|2) \times \mathrm{PSU}(2|2)] .$$

In an appropriate basis $\Gamma^{0'0}, \Gamma^{12}$ can be taken to be diagonal 4×4 matrices, each with two $+1$ and two -1 eigenvalues, therefore $\hat{\Gamma}$ is the diagonal matrix $diag\left(1_2, -1_2, -1_2, 1_2\right)$. Any PSU(2,2|4) generator that commutes with this matrix is a remaining gauge symmetry of this action. Thus, there is still the gauge symmetry [PSU(2|2)xPSU(2|2)]. The first PSU(2|2) acts on rows 1,2,7,8 of g ($+1$ eigenvalues of $\hat{\Gamma}$) while the second PSU(2|2) acts on rows 3,4,5,6 (-1 eigenvalues of $\hat{\Gamma}$). Therefore, after removing the gauge degrees of freedom the g in Eq. (18) belongs only to the coset $g \in$ PSU(2,2|4) / [PSU(2|2)xPSU(2|2)]. We can count the number of physical degrees of freedom as follows. The full PSU(2,2|4) supergroup contains $15+15$ real bosons and 16 complex fermions, thus altogether $30_B + 32_F$. The gauge subgroup PSU(2|2) contains 3+3 bosons and 4 complex fermions. Therefore PSU(2|2)xPSU(2|2) has $12_B + 16_F$ real gauge degrees of freedom. The physical degrees of freedom in the coset is the difference, which amounts to $18_B + 16_F$. As expected this is the same number as the $9x$, $9p$, 16 θ we counted for the $AdS_5 \times S^5$ superparticle above. The hidden SU(2,2|4) symmetries of the superparticle are also present here and made evident in this twistor gauge. Hence, one alternative description of $AdS_5 \times S^5$ is the coset given above. This is a new observation.

We can now rewrite this in terms of constrained twistors, as we did for $d = 4,6$ above. Rather than removing all of the gauge degrees of freedom we allow some of it to remain. Then from g we can extract four twistors which we call Z_{Aa} with the following properties[6]

$$\begin{array}{|l|} \hline Z_{Aa} = \overset{A=1,\cdots,8 \quad a=1,2 \ \ a=3,4}{\begin{pmatrix} bose & fermi \\ fermi & bose \end{pmatrix}} \quad \begin{array}{l} \text{8x4 rectangular matrix} \\ \text{4 fundamental reps of PSU(2,2|4)} \end{array} \\ \hline Z_{Aa} \ = \ (8,4) \ \text{of} \ PSU(2,2|4)_{global} \times [PSU(2|2) \times U(1)]_{local} \\ \hline L = \bar{Z}^{Aa}\left((\partial + A)Z\right)_{Aa} \rightarrow \bar{Z}^{aA} Z_{Ab} = 0, \quad \begin{array}{l} \text{take gauge invariants} \\ \text{in Fock space} \end{array} \\ \hline \end{array} \tag{19}$$

---

[6] These correspond to the four middle columns of g^{-1}, or equivalently the first two and last two columns of g^{-1}, as discussed in [9].

The first two twistors $a = 1,2$ each has four bosons in 4 of $SU(2,2)$ and four fermions in 4 of $SU(4)$. The last two twistors $a = 3,4$ each has four fermions in 4 of $SU(2,2)$ and four bosons in 4 of $SU(4)$. On the basis $a = 1,2,3,4$ we act with a gauge symmetry $PSU(2|2)xU(1)$, hence the Lagrangian is invariant under $PSU(2,2|4)_{global} \times [PSU(2|2) \times U(1)]_{local}$. Let us count the degrees of freedom. The complex Z_{Aa} has (16_B+16_F)x$2_{(\text{complex})}$, therefore 32_B+32_F. PSU(2|2)$\times$U(1) has $(3+3+1)_B+8_F$ real gauge parameters. The gauge parameters together with the corresponding constraints remove 14_B+16_F. Therefore the physical degrees of freedom in Z_{Aa} is the difference, namely 18_B+16_F, which is the same as the $9x$, $9p$, 16 θ we counted for the $AdS_5 \times S^5$ superparticle above. The global symmetry is still $PSU(2,2|4)$, and it has become evident rather than hidden in the twistor version we have just described. Hence, a new alternative description of $AdS_5 \times S^5$ is the constrained twistors Z_{Ai} given above.

The quantum theory for this case can again be described by using the BG oscillator approach [12, 11]. But now the "color" group is the supergroup $PSU(2|2) \times U(1)$, and the discussion in [11] should be modified accordingly. This is mathematically a new case in the BG approach, and will be further analyzed in [3].

3.6. Bosonic twistors in any d

The methods above can be applied in any dimension with the purely bosonic group $G =SO(d,2)$ as listed in (8). In this case the analogs of the twistors Z_{Aa} (1/4 of the columns of g^{-1}) come out with lots of constraints among the entries in these rectangular matrices, as dictated by the spinor representation of the group $SO(d,2)$. The independent parameters correspond to the coset t_Γ as in Eqs. (12,13). These give the correct generalizations of twistors in the sense that they provide an alternative (twistor) description of the massless relativistic particle in d dimensions. It is a holographic image in the 2T structure. As d gets larger and larger beyond $d = 6$ it becomes cumbersome to try to describe these in terms of oscillators, because of the complexity of the constraints on the oscillators. However, if we relax the requirement of only particles, and admit also D-brane degrees of freedom, as in footnote (2), then the oscillator (or twistor) approach becomes again a very efficient tool. The detailed discussion will appear in [3].

4. 2T SUPERSTRINGS D=3,4,5,6,10

So far in this lecture we discussed superparticles and the associated supertwistors, and their physical spectra. These have a direct generalization to superstrings via the 2T superstring formalism given in [9]. Briefly, the action is

$$L_{2T}^{\pm} = \begin{array}{c} \partial_\pm X \cdot P^\pm - \frac{1}{2} A X \cdot X - \frac{1}{2} B_{\pm\pm} P^\pm \cdot P^\pm - C_\pm P^\pm \cdot X \\ -\frac{1}{2^{d/2-1}} Str \left(\partial_\pm g \bar{g} \begin{pmatrix} L_{MN}^{\pm} \Gamma^{MN} & 0 \\ 0 & 0 \end{pmatrix} \right) + L_G \end{array}$$

$X_M(\tau,\sigma), P_M^{\pm}(\tau,\sigma), L_{MN}^{\pm} = X_{[M} P_{M]}^{\pm}$, $g(\tau,\sigma)$ are now string fields, and $\partial_\pm = \frac{1}{2}(\partial_\sigma \pm \partial_\tau)$. Here $L_{2T}^{\pm}$ represent left/right movers, and similar to [6] there is open string boundary conditions. L_G is an additional degree of freedom that describes an internal current algebra for some SYM group G. The local and global symmetries are similar to those of the particle and are described in [9]. The global symmetry is G_d chosen for various d as before, G_d=OSp$(8|4)_3$,SU$(2,2|4)_4$,F$(4)_5$,OSp$(8|4)_6$. The particle gauge for these give usual type superstrings and the twistor gauge gives twistor superstrings, with the twistors described above. In the 2T philosophy each one of these have many duals that can be found and investigated by choosing various gauges. This is a completely open field of investigation at this time.

Similarly to the $d=10$ particle case we also consider the d=10 2T superstring

$$L_{2T}^{\pm}=\begin{array}{c}\partial_{\pm}\hat{X}\cdot\hat{P}^{\pm}-\frac{1}{2}A\hat{X}\cdot\hat{X}-\frac{1}{2}B_{\pm\pm}\hat{P}^{\pm}\cdot\hat{P}^{\pm}-C_{\pm}\hat{P}^{\pm}\cdot\hat{X}\\ -\frac{1}{8}Str\left(\partial_{\pm}g\bar{g}\begin{pmatrix}L_{MN}^{\pm}\Gamma^{MN} & 0\\ 0 & -L_{IJ}^{\pm}\Gamma^{IJ}\end{pmatrix}\right)\end{array}$$

where SO(10,2)→SO(4,2)×SO(6), $\hat{X}^M=\left(X^m,X^I\right)$, $\hat{P}_M^{\pm}=\left(P_m^{\pm},P_I^{\pm}\right)$, $g(\tau,\sigma)\in SU(2,2|4)$, and $L_{MN}^{\pm}=X_{[M}P_{M]}^{\pm}$, $L_{IJ}^{\pm}=X_{[I}P_{J]}^{\pm}$. The local and global symmetries are discussed in detail in [10], [9]. In the particle-type gauge, the spectrum in the particle limit is the same as linearized type IIB SUGRA compactified on $AdS_5\times S^5$. In the twistor gauge this theory is currently being investigated by using the twistors in Eq. (19). As usual , being a 2T theory we expect dual versions of the theory in other gauges. This could be very interesting in terms of M-theory.

5. CLOSING REMARKS

I have described the following established facts about 2T-physics

- 2T-physics, with local Sp(2,R) & generalizations, gives emergent dynamics/spacetimes via d+2 →d holography.
- 1T-physics corresponds to d-dimensional holographs of the $d+2$ theory. Various holographs are dual; and the duality group is Sp$(2,R)$ & generalizations.
- When $d+2$ is in flat space each holograph has hidden SO$(d,2)$ symmetry. Its quantum Hilbert space forms a basis for same eigenvalues of the Casimirs of SO$(d,2)$. This applies with or without spin or supersymmetry.
- Twistor space is a particular hologram of the $d+2$ theory. In the twistor gauge the SO$(d,2)$ becomes more manifest as compared to other holograms.
- The $10+2$ twistor string, and its particle version show hidden $(10,2)$ holographic structures in $AdS_5\times S^5$ superstring and supergravity. These are strong indications that other aspects of M theory also has a 2T description. See related remarks in [9] that connect to [27, 29].

At this point it is hard to resist to also make some speculations, as follows

- The currently known corners of M-theory are very likely holograms of the same nature. The known M-dualities appear to be analogs of the Sp(2,R) & generalizations. This provides hints for the underlying gauge symmetry.

- Together with the earlier indications described in the introduction, it seems now even more likely that M-theory would be most clearly formulated as a 13D theory with signature $(11,2)$ and global supersymmetry $\mathrm{OSp}(1|64)$.

6. APPENDIX: TWISTOR STRING WITH SO$(3,1)$ SIGNATURE

Many people in this conference raised the question of the analytic continuation of $SO(3,1)$ to $SO(2,2)$ in the formulation as well as in the computations involving the twistor superstring. Since I gave some thought to this point, I outline below what the differences are when one uses the correct signature $SO(3,1)$. There are definite *changes in the formulation of the theory*, beyond the naive analytic continuation that inserts i in appropriate places, as follows. I work in the Berkovits formulation as it appears in [8], since this is the form of the 2T theory when we gauge fix the 2T superstring to the twistor gauge [9].

When the signature is (3,1), it requires the twistor Y^A in [8] to be the complex conjugate of the twistor Z_A, except for the metric, namely $Y^A = \bar{Z}^A$ and seen in Eq. (14). Therefore, Z_A and $\bar{Z}^A$ must have the same worldsheet conformal dimension $1/2$. This differs from Berkovits's $\dim(Z) = 0$ and $\dim(Y) = 1$. There is a definite consequence: There must be a shift in the stress tensor T, and in all the dimensions of the wavefunctions of physical vertices, as remarked in [9]. The following table gives the shifts

signature	(2,2)	(3,1)
stress tensor dimensions	$T = Y^A \partial Z_A$ 2 1 1 0	$\tilde{T} = \frac{1}{2}\bar{Z}^A \partial Z_A - \frac{1}{2}\partial \bar{Z}^A Z_A$ 2 1/2 1 1/2
SYM vertex op dimensions	$V_\phi = j^a \phi_a(Z)$ 1 1 0	$V_\phi = j^a \phi_a(Z)$ no changes 1 1 0 Z↔tZ dim 1/2 OK
Conf. SUGRA helicity +2, dims	$V_f = Y^A f_A(Z)$ 1 1 0	$V_f = \bar{Z}^A f_A(Z)$ 1 1/2 1/2 Z↔tZ
Conf. SUGRA helicity -2, dims	$V_g = \partial Z_A g^A(Z)$ 1 1 0 0	$V_g = \partial Z_A g^A(Z)$ 1 1 $\frac{1}{2}$ $-\frac{1}{2}$ Z↔tZ
Amplitudes instanton number	$T_n = T + n\partial(YZ)$ n	$\tilde{T}_n = \tilde{T} + n\partial(\bar{Z}Z)$ $=T_n - \frac{1}{2}\partial(YZ) \leftrightarrow n \to n-1/2$

Note that the SYM vertex $\phi_a(Z)$ has still dimension zero, but now it must be constructed from Z that has dimension $1/2$ instead of Z that has dimension 0. However since $\phi_a(Z)$ is homogeneous it means it is constructed from ratios of components of Z_A. Therefore the same wavefunctions will still appear, and so it seems that nothing changes in the SYM sector. This seems to be good news, at least on the surface.

Although the SYM wavefunction $\phi_a(Z)$ has the same dimension for either $SO(3,1)$ or $SO(2,2)$, the conformal supergravity wavefunctions $f_A(Z), g^A(Z)$ must have different dimensions, as shown in the table. I had hoped that $f_A(Z)$ with $\dim(f) = \frac{1}{2}$ and $g^A(Z)$ with $\dim(g) = -\frac{1}{2}$ would not exist (and therefore get rid of the "conformal gravity pollution" in the theory); but apparently they do exist without much modification also.

Another change is the stress tensor itself. The shift is equivalent to a shift in the instanton number $n \to n - 1/2$. I have not checked the details if this is cancelled by additional modifications, and whether this is harmless as well.

REFERENCES

1. I. Bars, C. Deliduman and O. Andreev, Phys. Rev. **D58**, 066004 (1998), [arXiv:hep-th/9803188]. For reviews of subsequent work see: I. Bars, *"Two-Time Physics"*, in the Proc. of the 22nd Intl. Colloq. on Group Theoretical Methods in Physics, Eds. S. Corney at. al., World Scientific 1999, [arXiv:hep-th/9809034];
Class. Quant. Grav. **18**, 3113 (2001), [arXiv:hep-th/0008164];
"2T-physics 2001," AIP Conf. Proc. **589** (2001), pp. 18–30; AIP Conf. Proc. **607** (2001), pp. 17–29 [arXiv:hep-th/0106021].
2. I. Bars, Phys. Lett. **B483**, 248 (2000), [arXiv:hep-th/0004090].
3. I. Bars and M. Picon, in preparation.
4. E. Witten, Commun. Math. Phys. **252**, 189 (2004), [arXiv:hep-th/0312171]; *"Parity invariance for strings in twistor space"*, hep-th/0403199.
5. F. Cachazo, P. Svrcek and E. Witten, JHEP **0409**, 06 (2004), [arXiv:hep-th/0403047]; JHEP **0410**, 074 (2004), [arXiv:hep-th/0406177]; JHEP **0410**, 077 (2004), [arXiv:hep-th/0409245].
6. N. Berkovits, Phys. Rev. Lett. **93**, 011601 (2004), [arXiv:hep-th/0402045].
7. N. Berkovits and L. Motl, JHEP **0404**, 56 (2004), [arXiv:hep-th/0403187].
8. N. Berkovits and E. Witten, JHEP **0408**, 009 (2204), [arXiv:hep-th/0406051].
9. I. Bars, Phys. Rev. **D70**, 104022 (2004), [arXiv:hep-th/0407239].
10. I. Bars, Phys. Rev. **D66**, 105024 (2002), [arXiv:hep-th/0208012].
11. I. Bars, Phys. Rev. **D66**, 105023 (2002), [arXiv:hep-th/0205194].
12. I. Bars and M. Günaydin, Comm. Math. Phys. **91**, 31 (1983).
13. M. Günaydin, P. van Nieuwenhuizen, N.P. Warner, Nucl. Phys. **B255**, 63 (1985).
14. E. Witten, *"Conformal field theory in four and six dimensions"*, In *Oxford 2002, Topology, geometry and quantum field theory*, p.405.
15. I. Bars, *"Duality and Hidden Dimensions"*, hep-th/9604200, published in Frontiers in Quantum Field Theory, Eds. H. Itoyama et. al., World Scientific (1996), page 52;
See also, Phys. Rev. **D54**, 5203 (1996), [=hep-th/9604139].
16. I. Bars, S-theory, Phys. Rev. **D55**, 2373 (1997), [arXiv:hep-th/9607112];
"Algebraic structure of S-theory", hep-th/9608061;
Phys. Rev. **D55**, 3633 (1997), [arXiv:hep-th/9610074];
Phys. Lett. **B403**, 257 (1997), [arXiv:hep-th/9704054].
17. C. Vafa, Phys. **B469**, 403 (1996), [arXiv:hep-th/9602022].
18. I. Bars and C. Kounnas, Phys. Lett. **B402**, 25 (1997), [arXiv:hep-th/9703060];
Phys. Rev. **D56**, 3664 (1997), [arXiv:hep-th/9705205];
I. Bars and C. Deliduman, Phys. Rev. **D56**, 6579 (1997), [arXiv:hep-th/9707215];
Phys. Lett. **B417**, 240 (1998), [arXiv:hep-th/9710066].
19. I. Bars, C. Deliduman and D. Minic, Phys. Lett. **B457**, 275 (1999), [arXiv:hep-th/9904063].
20. I. Bars, *"Toy M-model"*, unpublished. For a brief outline of the model see [19], [2] and third reference in [1].
21. I. Bars, Phys. Rev. **D58**, 066006 (1998), [arXiv:hep-th/9804028];
Phys. Rev. **D59**, 045019 (1999), [arXiv:hep-th/9810025].
22. I. Bars, Phys. Rev. **D62**, 085015 (2000), [arXiv:hep-th/0002140];
I. Bars and C. Deliduman, Phys. Rev. **D64**, 045004 (2001), [arXiv:hep-th/0103042].
23. I. Bars and C. Deliduman, Phys. Rev. **D58**, 106004 (1998), [arXiv:hep-th/9806085];
For additional quantum properties of the spinning theory in covariant quantization, including interactions, see the first reference in [26].
24. I. Bars, C. Deliduman and D. Minic, Phys. Rev. **D59**, 125004 (1999), [arXiv:hep-th/9812161].
25. I. Bars, C. Deliduman and D. Minic, Phys. Lett. **B466**, 135 (1999), [arXiv:hep-th/9906223].

26. I. Bars, Phys. Rev. **D62**, 046007 (2000), [arXiv:hep-th/0003100];
Phys. Rev. **D64**, 126001 (2001), [arXiv:hep-th/0106013];
I. Bars and S. J. Rey, Phys. Rev. **D64**, 046005 (2001), [arXiv:hep-th/0104135].
27. M.A. Vasiliev, " Higher spin superalgebras in any dimension and their representations", arXiv:hep-th/0404124.
28. I. Bars, Phys. Rev. **D52**, 3567 (1995), [arXiv:hep-th/9503228].
29. M. Bianchi, J. F. Morales and H. Samtleben, JHEP **0307**, 062 (2003), [arXiv:hep-th/0305052];
N. Beisert, M. Bianchi, J. F. Morales and H. Samtleben, JHEP **0402**, 001 (2004), [arXiv:hep-th/0310292];
"Higher spin symmetry and N = 4 SYM", [arXiv:hep-th/0405057].

Relativistic Spinor Dynamics Inducing the Extended Lorentz–Force–Like Equation

A. Bette* and J. Buitrago†

*The Royal Institute of Technology, KTH Syd, Campus Telge,
Mariekällgatan 3, S-151 81 Södertälje, Sweden
e-mail: bette@kth.se
†Department of Astrophysics of the University of La Laguna,
C/ Via Lactea, s/n, 38205, La Laguna, Tenerife, Spain.
e-mail: jbg@ll.iac.es

INTRODUCTION

Some years ago it has been noticed by J. Buitrago in [7] that Lorentz force equation may be regarded as a consequence of the action of infinitesimal Lorentz transformations on the velocity four-vector of a relativistic particle, where the parameters of the infinitesimal Lorentz transformations (i.e. of infinitesimal boosts and infinitesimal rotations) are regarded as functions of the position of the particle and not just constants. If these infinitesimal parameters are identified with the components of an external electromagnetic field (evaluated at the four-position x of the particle) multiplied by the infinitesimal lapse of the particle's proper time then the Lorentz force equation will follow automatically.

Let us reassume Buitrago's contribution. Consider the infinitesimal Lorentz transformation in the Minkowski four-vector space M_v:

$$u^a(s+\Delta s) = [\delta_b{}^a + \Delta\, L^a_b(x(s))]u^b(s), \tag{1}$$

where u^a is a time-like or space-like Lorentz four-vector and where $x(s)$, in the Minkowski space-time M, denotes a trajectory labelled by a parameter s.

The infinitesimal Lorentz transformation in (1) is defined by:

$$\Delta\, L^a_b(x(s)) := [\frac{1}{2}\alpha^{cd}(x^k(s))M_{cd}{}^a{}_b]\Delta s, \tag{2}$$

where a,b,c,d indices refer to the Lorentz four-vector and four-tensor character of the introduced quantities, so that e.g. $\delta_b{}^a$ is the Lorentz identity operator (matrix) etc.. The six generators of the Lorentz transformations $M_{cd}{}^a{}_b$ are, in the Minkowski four-vector

CP767, *Fundamental Interactions and Twistor-Like Methods,* edited by J. Lukierski and D. Sorokin

space, defined by the following constant Lorentz tensor of fourth rank[1]:

$$M_{cd\ b}^{\ \ a} = -M_{dc\ b}^{\ \ a} := \delta_c^{\ a}\ \eta_{db} - \delta_d^{\ a}\ \eta_{cb}, \tag{3}$$

where η_{kl} denotes the Minkowski metric. $\alpha^{cd}(x^a(s))\Delta s = -\alpha^{dc}(x^a(s))\Delta s$ represent the infinitesimal parameters of the infinitesimal Lorentz transformations changing continuously along the trajectory $x(s)$, while Δs measures the lapse of the parameter s along the trajectory. Using (3) it is easy to see that:

$$\frac{1}{2}\alpha^{cd}(x^k(s))M_{cd\ b}^{\ \ a} = \frac{1}{2}\alpha^{cd}(x^k(s))(\delta_c^{\ a}\ \eta_{db} - \delta_d^{\ a}\ \eta_{cb}) = \alpha^a{}_b(x^k(s)). \tag{4}$$

Therefore the equation in (1) may be now written as:

$$\frac{u^a(s+\Delta s) - u^a(s)}{\Delta s} = \alpha^a{}_b(x^k(s))u^b, \tag{5}$$

where the identity in (4) has been used. Taking the limit $\Delta s \to 0$ in (5) gives:

$$\frac{du^a}{ds} = \alpha^a{}_b(x^k(s))u^b. \tag{6}$$

It is easy to see from (6) that the necessarily non-zero Lorentz norm of the four-vector u^a is preserved[2] along the trajectory $x(s)$.

Two additional assumptions about u^a namely that it is a time-like four-vector and that it represents a four-velocity of a physical massive system with a well-defined four-position x in M amounts to[3] the following conditions:

$$u^a = \frac{dx^a}{ds}, \quad u^a u_a = 1. \tag{7}$$

The two equations in (6)–(7) can be regarded as a second order Lorentz-force-like differential equation:

$$\frac{d^2x^a}{ds^2} = \alpha^a{}_b(x^k)\frac{dx^b}{ds}, \tag{8}$$

whose solutions are the assumed trajectories $x(s)$ in M, while the parameter s is then recognized as the proper time of the system with the four-velocity u^a.

---

[1] as well-known from quantum mechanics the same generators multiplied by a purely imaginary number e.g. $i := \sqrt{-1}$ are identified with the intrinsic spin one angular four-momentum operator.

[2] because of α's antisymmetric Lorentz tensor character.

[3] in this paper the signature convention of the Lorentz metric in the Minkowski space M and the corresponding Minkowski vector space M_v is $+---$.

If we, in addition, assume that the function α, defining the infinitesimal parameters of the infinitesimal Lorentz transformations as given in (1) and (2), is proportional to the external electromagnetic field present along the system's trajectory in the following simplest possible way:

$$\alpha^{cd}(x^m(s)) := \frac{e}{m} F^{cd}(x^m(s)), \tag{9}$$

then the equation in (8) is not only Lorentz-force-like but becomes exactly the Lorentz force equation.

The classical dynamical principle leading to the equation in (6) where u^a is a time-like or a space-like Lorentz four-vector following a trajectory in M is very simple and fully geometrical. The key point for its validity is the identity proved in (4).

At the classical relativistic level that we discuss in this paper there is in this dynamical principle nothing to tell us how to choose a second rank antisymmetric Lorentz tensor valued function α in (2). We made a choice in (9) and that together with the two assumptions about u^a in (7) produced the Lorentz force equation. However, whatever choice and assumptions we make, quite generally, we get (6) from purely geometrical considerations.

In order to be able to follow the ideas presented in the sequel, the reader must have some basic knowledge about Weyl spinors and their relation to the Lorentz tensors at least to the extent as presented on the first pages of e.g. [1, 10, 11, 12, 13, 15]. Basic knowledge of the philosophy behind Penrose's Twistor Theory can also be of value when trying to understand the ideas that led us to the results obtained in this paper.

EXTENDED LORENTZ-FORCE-LIKE EQUATION

Consider the four Lorentz invariant, geometrically induced equations such as in (6):

$$\dot{P}^a := \frac{dP^a}{ds} = \alpha^{ab}(x^c(s))\, P_b, \tag{10}$$

$$\dot{S}^a := \frac{dS^a}{ds} = \alpha^{ab}(x^c(s))\, S_b, \tag{11}$$

$$\dot{V}^a := \frac{dV^a}{ds} = \alpha^{ab}(x^c(s))\, V_b, \tag{12}$$

$$\dot{W}^a := \frac{dW^a}{ds} = \alpha^{ab}(x^c(s))\, W_b, \tag{13}$$

where P is time-like while S, V and W are three space-like Lorentz four-vectors following a trajectory $x(s)$ in M and fulfilling the following conditions along the trajectory:

$$m^2 := P^a P_a = -S^a S_a = -V^a V_a = -W^a W_a \neq 0, \tag{14}$$

$$S^a P_a = 0, \;\; V^a P_a = 0, \;\; W^a P_a = 0, \;\; S^a W_a = 0, \;\; S^a V_a = 0, \;\; W^a V_a = 0. \tag{15}$$

Note that m^2, the square of the norm of the time-like four-vector P, is a constant of motion. In (10)–(15) we thus defined dynamics of an orthogonal tetrad of Lorentz four-vectors P, S, V and W, following an, as yet unspecified, trajectory in M, along which it is infinitesimally Lorentz transformed by an external α field.

Now we make an additional assumption about the time-like four-vector P identifying it with the linear four-momentum of an object following the trajectory $x(s)$ in M. Therefore we require additionally:

$$\dot{x}^a := \frac{dx^a}{ds} = \frac{P^a}{\sqrt{P^b P_b}} = \frac{P^a}{m}, \tag{16}$$

so that the parameter s may again be recognized as the proper time of the object moving along the trajectory $x(s)$. This fact follows trivially from (16) because it implies that $\dot{x}^a \dot{x}_a = 1$.

The equations in (10) and in (16) define together the (usual) Lorentz-force like equation while the additional equations in (11)–(13) obeying the conditions displayed in (14)–(15) form the extension of the Lorentz-force-like equation alluded to in the title of this section. This extension defines new degrees of freedom of the object following the trajectory $x(s)$ in M. These degrees of freedom may be associated with intrinsic spin of the object. See a short discussion concerning this issue below.

We wish also to stress the fact that the dynamical equations in (10)–(13), describing dynamics of the four four-vectors P, S, V and W, are a consequence of the geometrical considerations as briefly discussed in the introduction [7].

To represent the four-vectors P, S, V and W we will use a pair of non-proportional spinors π and η. By doing so we will *automatically* fulfil the conditions in (14)–(15). Let therefore the time-like four-vector P and the three space-like four-vectors S, V, W be defined spinorially [13] in the standard way (abstract index notation in the sense of Penrose [10] is used when appropriate):

$$P_a := P_{AA'} = \pi_{A'}(s)\bar{\pi}_A(s) + \bar{\eta}_A(s)\eta_{A'}(s), \tag{17}$$
$$S_a := S_{AA'} = \pi_{A'}(s)\bar{\pi}_A(s) - \bar{\eta}_A(s)\eta_{A'}(s), \tag{18}$$
$$V_a := V_{AA'} = \pi_{A'}(s)\bar{\eta}_A(s) + \bar{\pi}_A(s)\eta_{A'}(s), \tag{19}$$
$$W_a := W_{AA'} = i(\pi_{A'}(s)\bar{\eta}_A(s) - \bar{\pi}_A(s)\eta_{A'}(s)). \tag{20}$$

Using the definitions in (17)–(20), simple spinor algebra calculations show that the conditions in (14)–(15) are automatically fulfilled. Certain attempts, to formulate the traditional Lorentz force equation by the use of spinor representation of the linear four-momentum P as in (17), have been put forward previously [14].

Now we claim that besides this automatic fulfillment of (14)–(15), all the equations in (10)–(13) are induced by the following relatively simple dynamical spinor equations:

$$\frac{d\pi^{A'}}{ds} \equiv \dot{\pi}^{A'} = -c\,\pi^{A'} - b\,\eta^{A'}, \quad \frac{d\eta^{A'}}{ds} \equiv \dot{\eta}^{A'} = a\,\pi^{A'} + c\,\eta^{A'}, \tag{21}$$

where the complex valued Lorentz scalar functions a, b and c are given by:

$$a = \frac{\alpha^{S'T'}(x)\eta_{S'}\eta_{T'}}{(\pi^{K'}\eta_{K'})}, \quad b = \frac{\alpha^{S'T'}(x)\pi_{S'}\pi_{T'}}{(\pi^{K'}\eta_{K'})}, \quad c = -\frac{\alpha^{S'T'}(x)\pi_{S'}\eta_{T'}}{(\pi^{K'}\eta_{K'})}, \tag{22}$$

and where the symmetric second rank spinor $\alpha^{S'T'}(x)$, in the standard manner, represents the given external field α and where x represent events on the object's trajectory in M.

If this claim is true then this implies that the equation in (21) extends the geometrical principle as described by J. Buitrago in [7] to the space of the two spinors π, η. We now sketch the main parts of the proof showing that (21) induces the equations in (10)–(13):

Multiplying the first equation in (21) by $\bar{\pi}^A$ and the second by $\bar{\eta}^A$ gives:

$$\bar{\pi}^A\dot{\pi}^{A'} = \bar{\pi}^A(-c\,\pi^{A'} - b\,\eta^{A'}), \tag{23}$$

$$\bar{\eta}^A\dot{\eta}^{A'} = \bar{\eta}^A(a\,\pi^{A'} + c\,\eta^{A'}). \tag{24}$$

Taking the complex conjugates of (23) and (24) gives:

$$\pi^{A'}\dot{\bar{\pi}}^A = \pi^{A'}(-\bar{c}\,\bar{\pi}^A - \bar{b}\,\bar{\eta}^A), \tag{25}$$

$$\eta^{A'}\dot{\bar{\eta}}^A = \eta^{A'}(\bar{a}\bar{\pi}^A + c\,\bar{\eta}^A). \tag{26}$$

Adding the four equation in (23), (24), (25), (26) sidewise to each other gives:

$$\bar{\pi}^A\dot{\pi}^{A'} + \bar{\eta}^A\dot{\eta}^{A'} + \text{c.c.} =$$

$$= -[b\bar{\pi}^A\eta^{A'} + \bar{b}\bar{\eta}^A\pi^{A'} + (c+\bar{c})\bar{\pi}^A\pi^{A'}] + [a\bar{\eta}^A\pi^{A'} + \bar{a}\bar{\pi}^A\eta^{A'} + (c+\bar{c})\bar{\eta}^A\eta^{A'}]. \tag{27}$$

On the other hand we note that using spinor representation of the equation in (10) gives:

$$\dot{\bar{\pi}}^A\pi^{A'} + \bar{\pi}^A\dot{\pi}^{A'} + \dot{\bar{\eta}}^A\eta^{A'} + \bar{\eta}^A\dot{\eta}^{A'} = \alpha^{AA'BB'}(x)(\bar{\pi}_B\pi_{B'} + \bar{\eta}_B\eta_{B'}), \tag{28}$$

where $\alpha^{AA'BB'}(x)$ is the spinor equivalent of the antisymmetric Lorentz-tensor α^{ab} in (10). Spinor manipulating (28) further we note that the external four-force field α in (28) may, in the standard way, be represented by[4]:

$$\alpha^{ab}(x) = -\alpha^{ba}(x) = \alpha^{AA'BB'}(x) = -\alpha^{BB'AA'}(x) := \alpha^{A'B'}(x)\ \varepsilon^{AB} + \text{c.c..} \tag{29}$$

Now we decompose the spinor $\alpha^{A'B'}$ as follows:

$$\alpha^{A'B'}(x) = \frac{a}{\pi^{C'}\eta_{C'}}\pi^{A'}\pi^{B'} + \frac{c}{\pi^{C'}\eta_{C'}}\pi^{A'}\eta^{B'} + \frac{c}{\pi^{C'}\eta_{C'}}\eta^{A'}\pi^{B'} + \frac{b}{\pi^{C'}\eta_{C'}}\eta^{A'}\eta^{B'}, \tag{30}$$

with a, c, and b being defined as in (22). Using the decomposition in (30) we may rewrite (28) according to[5]:

$$\bar{\pi}^A\dot{\pi}^{A'} + \bar{\eta}^A\dot{\eta}^{A'} + \text{c.c.} =$$

$$= -[b\bar{\pi}^A\eta^{A'} + \bar{b}\bar{\eta}^A\pi^{A'} + (c+\bar{c})\bar{\pi}^A\pi^{A'}] + [a\bar{\eta}^A\pi^{A'} + \bar{a}\bar{\pi}^A\eta^{A'} + (c+\bar{c})\bar{\eta}^A\eta^{A'}], \tag{31}$$

proving our assertion that (10) is induced by (21). This follows simply from the fact that (31) and (27) are identical.

By imitating the above steps, using the spinor representations of S, V, and W, as defined in (18), (19), (20) it may be shown that also the equations in (11), (12), (13) are all induced by the *same* dynamical spinor equations in (21). We call therefore these equations the "master equations". This completes the proof that (10), (11), (12), (13) are induced by (21) while the conditions in (14)–(15) are automatically fulfilled due to the definitions in (17)–(20).

If we require in addition that (16), which when written spinorially, reads :

$$\dot{x}^{AA'} := \frac{dx^{AA'}}{ds} = \frac{\pi^{A'}\bar{\pi}^A + \bar{\eta}^A\eta^{A'}}{\sqrt{2}\,(\bar{\pi}^B\bar{\eta}_B)\,(\pi^{B'}\eta_{B'})} = (\frac{P^{AA'}}{m}), \tag{32}$$

is valid with x, as always, denoting points (events) along the object's trajectory in space-time, then the parameter s is once again recognized as the object's proper time parameter while P denotes its linear momentum four-vector.

---

[4] algebraically in (29) the force field is represented by either an antisymmetric Lorentz tensor of second rank or an hermitian spinor of fourth rank twice primed and twice unprimed or equivalently by a symmetric spinor of second rank either unprimed or primed, all these representations being physically equivalent. We use therefore the same generic letter α for these quantities.

[5] *c.c.* is short hand notation for complex conjugation.

The S, V and W in (18)–(20) may be thought of as defining the axis of inertia rigidly attached to an object with the linear four-momentum P. If so, then any space-like four-vector, formed as a linear combination of S, V and W (divided by m by obvious dimensional requirements) defines the, so called, Pauli-Lubański spin four-vector. Now, quick glance at any equation arising as any such linear combination of the equations in (18)–(20) and at the equation (11.164) in [8] reveals that the two equations are proportional to each other only if the gyromagnetic ratio g is equal to two. The "master equations" in (21) together with the requirement in (32) may thus be regarded as a classical relativistic limit of the equations of motion that describe dynamics of the (classical limit of the) spinning electron ($g = 2$) with any constant real value of its spin i.e. with any value of the norm of its Pauli-Lubański spin four-vector.

Note that the function defined by the positive real valued Lorentz scalar function:

$$P^a P_a = 2\,(\bar{\pi}^A \bar{\eta}_A)\,(\pi^{A'} \eta_{A'}) = m^2, \tag{33}$$

should then be identified with the square of the rest mass of the object (particle) while the function:

$$m = \sqrt{2}\,|\bar{\pi}^A \bar{\eta}_A| = \sqrt{2}\,|\pi^{A'} \eta_{A'}|, \tag{34}$$

defines its positive rest mass.

Note also that "master equations" in (21) imply that not only the function[6] $m \neq 0$, in (34), is a constant of motion but that entire complex valued Lorentz scalar function:

$$f := \pi^{A'} \eta_{A'} = f_0 \neq 0, \tag{35}$$

is also a constant of motion. This may be easily proved because by contracting the first equation in (21) with the spinor η and the second with the spinor π:

$$\eta_{A'}\, \dot{\pi}^{A'} = -c\, f, \tag{36}$$

$$\pi^{A'}\, \dot{\eta}_{A'} = c\, f, \tag{37}$$

and by adding (36) and (37) to each other, we find that the nonvanishing complex valued function f in (35) fulfils:

$$\dot{f} = 0 \;\text{ i.e. }\; f = f_0 = \text{const.}\; |f_0| > 0. \tag{38}$$

In the next section we show explicitly how the "master equations" (21) can be integrated in the case when the "magnetic" and "electric" fields are constant, equally valued and

[6] and thereby the function m^2 in (33).

perpendicular to each other and in the case of constant "electric" field or constant "magnetic" field or both of them constant and being parallel to each other (in some laboratory frame).

The obtained trajectories will, off course, be the very well known ones, see e.g. [9]. However we get some additional information about the motion of the remaining legs of the tetrad attached to our dynamical object. This may be interpreted as precession of the intrinsic spin vector (including the kinematic Thomas precession [8]) attached to the object[7].

Coping with the non-constant external α field is much more difficult and we present spinor equations for this general case in the last section of this paper.

TWO EXAMPLES AND THEIR SOLUTIONS

Concrete constructions of solutions require a choice of the external field and a choice of a suitable frame (corresponding to a laboratory frame in the Minkowski space-time) in the spinor space S. The basis in such a fixed (inertial) spin-frame (see e.g. [13, 15]) will be denoted by (ι, o):

$$\iota_A,\ o_A \ \text{and}\ \bar{\iota}_{A'},\ \bar{o}_{A'}\ , \tag{39}$$

and normalized by the requirement:

$$\iota^A o_A = 1,\ \bar{\iota}^{A'}\bar{o}_{A'} = 1. \tag{40}$$

The two dynamical spinors π, η may now, with respect to the chosen fixed spin frame, be expressed by means of their components:

$$\pi^{A'} = u\bar{\iota}^{A'} - z\bar{o}^{A'},\quad \eta^{A'} = v\bar{\iota}^{A'} - w\bar{o}^{A'}, \tag{41}$$

$$\bar{\pi}^{A} = \bar{u}\iota^{A} - \bar{z}o^{A},\quad \bar{\eta}^{A} = \bar{v}\iota^{A} - \bar{w}o^{A}, \tag{42}$$

where, in order to diminish the abundance of the indices, for the components of the two spinors π and η we introduced the notation:

$$\pi^{A'}\bar{o}_{A'} = u,\quad \pi^{A'}\bar{\iota}_{A'} = z,\quad \eta^{A'}\bar{o}_{A'} = v,\quad \eta^{A'}\bar{\iota}_{A'} = w. \tag{43}$$

$$\bar{\pi}^{A}o_{A} = \bar{u},\quad \bar{\pi}^{A}\iota_{A} = \bar{z},\quad \bar{\eta}^{A}o_{A} = \bar{v},\quad \bar{\eta}^{A}\iota_{A} = \bar{w}. \tag{44}$$

---

[7] an alternative classical limit of the dynamics (of a "charged" relativistic spinning and massive object) that starts from a twistor phase space formulation and the second order formulation of the minimally coupled Dirac equation has been introduced in [1, 2].

Using these coordinates the “master equations” in (21) become:

$$\dot{z} = -cz - bw, \quad \dot{w} = az + cw, \quad \dot{u} = -cu - bv, \quad \dot{v} = au + cv, \tag{45}$$

while the spinor product defining the function f (see (35)–(38)) which is a constant of motion reads:

$$f = zv - uw = f_0 = \text{constant} \neq 0. \tag{46}$$

The external Lorentz-force-like field α is, with respect to the chosen spin frame (ι, o) given by (see e.g. [15] p. 91):

$$\alpha^{A'B'}(x) = \alpha_0(x)\bar{\iota}^{A'}\bar{\iota}^{B'} - 2\alpha_1(x)\bar{\iota}^{(A'}\bar{o}^{B')} + \alpha_2(x)\bar{o}^{A'}\bar{o}^{B'}, \tag{47}$$

where we have:

$$\alpha_0(x) = \frac{(E_x - B_y) + i(E_y + B_x)}{2},$$

$$\alpha_1(x) = -\frac{E_z + iB_z}{2}, \quad \alpha_2(x) = -\frac{(E_x + B_y) - i(E_y - B_x)}{2}, \tag{48}$$

with $\vec{E}(x) = (E_x,\ E_y,\ E_z)$ and $\vec{B}(x) = (B_x,\ B_y,\ B_z)$ representing the applied Lorentz-force-like field[8]. Therefore from (47) it follows that the functions in (22) are, in the chosen frame, given by:

$$a = \frac{\alpha_0 w^2 - 2\alpha_1 vw + \alpha_2 v^2}{f_0}, \tag{49}$$

$$b = \frac{\alpha_0 z^2 - 2\alpha_1 uz + \alpha_2 u^2}{f_0}, \tag{50}$$

$$c = -\frac{\alpha_0 zw - \alpha_1 uw - \alpha_1 vz + \alpha_2 uv}{f_0}, \tag{51}$$

where f_0 is the constant of motion obtained in (38).

Taking into the account the relation in (32) the components of the linear momentum four-vector P are, with respect to the chosen spinor frame (that in the standard way defines the constant tetrad of an inertial frame in the Minkowski four-vector space) given by:

[8] Note that, in this context, our ”electric” ($\vec{E}$) and ”magnetic” ($\vec{B}$) fields have absorbed the factor $\frac{e}{m}$. In that way the dimension of the field α, in natural units, becomes inverse of the time.

$$P^{AA'}\iota_A\bar{\iota}_{A'} = \frac{m(\frac{dt}{ds}+\frac{dz}{ds})}{\sqrt{2}} = \frac{E+p_z}{\sqrt{2}} = z\bar{z}+w\bar{w}\,, \tag{52}$$

$$P^{AA'}o_A\bar{o}_{A'} = \frac{m(\frac{dt}{ds}-\frac{dz}{ds})}{\sqrt{2}} = \frac{E-p_z}{\sqrt{2}} = u\bar{u}+v\bar{v}\,, \tag{53}$$

$$P^{AA'}o_A\bar{\iota}_{A'} = \frac{m(\frac{dx}{ds}+i\frac{dy}{ds})}{\sqrt{2}} = \frac{p_x+ip_y}{\sqrt{2}} = z\bar{u}+w\bar{v}\,, \tag{54}$$

where $m = \sqrt{2f_0\bar{f}_0}$ and where s is the proper time of the system with the linear four-momentum P.

The three space-like four vectors S, V and W in (18)–(20) are, with respect to the chosen spinor frame, given by:

$$S^{AA'}\iota_A\bar{\iota}_{A'} = \frac{S_0+S_z}{\sqrt{2}} = z\bar{z}-w\bar{w},\;\; S^{AA'}o_A\bar{o}_{A'} = \frac{S_0-S_z}{\sqrt{2}} = u\bar{u}-v\bar{v}, \tag{55}$$

$$S^{AA'}o_A\bar{\iota}_{A'} = \frac{S_x+iS_y}{\sqrt{2}} = z\bar{u}-w\bar{v}, \tag{56}$$

$$V^{AA'}\iota_A\bar{\iota}_{A'} = \frac{V_0+V_z}{\sqrt{2}} = z\bar{w}+w\bar{z},\;\; V^{AA'}o_A\bar{o}_{A'} = \frac{V_0-V_z}{\sqrt{2}} = u\bar{v}+v\bar{u}, \tag{57}$$

$$V^{AA'}o_A\bar{\iota}_{A'} = \frac{V_x+iV_y}{\sqrt{2}} = z\bar{v}+w\bar{u}, \tag{58}$$

$$W^{AA'}\iota_A\bar{\iota}_{A'} = \frac{W_0+W_z}{\sqrt{2}} = i(z\bar{w}-w\bar{z}),\;\; W^{AA'}o_A\bar{o}_{A'} = \frac{W_0-W_z}{\sqrt{2}} = i(u\bar{v}-v\bar{u}), \tag{59}$$

$$W^{AA'}o_A\bar{\iota}_{A'} = \frac{W_x+iW_y}{\sqrt{2}} = i(z\bar{v}-w\bar{u}). \tag{60}$$

If the external field α is constant i.e. does not depend on the position x in M then we may separate two cases: the first case when the "electric" and "magnetic" parts of the α field are equal in magnitude and perpendicular to each other (compare with the solution of the problem 2 on page 58 in [9]) and the second case when the only non-vanishing part of the α field is the "electric" part or only the "magnetic" part or both "electric" and the "magnetic part are non-vanishing and parallel to each other (the mathematics of these three, just mentioned, options is the same so we call it case number two).

We now proceed to consider the case number one. For that reason we note that (45) and (49)–(51) simplify if we introduce the following constants and variables:

$$\beta_0 := \frac{\alpha_0}{f_0},\;\; \beta_1 := \frac{\alpha_1}{\alpha_0},\;\; \beta_2 := \beta_0\left(\frac{\alpha_2}{\alpha_0}-\left(\frac{\alpha_1}{\alpha_0}\right)^2\right), \tag{61}$$

$$w_1 := w - \beta_1 v, \;\; z_1 := z - \beta_1 u\,. \tag{62}$$

The expressions in (49)–(51) and in (46) now read:

$$a = \beta_0 (w_1)^2 + \beta_2 v^2, \;\; b = \beta_0 (z_1)^2 + \beta_2 u^2, \;\; c = -\beta_0 z_1 w_1 - \beta_2 uv, \;\; f_0 = z_1 v - u w_1\,. \tag{63}$$

We note further that:

$$\beta_2 = 0 \text{ implies that } \vec{E}\cdot\vec{B} = |\vec{E}|^2 - |\vec{B}|^2 = 0. \tag{64}$$

By putting $\beta_2 = 0$ the expressions for the first three functions in (63) become considerably simplified and we obtain:

$$a = \beta_0 (w_1)^2, \;\; b = \beta_0 z_1^2, \;\; c = -\beta_0 z_1 w_1\,. \tag{65}$$

The dynamical equations in (45) acquire now a form that is very easy to solve:

$$\frac{d(z_1 + \beta_1 u)}{ds} = -\alpha_0 \beta_1\, z_1\,, \;\; \frac{d(w_1 + \beta_1 v)}{ds} = -\alpha_0 \beta_1\, w_1\,, \tag{66}$$

$$\frac{du}{ds} = -\alpha_0\, z_1\,, \;\; \frac{dv}{ds} = -\alpha_0\, w_1\,. \tag{67}$$

Inserting the equations in (67) into the equations in (66) yields:

$$\frac{dz_1}{ds} = 0, \;\; \frac{dw_1}{ds} = 0\,. \tag{68}$$

Therefore for the constant external α field fulfilling the condition in (64) the following simple solutions for the coordinates of the two spinors are obtained from (67)–(68):

$$v = v_0 - \alpha_0 w_{10}\cdot s\,, \;\; u = u_0 - \alpha_0 z_{10}\cdot s\,, \tag{69}$$

$$w = w_{10} + \beta_1 v_0 - \alpha_1 w_{10}\cdot s\,, \;\; z = z_{10} + \beta_1 u_0 - \alpha_1 z_{10}\cdot s\,, \tag{70}$$

where the four complex numbers z_{10}, w_{10}, u_0, v_0 are constants of the spinor motion.

We choose now the y axis along the "electric" field and the z axis along the "magnetic" field and denote their common value by $B = B_z = E_y$. This gives that:

$$\alpha_1 = -\frac{iB}{2}\,, \;\; \alpha_0 = \frac{iB}{2}\,, \;\; \alpha_2 = \frac{iB}{2}\,, \;\; \beta_1 = \frac{\alpha_1}{\alpha_0} = -1\,, \tag{71}$$

while the solutions in (69)–(70) may now be written as follows:

$$v = v_0 - \frac{iB}{2} w_{10} \cdot s\,, \quad u = u_0 - \frac{iB}{2} z_{10} \cdot s\,, \tag{72}$$

$$w = w_{10} - v_0 + \frac{iB}{2} w_{10} \cdot s, \quad z = z_{10} - u_0 + \frac{iB}{2} z_{10} \cdot s\,. \tag{73}$$

Forming the components of the linear momentum P in (17) in the chosen frame (using (52)–(54)) out of the spinor components in (72)–(73) gives:

$$\frac{E + p_z}{\sqrt{2}} = (z_{10} - u_0)(\bar{z}_{10} - \bar{u}_0) + (w_{10} - v_0)(\bar{w}_{10} - \bar{v}_0) + \tag{74}$$

$$+ \frac{iB}{2}(\bar{z}_{10} u_0 - z_{10}\bar{u}_0 + \bar{w}_{10} v_0 - w_{10}\bar{v}_0) \cdot s + \frac{B^2}{4}(\bar{z}_{10} z_{10} + w_{10}\bar{w}_{10}) \cdot s^2,$$

$$\frac{E - p_z}{\sqrt{2}} = (v_0\bar{v}_0 + u_0\bar{u}_0) + \frac{iB}{2}(\bar{z}_{10} u_0 - z_{10}\bar{u}_0 \bar{w}_{10} v_0 - w_{10}\bar{v}_0) \cdot s + \frac{B^2}{4}(\bar{z}_{10} z_{10} + w_{10}\bar{w}_{10}) \cdot s^2, \tag{75}$$

$$\frac{p_x + ip_y}{\sqrt{2}} = (\bar{u}_0 z_{10} - u_0\bar{u}_0 + \bar{v}_0 w_{10} - v_0\bar{v}_0) + \frac{iB}{2}(\bar{z}_{10} z_{10} + w_{10}\bar{w}_{10} - u_0\bar{z}_{10} + \tag{76}$$

$$+ z_{10}\bar{u}_0 - \bar{w}_{10} v_0 + w_{10}\bar{v}_0) \cdot s - \frac{B^2}{4}(\bar{z}_{10} z_{10} + w_{10}\bar{w}_{10}) \cdot s^2.$$

Substracting (75) from (74) gives that p_z is a constant of motion:

$$p_z = \frac{(z_{10} - u_0)(\bar{z}_{10} - \bar{u}_0) + (w_{10} - v_0)(\bar{w}_{10} - \bar{v}_0) - (v_0\bar{v}_0 + u_0\bar{u}_0)}{\sqrt{2}} = \text{const.}$$

Adding (75) to (74) and to (76) and to its complex conjugate reveals that $E + p_x$ is also a constant of motion:

$$E + p_x = \frac{2\Re(\bar{u}_0 z_{10} + \bar{v}_0 w_{10}) - (\bar{u}_0 u_0 + \bar{v}_0 v_0) + (z_{10} - u_0)(\bar{z}_{10} - \bar{u}_0)}{\sqrt{2}} = \text{const.}$$

$\Re$ is short notation for the real part of. The trajectory in space-time is now obtained by a simple integration of (74)–(76) using the relation between the position four-vector and the linear four-momentum as displayed in (52)–(54). The trajectories coincide qualitatively with the ones obtained in [9] as they off course should.

Inserting the solutions in (72)–(73) into the definitions in (18)–(20) describes the motion (precession) of the tetrad attached to the object. It solves thereby the dynamical equation in (11)–(13) subject to the condition in (14)–(15). The precession consists of two parts one purely kinematical (the Thomas precession) and one dynamical induced by the "master equation".

We conclude that the first special case has a simple explicit analytical solution.

Now we proceed to analyze the second case when the "magnetic" part and the "electric" part of the α field are parallel to each other or any of them is vanishing. At the end we specialize to the case of constant "magnetic" field that gives a very well-known circular trajectory, however, the "master equations" imply also the precession of the legs of the tetrad in (18)–(20). Now we proceed to the construction of the explicit solution. Choose therefore the z axis as the common line along which the (possibly) two fields ("magnetic" and "electric") are directed. Then the formulas for the functions in (49)–(50) simplify again and read:

$$a = -\frac{2\alpha_1 vw}{f_0}\,, \quad b = -\frac{2\alpha_1 uz}{f_0}\,, \quad c = \frac{\alpha_1 uw + \alpha_1 vz}{f_0}\,. \tag{77}$$

Inserting the functions in (77) into the "master equation" in (45) and using (46) give a trivially simple set of four first order equations for the four components of the two spinors:

$$\dot{u} = \alpha_1 u\,, \;\; \dot{z} = -\alpha_1 z\,, \;\; \dot{v} = \alpha_1 v\,, \;\; \dot{w} = -\alpha_1 w\,. \tag{78}$$

The solutions of (78) are of course given by:

$$u = u_0 e^{\alpha_1 s}, \;\; z = z_0 e^{-\alpha_1 s}\,, \;\; v = v_0 e^{\alpha_1 s}, \;\; w = w_0 e^{-\alpha_1 s}\,, \tag{79}$$

where u_0, z_0, v_0 and w_0 are complex valued constants. Inserting (79) into the expressions in (52)–(60) gives final solution for this case. Therefore we may conclude that the second case with the constant external "magnetic" and "electric" field is also explicitly solved.

Exploring the second solution a bit further we note that in the special case of pure constant "magnetic" field along the z axis, since $\alpha_1 = -iB_z/2$, the integration constants u_0, z_0, v_0 and w_0 may be chosen as real. From (56) we find that S_x and S_y precess around the z axis according to

$$S_x = \sqrt{2}(z_0 u_0 - w_0 v_0)\cos B_z s\,, \tag{80}$$

$$S_y = \sqrt{2}(z_0 u_0 - w_0 v_0)\sin B_z s\,, \tag{81}$$

with similar behaviour for V_x, V_y, W_x and W_y. On the other hand, the S_z , V_z and W_z components remain constant, independent of the strength of the magnetic field, given by

$$S_z = \frac{\sqrt{2}}{2}(z_0^2 + v_0^2 - w_0^2 - u_0^2), \tag{82}$$

$$V_z = \sqrt{2}(z_0 v_0 - u_0 v_0), \;\; W_z = 0 \, . \tag{83}$$

SPINOR DYNAMICS

If the external field α is not constant in space-time, the integration of the "master equations" in (21) cannot be performed directly, as done in the previous section, because the functions a, b and c depend explicitly on the object's position four-vector x. To remedy this, we let the position four-vector of the object be also translated into its spinorial form by the following construction[9]:

$$x^a := x^{AA'} = h_1(s)\, \pi^{A'}\bar{\pi}^A + h_2(s)\;\; \bar{\eta}^A\eta^{A'} + h_3(s)\, \pi^{A'}\bar{\eta}^A + \bar{h}_3(s)\, \bar{\pi}^A\eta^{A'}, \tag{84}$$

where h_1 and h_2 are real valued Lorentz, s parameter dependent, scalar functions and where h_3 is a complex valued Lorentz, s parameter dependent, scalar function.

Contracting[10] (32) with the two spinors $\bar{\eta}$ and $\bar{\pi}$ gives:

$$\bar{\eta}_A\dot{x}^{AA'} = \frac{\bar{f}\pi^{A'}}{\sqrt{2\,(\bar{\pi}^B\bar{\eta}_B)(\pi^{B'}\eta_{B'})}}, \tag{85}$$

$$\bar{\pi}_A\dot{x}^{AA'} = -\frac{\bar{f}\eta^{A'}}{\sqrt{2\,(\bar{\pi}^B\bar{\eta}_B)(\pi^{B'}\eta_{B'})}}. \tag{86}$$

Taking derivative of (84) with respect to s gives:

$$\begin{aligned} \dot{x}^{AA'} &= \dot{h}_1(s)\, \pi^{A'}\bar{\pi}^A + \dot{h}_2(s)\;\; \bar{\eta}^A\eta^{A'} + \dot{h}_3(s)\, \pi^{A'}\bar{\eta}^A + \dot{\bar{h}}_3(s)\, \bar{\pi}^A\eta^{A'} \\ &\quad + h_1(s)\, \dot{\pi}^{A'}\bar{\pi}^A + h_2(s)\;\; \bar{\eta}^A\dot{\eta}^{A'} + h_3(s)\, \dot{\pi}^{A'}\bar{\eta}^A + \bar{h}_3(s)\, \bar{\pi}^A\dot{\eta}^{A'} \\ &\quad + h_1(s)\, \pi^{A'}\dot{\bar{\pi}}^A + h_2(s)\;\; \dot{\bar{\eta}}^A\eta^{A'} + h_3(s)\, \pi^{A'}\dot{\bar{\eta}}^A + \bar{h}_3(s)\, \dot{\bar{\pi}}^A\eta^{A'} \, . \end{aligned} \tag{87}$$

Inserting the right hand sides of the "master equations" (21) (and their complex conjugates) into (87), contracting[11] the obtained result at first with $\bar{\eta}_A$ and thereafter with $\bar{\pi}_A$ yield two expressions that can be compared with right hand side of (85)–(86). By that

---

[9] this is simply a projection of the position four-vector on the four legs of the dynamical tetrad defined by the four four-vectors P, S, V and W.

[10] note that assuming the validity of (32) amounts to identification of the parameter s with the proper time of the object.

[11] it is easier first to contract and later to insert the "master equations".

we obtain the following four first order differential equations for the scalar functions $h_1(s), h_2(s), h_3(s)$:

$$\frac{dh_1}{ds} - h_1(c+\bar{c}) + h_3\bar{a} + \bar{h}_3 a = \frac{1}{m}\,, \quad \frac{dh_2}{ds} + h_2(c+\bar{c}) - (h_3 b + \bar{h}_3\bar{b}) = \frac{1}{m}\,,$$

$$\frac{dh_3}{ds} - h_1\bar{b} + h_2 a - h_3 c + h_3\bar{c} = 0\,. \tag{88}$$

"Master equations" in (21) and the equations in (88) imply that the parameter s is the proper time of the object. Together they form a closed system of first order differential equations that induces the equations in (10)–(13) fulfilling (14)–(15) and the physical requirement (16).

Note that space-time dynamics described in this abstract way (entirely by the equations in (21) and (88)) do not make use of the notion of the space-time manifold at all. Position events traced out by a relativistic system become secondary constructions.

SUMMARY AND REMARKS

According to the principle presented in [7] and summarized in the introduction, any Lorentz force-like equation can be regarded as a consequence of the geometry of the Minkowski four-vector space. This principle is here extended to the spinor space and allows us to discover a set of coupled spinor equations that describe dynamics of a massive and spinning classical object (with g=2). The solutions of these equations describe not only the world trajectories of the object under study but also the degrees of freedom that can be associated to its intrinsic classical (limit of its quantum mechanical discrete) continuous spin values. It would be very interesting to know whether it is possible to find a Lorentz invariant hamiltonian/lagrangian formulation from which the equations in (21) and in (88) can be derived. If it is possible then a subsequent quantization procedure would produce a Lorentz invariant first quantized relativistic theory describing a massive, spinning (for any quantized value of the spin) object acted upon by an applied external α. These speculations, however, we postpone to future investigations.

ACKNOWLEDGMENTS

One of the authors (A.B.) would like to thank "IAC" (Instituto de Astrofisica de Canarias in Tenerife, Spain) for the financial support. He also wishes to thank The Royal Institute of Technology-KTH Syd for providing him with the so called "fakir" funds that freed him from the teaching duties at the time of this research. We also thank Professors Evencio Mediavilla and the members of his group for their hospitality at IAC, where this work was partly carried out and finally accomplished within the frame of the IAC project number 6/88.

REFERENCES

1. A. Bette, *Twistors, Special relativity, conformal symmetry and minimal coupling - a review.*, Reporte de Investigacion No. 26, Septiembre 2003, Universidad de Sonora, Division de Ciencias Exactas y Naturales, Rosales y Blvd. Luis Encinas J., Edif. 3K1, 83000 Hermosillo, Sonora, Mexico (2003).
2. A. Bette, *Twistor Approach to Relativistic Dynamics and to the Dirac Equation.* in *Clifford Algebras and their Applications in Mathematical Physics, Volume 1: Algebra and Physics* Editors: Rafał Abłamowicz, Bernard Fauser, Progress in Physics **18**, Birkhäuser, Boston, Basel, Berlin, 75–92 (2000).
3. A. Bette, J. Math. Phys. **7** (4), 1724–1734 (1996).
4. A. Bette, J. Math. Phys. **4** (10), 4617–4627 (1993).
5. A. Bette and S. Zakrzewski, J. Phys. A: Math. and Gen.: **30** 195–209 (1997).
6. A. Bette and S. Zakrzewski, *Massive relativistic systems with spin and the two twistor phase space.* in *The proceedings of the XII-th workshop on "soft" physics, Hadrons-96: Confinement*, Novy Svet, Crimea, June 9-16, 1996, National Academy of Sciences of Ukraine, Bogoliubov Institute for Theoretical Physics (Kiev), Simferopol State University (Crimea), Université Claude Bernard de Lyon, Kiev, pp. 336–346 (1996).
7. J. Buitrago, Eur. J. Phys. **16**, 113–118 (1995).
8. J.D. Jackson, *Classical Electrodynamics.*, 3rd Edition ISBN: 0-471-30932-X (1998).
9. L. Landau and E.M Lifshitz, *Classical theory of fields*, Pergamon Press. (1987) (reprinted 2000, ISBN 0750627689).
10. R. Penrose, in: *Batelle rencontres* , Eds. C.M. de Witt, J.A. Wheeler Princeton University W.A. Benjamin inc. New York, Amsterdam, 135–149 (1968).
11. R. Penrose, J. Math. Phys. **8** (2), 345–366 (1967).
12. R. Penrose and M.A.H MacCallum, Phys. Rep. **6** (4), 241–316 (1972).
13. R. Penrose and W. Rindler, *Spinors and Space-Time*, Cambridge Monographs on Mathematical Physics vol.1 and 2, Cambridge University Press, (1984) and (1986) Cambridge.
14. J. Plebański, private communication, (1983) Mexico City.
15. J. Stewart, *Advanced General Relativity*, Cambridge University Press, (1990) Cambridge Monographs on Mathematical Physics.

Two-Twistor Space, Commuting Composite Minkowski Coordinates and Particle Dynamics

A. Bette*, J. Lukierski† and C. Miquel-Espanya**

* *Royal Institute of Technology*
S-151 81 Södertälje, Sweden
†*Institute for Theoretical Physics,*
University of Wrocław,
pl. M. Borna 9, 50-204 Wrocław, Poland
***Departament de Física Teòrica i IFIC,*
Universitat de València,
Dr. Moliner 50, 46100 Burjassot (València), Spain

Abstract. We employ the modification of the basic Penrose formula in twistor theory, which allows to introduce commuting composite space-time coordinates. It appears that in the course of such modification the internal symmetry $SU(2)$ of two-twistor system is broken to $U(1)$. We consider the symplectic form on two-twistor space, permitting to interpret its 16 real components as a phase-space. After a suitable change of variables such a two-twistor phase space is split into three mutually commuting parts, describing respectively the standard relativistic phase space (8 degrees of freedom), the spin sector (6 degrees of freedom) and the canonical pair angle-charge describing the electric charge sector (2 degrees of freedom). We obtain a geometric framework providing a twistor-inspired 18-dimensional extended relativistic phase space $\mathscr{M}^{18}$. In such a space we propose the action only with first class constraints, describing the relativistic particle characterized by mass, spin and electric charge.

1. INTRODUCTION

The choice of basic geometric variables that describe the dynamics at the most elementary level is an important issue extensively discussed in mathematical physics as well as in fundamental interactions theory. In standard relativistic $D = 4$ theory we assume that the basic geometry is described by the Minkowski space-time coordinates $x_\mu = (\overrightarrow{x}, x_0 = ct)$.

There are two ways of extending the notion of classical Minkowski space-time:

i) One adds additional geometric degrees of freedom, e.g. the anticommuting Grassman variables in supersymmetric theory or additional commuting continuous or discrete coordinates. In principle the replacement of elementary point particles by strings can be also described as the extension of Minkowski space by infinite set of auxiliary coordinates describing Fourier modes of an extended object. In all these approaches the space-time coordinates remain elementary.

ii) One can consider the space-time geometry as a derived notion, with composite space-time coordinates. Because the most elementary representation of the Lorentz

CP767, *Fundamental Interactions and Twistor-Like Methods*, edited by J. Lukierski and D. Sorokin

algebra is spinorial[1], natural candidates for new elementary coordinates are Lorentz spinors. Taking into consideration that the mass can be considered as a dynamical effect, these elementary spinorial coordinates describing primary kinematics should describe the geometry of massless world with basic conformal invariance. In such a way we arrive at the notion of twistors (see e.g. [1–5]) - fundamental representations of the conformal algebra - as describing the coordinates of primary geometry.

In four dimensions ($D=4$) the conformal algebra is $SU(2,2)=\overline{SO(4,2)}$, and its fundamental twistor representation $T^4=(Z_1,Z_2,Z_3,Z_4)$ is complex. We define the twistor space as a fourdimensional complex metric space $T^4=(c^4,h)$, where the Hermitean metric h has the signature $(+,+,-,-)$. Choosing

$$h=\begin{pmatrix}0 & I_2\\ I_2 & 0\end{pmatrix}, \tag{1}$$

one can represent a twistor by a pair of $D=4$ Weyl spinors[2] ($\alpha,\beta=1,2;\ A=1,2,3,4$)

$$Z_A=\left(\omega^{\alpha},\overline{\pi}_{\dot\beta}\right),\qquad \overline{Z}_A=(Z_A)^*=\left(\overline{\omega}^{\dot\alpha},\pi_{\beta}\right). \tag{2}$$

More explicitly, the $SU(2,2)$ norm can be written as follows

$$\langle T,T\rangle=Z_A h^{AB}\overline{Z}_B=Z_A\overline{Z}^A=\omega^{\alpha}\pi_{\alpha}+\overline{\omega}^{\dot\alpha}\overline{\pi}_{\dot\alpha}. \tag{3}$$

The link with space-time coordinates is obtained by imposing the Penrose incidence relations

$$\omega^{\alpha}=i z^{\alpha\dot\beta}\,\overline{\pi}_{\dot\beta},\quad \overline{\omega}^{\dot\alpha}=-i\overline{z}^{\beta\dot\alpha}\,\pi_{\beta}, \tag{4}$$

where

$$z^{\alpha\dot\beta}=\tfrac{1}{2}(\sigma_{\mu})^{\alpha\dot\beta}z^{\mu},\qquad \overline{z}^{\beta\dot\alpha}=\tfrac{1}{2}(\overline{\sigma}_{\mu})^{\beta\dot\alpha}\,\overline{z}^{\mu}, \tag{5}$$

describe complex Minkowski coordinates. From (4) follows that the complex Minkowski coordinates parametrize two-planes in T^4, i.e.

$$z_{\mu}\in G_{4;2}(c)=\frac{SU(2,2)}{S(U(2)\times U(2))}. \tag{6}$$

Any such plane is parametrized by a pair of nonparallel twistors

$$Z_{A;i}=\left(\omega^{\alpha}{}_{;i},\overline{\pi}_{\dot\beta;i}\right)\quad \overline{Z}_A{}^{;i}=(Z_{A;i})^*=\left(\overline{\omega}^{\dot\alpha;i},\pi_{\beta}{}^{;i}\right)\qquad i=1,2\,, \tag{7}$$

---

[1] By most elementary representation we mean that all other irreducible representation can be obtained by tensoring procedure.

[2] We define $\omega^{\alpha}=\varepsilon^{\alpha\beta}\omega_{\beta}$, $\overline{\omega}^{\dot\alpha}=\varepsilon^{\dot\alpha\dot\beta}\overline{\omega}_{\dot\beta}$, $\omega_{\alpha}=\omega^{\beta}\varepsilon_{\beta\alpha}$, $\overline{\omega}_{\dot\alpha}=\overline{\omega}^{\dot\beta}\varepsilon_{\dot\beta\dot\alpha}$ and $\varepsilon^{\alpha\beta}\varepsilon_{\beta\gamma}=-\delta^{\alpha}_{\gamma}$, $\varepsilon^{\dot\alpha\dot\beta}\varepsilon_{\dot\beta\dot\gamma}=-\delta^{\dot\alpha}_{\dot\gamma}$, where $\varepsilon^{\alpha\beta}=\varepsilon_{\alpha\beta}=\begin{pmatrix}0 & -1\\ 1 & 0\end{pmatrix}=-\varepsilon^{\dot\alpha\dot\beta}=-\varepsilon_{\dot\alpha\dot\beta}$.

where we observe that the complex-conjugated spinors are contravariant in the internal $U(2)$ index space, i.e. $(\pi_{\alpha;i})^* = \overline{\pi}_{\dot{\alpha}}{}^{;i}$ etc. In such a way the $U(2)$-invariant norm we denote $A_{;i}(A_{;i})^* = A_{;i}\overline{A}^{;i}$.[3] Writing down the relation (4) for two twistors

$$\omega^{\alpha}{}_{;i} = i\, z^{\alpha\dot{\beta}}\, \overline{\pi}_{\dot{\beta};i}\,, \qquad \overline{\omega}^{\dot{\alpha};i} = -i\, \overline{z}^{\beta\dot{\alpha}}\, \pi_{\beta}{}^{;i}\,, \tag{8}$$

one gets as the solution of (8) the complex composite Minkowski space-time coordinates

$$z^{\alpha\dot{\beta}} = x^{\alpha\dot{\beta}} + i y^{\alpha\dot{\beta}} = \frac{i}{f}\, \omega^{\alpha;i}\, \overline{\pi}^{\dot{\beta}}{}_{;i}\,, \tag{9}$$

where $A^{;i} = \varepsilon^{ij} A_{;j}$, the fourvectors x_μ, y_μ are real, and

$$f = \overline{\pi}^{\dot{\alpha}}{}_{;1}\, \overline{\pi}_{\dot{\alpha};2} = \tfrac{1}{2}\overline{\pi}_{\dot{\alpha};i}\overline{\pi}^{\dot{\alpha}}{}_{;j}\, \varepsilon^{ij} = \tfrac{1}{2}\,\overline{\pi}_{\dot{\alpha};i}\,\overline{\pi}^{\dot{\alpha};i}\,, \qquad \overline{f} = \tfrac{1}{2}\pi_{\alpha;i}\,\pi^{\alpha;i}\,, \tag{10}$$

consistently with the numerical equality $\varepsilon_{ij} = \varepsilon^{ij}$. From the relation (9) one can show that in order to embed the real Minkowski coordinates in twistor spaces one should consider pairs of twistors T_1, T_2 which span a null 2-plane, i.e.

$$t_i{}^j = \langle T_i, T_j \rangle = Z_{A;i}\, h^{AB}\, \overline{Z}_{B;}{}^{j} = 0\,. \tag{11}$$

In such a case one gets from (8) that $y_\mu = 0$, i.e. the formula (9) describes real Minkowski coordinates.

The Hermitean metric h generates the $SU(2,2)$-invariant symplectic two-form Ω. Using (1) and (7) one can write

$$\begin{aligned} \Omega &= i\, dZ_{A;i} \wedge d\overline{Z}_B{}^{;i}\, h^{AB} \\ &= i\left(d\,\omega^{\alpha}{}_{;i} \wedge d\,\pi_{\alpha}{}^{;i} + d\,\overline{\pi}_{\dot{\alpha};i} \wedge d\,\overline{\omega}^{\dot{\alpha};i} \right) \end{aligned} \tag{12}$$

and the Liouville one form Θ, satisfying the relation $\Omega = d\Theta$, looks as follows

$$\Theta = \tfrac{i}{2}\left(\omega^{\alpha}{}_{;i}\, d\pi_{\alpha}{}^{;i} + \overline{\pi}_{\dot{\alpha};i}\, d\overline{\omega}^{\dot{\alpha};i} - H.C \right). \tag{13}$$

It appears that the two-form (12) defines fundamental twistorial Poisson brackets (TPB) and the Liouville one-form (13) should be important for the construction of dynamical Lagrangean and Hamiltonian models in twistor space.

The problem which we firstly consider in Sect. 2 is the proper notion of composite space-time in twistor space. If we use the formula (9) it was shown [6] that the real composite Minkowski space coordinates $x_\mu = Re\, z_\mu$ are noncommuting. Following [7, 8, 9] we introduce the modification of the standard Penrose formula (9) which leads to commuting composite space-time coordinates X_μ. Further in Sect. 3 by considering the nonlinear transformations of sixteen real coordinates in two-twistor space we introduce an

[3] We define by analogy with the footnote 2 that $A^{;i} = \varepsilon^{ij} A_{;j}$, $A_{;i} = A^{;j}\varepsilon_{ji}$ where $\varepsilon^{ij} = \begin{pmatrix} 0 & -1 \\ 1 & 0 \end{pmatrix} = \varepsilon_{ij}$.

enlarged 18-dimensional relativistic phase space $\mathcal{M}^{18}$ with three mutually commuting sectors

i) Relativistic phase-space (X_μ, P_μ)

ii) Spinorial complex phase space $(\eta_\alpha, \sigma_\alpha, \overline{\eta}_{\dot\alpha}, \overline{\sigma}_{\dot\alpha})$

iii) Electric charge phase space (e, ϕ)

One can prove that the equivalence with the two-twistor space (7) implies the imposition of two constraints in the spinorial complex phase space [10].

In Sect. 4 we consider in the extended phase space $\mathcal{M}^{18} = (X_\mu, P_\mu, \eta_\alpha, \overline{\eta}_{\dot\alpha}, \sigma^\alpha, \overline{\sigma}_{\dot\alpha}, e, \phi)$ a free particle model. In [10] we considered such a model with two geometric second class constraints, and three physical first class constraints defining physical quantities: mass, spin and charge. It appears however that the classical Dirac brackets, obtained in the process of elimination of the second class constraints, are very difficult to quantize in a consistent way (Jacobi identities!). In this paper we propose an alternative model in $\mathcal{M}^{18}$, with the two second class constraints proposed in [10] replaced by one geometric first class constraints. In such a model with four first class constraints (one geometric, three physical) the quantization is straightforward.

In the last Section we shall present the concluding remarks.

2. STANDARD RELATIVISTIC PHASE SPACE FROM TWO-TWISTOR GEOMETRY

The two-twistor symplectic form (12) implies the following nonzero fundamental Poisson brackets

$$\begin{aligned} \{\pi_\alpha{}^{;i}, \omega^\beta{}_{;j}\} &= i\,\delta_\alpha^\beta\,\delta_j^i\,, \\ \{\overline{\pi}_{\dot\alpha;i}, \overline{\omega}^{\dot\beta;j}\} &= -i\,\delta_{\dot\alpha}^{\dot\beta}\,\delta_i^j\,. \end{aligned} \tag{14}$$

After quantization one obtains from (14) the basic twistorial canonical commutation relations (TCCR)

$$\begin{aligned} [\widehat{\omega}^\beta{}_{;j}, \widehat{\pi}_\alpha{}^{;i}] &= \hbar\,\delta_\alpha^\beta\,\delta_j^i\,, \\ [\widehat{\overline{\omega}}^{\dot\beta;j}, \widehat{\overline{\pi}}_{\dot\alpha;i}] &= -\hbar\,\delta_{\dot\alpha}^{\dot\beta}\,\delta_i^j\,. \end{aligned} \tag{15}$$

Using the relation (9) one can calculate the TPB of the real composite Minkowski coordinates x_μ. One gets [6]

$$\{x_\mu, x_\nu\} = -\frac{1}{m^4}\,\varepsilon_{\mu\nu\rho\tau}\,W^\rho\,P^\tau\,, \tag{16}$$

where

$$m^2 = 2\,|f|^2 = 2\left|\tfrac{1}{2}\overline{\pi}_{\dot\alpha;i}\,\pi^{\dot\alpha;i}\right|^2 = \tfrac{1}{2}\left|\overline{\pi}_{\dot\alpha;i}\,\pi^{\dot\alpha;i}\right|^2 = P_\mu\,P^\mu\,, \tag{17}$$

and $(r = 1,2,3)$

$$P^{\alpha\dot{\beta}} = \pi^{\alpha;i}\overline{\pi}^{\dot{\beta}}{}_{;i}, \qquad P^{\mu} = \sigma^{\mu}{}_{\alpha\dot{\beta}}P^{\alpha\dot{\beta}}, \tag{18}$$

$$W^{\alpha\dot{\beta}} = \pi^{\alpha;i}(\tau^r)_i{}^j\overline{\pi}^{\dot{\beta}}{}_{;j}t_r, \qquad W^{\mu} = \sigma^{\mu}{}_{\alpha\dot{\beta}}W^{\alpha\dot{\beta}}, \tag{19}$$

where $(a = 0,1,2,3)$

$$t_i{}^j = (\tau^a)_i{}^j t_a = \langle T_i, T_j \rangle = Z_{A;i}\overline{Z}^{A;j}, \tag{20}$$

and $(\tau^0)_i{}^j = \begin{pmatrix} 1 & 0 \\ 0 & 1 \end{pmatrix}$ and $(\tau^r)_i{}^j$ describe three Pauli isospin matrices. The internal isospin symmetry is represented by the following $su(2) \otimes u(1)$ Poisson algebra brackets $(r,s,u = 1,2,3)$

$$\{t_r, t_s\} = \varepsilon_{rsu}t_u, \qquad \{t_0, t_r\} = 0, \tag{21}$$

as it follows from (14) and (20).

Using the relation (18) one can extended the TPB (16) by the following two relations:

$$\{P_{\mu}, x_{\nu}\} = \eta_{\mu\nu}, \tag{22}$$

$$\{P_{\mu}, P_{\nu}\} = 0. \tag{23}$$

Replacing the TPB (14) and (22–23) by TCCR (15) one gets the quantized relativistic phase space $(\widehat{x}_{\mu}, \widehat{P}_{\mu})$ with noncommuting composite Minkowski coordinates $\widehat{x}_{\mu}$. Such noncommutativity in the presence of nonvanishing spin $(W_{\mu} \neq 0)$ can be traced back to earlier considerations by Souriau [11] and Casalbuoni [12]. Indeed, the composite four-vector (19) can be identified with the Pauli-Lubanski vector in arbitrary relativistic frame. It is orthogonal, as it should be, to the composite fourmomentum (18)

$$P_{\mu}W^{\mu} = 0, \tag{24}$$

and in the rest system[4] $P_{\mu} = (m,0,0,0)$ one can write the noncommutativity relations of quantized composite Minkowski coordinates as folows: $(\widehat{x}_{\mu} = (\widehat{x}_k, \widehat{x}_0 = c\widehat{t})$

$$[\widehat{x}_k, \widehat{x}_l] = -i\frac{\hbar}{m^2}\varepsilon_{klm}\widehat{S}_m, \tag{25}$$

where $\widehat{W}_k = m\widehat{S}_k$ and as follows from (19)

$$\left[\widehat{S}_k, \widehat{S}_l\right] = i\hbar\varepsilon_{klm}\widehat{S}_m. \tag{26}$$

In this lecture we would like to consider the composite relativistic phase space (X_{μ}, P_{μ}), satisfying the standard TPB

[4] In quantized systems such description implies the consideration of eigenstates of the four-momentum operator $\widehat{P}_{\mu}$.

$$\{X_\mu, X_\nu\} = \{P_\mu, P_\nu\} = 0,$$
$$\{X_\mu, P_\nu\} = \eta_{\mu\nu}. \quad (27)$$

In such a case one has to change the definition of composite Minkowski coordinates by the modification of the standard definition (9), which we redefine as follows [7, 8, 9, 10]

$$\begin{aligned} z^{\alpha\dot\beta} \longrightarrow Z^{\alpha\dot\beta} &= z^{\alpha\dot\beta} + \Delta z^{\alpha\dot\beta} = X^{\alpha\dot\beta} + iY^{\alpha\dot\beta} \\ &= z^{\alpha\dot\beta} + i(t_1 - it_2)\,\frac{\pi^{\alpha;1}\bar\pi^{\dot\beta}{}_{;2}}{|f|^2}, \end{aligned} \quad (28)$$

or

$$\begin{aligned} X^{\alpha\dot\beta} &= x^{\alpha\dot\beta} + \Delta x^{\alpha\dot\beta} = x^{\alpha\dot\beta} - \frac{1}{2|f|^2}\left[t_1\,\pi^{\alpha;i}(\tau_2)_i{}^j\,\bar\pi^{\dot\beta}{}_{;j} - t_2\,\pi^{\alpha;i}(\tau_1)_i{}^j\,\bar\pi^{\dot\beta}{}_{;j}\right] \\ &= x^{\alpha\dot\beta} - \frac{1}{2|f|^2}\,\varepsilon_{3rs}\,t_r\,\pi^{\alpha;i}(\tau_s)_i{}^j\,\bar\pi^{\dot\beta}{}_{;j}\,. \end{aligned} \quad (29)$$

One can show that the TPB of the composite coordinates (29) are given by the relation (27) i.e. after quantization we obtain commuting composite Minkowski coordinate. We see from the relations (29) that the commutative coordinates $X^\mu = (\sigma^\mu)_{\alpha\dot\beta}X^{\alpha\dot\beta}$ distinguish a third direction in the isospace $O(3) \simeq SU(2)$ i.e. break the isospin symmetry from $O(3)$ to $O(2)$.

3. EXTENDED RELATIVISTIC PHASE SPACE FROM TWO-TWISTOR SPACE.

Let us consider the symplectic 2-form (13) and insert the formula (9) for the composite complex Minkowski coordinates. Using the $SU(2)$-covariant notation[5] one gets

$$\Theta = \pi_\alpha{}^{;i}\bar\pi_{\dot\beta;i}\,dx^{\alpha\dot\beta} + iy^{\alpha\dot\beta}\left(\pi_\alpha{}^{;i}\,d\bar\pi_{\dot\beta;i} - \bar\pi_{\dot\beta;i}\,d\pi_\alpha{}^{;i}\right). \quad (30)$$

Using the formula for the imaginary part of the complex four-vector z_μ one gets

$$t^i{}_j = -2y^{\alpha\dot\beta}\,\pi_\alpha{}^{;i}\bar\pi_{\dot\beta;j}, \quad (31a)$$

---

[5] The advantage of such a notation has been pointed out to us by S. Fedoruk. In such a framework the "half-twistors" $\pi_{\alpha;i}$, $\bar\pi_{\dot\alpha;i}$ can be treated as a pair of $Sl(2;C)$ harmonic spinors (see e.g. [13, 14, 15]) with particular normalization.

$$y^{\alpha\dot\beta} = -\frac{1}{2|f|^2} t_i{}^j \pi^{\alpha;i} \overline{\pi}^{\dot\beta}{}_{;j}. \tag{31b}$$

One gets using

$$\begin{aligned} \overline{\pi}^{\dot\alpha}{}_{;i} \overline{\pi}_{\dot\alpha;j} = -\varepsilon_{ij} f, \qquad \pi^{\alpha;i} \pi_\alpha{}^{;j} = -\varepsilon^{ij} \overline{f}, \\ \overline{\pi}_{\dot\alpha;i} \overline{\pi}_{\dot\beta}{}^{;i} = \varepsilon_{\dot\alpha\dot\beta} f, \qquad \pi_{\alpha;i} \pi_\beta{}^{;i} = \varepsilon_{\alpha\beta} \overline{f}, \end{aligned} \tag{32}$$

that

$$\begin{aligned} y^{\alpha\dot\beta} \pi_\alpha{}^{;k} &= \frac{1}{2f} t_i{}^j \overline{\pi}^{\dot\beta}{}_{;j} \varepsilon^{ik}, \\ y^{\alpha\dot\beta} \overline{\pi}_{\dot\beta;k} &= \frac{1}{2\overline{f}} t_i{}^j \pi^{\alpha;i} \varepsilon_{jk}. \end{aligned} \tag{33}$$

Substituting in (30) the formula (18) and using (33) one gets $((t^i{}_j)^* = t_i{}^j)$

$$\begin{aligned} \Theta &= P_\mu dx^\mu + i\left[-\frac{1}{2\overline{f}} t_i{}^j \pi^{\alpha;i} \varepsilon_{jk} d\pi_\alpha{}^{;k} + \frac{1}{2f} t_i{}^j \overline{\pi}^{\dot\beta}{}_{;j} \varepsilon^{ik} d\overline{\pi}_{\dot\beta;k}\right] \\ &= P_\mu dx^\mu + i\left[\frac{1}{2\overline{f}} t_i{}^j \pi^{\alpha;i} d\pi_{\alpha;j} + \frac{1}{2f} t_i{}^j \overline{\pi}^{\dot\beta}{}_{;j} d\overline{\pi}_{\dot\beta}{}^{;i}\right]. \end{aligned} \tag{34}$$

Let us observe that[6]

$$\begin{aligned} \overline{\pi}^{\dot\beta}{}_{;i} d\overline{\pi}_{\dot\beta;j} &= \overline{\pi}^{\dot\beta}{}_{;[i} d\overline{\pi}_{\dot\beta;j]} + \overline{\pi}^{\dot\beta}{}_{;(i} d\overline{\pi}_{\dot\beta;j)}, \\ \pi^{\beta;i} d\pi_\beta{}^{;j} &= \pi^{\beta;[i} d\pi_\beta{}^{;j]} + \pi^{\beta;(i} d\pi_\beta{}^{;j)}, \end{aligned} \tag{35}$$

and observe that

$$\begin{aligned} \overline{\pi}^{\dot\beta}{}_{;[i} d\overline{\pi}_{\dot\beta;j]} &= -\frac{1}{2} \varepsilon_{ij} df, \\ \pi^{\beta;[i} d\pi_\beta{}^{;j]} &= -\frac{1}{2} \varepsilon^{ij} d\overline{f}. \end{aligned} \tag{36}$$

We obtain

$$\Theta = P_\mu dx^\mu + \frac{i}{2} t_i{}^j \left(\frac{1}{\overline{f}} \pi^{\alpha;(i} d\pi_{\dot\alpha}{}^{;k)} \varepsilon_{kj} + \frac{1}{f} \overline{\pi}^{\dot\beta}{}_{;(j} d\overline{\pi}_{\dot\beta;k)} \varepsilon^{ik}\right) + \frac{i}{4} t_k{}^k d\ln\left(\frac{\overline{f}}{f}\right). \tag{37}$$

[6] We define $A_{(i}B_{j)} = \frac{1}{2}(A_iB_j + A_jB_i)$ and $A_{[i}B_{j]} = \frac{1}{2}(A_iB_j - A_jB_i)$.

In the symplectic prepotential (34) the space-time coordinates after quantization are noncommutating. In order to introduce symplectic form with commuting space-time coordinates X_μ one should use the formulae (28–29). One gets $(\Delta x^{\alpha\dot\beta} = \mathrm{Re}\Delta z^{\alpha\dot\beta})$

$$\begin{aligned} P_\mu\, dx^\mu &= P_\mu\, dX^\mu - P_{\alpha\dot\beta}\, d(\Delta x^{\alpha\dot\beta}) \\ &= P_\mu\, dX^\mu - \varepsilon_{3rs} P_{\alpha\dot\beta}\, d\left(\frac{-1}{2|f|^2} t_r\, \pi^{\alpha;i}(\tau_s)_i{}^j\, \overline{\pi}^{\dot\beta}{}_{;j}\right). \end{aligned} \tag{38}$$

Using the formula (18) one obtains

$$\begin{aligned} P_\mu dx^\mu &= P_\mu dX^\mu + P_{\alpha\dot\beta} \frac{1}{2|f|^2} \varepsilon_{3rs} t_r (\tau_s)_i{}^j d(\pi^{\alpha;i}\overline{\pi}^{\dot\beta}{}_{;j}) \\ &= P_\mu dX^\mu + \varepsilon_{3rs} t_r (\tau_s)_i{}^j \left[\frac{1}{2\overline{f}} \pi_{\alpha;j} d\pi^{\alpha;i} - \frac{1}{2f} \overline{\pi}_{\dot\beta}{}^{;i} d\overline{\pi}^{\dot\beta}{}_{;j}\right]. \end{aligned} \tag{39}$$

Substituting (39) in (37) one gets $(r = 1,2,3)$

$$\Theta = P_\mu dX^\mu + \frac{i}{f} t_r (\tau_r)_1{}^j \overline{\pi}^{\dot\alpha}{}_{;j} d\overline{\pi}_{\dot\alpha;2} - \frac{i}{\overline{f}} t_r (\tau_r)_j{}^1 \pi^{\alpha;j} d\pi_\alpha{}^{;2} + \frac{i}{2}(t_0 + t_3)\left(\frac{d\overline{f}}{\overline{f}} - \frac{df}{f}\right) \tag{40}$$

or

$$\Theta = P_\mu dX^\mu - i\left(\overline{\sigma}^{\dot\alpha}{}_{;1} d\overline{\pi}_{\dot\alpha;2} - \sigma^{\alpha;1} d\pi_\alpha{}^{;2}\right) + e\, d\phi\,, \tag{41}$$

where $(r = 1,2,3)$

$$\begin{aligned} \sigma^{\alpha;i} &= -\frac{1}{\overline{f}} t_r (\tau_r)_j{}^i \pi^{\alpha;j}\,, \\ \overline{\sigma}^{\dot\alpha}{}_{;i} &= -\frac{1}{f} t_r (\tau_r)_i{}^j \overline{\pi}^{\dot\alpha}{}_{;j}\,, \\ e &= t_0 + t_3\,, \\ \phi &= \frac{i}{2} \ln \frac{\overline{f}}{f}\,, \end{aligned} \tag{43}$$

(42)

implying

$$d\phi = \frac{i}{2}\left(\frac{d\overline{f}}{\overline{f}} - \frac{df}{f}\right). \tag{44}$$

We see therefore that the second part of the symplectic form (41) is the second rank fixed spinor in the internal space, and represents in the internal three-space ($SU(2) \simeq O(3)$) the direction $a_1 + ia_2$. The third part, describing the electric charge sector, indicates the direction $a_0 + a_3$ (a_0 is a scalar), i.e. the $SU(2)$ symmetry is also broken. It can be mentioned that the formula (42) for the electric charge recalls the Gell-Mann-Nishijima formula $Q = I_3 + \frac{Y}{2}$, where in Y is a scalar from the point of view of the internal $SU(2)$ isospin symmetry.

Let us point out that the 18 variables $(X_\mu, P_\mu, \pi^{\alpha;2}, \overline{\pi}_{\dot{\alpha};2}, \sigma^{\alpha;1}, \overline{\sigma}_{\dot{\alpha};1}, e, \phi)$ as composites of two-twistor space coordinates are not independent. On can show from (18) and (42) (see also [10][7]) that they satisfy the following three kinematical constraints:

$$\begin{aligned} R_1 &= \sigma_\alpha{}^{;1} p^{\alpha\dot{\beta}} \overline{\sigma}_{\dot{\alpha};1} - \mathbf{t}^2 = 0\,, \qquad \mathbf{t}^2 = (t_1)^2 + (t_2)^2 + (t_3)^2 && (45) \\ R_2 &= \pi_\alpha{}^{;2} p^{\alpha\dot{\beta}} \overline{\pi}_{\dot{\beta};2} - \frac{1}{2} P^2 = 0\,, && (46) \\ R_3 &= \pi_\alpha{}^{;2} \sigma^{\alpha;1} - \overline{\pi}_{\dot{\alpha};2} \overline{\sigma}^{\dot{\alpha}}{}_{;1} = 0\,, && (47) \end{aligned}$$

where (we recall that $f = \sqrt{\frac{P^2}{2}} e^{i\phi}$)

$$\begin{aligned} t_1{}^2 = t_1 - it_2 &= -\frac{1}{f} \pi_\alpha{}^{;2} p^{\alpha\dot{\beta}} \overline{\sigma}_{\dot{\beta};1}\,, && (48a) \\ t_2{}^1 = t_1 + it_2 &= -\frac{1}{f} \sigma_\alpha{}^{;1} p^{\alpha\dot{\beta}} \overline{\pi}_{\dot{\beta};2}\,, && (48b) \\ t_1{}^1 - t_2{}^2 = 2t_3 &= \pi^{\alpha;2} \sigma_\alpha{}^{;1} - \overline{\pi}_{\dot{\alpha};2} \overline{\sigma}^{\dot{\alpha}}{}_{;1}\,. && (48c) \end{aligned}$$

The constraints (45)–(47) reduce the 18 independent variables in (41) to 16 degrees of freedom in two-twistor space because only two of them are independent.

The symplectic form (41) implies the following canonical Poisson brackets (CPB):

$$\{X_\mu, X_\nu\} = 0\,, \quad \{P_\mu, P_\nu\} = 0\,, \quad \{P_\mu, X_\nu\} = \eta_{\mu\nu}\,, \tag{49a}$$

$$\left\{\pi_\alpha{}^{;2}, \sigma^{\beta;1}\right\} = i\delta_\alpha^{\ \beta}\,, \qquad \left\{\overline{\pi}_{\dot{\alpha};2}, \overline{\sigma}^{\dot{\beta}}{}_{;1}\right\} = -i\delta_{\dot{\alpha}}^{\ \dot{\beta}}\,, \tag{49b}$$

$$\{e, \phi\} = 1\,. \tag{49c}$$

Let us observe that the three constraints (R_1, R_2, R_3) have the following CPB:

$$\begin{aligned} \{R_1, R_2\} &= -2i\sigma_\alpha{}^{;1} p^{\alpha\dot{\beta}} \overline{\sigma}_{\dot{\beta};1}\,, && (50a) \\ \{R_2, R_3\} &= 2i\pi_\alpha{}^{;2} p^{\alpha\dot{\beta}} \overline{\pi}_{\dot{\beta};2}\,, && (50b) \\ \{R_1, R_3\} &= -iR_2 \cdot R_3\,. && (50c) \end{aligned}$$

[7] We should mention that the constraints $R_1 \cdots R_6$ are linear combinations of the constraints given in [10]

We suplement them with the following additional three physical constraints:

$$\begin{aligned} R_4 &= \mathbf{t}^2 - s(s+1) = 0\,, && \text{(51a)} \\ R_5 &= P^2 - m^2 = 0\,, && \text{(51b)} \\ R_6 &= e - e_0 = 0\,. && \text{(51c)} \end{aligned}$$

Using the canonical PB (see (49a-c)) one gets the following relations: $(A,B = 3,4,5,6)$

$$\{R_A, R_B\} = 0\,. \tag{52}$$

We see therefore that one can consider the set (R_3, R_4, R_5, R_6) as four first class constraints.

The constraints $R_4 = R_5 = 0$ can be interpreted as determining the numerical value of the mass operator P^2 and the isospin square $\mathbf{t}^2$. If we observe further that [6, 10]

$$\mathbf{t}^2 = -\frac{1}{2|f|^2} W_{\alpha\dot{\beta}} W^{\alpha\dot{\beta}} = -\frac{1}{P^2} W^2\,, \tag{53}$$

where $W^2 \equiv W_{\alpha\dot{\beta}} W^{\alpha\dot{\beta}}$ describes the square of the composite Pauli-Lubański fourvector, we see that one can identify $\mathbf{t}^2$ with the relativistic spin square Casimir of the Poincare algebra.

It should be added that the constraints (51a-c) can be supplemented with another relation determining the projection of isospin on the third axis.

4. RELATIVISTIC PARTICLES WITH MASS, SPIN AND ELECTRIC CHARGES IN EXTENDED SPACE-TIME.

Let us consider the 18 coordinates $(X_\mu, P_\mu, \eta^\alpha \equiv \pi^{\alpha;2}, \overline{\eta}_{\dot{\alpha}} \equiv \overline{\pi}_{\dot{\alpha};2}, \sigma^\alpha \equiv \sigma^{\alpha;1}, \overline{\sigma}_{\dot{\alpha}} \equiv \overline{\sigma}_{\dot{\alpha};1}, e, \phi)$ occuring in the symplectic one-form (41) as primary ones. They shall be restricted by four constraints $R_3 = \cdots = R_6 = 0$. We propose the following action with four constraints introduced through Lagrangian multipliers:

$$S = \int d\tau \mathscr{L} = \int d\tau \left[P_\mu \dot{X}^\mu + i(\sigma^\alpha \dot{\eta}_\alpha - \overline{\sigma}^{\dot{\alpha}} \dot{\overline{\eta}}_{\dot{\alpha}}) + e\dot{\phi} + \sum_{i=2}^{i=6} \lambda_i R_i \right]\,, \tag{54}$$

where R_i $(i = 3, \ldots, 6)$ are given by the formulae (47), (51a–c). Using (49a-c) one can show that all four constraints are first class. One can quantize the model by canonical quantization of the PB (49a-c). Using the standard quantization rule $\left(i\hbar\{a,b\} \leftrightarrow [\hat{a},\hat{b}]\right)$ one gets the following canonical commutators:

$$\begin{aligned} &[\hat{X}_\mu, \hat{X}_\nu] = [\hat{P}_\mu, \hat{P}_\nu] = 0\,, && [\hat{P}_\mu, \hat{X}_\nu] = i\hbar\eta_{\mu\nu}\,, \\ &[\hat{\sigma}^\alpha, \hat{\eta}_\beta] = \hbar\delta^\alpha_\beta\,, && [\hat{\overline{\sigma}}^{\dot{\alpha}}, \hat{\overline{\eta}}_{\dot{\beta}}] = -\hbar\delta^{\dot{\alpha}}_{\dot{\beta}}\,, \\ &[\hat{e}, \hat{\phi}] = i\hbar\,. && \text{(55)} \end{aligned}$$

We introduce the Schrödinger representation in the extended momentum space $\mathscr{P}_k = (P_\mu, \eta_\alpha, \overline{\eta}_{\dot\alpha}, \phi)$ (k=1,...,9) as follows

$$\hat{X}_\mu = -i\hbar\frac{\partial}{\partial P^\mu}, \qquad \hat{\sigma}_\alpha = \hbar\frac{\partial}{\partial\eta^\alpha}, \qquad \hat{\overline{\sigma}}_{\dot\alpha} = -\hbar\frac{\partial}{\partial\overline{\eta}^{\dot\alpha}},$$
$$\hat{e} = i\hbar\frac{\partial}{\partial\phi}. \tag{56}$$

In our quantized model the dynamics is characterized by the following four wave equations describing the wave function $\Psi(\mathscr{P}_k) \equiv \Psi(P_\mu, \eta_\alpha, \overline{\eta}_{\dot\alpha}, \phi)$

$$R_3 = 0: \quad \left(\eta_\alpha\frac{\partial}{\partial\eta^\alpha} + \overline{\eta}_{\dot\alpha}\frac{\partial}{\partial\overline{\eta}^{\dot\alpha}}\right)\Psi(\mathscr{P}_k) = 0, \tag{57a}$$

$$R_4 = 0: \quad \left[\eta_\alpha\overline{\eta}_{\dot\alpha}\frac{\partial}{\partial\eta_\alpha}\frac{\partial}{\partial\overline{\eta}_{\dot\alpha}} + \frac{1}{P^2}P_{\alpha\dot\alpha}P_{\beta\dot\beta}\left(\eta^\alpha\varepsilon^{\dot\beta\dot\alpha}\frac{\partial}{\partial\eta_\beta} + \overline{\eta}^{\dot\beta}\varepsilon^{\alpha\beta}\frac{\partial}{\partial\overline{\eta}_{\dot\alpha}} + 2\eta^\alpha\overline{\eta}^{\dot\beta}\frac{\partial}{\partial\eta_\beta}\frac{\partial}{\partial\overline{\eta}_{\dot\alpha}}\right) + \frac{s(s+1)}{\hbar^2}\right]\Psi(\mathscr{P}_k) = 0, \tag{57b}$$

$$R_5 = 0: \quad (P^2 - m^2)\Psi(\mathscr{P}_k) = 0, \tag{57c}$$

$$R_6 = 0: \quad \left(\frac{\partial}{\partial\phi} + \frac{i}{\hbar}e_0\right)\Psi(\mathscr{P}_k) = 0. \tag{57d}$$

The set (57a-d) describes the quantized first class constraints. If we add fifth constraint $R_4' = t_3 - m_3 = 0$, where $-s \le m_3 \le s$, we obtain the description of a massive relativistic particle with spin s, isospin projection m_3 on the third axis and the electric charge e_0. One can observe that the differential form of the Lorentz-invariant third component of isospin t_3

$$t_3 = \frac{1}{2}\left(\eta_\alpha\frac{\partial}{\partial\eta_\alpha} - \overline{\eta}_{\dot\alpha}\frac{\partial}{\partial\overline{\eta}_{\dot\alpha}}\right) \tag{58}$$

corresponds to the twistorial helicity formula for massless particles [1]-[4] employed recently in Witten's tensorial string theory [16]

It should be stressed that in distinction with [10] in our model we do not have the constraints $R_1 = R_2 = 0$, i.e. our phase space can not be identified with the two-twistorial phase space. Let us observe, however, that the conditions $R_1 = 0$ or $R_2 = 0$ can be obtained as gauge-fixing conditions for the gauge freedom generated by the constraint R_3, and viceversa, the transition from second to first class constraints is obtained by so-called gauge unfixing procedure (see e.g. [17, 18]).

More detailed discussion of the quantization of the model (54) will be given in a subsequent publication [19].

5. FINAL REMARKS

In the present lecture it is described the "physical basis" for two-twistor phase space and considered the particle models based on the symplectic form (41).

We stress that for the description of massive relativistic particles with spin we consider both relativistic phase space coordinates (X_μ, P_μ) as composite (compare e.g. with [20] where the coordinates X_μ are elementary) as well as we do not introduce any additional degrees of freedom besides two-twistor space (compare with [21, 22] where additional so-called index spinor was introduced).

In comparison with the results given in [10] we presented here the following two new aspects:

i) In the process of introduction of a "physical basis" in two-twistor space, defining the enlarged coomposite relativistic phase space $\mathscr{M}^{18}$, we exhibited explicitely the covariance and the breaking of the internal symmetry $SU(2)$.

ii) We introduced the particle model only inspired by the two-twistor space geometry with entirely first class constraints. If we wish to link the phase space of our model with two-twistor manifold and composite twistor formulae for $P_\mu, X_\mu, \eta_\alpha, \overline{\eta}_{\dot\alpha}, \sigma_\alpha, \overline{\sigma}_{\dot\alpha}, e$ and ϕ one should consider the gauge-fixed version of the model. In such a way will appear the second class constraints, considered in [10].

In this lecture there is presented only the model in four dimensions ($D = 4$). It appears that one can extend our considerations to the pair of super-twistors (see e.g. [23, 24]) and consider the corresponding superparticle models with mass and superspin. Other generalization consists in the extension of our discussion to other dimensions D (e.g. $D = 10$ or $D = 11$; see e.g. [24]-[26]).

ACKNOWLEDGMENTS

We would like to thank J.A. de Azcárraga, S. Fedoruk and A. Frydryszak for valuable remarks. One of the authors (J.L.) would like to acknowledge the support of the KBN grant 1 P03B 01828 and one of us (C.M-E.) wishes to thank the Spanish M.E.C. for his research grant.

REFERENCES

1. R. Penrose, Int. Journ. Theor. Phys. **1**, 61 (1968); *Twistor theory, its aims and achievements*, in *Quantum Gravity*, C.J. Isham *et al.* (eds.), Clarendon, Oxford 1975, p. 268–407.
2. R. Penrose and M.A. MacCallum, Phys. Rep. **C6**, 241 (1972).
3. Z. Perjés, Rep. Math. Phys. **12**, 193 (1977).
4. L.P. Hughston, *Twistors and Particles*, Lecture Notes in Physics **97**, Springer Verlag, Berlin (1979).
5. P.A. Tod, Rep. Math. Phys. **11**, 339 (1977).
6. A. Bette, J. Math. Phys. **25**, 2456 (1984).
7. A. Bette and S. Zakrzewski, J. Phys. **A30**, 195 (1997).
8. A. Bette and S. Zakrzewski,*Massive relativistic particles with spin and the two twistor phase space*, in Proc. XIIth Workshop on *Soft Physics-Hadrons 96*, Crimea. Pub. Nat. Acad. of Sci. of Ukraine, Kiev 1996, p. 336, [hep-th/0404024].

9. A. Bette, *Twistors, special relativity, conformal symmetry and minimal coupling*, hep-th/0402150.
10. A. Bette, J. de Azcárraga, J. Lukierski and C. Miquel-Espanya, Phys. Lett. **B595**, 491 (2004).
11. J.M. Souriau, *Structure des systèmes dynamiques*, Dunod, Paris (1970).
12. R. Casalbuoni, Nuovo Cim. **33A**, 389 (1976).
13. I. Bandos, Sov. J. Nucl. Phys. **51**, 906 (1990).
14. I. Bandos and A. Zheltukhin, Class. Quant. Grav. **12**, 609 (1995).
15. V. Zima and S. Fedoruk, Theor. Math. Phys. **102**, 305 (1995).
16. E. Witten, Comm. Math. Phys. **252**, 189 (2004).
17. L. Faddeev, S. Shatashvili, Phys. Lett. **B167**, 225 (1986).
18. A.S. Vytheeswaran, Ann. Phys. (NY) **236**, 297 (1994).
19. A. Bette, J. de Azcárraga, J. Lukierski and C. Miquel-Espanya, in preparation.
20. S.L. Lyakhovich, A.A. Sharapov and K.M. Shekter, Nucl. Phys. **b537**, 640 (1999); Int. J. Mod. Phys. **A15**, 4287 (2000).
21. S. Fedoruk, V.G. Zima, Mod. Phys. Lett. **A15**, 2281 (2000).
22. S. Fedoruk, V.G. Zima, hep-th/0308154; hep-th/0401064.
23. A. Ferber, Nucl. Phys. **B132**, 55 (1978).
24. I. Bandos, J. Lukierski and D. Sorokin, Phys. Rev. **D61**, 045002 (2000).
25. I. Bars, Phys. Lett. **B483**, 248 (2000).
26. I. Bandos, J. A. de Azcárraga, J. M. Izquierdo and J. Lukierski, Phys. Lett. **86**, 4451 (2001).

Twistor Cosmology and Quantum Space–Time

D.C. Brody* and L.P. Hughston†

*Blackett Laboratory, Imperial College, London SW7 2BZ, UK
†Department of Mathematics, King's College London, The Strand, London WC2R 2LS, UK

Abstract. The purpose of this paper is to present a model of a 'quantum space-time' in which the global symmetries of space-time are unified in a coherent manner with the internal symmetries associated with the state space of quantum-mechanics. If we take into account the fact that these distinct families of symmetries should in some sense merge and become essentially indistinguishable in the unified regime, our framework may provide an approximate description of or elementary model for the structure of the universe at early times. The quantum elements employed in our characterisation of the geometry of space-time imply that the pseudo-Riemannian structure commonly regarded as an essential feature in relativistic theories must be dispensed with. Nevertheless, the causal structure and the physical kinematics of quantum space-time are shown to persist in a manner that remains highly analogous to the corresponding features of the classical theory. In the case of the simplest conformally flat cosmological models arising in this framework, the twistorial description of quantum space-time is shown to be effective in characterising the various physical and geometrical properties of the theory. As an example, a sixteen-dimensional analogue of the Friedmann-Robertson-Walker cosmologies is constructed, and its chronological development is analysed in some detail. More generally, whenever the dimension of a quantum space-time is an even perfect square, there exists a canonical way of breaking the global quantum space-time symmetry so that a generic point of quantum space-time can be consistently interpreted as a quantum operator taking values in Minkowski space. In this scenario, the breakdown of the fundamental symmetry of the theory is due to a loss of quantum entanglement between space-time and internal quantum degrees of freedom. It is thus possible to show in a certain specific sense that the classical space-time description is an emergent feature arising as a consequence of a quantum averaging over the internal degrees of freedom. The familiar probabilistic features of the quantum state, represented by properties of the density matrix, can then be seen as a by-product of the causal structure of quantum space-time.

INTRODUCTION

This article is concerned with a programme that has as its goal the development of a theory of quantum space-time. In this programme, an outline of which will be given in more detail shortly, an important role is played by certain higher-dimensional analogues of spinors and twistors. It will be useful to begin, therefore, by remarking that there are two distinct notions of how one extends the concept of spinor into higher dimensions. This fundamental dichotomy arises in association with the fact that in four-dimensional space-time there is a local isomorphism between the Lorentz group $SO(1,3)$ and the spin transformation group $SL(2,\mathbb{C})$. In higher dimensions, however, this relation breaks down and as a consequence we are left with two concepts of spinors—one for the group $SO(N,\mathbb{C})$, and one for the group $SL(r,\mathbb{C})$.

The spinors associated with $SO(N,\mathbb{C})$, where we allow also for various possible signatures in the quadratic form defining these orthogonal or pseudo-orthogonal transforma-

CP767, *Fundamental Interactions and Twistor-Like Methods,* edited by J. Lukierski and D. Sorokin

tions when we specialise to the real subgroup $SO(p,q)$ with $p+q=N$, are the so-called Cartan spinors [7, 8, 9, 46]. The study of Cartan spinors has a long and interesting history, and there is a beautiful geometry associated with these spinors. There are also various specific cases of great interest—for example, the Cartan spinors associated with the group $SO(2,4)$ are Penrose's twistors; and the Cartan spinors associated with $SO(8)$ are intimately linked with the Cayley numbers (octonions) and the exceptional Lie groups. There are also a number of interesting connections between Cartan spinors and massless fields in higher dimensions [25, 26, 28].

The spinors associated with $SL(r,\mathbb{C})$, which are usually now called 'hyperspinors', have the advantage of being more directly linked with quantum mechanics. In fact, we shall show later that a naturally relativistic model for hyperspin arises when one considers 'multiplets' of two-component spinors, i.e. expressions of the form $\xi^{\mathbf{A}i}$ and $\eta_i^{\mathbf{A}'}$, where $\mathbf{A}, \mathbf{A}'$ are standard spinor indices, and $i = 1,2,\ldots,n$ is an 'internal' index. In the general case ($n=\infty$) we then think of $\xi^{\mathbf{A}i}$ as an element of the tensor product space $\mathbb{S}^{\mathbf{A}i} = \mathbb{S}^{\mathbf{A}} \otimes \mathbb{H}^i$, where $\mathbb{S}^{\mathbf{A}}$ is the complex vector space of two-component spinors, and $\mathbb{H}^i$ is an infinite-dimensional complex Hilbert space.

There is also a link, arising through a further extension of this idea, between hyperspinor theory and the theory of multi-twistor (hypertwistor) systems. Indeed, we find that the theory of hyperspin constitutes a natural starting place for building up a theory of quantum geometry or, as we shall call it here, *quantum space-time*. In summary, we shall be taking the left-hand path in the following diagram:

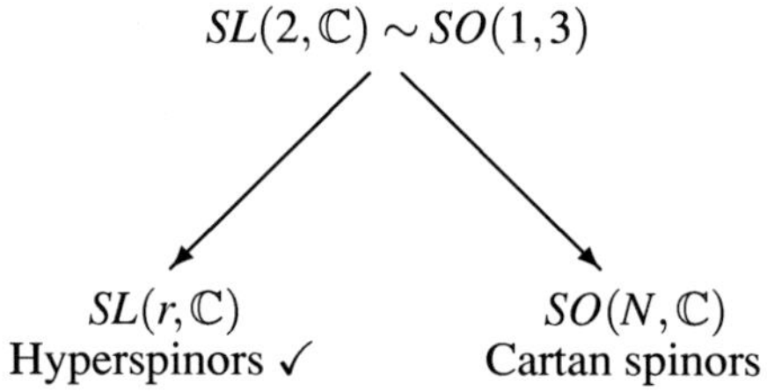

The hyperspinor route has the virtue that the resulting space-time has a rich causal structure associated with it, and as a consequence is unusually well-positioned to form the geometrical basis of a physical theory.

RELATIVISTIC CAUSALITY

To start, let us review briefly the role of two-component spinors in the description of four-dimensional Minkowskian space-time geometry. In what follows we use bold upright Roman letters to denote two-component spinor indices, and we adopt the standard conventions for the algebra of two-component spinors [37, 45, 46, 59]. Then we have the following correspondence between two-by-two Hermitian matrices and space-time points, relative to some origin:

$$x^{\mathbf{AA}'} \quad (\mathbf{A}, \mathbf{A}' = 1,2) \qquad \longleftrightarrow \qquad x^{\mathrm{a}} \quad (\mathrm{a} = 0,1,2,3) \,. \tag{1}$$

More explicitly, in a standard basis this correspondence is given by

$$\frac{1}{\sqrt{2}}\begin{pmatrix} t+z & x+\mathrm{i}y \\ x-\mathrm{i}y & t-z \end{pmatrix} \quad \longleftrightarrow \quad (t,x,y,z)\,. \tag{2}$$

We then have the fundamental relation

$$2\det(x^{\mathbf{A}A'}) = t^2 - x^2 - y^2 - z^2\,, \tag{3}$$

from which it follows that two-component spinors are connected both with quantum mechanics and with the causal structure of space-time. It is a peculiar aspect of relativistic physics that there is this link between (a) the spin degrees of freedom of spin one-half particles, and (b) the causal geometry of four-dimensional space-time.

Let us pursue this idea now in a little more detail, and then extend it to higher dimensions. For the interval $r^{\mathbf{A}A'}$ between a pair of points $x^{\mathbf{A}A'}$ and $y^{\mathbf{A}A'}$ in Minkowski space-time we write

$$r^{\mathbf{A}A'} = x^{\mathbf{A}A'} - y^{\mathbf{A}A'}\,, \tag{4}$$

from which it follows that

$$2\det(r^{\mathbf{A}A'}) = \varepsilon_{\mathbf{A}B}\varepsilon_{\mathbf{A'}\mathbf{B'}}r^{\mathbf{A}A'}r^{\mathbf{B}B'}\,, \tag{5}$$

where $\varepsilon_{\mathbf{A}B}$ is the antisymmetric spinor. Hence if we adopt the standard 'index clumping' convention and write $\mathrm{a} = \mathbf{A}A'$, $\mathrm{b} = \mathbf{B}B'$, and so on, according to which a pair of spinor indices, one primed and the other unprimed, corresponds to a lower case space-time vector index, then we can write

$$\varepsilon_{\mathbf{A}B}\varepsilon_{\mathbf{A'}\mathbf{B'}}r^{\mathbf{A}A'}r^{\mathbf{B}B'} = g_{\mathrm{a}b}r^{\mathrm{a}}r^{\mathrm{b}} \tag{6}$$

for the corresponding squared space-time interval, and thus we are able to identity

$$g_{\mathrm{a}b} = \varepsilon_{\mathbf{A}B}\varepsilon_{\mathbf{A'}\mathbf{B'}} \tag{7}$$

as the metric of Minkowski space.

There are essentially three different situations that can arise for the interval r^{a}, each of which represents a certain level of degeneracy. The first case is $g_{\mathrm{a}b}r^{\mathrm{b}} = 0$; the second case is $g_{\mathrm{a}b}r^{\mathrm{b}} \neq 0$ and $g_{\mathrm{a}b}r^{\mathrm{a}}r^{\mathrm{b}} = 0$; and the third case is $g_{\mathrm{a}b}r^{\mathrm{a}}r^{\mathrm{b}} \neq 0$. Each of these cases gives rise to a canonical form for the interval $r^{\mathbf{A}A'}$, with various sub-cases, which can be summarised as follows:

(i) $g_{\mathrm{a}b}r^{\mathrm{b}} = 0$:

$$r^{\mathbf{A}A'} = 0 \quad \text{zero separation}$$

(ii) $g_{ab}r^a r^b = 0$:

$$r^{\mathbf{A}A'} = \alpha^{\mathbf{A}}\bar{\alpha}^{\mathbf{A'}} \quad \text{future-pointing null separation}$$
$$r^{\mathbf{A}A'} = -\alpha^{\mathbf{A}}\bar{\alpha}^{\mathbf{A'}} \quad \text{past-pointing null separation}$$

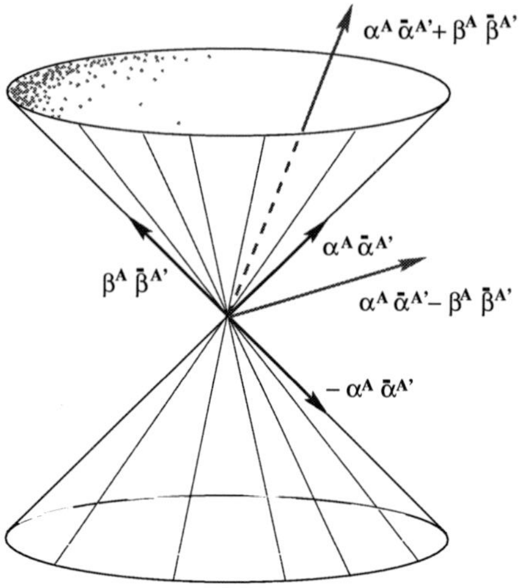

FIGURE 1. *Causal structure of four-dimensional space-time.* The canonical form of the spinor decomposition of the four-vector associated with a space-time interval determines its causal properties.

(iii) $g_{ab}r^a r^b \neq 0$:

$$r^{\mathrm{AA}'} = \alpha^{\mathrm{A}}\bar{\alpha}^{\mathrm{A}'} + \beta^{\mathrm{A}}\bar{\beta}^{\mathrm{A}'} \quad \text{future-pointing time-like separation}$$
$$r^{\mathrm{AA}'} = \alpha^{\mathrm{A}}\bar{\alpha}^{\mathrm{A}'} - \beta^{\mathrm{A}}\bar{\beta}^{\mathrm{A}'} \quad \text{space-like separation}$$
$$r^{\mathrm{AA}'} = -\alpha^{\mathrm{A}}\bar{\alpha}^{\mathrm{A}'} - \beta^{\mathrm{A}}\bar{\beta}^{\mathrm{A}'} \quad \text{past-pointing time-like separation}$$

Here it is understood that in case (iii) the spinors α^{A} and β^{A} do not coincide in direction. It is interesting to note that once the canonical form for $r^{\mathrm{AA}'}$ is specified, then so is the causal relationship that it determines on space-time. This correspondence is illustrated in Figure 1. On the other hand, the specification of $r^{\mathrm{AA}'}$ does not completely determine the spinors α^{A} and β^{A}. In general there is some freedom, and this is expressed by a group of transformations. In particular, if $r^{\mathrm{AA}'}$ is null, then this freedom is the phase shift $\alpha^{\mathrm{A}} \to \mathrm{e}^{\mathrm{i}\theta}\alpha^{\mathrm{A}}$, and the relevant group is $U(1)$. If $r^{\mathrm{AA}'}$ is time-like, then the group is $U(2)$, and if $r^{\mathrm{AA}'}$ is space-like, the group is $U(1,1)$.

HYPERSPIN SPACES

The terminology 'hyperspinor' is due to Finkelstein [13]. Essentially the same concept (although introduced for different purposes) is also touched on in [22]. The idea of a hyperspinor is a simple one—we replace the two-component spinors associated with four-dimensional space-time with r-component spinors. Thus we can regard hyperspin space as the vector space $\mathbb{C}^r$ with some extra structure. In particular, in addition to the original hyperspin space we have three other vector spaces—the dual hyperspin space, the complex conjugate hyperspin space, and the dual complex conjugate hyperspin space.

The theory of hyperspin has been pursued by a number of authors [3, 4, 6, 14, 15, 16, 18, 19, 20, 21, 62, 66], and the material we describe here builds on various aspects of this work. Let us write $\mathbb{S}^A$ and $\mathbb{S}^{A'}$, respectively, for the complex r-dimensional vector spaces of unprimed and primed hyperspinors. For hyperspinors we use italic indices to distinguish them from the boldface indices used exclusively for two-component spinors. It is assumed that $\mathbb{S}^A$ and $\mathbb{S}^{A'}$ are related by an anti-linear isomorphism under the operation of complex conjugation. Thus if $\alpha^A \in \mathbb{S}^A$, then under complex conjugation we have $\alpha^A \to \bar{\alpha}^{A'}$, where $\bar{\alpha}^{A'} \in \mathbb{S}^{A'}$. The dual spaces associated with $\mathbb{S}^A$ and $\mathbb{S}^{A'}$ are denoted $\mathbb{S}_A$ and $\mathbb{S}_{A'}$, respectively. If $\alpha^A \in \mathbb{S}^A$ and $\beta_A \in \mathbb{S}_A$, then their inner product is denoted $\alpha^A\beta_A$. Likewise if $\gamma^{A'} \in \mathbb{S}^{A'}$ and $\delta_{A'} \in \mathbb{S}_{A'}$ then their inner product is $\gamma^{A'}\delta_{A'}$.

We also introduce the totally antisymmetric hyperspinors of rank r associated with the spaces $\mathbb{S}^A$, $\mathbb{S}_A$, $\mathbb{S}^{A'}$, and $\mathbb{S}_{A'}$. These will be denoted $\varepsilon^{AB\cdots C}$, $\varepsilon_{AB\ldots C}$, $\varepsilon^{A'B'\cdots C'}$, and $\varepsilon_{A'B'\ldots C'}$, respectively. The choice of these antisymmetric hyperspinors is unique up to an overall scale factor. Once a choice has been made for $\varepsilon_{AB\cdots C}$, then the other epsilon hyperspinors are determined by the relations

$$\varepsilon^{AB\cdots C}\varepsilon_{AB\cdots C} = r!, \quad \varepsilon^{A'B'\cdots C'}\varepsilon_{A'B'\ldots C'} = r!, \quad \textit{and} \quad \varepsilon_{A'B'\ldots C'} = \bar{\varepsilon}_{A'B'\ldots C'}, \tag{8}$$

where $\bar{\varepsilon}_{A'B'\ldots C'}$ is the complex conjugate of $\varepsilon_{AB\cdots C}$.

Now let $\Lambda^m = \Lambda^{A_1A_2\cdots A_m}$ denote the skew tensor product space $\mathbb{S}^{A_1} \wedge \mathbb{S}^{A_2} \wedge \cdots \wedge \mathbb{S}^{A_m}$. Using an analogous notation, we introduce the spaces Λ_m, $\Lambda^{m'}$, and $\Lambda_{m'}$ for each $m = 0, 1, \ldots, r$. Then once the epsilon hyperspinors have been fixed we have a collection of maps of the form

$$\varepsilon_r : \Lambda^m \to \Lambda_{r-m}, \quad \varepsilon^r : \Lambda_m \to \Lambda^{r-m}, \quad \varepsilon_{r'} : \Lambda^{m'} \to \Lambda_{r'-m'}, \quad \textit{and} \quad \varepsilon^{r'} : \Lambda_{m'} \to \Lambda^{r'-m'}.$$

As a consequence, a wide range of algebraic theorems can be formulated, which are useful in calculations. For example, if $\alpha^A\beta_A = 0$ then β_A must be of the form

$$\beta_A = \varepsilon_{ABC\cdots D}\alpha^B\mu^{C\cdots D} \tag{9}$$

for some $\mu^{C\cdots D} \in \Lambda^{r-2}$.

QUANTUM SPACE-TIME

Now we introduce the complex matrix space $\mathbb{C}^{AA'} = \mathbb{S}^A \otimes \mathbb{S}^{A'}$. An element $x^{AA'} \in \mathbb{C}^{AA'}$ is said to be *real* if it satisfies the (weak) Hermitian property $x^{AA'} = \bar{x}^{A'A}$, where $\bar{x}^{A'A}$ is the complex conjugate of $x^{AA'}$. We shall have more to say about weak versus strong Hermiticity conditions in relation to the idea of symmetry breaking. We denote the vector space of real elements of $\mathbb{C}^{AA'}$ by $\mathbb{R}^{AA'}$. The elements of $\mathbb{R}^{AA'}$ constitute what we call the real quantum space-time $\mathscr{H}^{r^2}$ of dimension r^2. We then regard $\mathscr{C}\mathscr{H}^{r^2} = \mathbb{C}^{AA'}$ as the complexification of $\mathscr{H}^{r^2}$. Many problems in $\mathscr{H}^{r^2}$ are best first approached as problems in $\mathscr{C}\mathscr{H}^{r^2}$, and hence sometimes although we refer to $\mathscr{H}^{r^2}$ our operations are actually carried out in $\mathscr{C}\mathscr{H}^{r^2}$.

Let $x^{AA'}$ and $y^{AA'}$ be points in $\mathscr{H}^{r^2}$, and write $r^{AA'} = x^{AA'} - y^{AA'}$ for the corresponding separation vector, which is independent of the choice of origin. Using the index-clumping convention we set $x^{\mathrm{a}} = x^{AA'}$, $y^{\mathrm{a}} = y^{AA'}$, $r^{\mathrm{a}} = r^{AA'}$, and for the separation of x^{a} and y^{a} in $\mathscr{H}^{r^2}$ we write $r^{\mathrm{a}} = x^{\mathrm{a}} - y^{\mathrm{a}}$. There is a natural causal structure induced on such intervals by the so-called 'chronometric tensor'. Making use of the index-clumping convention, we define this fundamental tensor (introduced by Finkelstein [13]) by the following basic relation:

$$g_{\mathrm{ab}\cdots c} = \varepsilon_{AB\cdots C}\, \varepsilon_{A'B'\cdots C'}\ . \tag{10}$$

The chronometric tensor, which is of rank r, is totally symmetric and is nondegenerate in the sense that for any vector r^{a} the condition $r^{\mathrm{a}} g_{\mathrm{ab}\cdots c} = 0$ implies $r^{\mathrm{a}} = 0$. We say that x^{a} and y^{a} in $\mathscr{H}^{r^2}$ have a 'degenerate' separation if the chronometric form

$$\Delta(r) = g_{\mathrm{ab}\cdots c} r^{\mathrm{a}} r^{\mathrm{b}} \cdots r^{\mathrm{c}} \tag{11}$$

vanishes for $r^{\mathrm{a}} = x^{\mathrm{a}} - y^{\mathrm{a}}$. Degenerate separation is equivalent to the vanishing of the determinant of the matrix $r^{AA'}$, that is,

$$\varepsilon_{AB\cdots C}\, \varepsilon_{A'B'\cdots C'} r^{AA'} r^{BB'} \cdots r^{CC'} = 0\ . \tag{12}$$

If the hyperspin space has dimension $r = 2$, this reduces to the usual condition for x^{a} and y^{a} to be null-separated in Minkowski space. For $r > 2$, however, the situation is more complicated since there are various degrees of degeneracy that can arise between two points of quantum space-time, of which 'nullness' (in the Minkowskian sense) is only the most extreme.

As an example, consider $r = 3$. In this case the quantum space-time has dimension nine, and the chronometric form is given by

$$\Delta = g_{\mathrm{abc}} r^{\mathrm{a}} r^{\mathrm{b}} r^{\mathrm{c}}\ . \tag{13}$$

The different possibilities that can arise for the separation vector are as follows:

(i) $g_{\mathrm{abc}} r^{\mathrm{c}} = 0$:

$$r^{AA'} = 0 \quad \text{zero separation}$$

(ii) $g_{\mathrm{abc}} r^{\mathrm{b}} r^{\mathrm{c}} = 0$ and $g_{\mathrm{abc}} r^{\mathrm{c}} \neq 0$:

$$\begin{aligned} r^{AA'} &= \alpha^A \bar{\alpha}^{A'} && \text{future-pointing null separation} \\ r^{AA'} &= -\alpha^A \bar{\alpha}^{A'} && \text{past-pointing null separation} \end{aligned}$$

(iii) $\Delta = 0$ and $g_{\mathrm{abc}} r^{\mathrm{b}} r^{\mathrm{c}} \neq 0$:

$$\begin{aligned} r^{AA'} &= \alpha^A \bar{\alpha}^{A'} + \beta^A \bar{\beta}^{A'} && \text{degenerate time-like future-pointing separation} \\ r^{AA'} &= \alpha^A \bar{\alpha}^{A'} - \beta^A \bar{\beta}^{A'} && \text{degenerate space-like separation} \\ r^{AA'} &= -\alpha^A \bar{\alpha}^{A'} - \beta^A \bar{\beta}^{A'} && \text{degenerate time-like past-pointing separation} \end{aligned}$$

(iv) $\Delta \neq 0$ and $g_{ab} r^{a} r^{b} \neq 0$:

$r^{AA'} = \alpha^A \bar{\alpha}^{A'} + \beta^A \bar{\beta}^{A'} + \gamma^A \bar{\gamma}^{A'}$	future pointing time-like separation
$r^{AA'} = \alpha^A \bar{\alpha}^{A'} + \beta^A \bar{\beta}^{A'} - \gamma^A \bar{\gamma}^{A'}$	semi-space-like separation
$r^{AA'} = \alpha^A \bar{\alpha}^{A'} - \beta^A \bar{\beta}^{A'} - \gamma^A \bar{\gamma}^{A'}$	semi-space-like separation
$r^{AA'} = -\alpha^A \bar{\alpha}^{A'} - \beta^A \bar{\beta}^{A'} - \gamma^A \bar{\gamma}^{A'}$	past pointing time-like separation

When the separation of two points of quantum space-time is degenerate, we define the 'degree' of degeneracy by the rank of the matrix $r^{AA'}$. Null separation is the case for which the degeneracy is of the first degree, i.e. where $r^{AA'}$ is of rank one, and thus satisfies a system of quadratic relations of the following form:

$$\varepsilon_{AB\cdots C}\,\varepsilon_{A'B'\cdots C'} r^{AA'} r^{BB'} = 0\,, \tag{14}$$

or equivalently

$$g_{ab\cdots c} r^{a} r^{b} = 0\,. \tag{15}$$

This implies that $r^{AA'}$ can be expressed in the 'null' form

$$r^{AA'} = \pm \alpha^A \bar{\alpha}^{A'} \tag{16}$$

for some α^A. In the case of degeneracy of the second degree, $r^{AA'}$ is of rank two and satisfies a set of cubic relations given by

$$\varepsilon_{ABC\cdots D}\,\varepsilon_{A'B'C'\cdots D'} r^{AA'} r^{BB'} r^{CC'} = 0\,, \tag{17}$$

or equivalently

$$g_{abc\cdots d} r^{a} r^{b} r^{c} = 0\,. \tag{18}$$

In this situation $r^{AA'}$ can be put into one of the following three canonical forms:

(a) $r^{AA'} = \alpha^A \bar{\alpha}^{A'} + \beta^A \bar{\beta}^{A'}$,
(b) $r^{AA'} = \alpha^A \bar{\alpha}^{A'} - \beta^A \bar{\beta}^{A'}$,
(c) $r^{AA'} = -\alpha^A \bar{\alpha}^{A'} - \beta^A \bar{\beta}^{A'}$.

In case (a), x^{a} lies to the future of y^{a}, and r^{a} can be thought of as a degenerate future-pointing time-like vector. In case (b), r^{a} can be thought of as a degenerate space-like separation. In case (c), x^{a} lies to the past of y^{a}, and r^{a} is a degenerate past-pointing time-like vector. A similar analysis can be applied to degenerate separations of other 'intermediate' degrees.

If the determinant of the r-by-r weakly Hermitian matrix $r^{AA'}$ is nonvanishing, and $r^{AA'}$ is thus of maximal rank, then the chronometric form Δ is nonvanishing. In that case the matrix $r^{AA'}$ can be represented in the following canonical form:

$$r^{AA'} = \pm \alpha^A \bar{\alpha}^{A'} \pm \beta^A \bar{\beta}^{A'} \pm \cdots \pm \gamma^A \bar{\gamma}^{A'}\,, \tag{19}$$

with the presence of r nonvanishing terms, where the r hyperspinors $\alpha^A, \beta^A, \cdots, \gamma^A$ are all linearly independent.

Let us write (p,q) for the numbers of plus and minus signs appearing in the canonical form for the matrix $r^{AA'}$ given above. We call (p,q) the 'signature' of $r^{AA'}$. The hyperspinors $\alpha^A, \beta^A, \cdots, \gamma^A$ are determined by the specification of $r^{AA'}$ only up to an overall unitary (or pseudo-unitary) transformation of the form

$$\alpha_n^A \to U_n^m \alpha_m^A, \tag{20}$$

where $n, m = 1, 2, \ldots, r$, and

$$\alpha_n^A = \{\alpha^A, \beta^A, \cdots, \gamma^A\} . \tag{21}$$

The signature (p,q) is nevertheless an invariant of $r^{AA'}$. In the cases for which the signature is $(r,0)$ or $(0,r)$ we say that $r^{AA'}$ is future-pointing time-like or past-pointing time-like, respectively. Then recalling the definition (11) for the associated chronometric form, we define the 'proper time interval' between the events x^{a} and y^{a} by the formula

$$||x - y|| = |\Delta|^{\frac{1}{r}} . \tag{22}$$

In the case $r = 2$ we then recover the standard Minkowskian proper-time interval between the given events.

A remarkable feature of the causal structure of quantum space-time is that the essential physical features of the causal structure of Minkowski space are preserved. In particular, the space of future-pointing time-like vectors forms a convex cone. The same is true when we consider the structure of the associated momentum space, from which it follows that we can also give a good definition of what is meant by 'positive energy'.

EQUATIONS OF MOTION

Now suppose that $\lambda \mapsto x^{AA'}(\lambda)$ defines a smooth curve γ in $\mathscr{H}^{r^2}$ for $\lambda \in [a,b] \subset \mathbb{R}$. Then γ will be said to be time-like if the tangent vector along γ,

$$v^{AA'}(\lambda) = \frac{\mathrm{d}}{\mathrm{d}\lambda} x^{AA'}(\lambda), \tag{23}$$

is time-like and future-pointing. In that case we define the proper time s elapsed along γ by the integral

$$s = \int_a^b \left[g_{ab\cdots\mathrm{c}} v^{\mathrm{a}} v^{\mathrm{b}} \cdots v^{\mathrm{c}} \right]^{\frac{1}{r}} \mathrm{d}\lambda . \tag{24}$$

In the case of a very small interval, we can also write this in the 'pseudo-Finslerian' form

$$(\mathrm{d}s)^r = g_{ab\cdots\mathrm{c}} \mathrm{d}x^{\mathrm{a}} \mathrm{d}x^{\mathrm{b}} \cdots \mathrm{d}x^{\mathrm{c}} . \tag{25}$$

In the case $r = 2$ this clearly reduces to the standard pseudo-Riemannian expression for the line element.

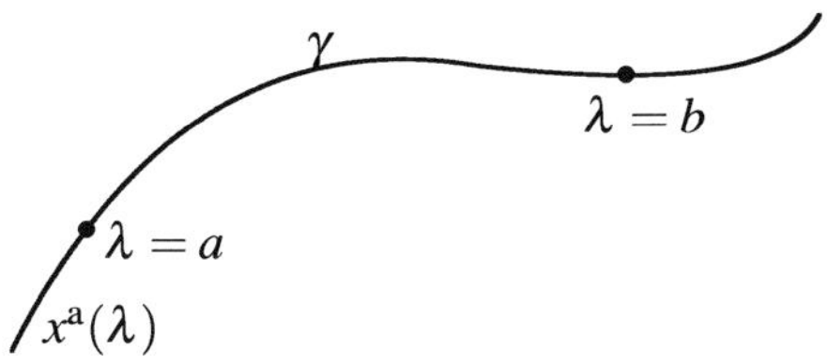

Now let us consider the condition γ must satisfy in order to be a geodesic in $\mathscr{H}^{r^2}$. Since the geometry is not Riemannian, the answer may not be entirely obvious. In the case of a time-like curve, we can choose the proper time as the parameter along the curve, in which case the resulting affine parameterisation of the curve is determined up to transformations of the form $s \to s+k$ where k is a constant. The equation of motion for the situation in which γ is a time-like geodesic is obtained by an application of the calculus of variations to formula (24). As usual, we assume the variation vanishes at the endpoints. Writing

$$L = \left(g_{abc\cdots d} v^a v^b v^c \cdots v^d \right)^{\frac{1}{r}}, \tag{26}$$

a standard argument shows that $x^a(\lambda)$ describes a geodesic only if the velocity vector v^a satisfies the Euler-Lagrange equation

$$\frac{\mathrm{d}}{\mathrm{d}\lambda}\left(\frac{\partial L}{\partial v^a}\right) = 0\,. \tag{27}$$

A calculation shows that this condition is given more explicitly by

$$g_{abc\cdots d}\frac{\mathrm{d}v^b}{\mathrm{d}\lambda} v^c \cdots v^d = \phi\, g_{abc\cdots d} v^b v^c \cdots v^d, \tag{28}$$

where

$$\phi = \frac{1}{L}\frac{\mathrm{d}L}{\mathrm{d}\lambda}\,. \tag{29}$$

If λ is chosen to be proper time, then $\phi = 0$ and the geodesic equation takes to form

$$g_{abc\cdots d}\dot{v}^b v^c \cdots v^d = 0, \tag{30}$$

where the dot denotes differentiation with respect to proper time.

In the case $r = 2$ the geodesic equation (30) reduces to the familiar relation $\dot{v}^a = 0$. We shall now show that the geodesic equation also implies $\dot{v}^a = 0$ in the case of a general quantum space-time with $r > 2$. It suffices to examine the case $r = 3$, which will indicate the relevant line of argument. For $r = 3$ the geodesic equation takes the form

$$g_{abc}\dot{v}^b v^c = 0, \tag{31}$$

which can be expressed in terms of hyperspinors in the form

$$\varepsilon_{ABC}\,\varepsilon_{A'B'C'}\,\dot{v}^{BB'}v^{CC'} = 0\,. \tag{32}$$

This relation can then be written

$$\dot{v}^{BB'}v^{CC'} - \dot{v}^{CB'}v^{BC'} - \dot{v}^{BC'}v^{CB'} + \dot{v}^{CC'}v^{BB'} = 0\,. \tag{33}$$

Because $\det(v^{AA'}) \neq 0$ we know that $v^{AA'}$ has an inverse $u_{AA'}$ satisfying

$$v^{AA'}u_{BA'} = \delta^{A}_{B} \quad \textit{and} \quad v^{AA'}u_{AB'} = \delta^{A'}_{B'}\,. \tag{34}$$

Therefore, contracting (33) with $u_{BB'}$ we obtain

$$(u_{BB'}\dot{v}^{BB'})v^{CC'} + (r-2)\dot{v}^{CC'} = 0\,. \tag{35}$$

This equation shows that if $\dot{v}^{CC'}$ were not zero, then it would have to be proportional to $v^{CC'}$. However, if that were so, then

$$\varepsilon_{ABC}\,\varepsilon_{A'B'C'}\,\dot{v}^{BB'}v^{CC'} = 0 \tag{36}$$

would imply $\det(v^{CC'}) = 0$, contrary to the assumption that $v^{CC'}$ is time-like. It follows that $\dot{v}^{\mathrm{a}} = 0$. A similar argument shows that for all $r \geq 2$ the geodesic equation (30) implies $\dot{v}^{\mathrm{a}} = 0$. Hence we have deduced the following result. Let y^{a} and z^{a} be quantum space-time points with the property that $y^{\mathrm{a}} - z^{\mathrm{a}}$ is time-like and future-pointing. Then the affinely parametrised geodesic γ connecting these points in $\mathscr{H}^{r^2}$ is given by

$$X^{\mathrm{a}}(s) = z^{\mathrm{a}} + \frac{y^{\mathrm{a}} - z^{\mathrm{a}}}{[\Delta(y,z)]^{1/r}}\,s \tag{37}$$

for $s \in (-\infty, \infty)$, where

$$\Delta(y,z) = g_{\mathrm{ab}\cdots\mathrm{c}}(y^{\mathrm{a}} - z^{\mathrm{a}})(y^{\mathrm{b}} - z^{\mathrm{b}})\cdots(y^{\mathrm{c}} - z^{\mathrm{c}})\,. \tag{38}$$

CONSERVED QUANTITIES

It is a straightforward exercise to verify that the chronometric form Δ for the separation between two points is invariant when the points of $\mathscr{H}^{r^2}$ are subjected to transformations of the following type:

$$x^{AA'} \to \lambda^{A}_{B}\bar{\lambda}^{A'}_{B'}x^{BB'} + \beta^{AA'}\,. \tag{39}$$

Here $\beta^{AA'}$ represents an arbitrary translation in quantum space-time, λ^{A}_{B} is an element of $SL(r,\mathbb{C})$, and $\bar{\lambda}^{A'}_{B'}$ is the complex conjugate of λ^{A}_{B}. The relation of this group of transformations to the Poincaré group in the case $r = 2$ should be apparent. Indeed, one of the attractive features of the extension of space-time geometry that we are putting forward here is that the hyper-Poincaré group allows such a description, which implies a wide range of intuitively plausible physical features.

More generally, we remark that the proper hyper-Poincaré group preserves the signature of any space-time interval, whether or not the interval is degenerate, and hence leaves the causal relations between events unchanged. We refer to a transformation of the form

$$r^{\mathrm{a}} \to L^{\mathrm{a}}_{\mathrm{b}} r^{\mathrm{b}} \tag{40}$$

as a 'hyper-Lorentz transformation' if

$$L^{\mathrm{a}}_{\mathrm{b}} = \lambda^{A}_{B} \bar{\lambda}^{A'}_{B'} \tag{41}$$

for some element $\lambda^{A}_{B} \in SL(r,\mathbb{C})$. In the case of Minkowski space ($r = 2$) it is well known that a two-component spinor $\xi^{\mathbf{A}}$ can be represented by a complex null bivector

$$\Lambda^{ab} = \xi^{\mathbf{A}} \xi^{\mathbf{B}} \varepsilon^{\mathbf{A}'\mathbf{B}'} \,. \tag{42}$$

Conversely, the bivector Λ^{ab} determines $\xi^{\mathbf{A}}$ up to the transformation

$$\xi^{\mathbf{A}} \to -\xi^{\mathbf{A}} \,. \tag{43}$$

This geometrical ambiguity is often referred to as the fundamental 'two-valuedness' of two-component spinors in relativity theory. In the case of a general quantum space-time of dimension r^2, a hyperspinor ξ^A can be represented by a complex null r-vector (antisymmetric tensor of rank r) of the form

$$\Lambda^{ab\cdots c} = \xi^{A} \xi^{B} \cdots \xi^{C} \varepsilon^{A'B'\cdots C'} \,. \tag{44}$$

The r-vector $\Lambda^{ab\cdots c}$ is 'null' in the sense that

$$g_{ab\cdots c} \Lambda^{ap\cdots q} \Lambda^{br\cdots s} = 0 \,. \tag{45}$$

Then $\Lambda^{ab\cdots c}$ determines ξ^A up to transformations of the form

$$\xi^{A} \to \mathrm{e}^{2\pi \mathrm{i}k/r} \xi^{A} \,, \tag{46}$$

where $k = 1, 2, \ldots, r-1$. Hence we can say that hyperspinors have a fundamental 'r-valuedness' (cf. Holm [21]).

The real dimension of the hyper-Lorentz group is $2r^2 - 2$, and thus the real dimension of the hyper-Poincaré group is $3r^2 - 2$. We observe that the dimension of the hyper-Poincaré group grows linearly with the dimension of the quantum space-time itself, which is given by r^2. This can be contrasted with the dimension of the group arising if we endow $\mathbb{R}^{r^2}$ with a standard Lorentzian metric with signature $(1, r^2 - 1)$. In that case the associated pseudo-orthogonal group has real dimension $\frac{1}{2}r^2(r^2 - 1)$, which together with the translation group gives a total dimension of $\frac{1}{2}r^2(r^2 + 1)$. The parsimonious dimensionality of the hyper-Poincaré group arises from the fact that it preserves the system of causal relations holding between pairs of points in quantum space-time.

In a flat four-dimensional space-time the symmetries of the Poincaré group are associated with a ten-parameter family of Killing vectors. Thus, for $r = 2$ we have the Minkowski metric (7), and the Poincaré group is generated by the ten-parameter family of vector fields ξ^{a} on $\mathfrak{M}^4$ satisfying

$$\mathscr{L}_\xi g_{ab} = 0, \tag{47}$$

where $\mathscr{L}_\xi$ denotes the Lie derivative with respect to ξ^{a}. For any vector field ξ^{a} and any symmetric tensor field g_{ab} we have

$$\mathscr{L}_\xi g_{ab} = \xi^{\mathrm{c}} \nabla_{\mathrm{c}} g_{ab} + 2 g_{\mathrm{c(a}} \nabla_{\mathrm{b)}} \xi^{\mathrm{c}} . \tag{48}$$

If g_{ab} is the metric and ∇_{a} denotes the associated covariant derivative satisfying

$$\nabla_{\mathrm{a}} g_{\mathrm{b}c} = 0, \tag{49}$$

we obtain the Killing equation

$$\nabla_{(\mathrm{a}} \xi_{\mathrm{b})} = 0, \tag{50}$$

where $\xi_{\mathrm{a}} = g_{ab} \xi^{\mathrm{b}}$. The condition $\mathscr{L}_\xi g_{ab} = 0$ therefore implies that ξ^{a} is a Killing vector.

For $r > 2$ we have no Riemannian metric, and the usual relations between symmetries and Killing vectors are lost. What survives, however, is of interest. Specifically, to generate a symmetry of the quantum space-time the vector field ξ^{a} has to satisfy

$$\mathscr{L}_\xi g_{ab\cdots c} = 0, \tag{51}$$

where $g_{ab\cdots c}$ is the chronometric tensor. For a general vector field ξ^{a} and a general symmetric tensor field $g_{ab\cdots c}$ we have

$$\mathscr{L}_\xi g_{ab\cdots c} = \xi^{\mathrm{d}} \nabla_{\mathrm{d}} g_{ab\cdots c} + r\, g_{\mathrm{d(a}\cdots b} \nabla_{\mathrm{c)}} \xi^{\mathrm{d}} . \tag{52}$$

In the case of the quantum space-time $\mathscr{H}^{r^2}$ we let ∇_{a} be the natural flat connection for which

$$\nabla_{\mathrm{a}} g_{\mathrm{bc}\cdots d} = 0 . \tag{53}$$

Then to generate a symmetry of the chronometric structure of $\mathscr{H}^{r^2}$ the vector field ξ^{a} has to satisfy

$$g_{\mathrm{d(a}\cdots b} \nabla_{\mathrm{c)}} \xi^{\mathrm{d}} = 0 . \tag{54}$$

Equation (54) can be written in an alternative form if we define a symmetric tensor $K_{ab\cdots c}$ of rank $r-1$ by setting

$$K_{ab\cdots c} = g_{ab\cdots cd} \xi^{\mathrm{d}} . \tag{55}$$

Then (54) says that $K_{ab\cdots c}$ satisfies the conditions for a symmetric Killing tensor:

$$\nabla_{(\mathrm{a}}K_{\mathrm{bc}\cdots d)} = 0\,. \tag{56}$$

Thus we see that $\mathscr{H}^{r^2}$ provides an example of a symmetry group generated by a family of Killing tensors. The symmetries of the quantum space-time are generated, more specifically, by a system of $3r^2-2$ irreducible symmetric Killing tensors of rank $r-1$. The significance of Killing tensors is that they are associated with conserved quantities. For other examples of Killing tensors arising in a physical context, see, e.g., [29, 30, 31, 46, 67]. In the present setting it follows that if the vector field v^{a} satisfies the geodesic equation, which on a quantum space-time of dimension r^2 is given, as we have seen, by

$$g_{abc\cdots d}\left(v^{\mathrm{e}}\nabla_{\mathrm{e}}v^{\mathrm{b}}\right)v^{\mathrm{c}}\cdots v^{\mathrm{d}} = 0\,, \tag{57}$$

and if $K_{ab\cdots c}$ is the Killing tensor of rank $r-1$ given by (55), then we have the following conservation law:

$$v^{\mathrm{e}}\nabla_{\mathrm{e}}\left(K_{ab\cdots c}v^{\mathrm{a}}v^{\mathrm{b}}\cdots v^{\mathrm{c}}\right) = 0\,. \tag{58}$$

In other words, the quantity

$$K = K_{ab\cdots c}v^{\mathrm{a}}v^{\mathrm{b}}\cdots v^{\mathrm{c}} \tag{59}$$

is a constant of the motion.

HYPER-RELATIVISTIC MECHANICS

It follows from the material of the previous section that in higher-dimensional quantum space-times the main conservation laws and symmetry principles of relativistic physics remain intact. In particular, the conservation of hyper-relativistic momentum and angular momentum for a system of interacting particles can be given a well-defined formulation, the basic principles of which are similar to those applicable in the Minkowskian case.

For this purpose it will be useful to introduce the notion of an 'elementary system' in hyper-relativistic mechanics. Such a system is defined by its hyper-relativistic momentum and angular momentum. The hyper-relativistic momentum of an elementary system is given by a momentum covector P_{a}. The associated mass m is given (cf. [13]) by the following natural expression:

$$m = \left(g^{\mathrm{ab}\cdots c}P_{\mathrm{a}}P_{\mathrm{b}}\cdots P_{\mathrm{c}}\right)^{\frac{1}{r}}\,. \tag{60}$$

The hyper-relativistic angular momentum of an elementary system is given by a tensor $L_{\mathrm{a}}^{\mathrm{b}}$ of the form

$$L_{\mathrm{a}}^{\mathrm{b}} = l_A^B\delta_{A'}^{B'} + \bar{l}_{A'}^{B'}\delta_A^B\,, \tag{61}$$

where the hyperspinor l_A^B is required to be trace-free: $l_A^A = 0$. The angular momentum is defined with respect to a choice of origin in such a manner that under a change of origin defined by a shift vector β^{a} we have

$$l_A^B \to l_A^B + P_{AC'}\beta^{BC'} . \tag{62}$$

In the case $r = 2$ these formulae reduce to the usual expressions for relativistic momentum and angular momentum in a Minkowskian setting. The real covector

$$S_{AA'} = \mathrm{i}m^{-1}\left(l_A^B P_{A'B} - \bar{l}_{A'}^{B'} P_{AB'}\right) \tag{63}$$

is invariant under a change of origin, and carries the interpretation of the intrinsic spin of the elementary system. The magnitude of the spin is then defined by

$$S = |g^{ab\cdots c} S_{\mathrm{a}} S_{\mathrm{b}} \cdots S_{\mathrm{c}}|^{\frac{1}{r}} . \tag{64}$$

In the case of a set of interacting hyper-relativistic systems we require that the total momentum and angular momentum should be conserved. This then implies conservation of the total mass and spin. In short, we see that the idea of 'relativistic mechanics' carries through nicely to the case of a general quantum space-time.

Now what is the interpretation of these conservation laws? We shall show later, once we introduce the idea of symmetry breaking, that hypermomentum can be interpreted as the momentum operator for a relativistic quantum system. Conservation of hypermomentum then can be thought of as conservation of four-momentum in relativistic quantum mechanics in the Heisenberg representation.

COMPLEX NULL DIRECTIONS

In four-dimensional space-time it is useful in many contexts to examine the geometry of the space of complex null vectors at a point in the space-time. This has the effect of giving us a vivid picture of the local causal relationships in space-time, and by sticking with a complex picture we also retain the link with quantum mechanics. Thus we consider complex vectors z^{a} satisfying the quadratic equation

$$g_{\mathrm{ab}} z^{\mathrm{a}} z^{\mathrm{b}} = 0 . \tag{65}$$

In spinor terms this relation implies that the corresponding complex matrix $z^{\mathrm{AA'}}$ is of the special form

$$z^{\mathrm{AA'}} = \alpha^{\mathrm{A}} \beta^{\mathrm{A'}} . \tag{66}$$

The space of complex vectors at a point in Minkowski space is $\mathbb{C}^4$. The space of complex directions (which results if we consider equivalence class of vectors modulo overall proportionality) is the complex projective 3-space $\mathbb{P}^3$. The null directions constitute a

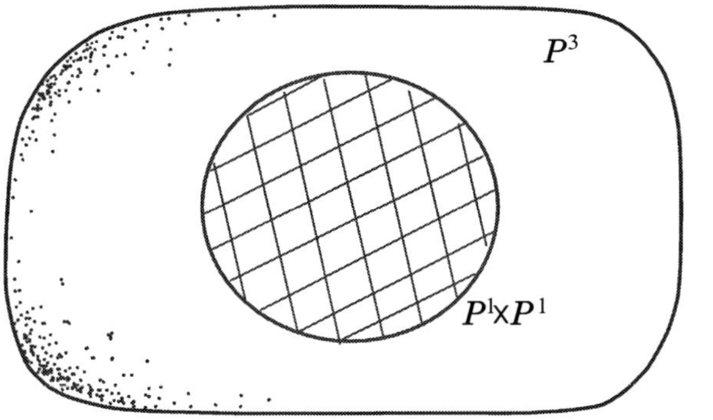

FIGURE 2. *Complex null directions in a four-dimensional space-time.* The space of directions at a point in complex Minkowski space has the geometry of a complex projective 3-space. The complex null directions constitute a quadric which has the structure of a doubly ruled surface $\mathbb{P}^1 \times \mathbb{P}^1$.

quadric $\mathbb{Q}^2$ in that space, which owing to the decomposition (66) has the structure of a doubly ruled surface

$$\mathbb{Q}^2 = \mathbb{P}^1 \times \mathbb{P}^1 \,. \tag{67}$$

We can identify the first set of lines (the α-lines) with the projective unprimed spinors, and the second set of lines (the β-lines) with the projective primed spinors. The quadric $\mathbb{Q}^2$ is ruled in such a way that two lines of the same type do not intersect, whereas two lines of the opposite type intersect at a point in $\mathbb{Q}^2$ (see Figure 2). This point corresponds to the null direction they jointly determine.

In the case of a general quantum space-time of dimension r^2, we consider the space of complex vectors at a point, and examine the corresponding space of directions, which has the structure of a complex projective space $\mathbb{P}^{r^2-1}$. The space of degenerate complex directions, which is given by the vanishing of the chronometric form $g_{ab\cdots c}z^{\mathrm{a}}z^{\mathrm{b}}\cdots z^{\mathrm{c}}$, is a hypersurface in $\mathbb{P}^{r^2-1}$, which we shall call $\mathscr{D}^{r^2-2}$.

The points of $\mathscr{D}^{r^2-2}$ correspond to degenerate directions of degree $r-1$. The null directions in $\mathscr{D}^{r^2-2}$ correspond to those directions for which the associated degenerate vectors are of minimal rank and hence of the form $z^{AA'} = \alpha^A\beta^{A'}$. These constitute a subvariety $\mathbb{Q}^{2r-2} \subset \mathscr{D}^{r^2-2}$ defined by the mutual intersection of a system of quadrics, given by the equation

$$g_{ab\cdots c}z^{\mathrm{a}}z^{\mathrm{b}} = 0 \,. \tag{68}$$

In this situation we have

$$\mathbb{Q}^{2r-2} = \mathbb{P}^{r-1} \times \mathbb{P}^{r-1} \,, \tag{69}$$

and we can identify the two systems of $(r-1)$-planes by which $\mathbb{Q}^{2r-2}$ is foliated, which we refer to as α-planes and β-planes, as the spaces of projective unprimed and primed hyperspinors, respectively. The various degenerate directions of intermediate degree

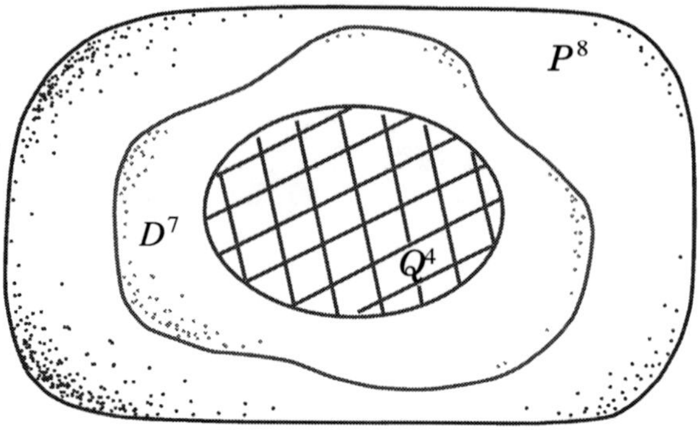

FIGURE 3. *The space of complex null directions at a point in a quantum space-time of dimension nine.* In this case the space of degenerate directions is given by a cubic surface $\mathscr{D}^7$ in the complex projective space $\mathbb{P}^8$. The complex null directions then constitute a four-dimensional doubly foliated subvariety $\mathbb{Q}^4 = \mathbb{P}^2 \times \mathbb{P}^2$ in $\mathscr{D}^7$.

correspond to points in $\mathscr{D}^{r^2-2}$ lying on the linear spaces spanned by the joins of d points in $\mathbb{Q}^{2r-2}$ $(d = 2, 3, \ldots, r)$. The degree of degeneracy is given by the integer d.

In the case $r = 3$, illustrated in Figure 3, the space of complex directions at a point in $\mathscr{C}\mathscr{H}^9$ is $\mathbb{P}^8$, and the degenerate directions constitute a cubic hypersurface $\mathscr{D}^7 \subset \mathbb{P}^8$. The null directions lie in the doubly foliated surface

$$\mathbb{Q}^4 = \mathbb{P}^2 \times \mathbb{P}^2 \tag{70}$$

in $\mathscr{D}^7$. The points of $\mathscr{D}^7$ all lie on the 'first join' of $\mathbb{Q}^4$ with itself; in other words, any point of $\mathscr{D}^7$ lies on a line joining two points of $\mathbb{Q}^4$. Thus we write $\mathscr{D}^7 = J_1(\mathbb{Q}^4)$. The space $\mathscr{D}^7 \backslash \mathbb{Q}^4$ then consists of degenerate directions that are strictly of the second degree. Note that any point of $\mathbb{P}^8$ can be represented as the join of three points in $\mathbb{Q}^4$, and hence $J_2(\mathbb{Q}^4) = \mathbb{P}^8$.

In the case $r = 4$, illustrated in Figure 4, the space of complex directions at a point in $\mathscr{C}\mathscr{H}^{16}$ is the complex projective space $\mathbb{P}^{15}$, and the degenerate directions constitute a quartic hypersurface $\mathscr{D}^{14} \subset \mathbb{P}^{15}$. The null directions (degenerate direction of the first degree) lie on the doubly foliated surface

$$\mathscr{D}_{(1)} = \mathbb{P}^3 \times \mathbb{P}^3 \tag{71}$$

in $\mathscr{D}^{14}$. The degenerate directions of the second degree lie on the first join of $\mathbb{Q}^6$ with itself:

$$\mathscr{D}_{(2)} = J_1(\mathbb{Q}^6)\,. \tag{72}$$

The degenerate directions of the third degree lie in

$$\mathscr{D}_{(3)} = J_2(\mathbb{Q}^6) \tag{73}$$

and constitute the general elements of $\mathscr{D}^{14}$.

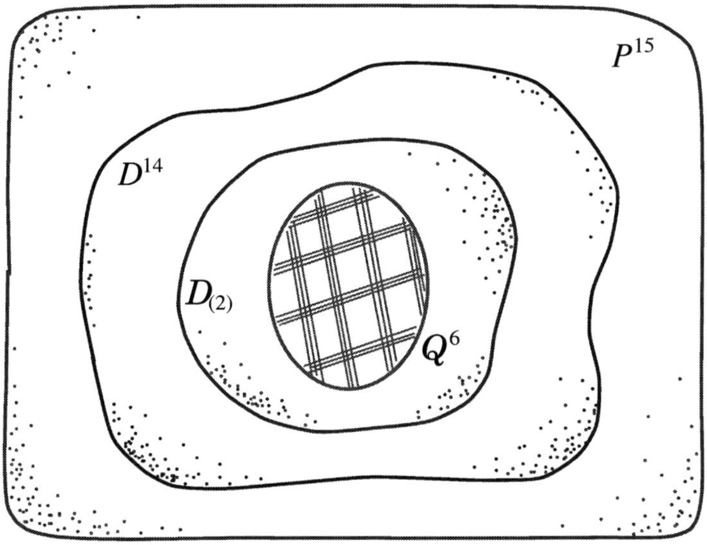

FIGURE 4. *The space of complex null directions at a point in the quantum space-time* $\mathscr{H}^{16}$. The degenerate directions of degree three constitute a quartic hypersurface $\mathscr{D}^{14}$ in $\mathbb{P}^{15}$. The null directions belong to a six-dimensional doubly foliated subvariety $\mathbb{Q}^6 = \mathbb{P}^3 \times \mathbb{P}^3$. Degenerate directions of degree two constitute a submanifold $\mathscr{D}_{(2)} = J_1(\mathbb{Q}^6) \subset \mathscr{D}^{14}$ consisting of points that lie on lines spanned by pairs of points in the subvariety $\mathbb{Q}^6$.

TWISTORS AND HYPERTWISTORS

The proceeding discussion of complex null directions can be seen as a 'warm-up' exercise for the introduction of the concept of hypertwistors. For this purpose we introduce a notation that closely parallels the standard notation for $r = 2$ Penrose twistors. Let us denote by $\mathbb{T}^\alpha$ the complex vector space of dimension $2r$ given by the pair $(\mathbb{S}^A, \mathbb{S}_{A'})$. We write

$$Z^\alpha = (\omega^A, \pi_{A'}) \tag{74}$$

for a typical element of $\mathbb{T}^\alpha$. Such elements will be referred to as 'hypertwistors' (also called 'generalised twistors' [11, 22]). Let $\mathbb{T}_\alpha = (\mathbb{S}_A, \mathbb{S}^{A'})$ denote the space of dual hypertwistors. A natural pseudo-Hermitian structure can be introduced on the geometry of hypertwistors by means of the complex conjugation operation that maps $(\omega^A, \pi_{A'}) \in \mathbb{T}^\alpha$ to $(\bar{\pi}_A, \bar{\omega}^{A'}) \in \mathbb{T}_\alpha$. The corresponding pseudo-Hermitian form is then given by

$$Z^\alpha \bar{Z}_\alpha = \omega^A \bar{\pi}_A + \pi_{A'} \bar{\omega}^{A'} . \tag{75}$$

It is straightforward to verify that the inner product $Z^\alpha \bar{Z}_\alpha$ is invariant under the action of the group $U(r,r)$. In the case $r = 2$, the elements of $\mathbb{T}^\alpha$ are standard Penrose twistors, and the relevant symmetry group is $U(2,2)$.

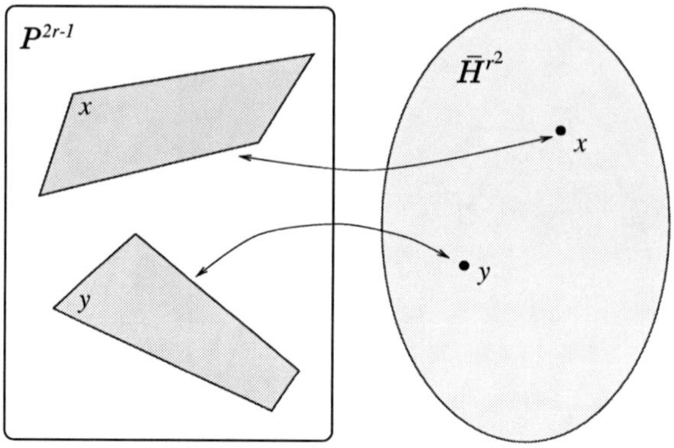

FIGURE 5. *The Klein representation for a general quantum space-time.* The aggregate consisting of all $(r-1)$-planes in the complex projective space $\mathbb{P}^{2r-1}$ forms a manifold of dimension r^2. This space is the complexified, compactified version of the quantum space-time $\mathscr{C}\mathscr{H}^{r^2}$.

The space $\mathbb{P}^{2r-1}$ of projective hypertwistors is a good starting point for the analysis of the conformal geometry of complex quantum space-time, which can be regarded as the Grassmannian variety of projective $(r-1)$-planes in $\mathbb{P}^{2r-1}$, as illustrated in Figure 5. More precisely, the aggregate of all projective $(r-1)$-planes in $\mathbb{P}^{2r-1}$ constitutes a compact manifold of dimension r^2, which we identify as the complexified, compactified quantum space-time $\mathscr{C}\bar{\mathscr{H}}^{r^2}$. The 'finite' points of $\mathscr{C}\bar{\mathscr{H}}^{r^2}$ correspond to those $(r-1)$-planes of $\mathbb{P}^{2r-1}$ that are determined by a linear relation of the form

$$\omega^A = \mathrm{i}x^{AA'}\pi_{A'} \tag{76}$$

for some fixed $x^{AA'}$. Thus for each (complex) point $x^{AA'}$ in $\mathscr{C}\mathscr{H}^{r^2}$ we obtain, according to equation (76), an $(r-1)$-plane in $\mathbb{P}^{2r-1}$. The aggregate of such $(r-1)$-planes constitute the points of $\mathscr{C}\mathscr{H}^{r^2}$. The $(r-1)$-planes for which $x^{AA'}$ is Hermitian then constitute the points of the real quantum space-time $\mathscr{H}^{r^2}$.

The conformal structure of a quantum space-time is implicit in the various possibilities arising for the intersections of $(r-1)$-planes in the projective hypertwistor space $\mathbb{P}^{2r-1}$. In the case $r=2$ (Penrose twistors) the projective space $\mathbb{P}^3$ contains a four-dimensional family of complex projective lines, the aggregation of which constitutes the associated space-time (see Figure 6). In this case, a pair of space-time points x and y are null-separated in the Minkowskian sense if and only if the corresponding lines in $\mathbb{P}^3$ intersect. The space of all complex projective lines in $\mathbb{P}^3$ constitutes a four-dimensional quadric hypersurface Q^4 in $\mathbb{P}^5$, which we identify as complexified, compactified Minkowski space-time $\mathscr{C}\bar{\mathscr{H}}^4$. The quadric Q^4 contains two systems of projective 2-planes, called α-planes and β-planes. The α-planes are in one-to-one correspondence with the points of $\mathbb{P}^3$, whereas the β-planes are in one-to-one correspondence with the 2-planes of $\mathbb{P}^3$, as illustrated in Figure 7.

The case $r=3$ has been studied closely by Finkelstein [13] and his collaborators. In this case, the nine-dimensional quantum space-time $\mathscr{C}\bar{\mathscr{H}}^9$ arises when we consider

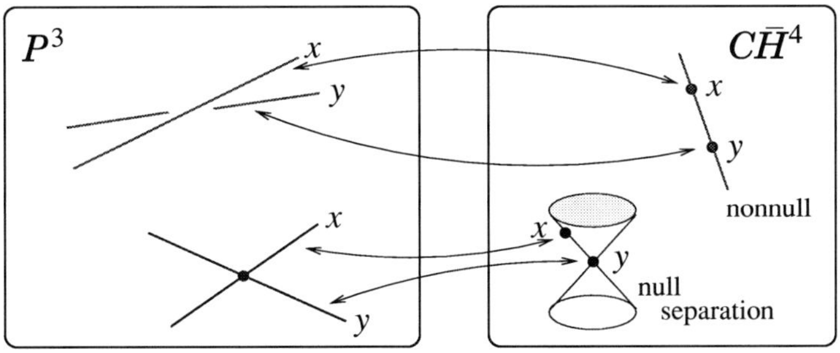

FIGURE 6. *Twistor theory.* The lines of the complex projective space $\mathbb{P}^3$ correspond to the points of the complex space-time $\mathscr{C}\mathscr{H}^4$. Two points in $\mathscr{C}\mathscr{H}^4$ are null-separated if and only if the corresponding lines in $\mathbb{P}^3$ intersect.

the Grassmanian of complex projective 2-planes in complex projective 5-space. A pair of planes in $\mathbb{P}^3$ in general will not intersect. If they do, however, then there are three possibilities: they can intersect in a point, a line, or a plane. These situations correspond to the various levels of degeneracy obtainable for the separation vector. In particular, if the two planes intersect in a line, then the separation vector is null (see Figure 8).

A pair of $(r-1)$-planes in $\mathbb{P}^{2r-1}$ will not in general intersect. This generic non-intersection property corresponds to the nonvanishing of the chronometric form for the separation vector of generic quantum space-time points. An $(r-1)$-plane in $\mathbb{P}^{2r-1}$ is represented by a simple skew hypertwistor $P^{\alpha\beta\cdots\gamma}$ of rank r. By a 'simple' skew hypertwistor we mean one of the form

$$P^{\alpha\beta\cdots\gamma} = A^{[\alpha}B^{\beta}\cdots C^{\gamma]} \tag{77}$$

for some set $A^{\alpha}, B^{\alpha}, \cdots, C^{\alpha}$ of r hypertwistors. Suppose that the simple skew hypertwistors $P^{\alpha\beta\cdots\gamma}$ and $Q^{\alpha\beta\cdots\gamma}$ represent, respectively, the $(r-1)$-planes P and Q in $\mathbb{P}^{2r-1}$. A necessary and sufficient condition for the vanishing of the chronometric form for the corresponding space-time points is

$$\varepsilon_{\alpha\beta\cdots\gamma\rho\sigma\cdots\tau}P^{\alpha\beta\cdots\gamma}Q^{\rho\sigma\cdots\tau} = 0, \tag{78}$$

where $\varepsilon_{\alpha\beta\cdots\gamma\rho\sigma\cdots\tau}$ is the skew hypertwistor of rank $2r$. We note that (78) is symmetric under the interchange of P and Q if r is even, and antisymmetric if r is odd. The vanishing of (78) is the condition that the planes P and Q contain a point in common. This means that the skew hypertwistors $P^{\alpha\beta\cdots\gamma}$ and $Q^{\rho\sigma\cdots\tau}$ contain at least one hypertwistor as a common factor. Thus, a necessary and sufficient condition for a pair of quantum space-time points to have a degenerate separation is that the corresponding $(r-1)$-planes in $\mathbb{P}^{2r-1}$ should intersect.

More generally, the degeneracy d of the separation of a pair of quantum space-time events is given by $d = r - m - 1$, where m is the dimensionality of the intersection of the corresponding $(r-1)$-planes in $\mathbb{P}^{2r-1}$. The possible degrees of degeneracy are $d = 1, 2, \ldots, r-1$. If we interpret the case of no intersection as an intersection of dimension -1, then a nondegenerate separation can be interpreted as a 'degeneracy of

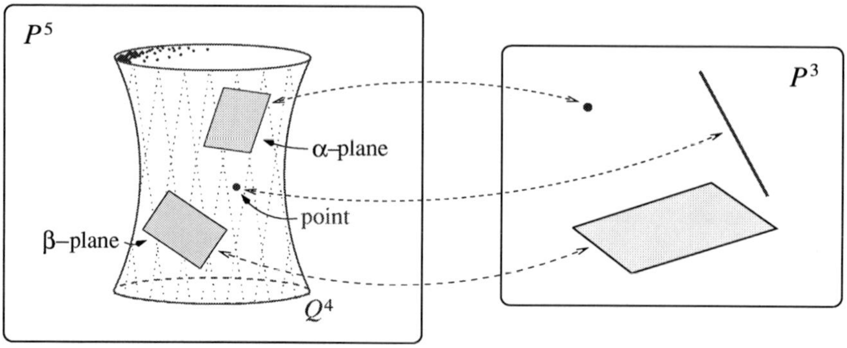

FIGURE 7. *Complexified, compactified Minkowski space-time as the Klein quadric in complex projective 5-space.* The aggregate of complex projective lines in $\mathbb{P}^3$ constitutes a nondegenerate quadric Q^4 in $\mathbb{P}^5$. The quadric contains two distinct systems of projective 2-planes, called α-planes and β-planes. Any two distinct planes of the same type in Q^4 intersect at a point. Two planes of the opposite type in Q^4 will in general not intersect, but if they do, they intersect in a line. The points of $\mathbb{P}^3$ correspond to α-planes in Q^4, and the 2-planes of $\mathbb{P}^3$ correspond to β-planes in Q^4.

degree r'. Thus separations of degree less than r are degenerate, whereas a separation of degree r is nondegenerate. The degree of degeneracy is given by the rank of the separation matrix $r^{AA'} = x^{AA'} - y^{AA'}$.

Alternatively, given two skew hypertwistors $P^{\alpha\beta\cdots\gamma}$ and $Q^{\alpha\beta\cdots\gamma}$, each with r indices, let us form the dual hypertwistor by

$$\mathcal{Q}_{\alpha\beta\cdots\gamma} = \varepsilon_{\alpha\beta\cdots\gamma\rho\sigma\cdots\tau} Q^{\rho\sigma\cdots\tau} \,. \tag{79}$$

Then d is the maximum number of index contractions we can make between $P^{\alpha\beta\cdots\gamma}$ and $\mathcal{Q}_{\alpha\beta\cdots\gamma}$ without obtaining the result zero. If a single index contraction gives zero, this corresponds to the case where $P^{\alpha\beta\cdots\gamma}$ is proportional to $Q^{\alpha\beta\cdots\gamma}$. Thus $d = 0$ (separation of degree zero) can be interpreted as the 'completely degenerate' case where the two space-time points coincide.

HYPERTWISTOR FIELDS

Now we present a few elementary applications of hypertwistor theory. In the case of a four-dimensional space-time, it is well known that twistors can be characterised as solutions of the differential equation

$$\nabla^{\mathbf{A}'(\mathbf{A}}\xi^{\mathbf{B})} = 0 \,, \tag{80}$$

the so-called 'twistor equation'. Indeed, the solution of this equation is

$$\xi^{\mathbf{A}} = \omega^{\mathbf{A}} - \mathrm{i}x^{\mathbf{A}\mathbf{A}'}\pi_{\mathbf{A}'} \,, \tag{81}$$

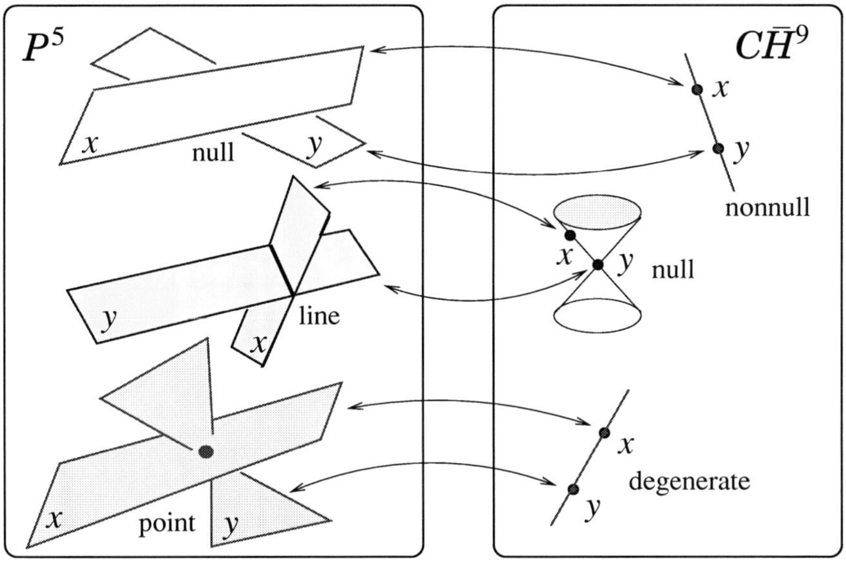

FIGURE 8. *Hypertwistor theory for* $r = 3$. The aggregate of complex projective 2-planes in $\mathbb{P}^5$ constitutes a nine-dimensional quantum space-time $\mathscr{C}\mathscr{H}^9$. In general, two such planes will not intersect. If two planes intersect in a point, then the separation vector of the corresponding space-time points is degenerate. If the intersection is in a line, then the corresponding points are null-separated.

and the associated twistor is determined by the fixed spinor pair $(\omega^{\mathbf{A}}, \pi_{\mathbf{A}'})$. More generally, in the case of a quantum space-time of dimension r^2 we consider the equation

$$r\nabla_{AA'}\xi^C = \delta_A^C \nabla_{DA'}\xi^D \, . \tag{82}$$

It is not difficult to verify that in the case $r = 2$ equation (82) reduces to the conventional twistor equation (80). The solutions of (82) can be expressed in terms of generalised twistors:

$$\xi^A = \omega^A - \mathrm{i}x^{AA'}\pi_{A'} \, , \tag{83}$$

where ω^A and $\pi_{A'}$ are constant. This result is established in [22].

We note that a necessary and sufficient condition for a hypertwistor field $\xi^A(x)$ to satisfy the hypertwistor equation (82) is that it should satisfy

$$\sigma_C \sigma_{[A} \nabla_{B]C'} \xi^C = 0 \tag{84}$$

for all values of σ_A. For $r = 2$ the twistor equation (80) is a special case of a more general nonlinear partial differential equation known as the geodesic shear-free condition:

$$\xi^{\mathbf{A}}\xi^{\mathbf{B}}\nabla_{\mathbf{A'A}}\xi_{\mathbf{B}} = 0 \, . \tag{85}$$

When $r > 2$, however, the connection between the twistor equation and the geodesic shear-free condition is lost, and the appropriate generalisation of the latter is

$$\left(\xi_{[A}\nabla_{B]B'}\xi_{[C}\right)\xi_{D]} = 0\,, \tag{86}$$

which reduces to the geodesic shear-free condition (85) when $r = 2$. Solutions of the equation (86) can be generated through the consideration of the intersections of projective $(r-1)$-planes in $\mathbb{P}^{2r-1}$ with certain types of complex analytic varieties in $\mathbb{P}^{2r-1}$.

Now let us consider the dynamical equations for fields of totally null momentum on a general quantum space-time. The relevant field equations take the form

$$\nabla_{A[A'}\nabla_{B']B}\phi = 0\,, \tag{87}$$

$$\nabla_{A[A'}\phi_{B']\cdots C'} = 0\,, \tag{88}$$

and

$$\nabla_{A'[A}\phi_{B]\cdots C} = 0\,. \tag{89}$$

These equations generalise the zero rest-mass equations for each helicity $s = \pm\frac{1}{2}n$ (where $n = 0, 1, 2, \ldots$) in four-dimensional space-time.

The solutions can be described briefly as follows. We write $Z^{\alpha} = (\omega^A, \pi_{A'})$ and let $\psi(Z^{\alpha})$ denote an analytic, homogeneous function of degree $-r+2s$ defined on some region of the hypertwistor space. For $s = 0$ we set

$$\phi(x) = \oint \psi(\mathrm{i}x^{AA'}\pi_{A'}, \pi_{A'})\,\Delta\pi\,, \tag{90}$$

where the differential form $\Delta\pi$ is given by

$$\Delta\pi = \varepsilon^{A'B'\ldots C'}\pi_{A'}\pi_{B'}\cdots\pi_{C'}\,. \tag{91}$$

Note that $\Delta\pi$ is homogeneous of degree r in $\pi_{A'}$, and thus that the quantity appearing in the integral sign in (90) is homogeneous of degree zero. Similarly, for helicity $s = -\frac{1}{2}$ we define

$$\phi_{A'}(x) = \oint \pi_{A'}\psi(\mathrm{i}x^{AA'}\pi_{A'}, \pi_{A'})\,\Delta\pi\,, \tag{92}$$

and for helicity $s = \frac{1}{2}$ we define

$$\phi_{A'}(x) = \oint \frac{\partial}{\partial\omega^A}\psi(\mathrm{i}x^{AA'}\pi_{A'}, \pi_{A'})\,\Delta\pi\,. \tag{93}$$

Here it is understood that first we differentiate $\psi(\omega^A, \pi_{A'})$ with respect to ω^A, and we set $\omega^A = \mathrm{i}x^{AA'}\pi_{A'}$ before integrating. A proposition established in [22] shows that the contour integrals (90), (92) and (93) satisfy the totally null momentum conditions (87), (88) and (89), respectively.

As an illustration, let us first take the simplest case, given by equation (90) for $r = 3$. For our hypertwistor function we take

$$\psi(Z^\alpha) = \frac{1}{P_\alpha Z^\alpha Q_\beta Z^\beta R_\gamma Z^\gamma}, \tag{94}$$

where P_α, Q_β, and R_γ denote fixed points in the dual hypertwistor space. For their hyperspinor decomposition we write

$$P_\alpha = (p_A, p^{A'}), \quad Q_\alpha = (q_A, q^{A'}), \quad \text{and} \quad R_\alpha = (r_A, r^{A'}), \tag{95}$$

and set

$$\tilde{p}^{A'}(x) = p^{A'} + \mathrm{i}x^{A'A} p_A, \quad \tilde{q}^{A'}(x) = q^{A'} + \mathrm{i}x^{A'A} q_A, \quad \text{and} \quad \tilde{r}^{A'}(x) = r^{A'} + \mathrm{i}x^{A'A} r_A, \tag{96}$$

where $\tilde{p}^{A'}$, $\tilde{q}^{B'}$, and $\tilde{r}^{C'}$ are solutions of the primed analogue

$$r\nabla_{AA'}\tilde{\eta}^{C'} = \delta^{C'}_{A'}\nabla_{AD'}\tilde{\eta}^{D'}, \tag{97}$$

of the hypertwistor equation (82). Inserting this expression into the contour integral formula, we obtain the result

$$\phi(x) = \frac{4\pi^2}{\varepsilon_{A'B'C'}\tilde{p}^{A'}\tilde{q}^{B'}\tilde{r}^{C'}}, \tag{98}$$

where the contour is taken to be $S^1 \times S^1$. It is then a straightforward exercise to verify directly that (98) satisfies the scalar hyper-wave equation (87). This example generalises the so-called 'elementary states' of standard Penrose twistor theory [44].

GEOMETRICAL STRUCTURES IN COSMOLOGY

One of the motivations for the idea of quantum space-time is that it provides a possible way forward towards the unification of cosmology and elementary particle physics. Clearly some such unification is required for a coherent discussion of the early stages of the universe—but in what mathematical framework should this unification be pursued? An added impetus to these considerations comes from the dark matter/energy problem (see, e.g., [12, 60] and references cited therein), which has had the effect of encouraging physicists to rethink the foundations of cosmology. If we are to consider new classes of models, then some criteria are required to limit the range of possibilities. In particular, it is reasonable to propose that: (a) the model should be geometrical in character; and (b) the model should be rich enough to admit within its space the possibility of a geometrisation of quantum mechanics (by 'geometrisation' we mean an approach in the spirit, e.g., of [1, 2, 5, 17, 34]). Our intention now is to put forward a tentative approach to 'quantum cosmology' in this spirit, based on the idea of quantum space-time.

In doing so, we bear in mind that there are a number of distinct inter-related geometrical structures arising in cosmology, that need to be taken into consideration as we proceed. These include: (a) reality structure (as in the distinction between a real and a complex space-time); (b) causality structure (identification of the light cones); (c) infinity structure (is the universe open or closed?); (d) chronometric structure (how are time, distance, and energy measured?); and (e) singularity structure (how does one characterise the beginning and end?).

The theory of hyperspin is well suited for addressing these issues in a mathematically compelling way (see [15, 19, 20] for earlier examples of models for hypercosmologies). It is convenient to begin the present analysis with a discussion of the structure of infinity in the case of a general flat quantum space-time. This will lead us to more general cosmological considerations.

As indicated above, for any skew hypertwistor $Q^{\alpha\beta\cdots\gamma}$ of rank r in a quantum space-time of dimension r^2 we define its dual $Q_{\alpha\beta\cdots\gamma}$ by the relation

$$Q_{\alpha\beta\cdots\gamma} = \varepsilon_{\alpha\beta\cdots\gamma\rho\sigma\cdots\tau} Q^{\rho\sigma\cdots\tau} \,. \tag{99}$$

Here $\varepsilon_{\alpha\beta\cdots\gamma\rho\sigma\cdots\tau}$ is the skew hypertwistor of rank $2r$, which is unique up to scale. Depending on whether r is even or odd, we have the following interchange relations:

$$\varepsilon_{\alpha\beta\cdots\gamma\rho\sigma\cdots\tau} = \pm\varepsilon_{\rho\sigma\cdots\tau\alpha\beta\cdots\gamma} . \tag{100}$$

Thus if r is even, then once the scale is fixed we obtain a symmetric inner product on the space of skew hypertwistors of rank r, which we denote

$$\langle P,Q\rangle = \varepsilon_{\alpha\beta\cdots\gamma\rho\sigma\cdots\tau} P^{\alpha\beta\cdots\gamma} Q^{\rho\sigma\cdots\tau} \,. \tag{101}$$

If r is odd then the product (101) gives us a symplectic structure.

Under complex conjugation the skew hypertwistor $P^{\alpha\beta\cdots\gamma}$ becomes $\bar{P}_{\alpha\beta\cdots\gamma}$. If $P^{\alpha\beta\cdots\gamma}$ is simple, thus corresponding to an $(r-1)$-plane P in $\mathbb{P}^{2r-1}$, then we say that P is real if $\bar{P}_{\alpha\beta\cdots\gamma}$ is proportional to $P_{\alpha\beta\cdots\gamma}$. The real $(r-1)$-planes of $\mathbb{P}^{2r-1}$ correspond to the real points of quantum space-time.

The structure at infinity in the compactified quantum space-time $\bar{\mathscr{H}}^{r^2}$ can be described as follows. In the hypertwistor space $\mathbb{P}^{2r-1}$ we choose a real $(r-1)$-plane I represented by a simple skew hypertwistor $I^{\alpha\beta\cdots\gamma}$. The point I in $\bar{\mathscr{H}}^{r^2}$ corresponding to the $(r-1)$-plane I in $\mathbb{P}^{2r-1}$ will be called the 'point at infinity'. The locus in $\bar{\mathscr{H}}^{r^2}$ consisting of all points that have a degenerate separation from I will be called 'infinity'. It should be evident that infinity has a rich structure, with various domains that can be classified according to the degree of degeneracy of their separation from the point I.

The 'finite' points of $\bar{\mathscr{H}}^{r^2}$ are those for which the separation from I is nondegenerate, i.e. those points P for which $\langle P,I\rangle \neq 0$. In the case of two finite quantum space-time points the chronometric form Δ is given as follows:

$$\Delta(P,Q) = \frac{\varepsilon_{\alpha\beta\cdots\gamma\rho\sigma\cdots\tau} P^{\alpha\beta\cdots\gamma} Q^{\rho\sigma\cdots\tau}}{(\varepsilon_{\alpha\beta\cdots\gamma\rho\sigma\cdots\tau} P^{\alpha\beta\cdots\gamma} I^{\rho\sigma\cdots\tau})(\varepsilon_{\alpha\beta\cdots\gamma\rho\sigma\cdots\tau} Q^{\alpha\beta\cdots\gamma} I^{\rho\sigma\cdots\tau})} \,. \tag{102}$$

Equivalently we can write

$$\Delta(P,Q) = \frac{\langle P,Q\rangle}{\langle P,I\rangle\langle Q,I\rangle} \,. \tag{103}$$

If P and I are not null-separated, then we can choose the scales of $P^{\alpha\beta\cdots\gamma}$ and $I^{\alpha\beta\cdots\gamma}$ such that $\langle P,I\rangle = 1$, without loss of generality, and similarly for $Q^{\alpha\beta\cdots\gamma}$ and $I^{\alpha\beta\cdots\gamma}$.

We note that $\Delta(P,Q)$ is independent of the scale of $P^{\alpha\beta\cdots\gamma}$ and $Q^{\alpha\beta\cdots\gamma}$. On the other hand, $\Delta(P,Q)$ does depend on the scale of $\varepsilon_{\alpha\beta\cdots\gamma\rho\sigma\cdots\tau}$ and the scale of $I^{\alpha\beta\cdots\gamma}$. It has an epsilon 'weight' of -1 and an I 'weight' of -2 (cf. [27]). If we form the ratio associated with four hypertwistors P, Q, R, and S, given by

$$\frac{\Delta(P,Q)}{\Delta(R,S)} = \frac{\|p-q\|^r}{\|r-s\|^r}, \tag{104}$$

where p, q, r, and s are the quantum space-time points corresponding to P, Q, R, and S, respectively, then we obtain an expression that is absolute—that is to say, a geometric invariant. This is because $\Delta(P,Q)$ has the 'dimensionality' of time raised to the power r; whereas the ratio (104) arises as a comparison of two such time intervals, and thus is dimensionless. The basic chronometric geometry, with infinity chosen as indicated above, admits no absolute or 'preferred' unit of time: in this geometry only ratios of time intervals have an absolute meaning.

HIGHER-DIMENSIONAL TWISTOR COSMOLOGY

There is, on the other hand, no reason *a priori* why just such a structure should apply at infinity. Other choices are available for $I^{\alpha\beta\cdots\gamma}$, and these have the effect of giving $\bar{\mathscr{H}}^{r^2}$ the structure of a cosmological model. In the case $r = 2$, for example, if $I^{\alpha\beta}$ is chosen to be real and non-simple, then the quadratic form

$$\lambda = \varepsilon_{\alpha\beta\gamma\delta} I^{\alpha\beta} I^{\gamma\delta}, \tag{105}$$

which has an epsilon weight of one and an I-weight of two, has the dimensionality of inverse squared-time. Hence in this case there *is* a preferred unit of time. Other time intervals can then be expressed in multiples of the preferred unit of time.

To pursue this point further, we recall that $\bar{\mathscr{H}}^4$ has the structure of a quadric Ω in $\mathbb{P}^5$. More specifically, for $r = 2$ the space of skew rank two twistors is $\mathbb{C}^6$, which is projectively $\mathbb{P}^5$, and $\bar{\mathscr{H}}^4$ is the locus defined by the homogeneous quadratic equation

$$\varepsilon_{\alpha\beta\gamma\delta} X^{\alpha\beta} X^{\gamma\delta} = 0. \tag{106}$$

Infinity in $\bar{\mathscr{H}}^4$ can then be defined by the intersection of $\bar{\mathscr{H}}^4$ in $\mathbb{P}^5$ with the projective 4-plane I^4 given by the equation

$$\varepsilon_{\alpha\beta\gamma\delta} I^{\alpha\beta} X^{\gamma\delta} = 0. \tag{107}$$

If $I^{\alpha\beta}$ is simple, then I^4 is tangent to $\bar{\mathscr{H}}^4$, and the intersection is a cone—the null cone at infinity. The geometry of this space plays a crucial role in determining the properties of time-like geodesics in Minkowski space, as discussed in Figure 9.

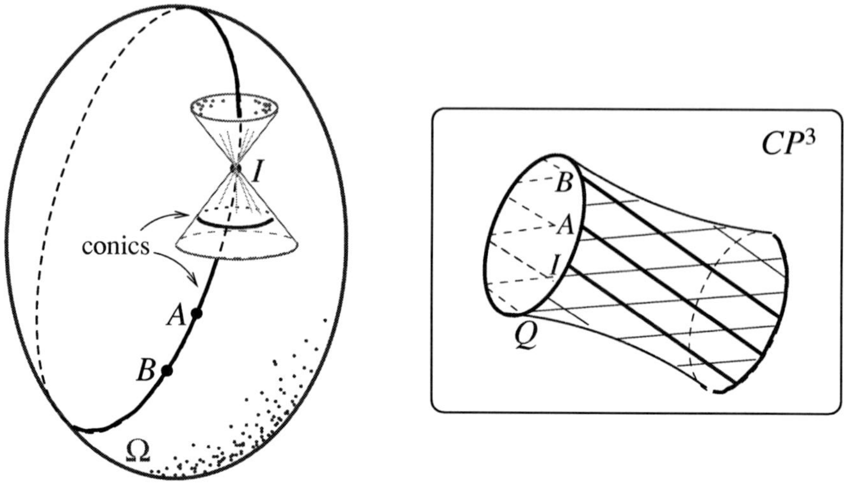

FIGURE 9. *Time-like geodesics in Minkowski space.* Each time-like geodesic in a flat four-dimensional space-time determines a quadric in projective twistor space. A quadric in $\mathbb{P}^3$ has two distinct systems of generators. In the case of a time-like geodesic, the line I belongs to one of the two systems. The corresponding geodesic is a conic curve in the Klein quadric Ω, and has the property that it passes through the point I. The second system of generators then corresponds to a polar conic that lies on the null cone infinity and does not go through I.

On the other hand, if $I^{\alpha\beta}$ is not simple, then the intersection $I^4 \cap \mathscr{H}^4$ is a 3-quadric. The resulting geometry, if $I^{\alpha\beta}$ is real, is that of de Sitter space. The metric on $I^4 \cap \mathscr{H}^4$ in this case is given by

$$\mathrm{d}s^2 = \frac{\varepsilon_{\alpha\beta\gamma\delta}\mathrm{d}X^{\alpha\beta}\mathrm{d}X^{\gamma\delta}}{(\varepsilon_{\alpha\beta\gamma\delta}I^{\alpha\beta}X^{\gamma\delta})^2}, \tag{108}$$

and the parameter λ defined by (105) has the interpretation of being the associated cosmological constant. We note that $\mathrm{d}s$ is independent of the scale of $X^{\alpha\beta}$, and has an I-weight of -1, as is appropriate for an element of time. The de Sitter group consists of transformations of $\mathbb{P}^5$ that preserve both the quadric Ω and the point I.

With the incorporation of some additional structure at infinity, the entire class of Robertson-Walker cosmological models can be represented in a similar way [32, 33, 36, 42, 45, 46]. The idea can be described briefly as follows. We start with the complex projective space $\mathbb{P}^5$ and in it the quadric Ω defined by (106). The points of space-time are given by a reality structure by requiring that $\overline{X}^{\alpha\beta} = X^{\alpha\beta}$. Next we introduce a pencil of 4-planes in $\mathbb{P}^5$ of the form

$$Z_{\alpha\beta} = pP_{\alpha\beta} + qQ_{\alpha\beta}, \tag{109}$$

where $(p,q) \in \mathbb{C}^2 - \{0,0\}$. For each such $Z_{\alpha\beta}$, the intersection of the 4-plane

$$Z_{\alpha\beta}X^{\alpha\beta} = 0 \tag{110}$$

with the real quadric given by

$$\varepsilon_{\alpha\beta\gamma\delta}X^{\alpha\beta}X^{\gamma\delta}=0 \quad and \quad \overline{X}^{\alpha\beta}=X^{\alpha\beta} \tag{111}$$

defines a certain subspace of the space-time. For certain choices of $\{P_{\alpha\beta}, Q_{\alpha\beta}, p, q\}$ the resulting subspace can be interpreted (in some cases with the deletion of certain elements) as a constant-time space-like hypersurface σ^3 in the space-time. The corresponding cosmological model is then obtained by selecting a one-parameter 'chronological family' of such surfaces, and choosing an appropriate conformal factor for the metric geometry.

In more detail, the construction is as follows. For $k=1$ (a closed universe) we require the skew twistors $P_{\alpha\beta}$ and $Q_{\alpha\beta}$ to be complex and to satisfy the following relations:

$$\varepsilon^{\alpha\beta\gamma\delta}P_{\alpha\beta}P_{\gamma\delta}=0, \quad \varepsilon^{\alpha\beta\gamma\delta}Q_{\alpha\beta}Q_{\gamma\delta}=0, \quad and \quad P_{\alpha\beta}=\overline{Q}_{\alpha\beta}\,. \tag{112}$$

We then let $p=\mathrm{e}^{\mathrm{i}\theta}$, $q=\mathrm{e}^{-\mathrm{i}\theta}$ for some $\theta\in[0,2\pi]$. The resulting one-parameter family of surfaces determined by

$$Z_{\alpha\beta}=\mathrm{e}^{\mathrm{i}\theta}P_{\alpha\beta}+\mathrm{e}^{-\mathrm{i}\theta}\overline{P}_{\alpha\beta} \tag{113}$$

has the property that each is a 3-sphere. For a $k=-1$ cosmology we require

$$\varepsilon^{\alpha\beta\gamma\delta}P_{\alpha\beta}P_{\gamma\delta}=0, \quad \varepsilon^{\alpha\beta\gamma\delta}Q_{\alpha\beta}Q_{\gamma\delta}=0, \quad P_{\alpha\beta}=\overline{P}_{\alpha\beta}, \quad and \quad Q_{\alpha\beta}=\overline{Q}_{\alpha\beta}\,. \tag{114}$$

In this case the relevant family of 4-planes is given by (109) with $(p,q)\in\mathbb{R}^2-\{0,0\}$, the overall scale of $Z_{\alpha\beta}$ being unimportant. For $k=0$ we set

$$\varepsilon^{\alpha\beta\gamma\delta}P_{\alpha\beta}P_{\gamma\delta}=0, \quad \varepsilon^{\alpha\beta\gamma\delta}Q_{\alpha\beta}Q_{\gamma\delta}=1, \quad P_{\alpha\beta}=\overline{P}_{\alpha\beta}, \quad and \quad Q_{\alpha\beta}=\overline{Q}_{\alpha\beta}\,, \tag{115}$$

with (p,q) as above in the $k=-1$ case.

In each case we can consider the 'axis' obtained by intersecting the elements of the given family of 4-planes. In the $k=+1$ case, the axis itself does not intersect the associated real space-time, and as a consequence the resulting hypersurfaces of constant time are topologically 3-spheres (see Figure 10). In the $k=0$ case the axis 'touches' the space-time at a point (given by $q=0$) common to all of the intersection spaces. If we remove this point (or treat it as a point at infinity), then the resulting constant-time surfaces are each topologically $\mathbb{R}^3$. In the $k=-1$ case, the common intersection region is a 2-sphere, and as a consequence an 'open' cosmological model results in this case as well.

Thus we see that the algebraic geometry of twistor theory gives us an essentially 'unified' point of view over the various standard cosmological models—this approach can be pursued at greater length, giving rise to a geometrical characterisation of the different types of situations that can occur, depending, in particular, on the global structure and topology of the space-time, on the equation of state of the fluid representation in the energy tensor, and on the type of cosmological constant (if any) in the model. Much more detail, along with specific examples for various choices of the equation of state, can be found in [32, 33, 42, 46].

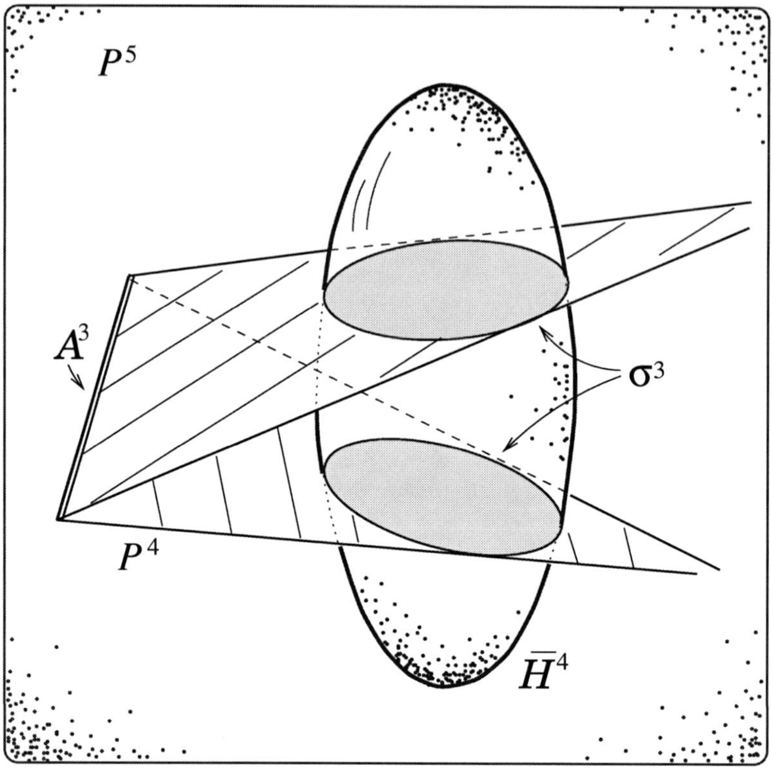

FIGURE 10. *Friedmann-Robertson-Walker cosmologies.* In the case $k = 1$ depicted here the chronology of the FRW cosmology is generated by a pencil of projective 4-planes hinged on an axis $\mathscr{A}^3$ in $\mathbb{P}^5$. The axis is chosen so that it does not impinge on any of the real points of the space-time. As a consequence the constant-time hypersurfaces in the cosmological model are compact.

It is interesting to note that more or less the same state of affairs prevails in higher dimensions (see Figure 11 for an example of a sixteen-dimensional class of 'quantum cosmologies' analogous to the $k = 1$ Friedmann-Robertson-Walker models). In other words, the choice of structure at infinity gives rise to various possible global structures for the quantum space-time, and in particular, to a chronometric form that is in general not flat, thus making $\mathscr{H}^{\bar{r}^2}$ a cosmological model. In the case of a standard four-dimensional cosmological model based on Einstein's theory, the existence of structure at (or 'beyond') infinity has a bearing on the geometry of space-time alone. In the case of a quantum cosmology, however, the structure at infinity also has implications for microscopic physics. For instance, whereas in the four-dimensional de Sitter cosmology the relevant structure at infinity contains the 'invariant' information of one dimensional constant (the cosmological constant), in the higher-dimensional situation there are in general a number of such constants that may emerge as geometrical invariants of the theory. Thus within a single geometric framework one has the scope for introducing

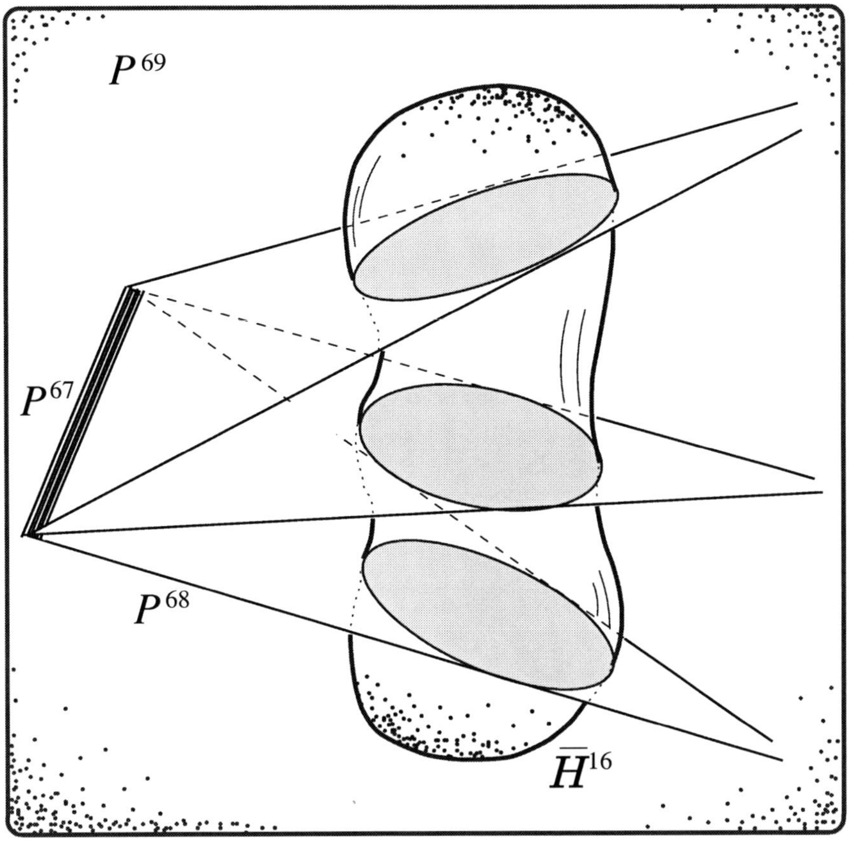

FIGURE 11. *Chronological foliation of an $r = 4$ hypercosmology.* The compactified quantum space-time $\bar{\mathscr{H}}^{16}$ is a real submanifold of the complex projective space $\mathbb{P}^{69}$. A pencil of $\mathbb{P}^{68}$-hyperplanes hinged on a complex axis $\mathbb{P}^{67}$ intersects the space-time in a set of hypersurfaces. Depending on the reality structure of the pencil, and its real subparameterisation, a variety of different possibilities can emerge for the global structure of the hypersurface family.

structure (or what amounts to the same thing—the breaking of symmetry) both on a global or cosmological scale, as well as on the microscopic scales of distance, time, and energy associated with the phenomenology of elementary particles. One might say that in these models the structure at infinity is playing the role of the Higgs fields. One can even envisage the possibility of explaining, in a purely geometrical language, the basis of the remarkable coincidences involving various fundamental constants of nature that have puzzled physicists for many decades.

WEAK AND STRONG HERMITICITY

As a prelude to our discussion of the idea of symmetry breaking in quantum space-time, we digress briefly to review the notions of weak and strong Hermiticity. This material is relevant to the origin of unitarity in quantum mechanics. Intuitively speaking, we observe that when the weak Hermiticity condition is imposed on a hyperspinor $x^{AA'}$ representing a space-time event, then $x^{AA'}$ belongs to the real subspace $\mathbb{R}^{AA'}$. As such, the hyper-relativistic symmetry of quantum space-time is not affected by the imposition of this condition. If, however, we break the hyper-relativistic symmetry by selecting a preferred time-like direction, then we can speak of a stronger reality condition whereby an isomorphism is established between the primed and unprimed hyperspin spaces.

We begin with the weak Hermitian property. Let $\mathbb{S}^A$ denote, as before, an r-dimensional complex vector space. We also introduce the associated spaces $\mathbb{S}_A$, $\mathbb{S}^{A'}$, and $\mathbb{S}_{A'}$. In general, there is no natural isomorphism between $\mathbb{S}^{A'}$ and $\mathbb{S}_A$. Therefore, there is no natural matrix multiplication law or trace operation defined for elements of $\mathbb{S}^A \otimes \mathbb{S}^{A'}$. Nevertheless, certain matrix operations are well defined. For example, the determinant of a generic element $\mu^{AA'}$ is given by

$$\det(\mu) = \frac{1}{r!}\,\varepsilon_{AB\cdots C}\,\varepsilon_{A'B'\cdots C'}\,\mu^{AA'}\,\mu^{BB'}\cdots\mu^{CC'}\,. \tag{116}$$

The weak Hermitian property is also well-defined. In particular, if $\bar{\mu}^{A'A}$ is the complex conjugate of $\mu^{AA'}$, then we say that $\mu^{AA'}$ is weakly Hermitian if $\mu^{AA'} = \bar{\mu}^{A'A}$. As we have observed, for many applications, weak Hermiticity suffices.

Now we consider the strong Hermitian property. In some situations there may exist a natural map $\mathbb{S}^{A'} \to \mathbb{S}_A$ defined by the context of the particular problem. Such a map is called a Hermitian correlation. In this case, the complex conjugate of an element $\alpha^A \in \mathbb{S}^A$ determines an element $\bar{\alpha}_A \in \mathbb{S}_A$. For any element $\mu^A_B \in \mathbb{S}^A \otimes \mathbb{S}_B$ we define the operations of determinant, matrix multiplication, and trace in the usual manner. The determinant is

$$\det(\mu) = \frac{1}{r!}\,\varepsilon_{AB\cdots C}\,\varepsilon^{PQ\cdots R}\,\mu^A_P\mu^B_Q\cdots\mu^C_R\,, \tag{117}$$

and the Hermitian conjugate of μ^A_B is $\bar{\mu}_A{}^B$. The Hermitian correlation is given by the choice of a preferred element $t_{AA'} \in \mathbb{S}_A \otimes \mathbb{S}_{A'}$. Then we write

$$\bar{\alpha}_A = t_{AA'}\bar{\alpha}^{A'}\,, \tag{118}$$

where $\bar{\alpha}_A$ is now called the complex conjugate of α^A. When there is a Hermitian correlation $\mathbb{S}^{A'} \leftrightarrow \mathbb{S}_A$, we call the condition $\mu^A_B = \bar{\mu}^A_B$ the strong Hermitian property.

Thus once we break the relativistic invariance by introducing a preferred element $t_{AA'}$ that determines a Hermitian correlation, we may carry out specific calculations in that frame. To gain a better understanding of this, consider an event in complex Minkowski space defined by its separation $x^{\mathbf{A}A'}$ from the origin. The complex conjugate of $x^{\mathbf{A}A'}$ is $\bar{x}^{\mathbf{A}A'}$, so $x^{\mathbf{A}A'}$ is real if $x^{\mathbf{A}A'} = \bar{x}^{\mathbf{A}A'}$. Let $t^{\mathbf{A}A'}$ be a fixed time-like vector satisfying

$g_{ab}t^{a}t^{b} = 2$. With respect to this choice of $t^{\mathbf{AA}'}$, the trace of $x^{\mathbf{AA}'}$ is defined by

$$\mathrm{tr}(x^{\mathbf{AA}'}) = x^{\mathbf{AA}'} t_{\mathbf{AA}'} \,. \tag{119}$$

Once $t^{\mathbf{AA}'}$ is fixed, we may represent $x^{\mathbf{AA}'}$ in terms of Pauli matrices, according to which $x^{\mathbf{AA}'}$ admits a matrix representation of the form (2). The time variable is then given by $t = \frac{1}{2}\mathrm{tr}(x^{\mathbf{AA}'})$ and the Minkowski metric is given by the determinant (3). We can also define a commutator for a pair of space-time position vectors $x^{\mathbf{AA}'}$ and $y^{\mathbf{AA}'}$ by setting

$$z^{\mathbf{AA}'} = \left(x^{\mathbf{AB}'} y^{\mathbf{BA}'} - y^{\mathbf{AB}'} x^{\mathbf{BA}'}\right) t_{\mathbf{BB}'} \,. \tag{120}$$

The geometrical meaning of (120) is that $z^{\mathbf{AA}'}$ corresponds to the ordinary 3-space cross-product of the projection of the vectors $x^{\mathbf{AA}'}$ and $y^{\mathbf{AA}'}$ onto the space-like hyperplane orthogonal to $t^{\mathbf{AA}'}$ that passes through the origin. Analogously, we can discuss the 'spectral' properties of vectors with respect to a given choice of $t^{\mathbf{AA}'}$ and a given choice of origin in space-time. The two eigenvalues of $x^{\mathbf{AA}'}$ are then given by $t \pm \sqrt{x^2 + y^2 + z^2}$. Thus two such vectors $x^{\mathbf{AA}'}$ and $y^{\mathbf{AA}'}$ are isospectral if and only if they lie on a common sphere about the origin lying in the given space-like hyperplane. Similar remarks apply to higher-dimensional quantum space-times.

SYMMETRY BREAKING MECHANISM

Now we proceed to introduce a natural mechanism for symmetry breaking that arises in the case of a standard 'flat' quantum space-time endowed with the canonical structures associated with reality and infinity. We shall make the point in particular that the breaking of symmetry in quantum space-time is intimately linked to the notion of quantum entanglement. According to this point of view the introduction of symmetry-breaking in the early stages of the universe can be understood as a phase transition, or a sequence of phase transitions, the ultimate consequence of which is an approximate disentanglement of a four-dimensional 'classical' space-time.

In practical terms the breaking of symmetry is represented in our framework by an 'index decomposition'. In particular, if the dimension r of the hyperspin space is not a prime number, then a natural method of breaking the symmetry arises by consideration of the decomposition of r into factors. The specific essential assumption that we shall make at this juncture will be that the dimension of the hyperspin space $\mathbb{S}^{A}$ is *even*. Then we write $r = 2n$, where $n = 1, 2, \ldots$, and set

$$\mathbb{S}^{A} = \mathbb{S}^{\mathbf{A}i}, \tag{121}$$

where $\mathbf{A}$ is a standard two-component spinor index, and i will be called an 'internal' index $(i = 1, 2, \ldots, n)$. Thus we can write $\mathbb{S}^{\mathbf{A}i} = \mathbb{S}^{\mathbf{A}} \otimes \mathbb{H}^{i}$, where $\mathbb{S}^{\mathbf{A}}$ is a standard spin space of dimension two, and $\mathbb{H}^{i}$ is a complex vector space of dimension n. The breaking of the symmetry then amounts to the fact that we can identify the hyperspin space with the tensor product of these two spaces.

We shall assume, moreover, that $\mathbb{H}^i$ is endowed with a strong Hermitian structure, i.e. we shall assume that there is a canonical anti-linear isomorphism between the complex conjugate of the internal space $\mathbb{H}^i$ and the dual space $\mathbb{H}_i$. If $\psi^i \in \mathbb{H}^i$, then we write $\bar{\psi}_i$ for the complex conjugate of ψ^i, where $\bar{\psi}_i \in \mathbb{H}_i$. We see therefore that $\mathbb{H}^i$ is a complex Hilbert space—and indeed although here we consider for technical simplicity the case for which n is finite, one should have in mind also the general infinite dimensional situation. For the other hyperspin spaces we write

$$\mathbb{S}_A = \mathbb{S}_{\mathbf{A}i}, \quad \mathbb{S}^{A'} = \mathbb{S}^{\mathbf{A}'}{}_i, \quad and \quad \mathbb{S}_{A'} = \mathbb{S}_{\mathbf{A}'}{}^i, \tag{122}$$

respectively. These equivalences preserve the duality between $\mathbb{S}^A$ and $\mathbb{S}_A$, and between $\mathbb{S}^{A'}$ and $\mathbb{S}_{A'}$; and at the same time are consistent with the complex conjugation relations between $\mathbb{S}^A$ and $\mathbb{S}^{A'}$, and between $\mathbb{S}_A$ and $\mathbb{S}_{A'}$. Hence if $\alpha^{\mathbf{A}i} \in \mathbb{S}^{\mathbf{A}i}$ then under complex conjugation we have $\alpha^{\mathbf{A}i} \to \bar{\alpha}^{\mathbf{A}'}{}_i$, and if $\beta_{\mathbf{A}i} \in \mathbb{S}_{\mathbf{A}i}$ then $\beta_{\mathbf{A}i} \to \bar{\beta}_{\mathbf{A}'}{}^i$.

In the case of a quantum space-time vector $r^{AA'}$ we have a corresponding structure induced by the identification

$$r^{AA'} = r^{\mathbf{AA}'i}{}_j . \tag{123}$$

When the quantum space-time vector is real, the weak Hermitian structure on $r^{AA'}$ is manifested in the form of a standard weak Hermitian structure on the two-component spinor index pair, together with a strong Hermitian structure on the internal index pair. In other words, the Hermitian condition on the space-time vector $r^{AA'}$ is given by

$$\bar{r}^{\mathbf{A}'\mathbf{A}}{}_j{}^i = r^{\mathbf{AA}'i}{}_j . \tag{124}$$

One consequence of these relations is that we can interpret each point in quantum space-time as being a space-time valued operator. The ordinary classical space-time then 'sits' inside the quantum space-time in a canonical manner—namely, as the locus of those points of quantum space-time that factorise into the product of a space-time point $x^{\mathbf{AA}'}$ and the identity operator on the internal space:

$$x^{\mathbf{AA}'i}{}_j = x^{\mathbf{AA}'} \delta^i{}_j . \tag{125}$$

Thus, in situations where special relativity is a satisfactory theory, we regard the relevant events as taking place on or in the immediate neighbourhood of this embedding of Minkowski space in $\mathcal{H}^{4n^2}$.

This picture can be presented in more geometric terms as follows. The hypertwistor space $\mathbb{P}^{2r-1}$ in the case $r = 2n$ admits a Segré embedding of the form

$$\mathbb{P}^3 \times \mathbb{P}^{n-1} \subset \mathbb{P}^{4n-1} . \tag{126}$$

Many such embeddings are possible, though they are all equivalent to one another under the action of the overall symmetry group $U(2n, 2n)$. If the symmetry is broken and one such embedding is selected out, then following the conventions discussed earlier we can introduce homogeneous coordinates and write $Z^{\alpha i}$ for the hypertwistor. Here the Greek letter α denotes an ordinary twistor index ($\alpha = 0,1,2,3$) and i denotes an internal index ($i = 1,2,\ldots,n$). These two indices, when clumped together, constitutes a hypertwister index. The Segré embedding consists of those points in $\mathbb{P}^{4n-1}$ for which we have a product decomposition of the associated hypertwistor given by

$$Z^{\alpha i} = Z^{\alpha}\psi^{i} . \tag{127}$$

The idea of symmetry breaking that we are putting forward here is related to the notion of disentanglement in standard quantum mechanics (cf. Gibbons 1992; Brody & Hughston 2001). That is to say, at the unified level the degrees of freedom associated with space-time symmetry are quantum mechanically entangled with the internal degrees of freedom associated with microscopic physics. The phenomena responsible for the breakdown of symmetry are thus analogous to the mechanisms of decoherence through which quantum entanglements are gradually diminished. Some readers may raise the objection that surely it is impossible to unify the unitary symmetries of elementary particle phenomenology with the symmetries of space-time (cf., e.g., [10]). It should be noted, however, that our approach is not to attempt to embed a relativistic symmetry group in a higher-dimensional unitary group, but rather to embed the unitary group in a higher-dimensional relativistic symmetry group. Our methodology is consistent with the point of view put forward by Penrose that for a coherent unification of general relativity and quantum mechanics, the rules of quantum theory must undergo 'profound modification' [43].

The compactified complexified quantum space-time $\mathscr{C}\bar{\mathscr{H}}^{4n^2}$ can be regarded as the aggregate of projective $(2n-1)$-planes in $\mathbb{P}^{4n-1}$. Now generically a $\mathbb{P}^{2n-1}$ in $\mathbb{P}^{4n-1}$ will not intersect the Segré variety

$$\mathscr{G}^{n+2} = \mathbb{P}^3 \times \mathbb{P}^{n-1} . \tag{128}$$

Such a generic $(2n-1)$-plane corresponds to a generic point in $\mathscr{C}\bar{\mathscr{H}}^{4n^2}$. The $(2n-1)$-planes that correspond to the points of compactified complexified Minkowski space can be constructed as follows. For each line L in $\mathbb{P}^3$ we consider the subvariety $\mathscr{G}^n_L \subset \mathscr{G}^{n+2}$ where $\mathscr{G}^n_L = \mathbb{P}^1_L \times \mathbb{P}^{n-1}$. For any algebraic variety $V^j \subset \mathbb{P}^l$ $(j \leq l-1)$ we define the *span* of V^j to be the projective plane spanned by the points of V^j. We say a point X in the ambient space $\mathbb{P}^l$ lies in the span of the variety V^j if and only if there exist m points in V^j for some $m \geq 2$ with the property that X lies in the $(m-1)$-plane spanned by those m points. The dimension k of the span of V^j satisfies $j \leq k \leq l$; however, the value of k depends on the geometry of V^j.

The linear span of the points in $\mathscr{G}^n_L$, for any given L, is a $(2n-1)$-plane. This is the $\mathbb{P}^{2n-1}_L$ in $\mathbb{P}^{4n-1}$ that represents the point in $\bar{\mathscr{H}}^{4n^2}$ corresponding to the line L in $\mathbb{P}^3$. The aggregate of such special $(2n-1)$-planes, defined by their intersection properties with the Segré variety $\mathscr{G}^{n+2}$, constitutes a submanifold of $\mathscr{C}\bar{\mathscr{H}}^{4n^2}$, and this submanifold is compactified complexified Minkowski space $\mathscr{C}\bar{\mathscr{H}}^{4}$. Thus we see that once the symmetry of quantum space-time $\mathscr{C}\mathscr{H}^{4n^2}$ has been broken in the particular way we have discussed, then ordinary Minkowski space can be identified as a submanifold.

Let us now consider the implications of our symmetry breaking mechanism for fields defined on quantum space-time. As an example, let $\phi(x^{AA'})$ be a scalar field on quantum space-time. After we break the symmetry by writing $x^{AA'} = x^{\mathbf{AA'}i}{}_j$, we consider a Taylor expansion of the field around the embedded Minkowski space-time. Specifically, for such an expansion we have

$$\phi(x^{AA'}) = \phi^{(0)}(x^{\mathbf{AA'}}) + \phi^{(1)\,i}_{\mathbf{AA'}\,j}(x^{\mathbf{AA'}})\left(x^{\mathbf{AA'}j}{}_i - x^{\mathbf{AA'}}\delta^j_i\right) + \cdots\,, \tag{129}$$

where

$$\phi^{(0)}(x^{\mathbf{AA'}}) = \phi(x^{AA'})\Big|_{x^{AA'}=x^{\mathbf{AA'}}\delta^j_i}\,, \tag{130}$$

and

$$\phi^{(1)\,i}_{\mathbf{AA'}\,j}(x^{\mathbf{AA'}}) = \frac{\partial\phi}{\partial x^{\mathbf{AA'}j}{}_i}\Bigg|_{x^{\mathbf{AA'}j}{}_i=x^{\mathbf{AA'}}\delta^j_i}\,. \tag{131}$$

Therefore, the order zero term has the character of a classical field on Minkowski space, and the first order term can be interpreted as a 'multiplet' of fields, transforming according to the adjoint representation of the internal symmetry group $U(n)$.

In this connection we note that the symmetry breaking mechanism that we have proposed here has yet another representation—namely, the expression of a hypertwistor as a multi-twistor system, i.e. as a multiplet of Penrose twistors. The physical and geometrical characteristics of such n-twistor systems have been analysed at great length by a number of authors (see, e.g., [23, 24, 38, 39, 47, 63, 64] and references cited therein), and it is interesting therefore to see the direct link with hypertwistor theory and quantum space-time geometry. It is fitting also to make a tribute here to the work of Zoltan Perjés, whose extensive contributions to relativity theory include, in particular, a number of important studies concerning the properties of n-twistor systems and their symmetries [35, 48, 49, 50, 51, 52, 53, 54, 55, 56, 57, 58, 65].

It is tempting to speculate that even in a more dynamic context some version of the symmetry breaking mechanism provided here will manifest itself. In this picture we would envisage the earliest stages of the universe as being highly symmetrical, rather in the spirit of Penrose's Weyl curvature hypothesis [40, 41], with no appreciable distinction between the conventional space-time degrees of freedom and the internal degrees of freedom associated with quantum theory. Nevertheless, the causal geometry of the universe remains well defined, and it is interesting to ask whether there might be some scenario within the rather rich causal structure of a quantum space-time that would allow us to account for the so-called 'horizon problem'. In any event, once symmetry breaking takes place—and this may happen in stages, corresponding to a successive factorisation of the underlying hypertwistor space—then it makes sense to think of ordinary four-dimensional space-time as becoming more or less disentangled from the rest of the universe, and behaving in a way that is to some extent autonomous. Nonetheless, we might reasonably expect its global dynamics, on a cosmological scale, to be affected by the distribution of mass and energy elsewhere in the quantum space-time as well (see, e.g., Figure 12).

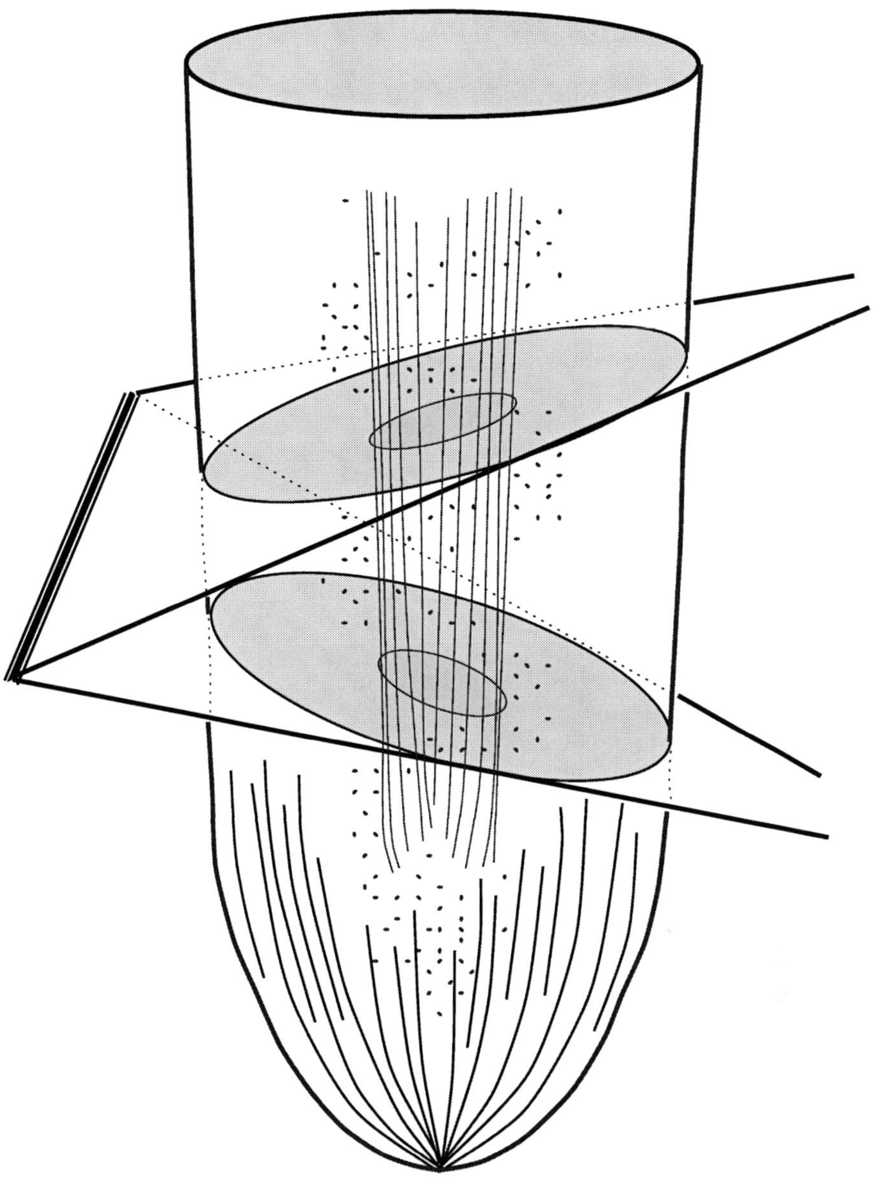

FIGURE 12. *Symmetry breaking and matter formation in a hypertwistorial quantum cosmology.* In its earliest stages the universe is highly symmetrical. Eventually, symmetry is broken, and a four-dimensional 'classical' cosmology freezes out, becoming largely disentangled from the reminder of the quantum space-time. The formation of matter during the disentanglement process may not be confined to the four-dimensional subspace. If such a scenario prevails, the bulk of this 'dark material' is likely to remain outside the four-dimensional subspace, though nevertheless having an impact on its dynamics.

The embedding of Minkowski space in the quantum space-time $\mathscr{H}^{4n^2}$ given by (125) implies a corresponding embedding of the Poincaré group in the hyper-Poincaré group. This can be seen as follows. The standard Poincaré group in $\mathscr{H}^4$ consists of transformations of the form

$$x^{\mathbf{A}A'} \longrightarrow \lambda^{\mathbf{A}}_{\mathbf{B}} \bar{\lambda}^{\mathbf{A}'}_{\mathbf{B}'} x^{\mathbf{B}B'} + \beta^{\mathbf{A}A'}, \tag{132}$$

and the hyper-Poincaré transformations in $\mathscr{H}^{4n^2}$ are of the form

$$x^{\mathbf{A}A'i}{}_j \longrightarrow \lambda^{\mathbf{A}}_{\mathbf{B}} \bar{\lambda}^{\mathbf{A}'}_{\mathbf{B}'} x^{\mathbf{B}B'i}{}_j + \beta^{\mathbf{A}A'} \delta^i_j\,. \tag{133}$$

With the identification $A = \mathbf{A}i$, the general hyper-Poincaré transformation in the broken symmetry phase can be expressed in the form

$$x^{\mathbf{A}A'i}{}_j \longrightarrow L^{\mathbf{A}i}_{\mathbf{B}k} \bar{L}^{\mathbf{A}'l}_{\mathbf{B}'j} x^{\mathbf{B}B'k}{}_l + \beta^{\mathbf{A}A'i}{}_j\,. \tag{134}$$

Thus the embedding of the Poincaré group as a subgroup of the hyper-Poincaré group is given by

$$L^{\mathbf{A}i}_{\mathbf{B}j} = \lambda^{\mathbf{A}}_{\mathbf{B}} \delta^i{}_j \quad \textit{and} \quad \beta^{\mathbf{A}A'i}{}_j = \beta^{\mathbf{A}A'} \delta^i_j\,. \tag{135}$$

Bearing this in mind, we now construct a class of maps from the general even-dimensional quantum space-time $\mathscr{H}^{4n^2}$ to Minkowski space $\mathscr{H}^4$. It turns out [6] that under rather general physical assumptions such maps are necessarily of the form

$$x^{\mathbf{A}A'i}{}_j \longrightarrow x^{\mathbf{A}A'} = \rho^j_i\, x^{\mathbf{A}A'i}{}_j, \tag{136}$$

where ρ^j_i is a density matrix. As usual, by a density matrix we mean a positive semi-definite Hermitian matrix with unit trace. Thus the maps arising here can be regarded as quantum expectations.

In particular, let $\rho:\ \mathscr{H}^{4n^2} \to \mathscr{H}^4$ satisfy the following conditions: (i) ρ is linear and maps the origin of $\mathscr{H}^{4n^2}$ to the origin of $\mathscr{H}^4$; (ii) ρ is Poincaré invariant; and (iii) ρ preserves causal relations. Then ρ is given by a density matrix on the internal space.

The general linear map from $\mathscr{H}^{4n^2}$ to $\mathscr{H}^4$ preserving the origin is given by

$$x^{\mathbf{A}A'i}{}_j \longrightarrow x^{\mathbf{A}A'} = \rho^{\mathbf{A}A'j}_{\mathbf{B}B'i}\, x^{\mathbf{B}B'i}{}_j, \tag{137}$$

where $\rho^{\mathbf{A}A'j}_{\mathbf{B}B'i}$ is weakly Hermitian. Now suppose that we subject $\mathscr{H}^{4n^2}$ to a Poincaré transformation of the form (133), and require the corresponding transformation of $\mathscr{H}^4$ should be of the form (132). If ρ satisfies these conditions then we shall say that the map ρ is Poincaré invariant. Clearly, Poincaré invariance holds if and only if

$$\rho^{\mathbf{A}A'j}_{\mathbf{B}B'i}\left(\lambda^{\mathbf{B}}_{\mathbf{C}} \bar{\lambda}^{\mathbf{B}'}_{\mathbf{C}'} x^{\mathbf{C}C'i}{}_j + b^{\mathbf{B}B'} \delta^i_j\right) = \lambda^{\mathbf{A}}_{\mathbf{B}} \bar{\lambda}^{\mathbf{A}'}_{\mathbf{B}'}\left(\rho^{\mathbf{B}B'j}_{\mathbf{C}C'i}\, x^{\mathbf{C}C'i}{}_j\right) + \beta^{\mathbf{A}A'} \tag{138}$$

for all $\lambda_{\mathbf{B}}^{\mathbf{A}} \in SL(2n,\mathbb{C})$, for all $\beta^{\mathbf{A}A'} \in \mathbb{S}^{\mathbf{A}A'}$, and for all $x^{\mathbf{A}A'i}{}_{j} \in \mathscr{H}^{4n^2}$. Thus we have

$$\rho_{\mathbf{B}B'i}^{\mathbf{A}A'j} \lambda_{\mathbf{C}}^{\mathbf{B}} \bar{\lambda}_{\mathbf{C}'}^{\mathbf{B}'} = \lambda_{\mathbf{B}}^{\mathbf{A}} \bar{\lambda}_{\mathbf{B}'}^{\mathbf{A}'} \rho_{\mathbf{C}C'i}^{\mathbf{B}B'j} \tag{139}$$

for all $\lambda_{\mathbf{B}}^{\mathbf{A}}$, and

$$\rho_{\mathbf{B}B'i}^{\mathbf{A}A'j} \delta_j^i \beta^{\mathbf{B}B'} = \beta^{\mathbf{A}A'} \tag{140}$$

for all $\beta^{\mathbf{A}A'}$. Equation (139) implies that ρ is of the form

$$\rho_{\mathbf{B}B'i}^{\mathbf{A}A'j} = \delta_{\mathbf{B}}^{\mathbf{A}} \delta_{\mathbf{B}'}^{\mathbf{A}'} \rho_i^j \tag{141}$$

for some ρ_i^j. Then (140) implies that ρ must satisfy the trace condition $\rho_i^i = 1$. Finally we require that if $x^{\mathbf{A}A'i}{}_{j}$ and $y^{\mathbf{A}A'i}{}_{j}$ are quantum space-time points with the property that the interval

$$r^{\mathbf{A}A'i}{}_{j} = x^{\mathbf{A}A'i}{}_{j} - y^{\mathbf{A}A'i}{}_{j} \tag{142}$$

is future-pointing then $r^{\mathbf{A}A'} = x^{\mathbf{A}A'} - y^{\mathbf{A}A'}$ is also future pointing, where

$$r^{\mathbf{A}A'} = \rho_i^j \, r^{\mathbf{A}A'i}{}_{j} \, . \tag{143}$$

This is the requirement that ρ should be a 'causal' map. This condition implies that ρ must be positive semi-definite. In particular, if $r^{\mathbf{A}A'i}{}_{j}$ is future-pointing then it must be of the form

$$r^{\mathbf{A}A'i}{}_{j} = \xi^{\mathbf{A}i} \bar{\xi}_j^{\mathbf{A}'} + \eta^{\mathbf{A}i} \bar{\eta}_j^{\mathbf{A}'} + \cdots \, . \tag{144}$$

Consider therefore the case for which $r^{\mathbf{A}A'i}{}_{j}$ is null. Then we require that the expression $\xi^{\mathbf{A}i} \bar{\xi}_j^{\mathbf{A}'} \rho_i^j$ should be future-pointing (or vanish) for any choice of $\xi^{\mathbf{A}i}$. In particular, we require that the vector $\xi^{\mathbf{A}i} \bar{\alpha}_j^{\mathbf{A}'} \rho_i^j$ should be future-pointing if $\xi^{\mathbf{A}i}$ is of the form

$$\xi^{\mathbf{A}i} = \alpha^{\mathbf{A}} \psi^i \tag{145}$$

for any choice of $\xi^{\mathbf{A}}$ and ψ^i. This means that the inequality

$$\rho_i^j \psi^i \bar{\psi}_j \geq 0 \tag{146}$$

holds for all ψ^i, which shows that ρ_i^j is positive semi-definite. Since we have shown that the trace of ρ_i^j is unity, it follows that ρ_i^j is a density matrix.

This result shows how the causal structure of quantum space-time is linked with the probabilistic structure of quantum mechanics. The concept of a quantum state emerges when we ask for consistent ways of 'averaging' over the geometry of quantum space-time in order to obtain a reduced description of physical phenomena in terms of the geometry of Minkowski space. We see that a probabilistic interpretation of the map from a general quantum space-time to Minkowski space arises as a consequence of elementary causality requirements. We can thus view the space-time events in $\mathscr{H}^{4n^2}$ as representing quantum observables, the expectations of which correspond to points of $\mathscr{H}^4$.

ACKNOWLEDGMENTS

DCB gratefully acknowledges financial support from The Royal Society. The work described here is based, in part, on ideas and suggestions arising in discussions with E. J. Brody. The authors are grateful to participants at the XIXth Max Born Symposium, Institute of Theoretical Physics, Wrocław, Poland, for helpful comments.

REFERENCES

1. J. Anandan and Y. Aharonov, Phys. Rev. Lett. **65**, 1697–1700 (1990).
2. A. Ashtekar and T. Schilling, *Geometrical formulation of quantum mechanics*, in *On Einstein's Path* (A. Harvey, ed.) Berlin: Springer-Verlag.
3. A. Borowiec, Int. J. Theor. Phys. **28**, 1229–1232 (1989).
4. A. Borowiec, *G*-structure for hypermanifold, in *Spinors, twistors, Clifford algebras and quantum deformations* (Sobótka Castle, Z. Oziewicz, B. Jancewicz, & A. Borowiec, Eds.) Fund. Theories Phys. **52**, 75–79 (1992), Dordrecht: Kluwer.
5. D.C. Brody and L.P. Hughston, J. Geom. Phys. **38**, 19–53 (2001).
6. D.C. Brody and L.P. Hughston, *Theory of quantum space-time*, Proc. R. Soc. London **A461**, (to appear).
7. P. Budinich and A. Trautman, *The spinorial chessboard*, Berlin: Springer (1988).
8. Cartan, É. 1937 *Leçons sur la théorie des Spineurs* Paris: Hermann & Cie. (English translation: *The theory of spinors*, Paris: Hermann & Cie., 1966 and New York: Dover, 1981).
9. Chevalley, C. 1954 *The algebraic theory of spinors*, New York: Columbia University Press.
10. S. Coleman, Phys. Rev. **138**, B1262–B1267 (1965).
11. M.G. Eastwood, R. Penrose and R.O. Wells, Commun. Math. Phys. **78**, 305–351 (1981).
12. J. Ellis, Phil. Trans. R. Soc. London **A361**, 2607–2627 (2003).
13. D. Finkelstein, Phys. Rev. Lett. **56**, 1532–1533 (1986).
14. D. Finkelstein, S.R. Finkelstein and C. Holm, Int. J. Theor. Phys. **25**, 441–463 (1986).
15. D. Finkelstein, S.R. Finkelstein and C. Holm, Phys. Rev. Lett. **59**, 1265–1266 (1987).
16. S.R. Finkelstein, Int. J. Theor. Phys. **27**, 251–272 (1988).
17. G.W. Gibbons, J. Geom. Phys. **8**, 147–162 (1992).
18. C. Holm, Int. J. Theor. Phys. **25**, 1209–1213 (1986).
19. C. Holm, J. Math. Phys. **29**, 978–986 (1988), Erratum *ibid.* **30**, 2451 (1989).
20. C. Holm, J. Math. Phys. **29**, 2273–2279 (1988).
21. C. Holm, Int. J. Theor. Phys. **29**, 23–36 (1990).
22. L.P. Hughston, *Some new contour integral formulae*, in *Complex Manifold Techniques in Theoretical Physics* (D. Lerner and P. D. Sommers, Eds.) (1979) San Francisco: Pitman.
23. L.P. Hughston, *A derivation of the twistor internal symmetry group*, in *Advances in twistor theory* (L.P. Hughston and R.S. Ward, Eds.) (1979) San Francisco: Pitman.
24. L.P. Hughston, *Twistors and Particles*, (1979) Heidelberg: Springer.
25. L.P. Hughston, *Applocations of SO(8) spinors*, in *Gravitation and Geometry* (W. Rindler and A. Trautman, Eds.) (1987) Naples: Bibliopolis.
26. L.P. Hughston, *A remarkable connection between the wave equation and pure spinors in higher dimensions*, in: *Further advances in twistor theory, vol. I: The Penrose transform and its applications* (L.J. Mason and L.P. Hughston, Eds.) (1990) Harlow: Longman.
27. L.P. Hughston, and T.R. Hurd, Phys. Rep. **100**, 273–326 (1982).
28. L.P. Hughston and L.J Mason, Class. Quantum Grav. **5**, 275–285 (1988).
29. L.P. Hughston, R. Penrose, P. Sommers and M. Walker, Commun. Math. Phys. **27**, 303–308 (1971).
30. L.P. Hughston and P. Sommers, Commun. Math. Phys. **32**, 147–152 (1973).
31. L.P. Hughston and P. Sommers, Commun. Math. Phys. **33**, 129–133 (1973).
32. T.R. Hurd, Proc. R. Soc. London **A397**, 233–243 (1985).
33. T.R. Hurd, *Cosmological models in* $\mathbb{P}^5$, in *Further advances in twistor theory, vol. II: Integrable systems, conformal geometry and gravitation* (L.J. Mason, L.P. Hughston and P.Z. Kobak, Eds.)

(1995) Harlow: Longman.
34. T.W.B> Kibble, Commun. Math. Phys. **65**, 189–201 (1979).
35. B. Lukács, Z. Perjés, A. Sebestyén, E.T. Newman and J. Porter, J. Math. Phys. **23**, 2108–2115 (1982).
36. R. Penrose, J. Math. Phys. **8**, 345–366 (1967).
37. R. Penrose, *Structure of space-time*, in: *Battelle Rencontres* (C.M. DeWitt and J.A. Wheeler, Eds.) (1968) New York: W. A. Benjamin.
38. R. Penrose, *Twistors and particles*, in: *Quantum theory and the structure of time and space* (L. Castell, M. Drieschner and C.F. von Weizsacker, Eds.) (1975) München: Carl Hanser Verlag.
39. R. Penrose, Rep. Math. Phys. **12**, 65–76 (1977).
40. R. Penrose, *Singularities and time-asymmetry*, in: *General relativity: an Einstein centenary survey* (S.W. Hawking and W. Israel, Eds.) (1979) Cambridge: Cambridge University Press.
41. R. Penrose, *Time-asymmetry and quantum gravity*, in: *Quantum Gravity 2* (C.J. Isham, R. Penrose and D.W. Sciama, Eds.) (1981) Oxford: Oxford University Press.
42. R. Penrose, *Twistors for cosmological models*, in: *Further advances in twistor theory, vol. II: Integrable systems, conformal geometry and gravitation* (L.J. Mason, L.P. Hughston and P.Z. Kobak, Eds.) (1995) Harlow: Longman.
43. R. Penrose, Chaos, Solitons & Fractals **10**, 581–611 (1999).
44. R. Penrose and M.A.H MacCallum, Phys. Rep. **6C**, 241–315 (1973).
45. R. Penrose and W. Rindler, *Spinors and Space-time* vol. 1, (1984) Cambridge: Cambridge University Press.
46. R/ Penrose and W. Rindler, *Spinors and Space-time* vol. 2 (1986) Cambridge: Cambridge University Press.
47. R. Penrose and G.A.J. Sparling, *A note on the n-twistor internal symmetry group*, in *Advances in twistor theory*, (L.P. Hughston and R.S. Ward, Eds.) (1979) San Francisco: Pitman.
48. Z. Perjés, Phys. Rev. D**11**, 2031–2041 (1975).
49. Z. Perjés, Rep. Math. Phys. **12**, 193-211 (1977).
50. Z. Perjés and G.A.J. Sparling, *The twistor structure of hadrons*, in: *Advances in twistor theory* (L.P. Hughston and R.S. Ward, Eds.) (1979) San Francisco: Pitman.
51. Z. Perjés, Phys. Rev. D**20**, 1857–1876 (1979).
52. Z. Perjés, *Twistors and unitary space*, in: *85th Summer Meeting of AMS*, (1981) American Mathematical Society, 457.
53. Z. Perjés, *Twistor theory—a particle physicist attitude*, *Proc. 2nd Marcel Grossmann Conference* (R. Ruffini, Ed.) (1982) Amsterdam: North-Holland.
54. Z. Perjés, Czech. J. Phys. **B32**, 540–548 (1982).
55. Z. Perjés and G.A.J. Sparling, Surveys High Energ. Phys. **3**, 27–37 (1982).
56. Z. Perjés, *Twistor internal symmetry groups*, in: *Group theoretical methods in physics* (M.A. Markov, Ed.) (1983) Chur: Harwood Acad. Publ.
57. Z. Perjés, *Twistors and unitary space*, in: *Global analysis—analysis on manifolds* (T. Rassias, Ed.) (1983) Leipzig: Teubner.
58. Z. Perjés, *Twistor theory*, in: *Quantum Gravity* (M.A. Markov and P.C. West, Eds.) (1984) New York: Plenum.
59. F.A.E. Pirani, *Introduction to gravitational radiation*, in: Lectures on General Relativity: 1964 Brandeis Summer Institute in Theoretical Physics, vol. 1 (S. Deser and K. W. Ford, Eds.) (1965) Englewood Cliffs, NJ: Prentice-Hall.
60. M. Rees, Phil. Trans. R. Soc. London **A361**, 2427–2434 (2003).
61. J.G. Semple and G.T. Kneebone, *Algebraic projective geometry*, (1952) Oxford: Claredon Press.
62. A.V. Solov'yov and Yu.S. Vladimirov, Int. J. Theor. Phys. **40**, 1511–1523 (2001).
63. G.A.J. Sparling, Phil. Trans. R. Soc. London **A301**, 27–74 (1981).
64. K.P. Tod, Rep. Math. Phys. **11**, 339–346 (1977).
65. K.P. Tod and Z. Perjés, Gen. Rel. Grav. **7**, 903–913 (1976).
66. H. Urbantke, Int. J. Theor. Phys. **28**, 1233–1235 (1989).
67. M. Walker and R. Penrose, Commun. Math. Phys. **18**, 265–274 (1970).

AdS Twistors for Higher Spin Theory

M. Cederwall

Department of Theoretical Physics Göteborg University and Chalmers University of Technology
S-412 96 Göteborg, Sweden

Abstract. We construct spectra of supersymmetric higher spin theories in $D = 4,5$ and 7 from twistors describing massless (super-)particles on AdS spaces. A massless twistor transform is derived in a geometric way from classical kinematics. Relaxing the spin-shell constraints on twistor space gives an infinite tower of massless states of a "higher spin particle", generalizing previous work of Bandos *et* al.. This can generically be done in a number of ways, each defining the states of a distinct higher spin theory, and the method provides a systematic way of finding these. We reproduce known results in $D = 4$, minimal supersymmetric 5- and 7-dimensional models, as well as supersymmetrisations of Vasiliev's Sp-models as special cases. In the latter models a dimensional enhancement takes place, meaning that the theory lives on a space of higher dimension than the original AdS space, and becomes a theory of doubletons. This talk was presented at the XIX-th Max Born Symposium "Fundamental Interactions and Twistor-Like Methods", September 2004, in Wrocław, Poland.

INTRODUCTION

It has been known for some time that higher spin theory, *i.*e, theories of interacting massless higher spin fields, should be formulated in anti-de Sitter space, or more precisely, allow AdS space as a vacuum solution [1]. The theory of higher spin fields was subsequently developed in a series of papers [2]. For excellent recent reviews, see refs. [3, 4, 5], which also give an account to the earlier history of higher spin.

In recent years, there has been a growing interest in the theory of massless higher spins. There is some hope that such a theory may provide a geometric framework and a symmetry principle underlying string theory, which in that case would be interpreted as a broken phase of higher spin theory, where the higher spin fields have become massive. For the possible connection between string theory and higher spin theory, see refs. [6, 7, 8, 9, 10].

Higher spin theory has mostly been constructed using spinorial oscillators. These are the kinds of models that will be discussed in the present talk. Recently, constructions with vectorial variables, for any dimension, have been performed [11], and we will have nothing to say about these. The purpose of the work presented in this talk is to present a unified framework for obtaining spinorial (twistorial) variables for higher spin theory by relaxation of spin-shell constraints for ordinary bosonic or supersymmetric particles. The discussion is performed entirely at the kinematic, non-interacting, level.

CP767, *Fundamental Interactions and Twistor-Like Methods,* edited by J. Lukierski and D. Sorokin

TWISTOR TRANSFORM FOR MASSLESS PARTICLES ON ADS

Consider AdS_{d+1} space with radius R as the hyperboloid

$$x_M x^M \equiv -(x^0)^2 - (x^{0'})^2 + \sum_{i=1}^{d} (x^i)^2 = -R^2 \tag{1}$$

in flat space with signature $(2,d)$. The trajectory of a massless particle, a light-like geodesic, is the intersection of the hyperboloid with a plane through the origin spanned by one light-like and one time-like vector, *i.e.*, $x = \alpha X + \beta P$, where $P^2 = 0$, $X^2 < 0$ and $X \cdot P = 0$. X may then be seen as the coordinate for the location of the particle, fulfilling Eq. (1), and P as its momentum, being light-like and directed along the hyperboloid. A plane is also defined by a bi-vector Π^{MN} which is *simple*, meaning that it can be expressed in terms of two vectors as $\Pi^{MN} = X^{[M}P^{N]}$.

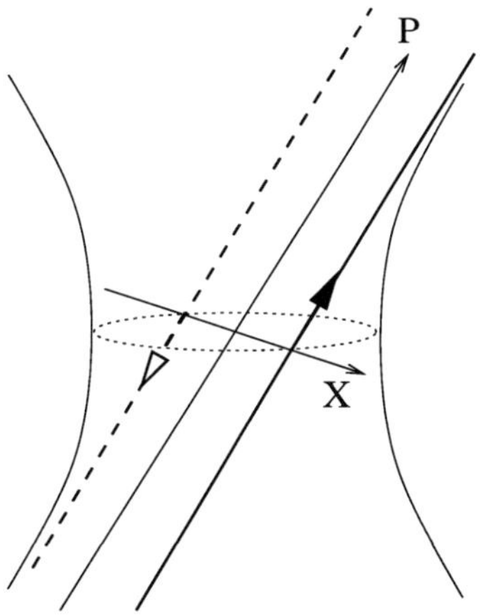

FIGURE 1. A light-like geodesic as the intersection of a plane and a hyperboloid.

The condition that Π is simple can be expressed as

$$\Pi^{[MN}\Pi^{PQ]} = 0. \tag{2}$$

The properties of the vectors X and P imply in addition that

$$\Pi_{MN}\Pi^{MN} = 0. \tag{3}$$

We want to find a twistor transform for a massless particle in AdS space. This means finding an expression Π fulfilling Eqs. (2) and (3) as a spinor bilinear modulo some well defined constraints and transformations. The spinor should be a spinor under the AdS_{d+1} group $\mathrm{Spin}(2,d)$. We restrict ourselves to the cases $d+1 = 4,5,7$, which are naturally related to the division algebras $\mathbb{K}_1 = \mathbb{R}$, $\mathbb{K}_2 = \mathbb{C}$ and $\mathbb{K}_4 = \mathbb{H}$. Using the isomorphisms $Sp(4;\mathbb{K}_\nu) \approx \mathrm{Spin}(2,\nu+2)$, the spinors are 4-component with elements in $\mathbb{K}_\nu$.

The form of the twistor transform is

$$\Pi^{MN} = \frac{1}{8}\Lambda^{\dagger I}\Gamma^{MN}\Lambda^{I}, \tag{4}$$

where I is an internal index that will turn out to run over two values, or equivalently, since the adjoint of $\mathrm{Spin}(2,\nu+2)$ can be represented as a hermitean 4×4-matrix, $\Pi^{A\dot{B}}=\frac{1}{2}\Lambda^{AI}\Lambda^{\dagger I\dot{B}}$, or simply

$$\Pi=\frac{1}{2}\Lambda\Lambda^{\dagger}\,. \tag{5}$$

Π contains the generator of AdS transformations, if Λ has canonical Poisson brackets with itself.

It is not possible to form a simple bivector from a single AdS spinor, the minimum number is two. That this is true can be checked explicitly for some simple specific choice of the plane. Take $\Pi^{\oplus 0}\neq 0$ (with the light-like $\oplus$ direction and the time-like 0 direction orthogonal) and the rest of the components vanishing. Using the gamma matrices of the Appendix, and denoting the spinor

$$\Lambda=\begin{bmatrix}\lambda^{\alpha}\\ \mu_{\dot{\alpha}}\end{bmatrix}=\begin{bmatrix}k\\ l\\ m\\ n\end{bmatrix} \tag{6}$$

(suppressing the index I), one needs

$$\begin{aligned}
\Lambda^{\dagger}\Gamma^{\oplus 0}\Lambda &\sim \bar{k}k+\bar{l}l\neq 0\\
\Lambda^{\dagger}\Gamma^{\oplus,\nu+1}\Lambda &\sim \bar{k}k-\bar{l}l=0\\
\Lambda^{\dagger}\Gamma^{\oplus i}\Lambda &\sim \bar{k}\bar{e}_i l+\bar{l}e_i k=0\,,\quad i=1,\ldots\nu\\
\\
\Lambda^{\dagger}\Gamma^{\ominus 0}\Lambda &\sim \bar{m}m+\bar{n}n=0\\
\Lambda^{\dagger}\Gamma^{\ominus,\nu+1}\Lambda &\sim \bar{m}m-\bar{n}n=0\\
\Lambda^{\dagger}\Gamma^{\ominus i}\Lambda &\sim \bar{m}\bar{e}_i n+\bar{n}e_i m=0\,,\quad i=1,\ldots\nu
\end{aligned} \tag{7}$$

In order to satisfy the first three equations, two spinors are needed. Then the last three equations imply that $\mu=0$. The rest of the components of Π mix λ and μ and will thus vanish.

A pair of spinors has an "R-symmetry", acting from the right on the twistor $\Lambda\in\mathbb{K}_{\nu}{}^{8}$ of the form

$$\Lambda\sim\begin{bmatrix}\cdot&\cdot\\ \cdot&\cdot\\ \cdot&\cdot\\ \cdot&\cdot\end{bmatrix}, \tag{8}$$

generated by anti-hermitean 2×2-matrices with entries in $\mathbb{K}_{\nu}$. The corresponding groups are[1]

---

[1] Here it becomes clear why AdS_{11} is not included, although one has the isomorphism $\mathrm{Sp}(4;\mathbb{O})\approx\mathrm{Spin}(2,10)$. "$A_2(\mathbb{O})$" does not form a Lie algebra, not even in the soft sense that $A_1(\mathbb{O})\approx S^7$ does [12, 13, 14, 15, 16]; it forms a set of generators in a $\mathrm{Spin}(9)/G_2$ coset of $\mathrm{Spin}(9)$.

$$A_2(\mathbb{K}_\nu) = \begin{cases} \mathrm{U}(1) \\ \mathrm{SU}(2) \times \mathrm{U}(1) \ . \\ \mathrm{Spin}(5) \end{cases} \tag{9}$$

Note that although these symmetries are "internal", and commute with the AdS generators, they are isomorphic to the groups of transverse rotations for a massless particle (except for the extra U(1), whose rôle is commented on below). Denote the corresponding generators T. The specific spinor Λ above clearly has $T = 0$, since all generators contain "$\mu\lambda$". The rest of the conditions turn out to be consequences of the choice of frame. It is also clear that since Π commutes with T, spinors should be counted modulo T-transformations. In fact,

$$T = \frac{1}{2}\Lambda^\dagger\Lambda \approx 0 \tag{10}$$

is the gauge invariance of the twistor. The generators T are anti-hermitean due to the anti-hermiticity of the "spinor metric" ε (see the Appendix). The counting of physical degrees of freedom is straightforward:

$$2 \times 4\nu - 2(3\nu - 2) = 2\nu + 4 = 2d\,, \tag{11}$$

which is the correct number for the phase space of a massless particle in $d+1$ dimensions.

A twistor construction for massive particles was performed in refs. [17, 18, 19]. It was noted that four spinors were needed to form a plane whose intersection with the hyperboloid is a massive geodesic for AdS_4 and AdS_7, while two spinors were sufficient for AdS_5. In the last of these cases, the U(1) generator was identified with the mass, and the SU(2) were gauge generators. In the 4- and 7-dimensional cases, it was necessary to break the "R-symmetry" $A_4(\mathbb{K}_\nu)$ by identifying a non-vanishing U(1) generator as the mass, which lead to a mixture of first and second class constraints. The present case of massless twistors is more uniform in the different dimensionalities, and only first class constraints are present.

We should stress a difference between twistor transforms on AdS space and Minkowski space. The division algebra construction [20, 14] is natural in Minkowski space of dimensions $\nu + 2 = 3$, 4, 6 and 10. The basic relation is

$$p \sim \lambda\lambda^\dagger\,, \tag{12}$$

which due to Fierz identities in these dimensionalities ensure that p is light-like. Such a construction is not possible in AdS, where the isometry group is semi-simple and does not contain translations. Any natural twistor transform will involve the whole twistor phase space, as in Eq. (5), not only a commuting half of it as in Eq. (12). Also in Minkowski space, however, Eq. (4) can be used as a definition of the twistor transform, when the spinor λ^α is complemented with its canonical conjugate $\mu_{\dot\alpha}$ and Π is interpreted as generators of conformal transformations. This interpretation is relevant also in AdS space, where Eq. (12) is seen as the twistor transform on the boundary, involving a commuting half of the AdS twistor, which will be utilized below.

We finally note that the division algebra language is well suited to display certain Fierz identities relating the behavior under AdS transformations and under the "R-symmetry". We note that a 4×4 $\mathbb{K}_\nu$-valued matrix carrying two spinor indices A and $\dot{B}$ decomposes into three different irreducible representations: hermitean (the adjoint, and in the complex case a singlet), anti-hermitean traceless (a 4-form, which in the real case is a vector, in the complex case the adjoint and in the quaternionic case self-dual) and the trace (singlet). If we form the twistor quadrilinear

$$\Pi\Pi = \Lambda\Lambda^\dagger\Lambda\Lambda^\dagger\,, \tag{13}$$

it will be anti-hermitean and vanish due to $T = 0$. The vanishing of the 4-form part shows that Π is simple as in Eq. (2), and the singlet that $\Pi_{MN}\Pi^{MN} = 0$. When we later relax the constraint $T = 0$ to obtain "higher spin particles", such Fierz identities will relate Casimirs of the AdS group to Casimirs of R-symmetry.

The construction is straightforwardly extended to superparticles by introducing $2N$ anticommuting scalar variables Θ arranged in an $N \times 2$ matrix, transforming from the left by $O(N;\mathbb{K}_\nu)$ and from the right by $A_2(\mathbb{K}_\nu)$. The supertwistor

$$\Xi = \begin{bmatrix} \Lambda \\ \Theta \end{bmatrix} \tag{14}$$

transforms in the fundamental of the N-extended AdS$_{\nu+3}$ supergroup $OSp(N|4;\mathbb{K}_\nu)$. The constraint structure is not affected, we now have $T = \frac{1}{2}\Xi^\dagger\Xi = \frac{1}{2}(\Lambda^\dagger\Lambda + \Theta^\dagger\Theta) \approx 0$, still generating $A_2(\mathbb{K}_\nu)$.

TWISTORS FOR HIGHER SPIN

We will now relax the constraint $T = 0$ in order to incorporate higher spin massless states in the model. Since four spinors are needed in the 4- and 7-dimensional models to describe massive states, it is essentially clear that such states will not appear. In a twistor model, the on-shell constraint is not a mass-shell constraint, but rather a spin-shell constraint, so relaxing it will lead to a multitude of spins. In the 5-dimensional case we need to be careful—since the U(1) generator measures the mass we are only to relax the SU(2) constraint (which matches nicely with the observation that this is isomorphic to transverse rotations for a massless particle in five dimensions).

It is possible to relax only part of the spin-shell constraints. In AdS$_4$ with gauge group U(1), there is no choice, but in the other cases keeping as gauge group any subgroup of SU(2) or Spin(5) (smaller than the group itself) should define a distinct higher spin theory. Since the generators T are AdS scalars, this procedure is covariant, and will result in restrictions on the massless representations occurring. In this sense, the "smallest" higher spin theory should be given by the "biggest" subgroup. For example, one could consider a higher spin theory on AdS$_5$ with internal "spin manifold" (the non-gauged part of the R-symmetry) $\mathrm{SU}(2)/U(1) = S^2$ or a theory on AdS$_7$ with spin manifold $Spin(5)/(\mathrm{SU}(2) \times \mathrm{SU}(2))$. These choices give the minimal models considered by Sezgin and Sundell [21, 22]. If, on the other hand, the subgroup chosen is the trivial one, one obtains the "Sp space-times" of Ref. [23].

Quantization is straightforward and goes as follows. States in unitary representations of the AdS group are formed by letting an (anti-)commuting subset of variables in Ξ (a configuration space) act on a (preferably scalar) vacuum. States are thus obtained as polynomials in half of the supertwistor variables. For an ordinary superparticle, where $T \approx 0$, the constraints are implemented by only considering states that are R-symmetry singlets. If a subset of generators are kept as gauge generators, as in the minimal models, they are treated accordingly. A natural choice of configuration space for the bosonic variables is the upper half λ of Λ, which is a pair of spinors under the Lorentz group acting on the boundary of AdS space.

The minimal models in 4, 5 and 7 dimensions have gauge groups $\emptyset$, U(1)$\times$U(1) and SU(2)$\times$SU(2), respectively. The two factors act each on one column of Ξ and do not mix them—the two spinors decouple with this choice of spin constraints. The states of the minimal theories are thus obtained as the tensor product of two representations obtained from a single AdS spinorial oscillator, *i.e.*, as the tensor product of two singletons/doubletons, as in refs. [21, 22].

Let us for a moment dwell on the "maximal" models, the supersymmetric versions of the Sp-models. It is trivial to write an action for the higher spin particle,

$$S = \int d\tau \Xi^{\dagger}\dot{\Xi} \,, \tag{15}$$

(for AdS_5 we should still include a Lagrange multiplier for the mass U(1)). From just inspecting the action, we see that the higher spin particle is invariant under a much larger symmetry than the AdS group. In 4 and 7 dimensions, where there are no constraints, this is the real orthosymplectic supergroup acting on the number of real fermionic and bosonic components of the supertwistor, *i.e.*, $OSp(2N|8)$ (as noted in Ref. [24]) and $OSp(8N|32)$, respectively. In AdS_5 we have the subgroup of $OSp(4N|16)$ that commutes with U(1), which is $SU(2N|4,4)$. The corresponding groups acting on the (bosonic part of) configuration space, and thus extending the Lorentz symmetry on the boundary, are SL(4), SL(4,$\mathbb{C}$) and SL(16).

As mentioned earlier, the twistor parametrization of the AdS generators Π imply a direct relation between Casimirs of the AdS group and of the internal symmetry. This relation follows immediately from equations like (13) and reads

$$\mathrm{tr}\Pi^n = \mathrm{tr}T^n \,. \tag{16}$$

One might worry that once the spin-shell constraint is relaxed, the relation to spacetime is lost. The following consideration shows that this is not the case. A simple bivector, describing a plane, can be written as $\Pi^{MN} = X^{[M}P^{N]}$, where $X^2 = -R^2, X \cdot P = 0$ (P is tangent to the hyperboloid) and $P^2 = 0$. When $T \neq 0$, the best one can do is $\Pi^{MN} = X^{[M}P^{N]} + S^{MN}$. In order to still have $\mathrm{tr}\Pi^2 = \mathrm{tr}T^2$, one should demand that $X_M S^{MN} = 0$ (spin is in the AdS tangent space) and $P_M S^{MN} = 0$ (spin is transverse). In addition, S is defined modulo $\delta S^{MN} = V^{[M}P^{N]}$, with $V_M X^M = V_M P^M = 0$, which can be absorbed in a redefinition of X (a reflection of gauge invariance). Such an S is restricted to lie in the transverse rotations, and one will have $\mathrm{tr}\Pi^n = \mathrm{tr}S^n$. This also shows that although T are the generators of an algebra of *internal* rotations, the parametrization of the AdS generators in terms of spinors (the twistor transform) implies the identification of all spin quantum numbers with quantum numbers with respect to T, and the isomorphism between the internal algebra and the algebra of transverse spin degrees of freedom becomes a physical identification.

In the spirit of Ref. [23], we can ask questions about the causal structure in such a theory. Vasiliev showed in Ref. [23] that the causal structure is determined by the maximal Clifford algebras contained in the representations of the bilinears in λ. They must contain matrices acting on 4-component real, 4-component complex and 16-component real spinors, respectively, considering that the subalgebras acting linearly on the configuration space of boundary spinors are SL(4), SL(4,$\mathbb{C}$) and SL(16). We note that these groups are large enough to contain as subgroups the Lorentz groups in higher dimensions, and in fact, since we consider pairs of division algebra spinors, the dimensions corresponding to choosing the next larger division algebra. This means that the boundary in the maximal higher spin model in dimensions 4, 5 and 7 may be considered as 4-, 6- and 10-dimensional. Interpreted in the higher dimensionalities, the pair of spinors becomes a single spinor, and the corresponding states are those of a doubleton in 5, 7 and 11 dimensions.

The maximal dimensionality of the Minkowski space is then 4, 6 and 10, respectively, where the boundary spinors are 2-component complex, quaternionic and octonionic. When we thus consider the $SL(2,\mathbb{K}_{2\nu}) \approx Spin(1,2\nu+1)$ subgroup of the boundary group, the bilinears in λ decompose into a vector and a set of tensorial charges. In the theory originating from AdS_4, where the boundary has now become 4-dimensional, we have the vectorial coordinates and a 2-form (this is the "tensorial" model considered by Bandos *et* al.. [24, 25, 26, 27]). In the theory on AdS_5, where the boundary is enhanced to be 6-dimensional, we have the vector and an SU(2)-triplet of self-dual 3-forms, and in the AdS_7 model, with a 10-dimensional boundary, there is the vector and a self-dual 5-form.

Starting out from a theory in AdS_{d+1} space, with $d=4,5,7$, we end up with a different space-time interpretation, where the "coordinates" are the ordinary coordinates together with a set of forms, "central charge coordinates". The latter provide an alternative picture of the spin degrees of freedom in higher spin theory. However, the general picture is more intricate than in 4 dimensions, where the alternative space-time also is 4-dimensional. We see that the alternative descriptions of the 5- and 7-dimensional AdS theories are theories in 6- and 10-dimensional Minkowski space.

Some more things can be made explicit about the relation between the two descriptions/interpretations. Let us take the AdS_7 theory as an example. The original AdS group is $Spin(2,6) \approx Sp(4;\mathbb{H})$, and the internal symmetry group is $Spin(5) \approx A_2(\mathbb{H})$. $Sp(4;\mathbb{H}) \times A_2(\mathbb{H})$ is a subgroup of the group of all symplectic transformations on the bispinor, which is Sp(32). Restricting to transformations on configuration space, *i*.e., on half the spinor, breaks $Sp(4;\mathbb{H}) \times A_2(\mathbb{H})$ to $SL(2;\mathbb{H}) \times A_2(\mathbb{H}) \approx Spin(1,5) \times Spin(5)$ and Sp(32) to SL(16). When we looked for the maximal Clifford algebra above, that procedure singled out $Spin(1,9) \approx SL(2;\Omega)$. The Spin(1,5) group is smaller than the Lorentz group on the tangent space of AdS_7. It can be identified with the Lorentz group on the conformally Minkowski boundary of AdS_7. This gives a picture of what happens: the boundary Lorentz group is a common subgroup of the AdS group and the boundary symmetry group (in this case Sp(16)). While adjoining a radial coordinate gives the AdS picture, the alternative picture is obtained by supplementing the boundary coordinates with a number of "spin degrees of freedom", manifested as tensorial variables. While this discussion refers to the configuration space, or the boundary variables, one may equally well consider the dimensionally enhanced model as a bulk theory on AdS_{11}. There the states are those obtained from a single supertwistor, which means that it is a doubleton in the sense of Ref. [28, 29].

CONCLUDING REMARKS

We have given a systematic way of deriving different versions of bosonic and supersymmetric higher spin theory on anti-de Sitter spaces of dimension 4, 5 and 7, *i.e.*, the dimensionalities related to the real, complex and quaternionic division algebras. The construction, as it stands is first-quantized, and can be understood as describing the dynamics of a (free) higher spin particle.

The main message is that starting with a twistorial description of particle mechanics, the spin-shell constraints may be relaxed in a systematic and controlled way to yield higher spin degrees of freedom. Using this prescription produces all known spinorial descriptions of higher spin kinematics as special cases. So far, nothing really new has come out of the present work, although we think it provides a nice and unified framework for known models, and a better understanding of their respective rôles and relations.

It would be interesting, especially considering the potential relation of higher spin theory to string theory [6, 7, 8, 9, 10], to see if one can use the particle action and couple it to background fields describing exactly the states produced by the first-quantized action itself, and what information about the interacting theory may be obtained this way.

APPENDIX A: SPINORS AND GAMMA MATRICES

Here, we set the conventions for the gamma matrices used in the twistor transform (4). They can be given in a unified notation for the three cases. We denote by e_i, $i = 1,\ldots,\nu$, the standard orthonormal basis for the division algebra $\mathbb{K}_\nu$. For the flat embedding space with signature $(2,\nu+2)$ we use light-cone coordinates $M = (\oplus,\ominus,\mu) = (\oplus,\ominus,+,-,i)$ and scalar product $V \cdot W = -V^{\oplus}W^{\ominus} - V^{\ominus}W^{\oplus} - V^{+}W^{-} - V^{-}W^{+} + V^{i}W^{i}$.

A spinor under the AdS group belongs to the fundamental representation of $Sp(4;\mathbb{K}_\nu)$, *i.e.*, it is a 4-component column with entries in $\mathbb{K}_\nu$. We use a dotted/undotted notation for spinors, and in addition primed and unprimed spinor indices (since there generically are two chiralities). The two spinor representations, with indices A and A', both decompose into as ${}^{A} \rightarrow ({}^{\alpha},{}_{\dot\alpha}) \leftarrow {}^{A'}$ under the subgroup $SL(2;\mathbb{K}_\nu) \approx Spin(1,\nu+1)$. The gamma matrices (or, strictly speaking, sigma matrices) acting on one chirality (the unprimed one that is chosen for the twistors) are

$$\Gamma^{\oplus A'}{}_{=}\begin{bmatrix}\sqrt{2}\,\mathbb{1}^{\alpha}{}_{\beta} & 0\\ 0 & 0\end{bmatrix}, \quad \Gamma^{\ominus A'}{}_{B} = \begin{bmatrix}0 & 0\\ 0 & \sqrt{2}\,\mathbb{1}_{\dot\alpha}{}^{\dot\beta}\end{bmatrix}, \quad \Gamma^{\mu A'}{}_{B} = \begin{bmatrix}0 & \tilde\gamma^{\mu\alpha\dot\beta}\\ \gamma^{\mu}{}_{\dot\alpha\beta} & 0\end{bmatrix}, \qquad (17)$$

and on the other one

$$\tilde{\Gamma}^{\oplus A}{}_{B'} = \begin{bmatrix} 0 & 0 \\ 0 & -\sqrt{2}\mathbb{1}_{\dot\alpha}{}^{\dot\beta} \end{bmatrix}, \quad \tilde{\Gamma}^{\ominus A}{}_{B'} = \begin{bmatrix} -\sqrt{2}\mathbb{1}^{\alpha}{}_{\beta} & 0 \\ 0 & 0 \end{bmatrix}, \quad \tilde{\Gamma}^{\mu A}{}_{B'} = \begin{bmatrix} 0 & \tilde{\gamma}^{\mu\alpha\dot\beta} \\ \gamma^{\mu}{}_{\dot\alpha\beta} & 0 \end{bmatrix}, \tag{18}$$

where γ^{μ}, $\tilde{\gamma}^{\mu}$ are $\mathrm{SL}(2;\mathbb{K}_{\nu}) \approx \mathrm{Spin}(1,\nu+1)$ gamma matrices:

$$\gamma^{+}{}_{\dot\alpha\beta} = \begin{bmatrix} \sqrt{2} & 0 \\ 0 & 0 \end{bmatrix}, \qquad \gamma^{-}{}_{\dot\alpha\beta} = \begin{bmatrix} 0 & 0 \\ 0 & \sqrt{2} \end{bmatrix}, \qquad \gamma^{i}{}_{\dot\alpha\beta} = \begin{bmatrix} 0 & \bar{e}_i \\ e_i & 0 \end{bmatrix},$$

$$\tilde{\gamma}^{+\alpha\dot\beta} = \begin{bmatrix} 0 & 0 \\ 0 & -\sqrt{2} \end{bmatrix}, \qquad \tilde{\gamma}^{-\alpha\dot\beta} = \begin{bmatrix} -\sqrt{2} & 0 \\ 0 & 0 \end{bmatrix}, \qquad \tilde{\gamma}^{i\alpha\dot\beta} = \begin{bmatrix} 0 & \bar{e}_i \\ e_i & 0 \end{bmatrix}. \tag{19}$$

The matrices Γ^{MN} used in the construction of the plane defining the geodesics are constructed as

$$\Gamma^{MNA}{}_{B} = \frac{1}{2}\left(\tilde{\Gamma}^{M}\Gamma^{N} - \tilde{\Gamma}^{N}\Gamma^{M}\right)^{A}{}_{B}, \tag{20}$$

and the twistor bilinear is given by

$$\Pi^{MN} = \frac{1}{2}\Lambda^{\dagger}\Gamma^{MN}\Lambda \equiv \frac{1}{2}\Lambda^{\dagger}_{\dot A}\Gamma^{MNA}{}_{B}\Lambda^{B} \equiv \frac{1}{2}\Lambda^{\dagger\dot A}\varepsilon_{\dot AB}\Gamma^{MNB}{}_{C}\Lambda^{C}, \tag{21}$$

where the anti-hermitean "spinor metric"

$$\varepsilon_{\dot AB} = \begin{bmatrix} 0 & \mathbb{1}_{\dot\alpha}{}^{\dot\beta} \\ -\mathbb{1}^{\alpha}{}_{\beta} & 0 \end{bmatrix} \tag{22}$$

is used to lower the spinor index, and where † implies division algebra hermitean conjugation. In the real case, dots are of course superfluous, and there is only one chirality. The above formulæ are still correct, and primed and unprimed indices are then related via

$$E^{A'}{}_{B} = \begin{bmatrix} 0 & \varepsilon^{\alpha\beta} \\ \varepsilon_{\alpha\beta} & 0 \end{bmatrix}. \tag{23}$$

ACKNOWLEDGMENTS

The author is grateful to Jerzy Lukierski, Per Sundell, Dmitri Sorokin and, in particular, to Michail Vasiliev for discussions and explanations of their work, and to Murat Günaydin for comments. He would also like to thank the organizers of the XIXth Max Born Symposium in Wrocław, September 2004, where this work was presented, as well as the organizers of the 9th Adriatic Meeting, Dubrovnik, September 2003, where most of the present work was done, for their hospitality.

REFERENCES

1. E. Fradkin and M.A. Vasiliev, Phys. Lett. **B189**, 89 (1987).
2. M.A. Vasiliev, Phys. Lett. **B209**, 491 (1988);
   Ann. Phys. **1**90, 59 (1989) and references therein.
3. M.A. Vasiliev, *"Progress in higher spin gauge theories"*, hep-th/0104246.
4. D. Sorokin, *"Introduction to the classical theory of higher spins"*, hep-th/0405069.
5. N. Bouatta, G. Compere and A. Sagnotti, *"An introduction to free higher-spin fields"*, Lectures given at the Workshop on Higher Spin Gauge Theories, Brussels, May 2004, hep-th/0409068.
6. B. Sundborg, Nucl. Phys. Proc. Suppl. **102**, 113 (2001), hep-th/0103247.
7. E. Sezgin and P. Sundell, Nucl. Phys. **B644**, 303 (2002); erratum Nucl. Phys. **B660**, 403 (2003), hep-th/0205131.
8. G. Bonelli, Nucl. Phys. **B669**, 159 (2003), hep-th/0305155.
9. A. Sagnotti and M. Tsulaia, Nucl. Phys. **B682**, 83 (2004), hep-th/0311257.
10. N. Beisert, M. Bianchi, J.F. Morales and H. Samtleben, J. High Energy Phys. **0407**, 058 (2004), hep-th/0405057.
11. M.A. Vasiliev, Phys. Lett. **B567**, 139 (2003), hep-th/0304049.
12. F. Englert, A. Sevrin, W. Troost, A. Van Proeyen and P. Spindel, J. Math. Phys. **29**, 281 (1988).
13. N. Berkovits, Nucl. Phys. **B350**, 193 (1991).
14. M. Cederwall, J. Math. Phys. **33**, 388 (1992).
15. L. Brink, M. Cederwall and C. Preitschopf, Phys. Lett. **B311**, 76 (1993), hep-th/9303172.
16. M. Cederwall and C. Preitschopf, Commun. Math. Phys. **167**, 273 (1995), hep-th/9309030.
17. P. Claus, M. Günaydin, R. Kallosh, J. Rahmfeld and Y. Zunger, J. High Energy Phys. **9905**, 019 (1999), hep-th/9905112.
18. P. Claus, R. Kallosh and J. Rahmfeld, Phys. Lett. **B462**, 285 (1999), hep-th/9906195.
19. M. Cederwall, Phys. Lett. **B483**, 257 (2000), hep-th/0002216.
20. I. Bengtsson and M. Cederwall, Nucl. Phys. **B302**, 81 (1988).
21. E. Sezgin and P. Sundell, J. High Energy Phys. **0109**, 036 (2001), hep-th/0105001.
22. E. Sezgin and P. Sundell, Nucl. Phys. **B634**, 120 (2002), hep-th/0112100.
23. M.A. Vasiliev, *"Relativity, causality, locality, quantization and duality in the Sp(2M) invariant generalized space-time"*, hep-th/0111119;
    "Higher spin theories and Sp(2M)-invariant space-times", hep-th/0301235.
24. I. Bandos, J. Lukierski, C. Preitschopf and D. Sorokin, Phys. Rev. **D61**, 065009 (2000), hep-th/9907113.
25. I. Bandos, E. Ivanov, J. Lukierski and D. Sorokin, J. High Energy Phys. **0206**, 040 (2002), hep-th/0205104.
26. I. Bandos, P. Pasti, D. Sorokin and M. Tonin, *"Superfield theories in tensorial superspaces and the dynamics of higher spin fields"*, hep-th/0407180.
27. M. Plyushchay, D. Sorokin and M. Tsulaia, *"GL flatness of OSp(1|2N) and higher spin field theory from dynamics in tensorial spaces"*, in the Proceedings of the International Seminar on Supersymmetries and Quantum Symmetries SQS '03, Dubna, July 2003, hep-th/0310297.
28. M. Günaydin, Nucl. Phys. **B528**, 432 (1998), hep-th/9803138.
29. M. Günaydin and Dj. Minić, Nucl. Phys. **B523**, 145 (1998), hep-th/9802047.

Twistor Transform for Spinning Particle

S. Fedoruk

Ukrainian Engineering–Pedagogical Academy, 61003 Kharkiv, 16 Universitetska Str., Ukraine
Kharkiv National University, 61077 Kharkiv, 4 Svobody Sq., Ukraine
e-mail: fed@postmaster.co.uk

Abstract. Twistorial formulation of a particle of arbitrary spin has been constructed. The twistor formulation is deduced from a space–time formulation of the spinning particle by introducing pure gauge Lorentz harmonics in this system. Canonical transformations and gauge fixing conditions, excluding space–time variables, produce the fundamental conditions of twistor transform relating the space–time formulation and twistor one. Integral transformations, relating massive twistor fields with usual space–time fields, have been constructed.

1. INTRODUCTION

Twistor theory [1]-[6] plays important role in solution of different tasks of theoretical physics. Besides fundamental role, the twistors are also powerful tool in analysis of supersymmetric models of point-like and extended objects. In contrast to space-time approach, in which space-time translations of Poincare group are realized be inhomogeneous transformations and actions of conformal group boosts are nonlinear, in twistor space conformal group are realized by homogeneous transformations of $SU(2,2)$ group. Twistor space is four–dimensional complex vector space provided with nongenerate Hermitian form of signature 0 while the twistors are elements of fundamental representation of $SU(2,2)$ group. Incidence conditions determine the relation between objects of the twistor space and ones of complexified Minkowski space–time. So the points of complexified space–time are represented by complex projective lines of projective twistor space having structure of CP^3. On the other hand the points of projective twistor space are realized by totally isotropic two-dimensional planes of complexified space–time. The relation of the twistor fields and space–time ones is realized by integro–differential transform which is, in fact, a generalization of classical Radon transform translating complex–analytic dates on twistor space in solutions of linear equations for massless field.

Conformal invariance, inherent in twistor description, assumes massless particles or its supersymmetric analogs [7]-[20] as basic objects of the consideration. Twistor formulation makes possible to carry out covariant quantization of massless superparticle and also clarifies geometric sense of κ–symmetry as local supersymmetry. Superembedding method [21] used twistor approach allows to formulate twice supersymmetric description of extended superobjects and gives additional information about its dynamics.

The relation of twistor description with space–time formulation of massless particle is determined equations of twistor transform among them Cartan resolution of light–like particle momentum in term of product of two spinors and incidence conditions

CP767, *Fundamental Interactions and Twistor-Like Methods,* edited by J. Lukierski and D. Sorokin

establishing the relation between twistors and space–time coordinates. As we shall have discussed in sect.2, on classical level, namely on level action and equations of motion, incidence conditions lead to zero helicity of particle if we use only one twistor (it is sufficient for twistor description of massless particle) and real space–time coordinates. Classical relation of the formulations in cases of nonzero helicity assumes a modification of incidence conditions at corresponding choice of space–time formulation for spinning particle.

Twistor description of massive particle requires with necessity more than one twistor [1]-[4], [22]-[27] since time–like momentum of massive particle can be resolved only with using two or more twistor spinors. But the complete twistor description of the arbitrary spin massive particles including interaction of them has not been accomplished yet [28]. It should include the constraints and the Lagrangian of the massive spinning particle and also a correct chooses of corresponding variables in twistor formulation. In contrast to the massless case where, in principle, the twistor description of a particle of arbitrary helicity is achieved by using only one twistor in the massive case it is necessary to use some spinning variables in addition to the twistor ones. As in case massless particle it is necessary to have corresponding space–time formulation of massive spinning particle connected with twistor formulation after using of fundamental relations of twistor transform. It is desirable also that massless particle and massive one have been considered uniformly and massless case can be obtained from massive one in level zero mass. This moment is especially important in last time in connection with intensive development of twistor superstring theory [29], full spectrum of which includes both massless modes and massive ones. In present paper we follow the constructive way to finding of twistor formulation of spinning particle implies using the appropriate space–time formulation.

For this aim from all space–time formulations the more appropriate formulation are those in which the spin degrees of freedom are described by means of commuting variables. Also from such formulations there are appropriate ones in which spin variables are spinors (for obtaining of half–integer spins) and the description of arbitrary spins is realized in uniform way. The formulation spinning particle with index spinor [30]-[32] is more appropriate formulation for these aims. In this formulation uniform Lagrangian describes massive particle ($m \neq 0$) or massless one ($m = 0$). Also spinning particle with index spinor has a some analogy with usual superparticle which use Grassmannian spinor variable and we can exploit the some elements of transition from space–time formulation of superparticle to twistor one. A brief review of spinning particle in space–time formulation with index spinor [30]-[32] is presented in sect.3.

Twistor formulation of spinning particle is constructed in sect.4 ($m \neq 0$) and sect.5 ($m = 0$). We introduce pure gauge variables in the space–time formulation of the spinning particle which are Lorentz harmonics. In what follows after canonical transformation we exclude space–time variables by means of gauge fixing and remain only with twistor variables. Canonical transformations and gauge fixing conditions produce the fundamental conditions of twistor transform relating the space–time formulation and twistor one. It is important that standard twistor formulation of massless particle [7]-[20] is obtained from the constructed model by excluding of auxiliary variables. Sect. 6 contains a detail analysis of the twistor transform for massive fields. In sect. 7 we discuss of the results obtained.

2. TWISTOR FORMULATION OF SPINLESS PARTICLE

In space–time formulation a massless particle is described by the Lagrangian

$$L = -\tfrac{1}{2} p_{\alpha\dot{\alpha}} \dot{x}^{\dot{\alpha}\alpha} + \tfrac{1}{2} V p_{\alpha\dot{\alpha}} p^{\dot{\alpha}\alpha} \tag{1}$$

where the coordinate x and the momentum p are real space–time vectors and the scalar V is a Lagrange multiplier for the mass constraint $p_{\alpha\dot{\alpha}} p^{\dot{\alpha}\alpha} = -2p^2 \approx 0$. In the paper we use spinor notation of [33]. In particular, $p_{\alpha\dot{\alpha}} = p_\mu \sigma^\mu_{\alpha\dot{\alpha}}$, $x^{\dot{\alpha}\alpha} = x^\mu \sigma_\mu^{\dot{\alpha}\alpha}$ where matrices σ_μ satisfy $\sigma^{\dot{\alpha}\alpha}_{(\mu} \sigma_{\nu)\alpha\dot{\beta}} = -\eta_{\mu\nu} \delta^{\dot{\alpha}}_{\dot{\beta}}$ and the metric will be $\eta_{\mu\nu} = \mathrm{diag}(-,+,+,+)$.

The model (1) of massless particle possesses the invariance under the global conformal transformations

$$\delta x^{\dot{\alpha}\alpha} = a^{\dot{\alpha}\alpha} + x^{\dot{\alpha}\beta} l_\beta{}^\alpha + \bar{l}^{\dot{\alpha}}{}_{\dot{\beta}} x^{\dot{\beta}\alpha} + x^{\dot{\alpha}\alpha} d + x^{\dot{\alpha}\beta} k_{\beta\dot{\beta}} x^{\dot{\beta}\alpha} , \tag{2}$$

$$\delta p_{\alpha\dot{\alpha}} = -l_\alpha{}^\beta p_{\beta\dot{\alpha}} - p_{\alpha\dot{\beta}} \bar{l}^{\dot{\beta}}{}_{\dot{\alpha}} - p_{\alpha\dot{\alpha}} d - k_{\alpha\dot{\beta}} x^{\dot{\beta}\beta} p_{\beta\dot{\alpha}} - p_{\alpha\dot{\beta}} x^{\dot{\beta}\beta} k_{\beta\dot{\alpha}} , \tag{3}$$

$$\delta V = 2Vd + 2(x^{\dot{\beta}\beta} k_{\beta\dot{\beta}}) V . \tag{4}$$

Here the infinitesimal parameters of translations $a^{\dot{\alpha}\alpha}$ and boosts $k_{\alpha\dot{\alpha}}$ are Hermitian, the dilatation parameter d is real and $l_\alpha{}^\beta$, $\bar{l}^{\dot{\alpha}}{}_{\dot{\beta}} = (\overline{l_\beta{}^\alpha})$, $l_\alpha{}^\alpha = 0$, $\bar{l}^{\dot{\alpha}}{}_{\dot{\alpha}} = 0$ are the infinitesimal parameters of the Lorentz group $SL(2,C)$.

In the twistor formulation the phase space of a massless particle is described by two Weyl spinors λ_α, $\bar{\lambda}_{\dot{\alpha}} = (\overline{\lambda_\alpha})$ and $\bar{\omega}^\alpha$, $\omega^{\dot{\alpha}} = (\overline{\bar{\omega}^\alpha})$ which are canonically conjugated to each other. The four–component Penrose twistor is constructed out of these two Weyl spinors $Z_a = \left(\lambda_\alpha,\, \omega^{\dot{\alpha}}\right)$. Its conjugate twistor is defined by $\bar{Z}_{\dot{a}} = (\overline{Z_a}) = \left(\bar{\lambda}_{\dot{\alpha}},\, \bar{\omega}^\alpha\right)$, $\bar{Z}^a = g^{a\dot{a}} \bar{Z}_{\dot{a}} = \left(\bar{\omega}^\alpha,\, -\bar{\lambda}_{\dot{\alpha}}\right)$ where $g^{a\dot{b}}$ is the metric of twistor space. In terms of twistor variables the Lagrangian of massless particle is described in form

$$L = -\tfrac{1}{2}(\dot{\bar{Z}}^a Z_a - \bar{Z}^a \dot{Z}_a) - N(\tfrac{i}{2}\bar{Z}^a Z_a + s) , \tag{5}$$

where N is Lagrange multiplier for twistor spin constraint $\frac{i}{2}\bar{Z}^a Z_a + s = \frac{i}{2}(\bar{\omega}^\alpha \lambda_\alpha - \bar{\lambda}_{\dot{\alpha}} \omega^{\dot{\alpha}}) + s \approx 0$.

Twistor formulation (5) is invariant under linear transformations

$$\delta \begin{pmatrix} \lambda_\alpha \\ \omega^{\dot{\alpha}} \end{pmatrix} = \begin{pmatrix} -l_\alpha{}^\beta - \frac{1}{2} d \delta_\alpha{}^\beta & -2k_{\alpha\dot{\beta}} \\ \frac{1}{2} a^{\dot{\alpha}\beta} & \bar{l}^{\dot{\alpha}}{}_{\dot{\beta}} + \frac{1}{2} d \delta^{\dot{\alpha}}{}_{\dot{\beta}} \end{pmatrix} \begin{pmatrix} \lambda_\beta \\ \omega^{\dot{\beta}} \end{pmatrix} , \tag{6}$$

which form group $SU(2,2)$ being twice covering of group $SO(2,4)$ which is in turn locally isomorphic to conformal group (2)-(4) of $D = 4$ Minkowski space.

By Noether procedure corresponding to transformations (6) conserved charges $P_{\alpha\dot{\alpha}} = \lambda_\alpha \bar{\lambda}_{\dot{\alpha}}$, $M_{\alpha\beta} = i\lambda_{(\alpha} \bar{\omega}_{\beta)}$, $\bar{M}_{\dot{\alpha}\dot{\beta}} = i\bar{\lambda}_{(\dot{\alpha}} \omega_{\dot{\beta})}$, $K^{\dot{\alpha}\alpha} = \omega^{\dot{\alpha}} \bar{\omega}^\alpha$, $D = \frac{1}{2}(\bar{\omega}^\alpha \lambda_\alpha + \bar{\lambda}_{\dot{\alpha}} \omega^{\dot{\alpha}})$ give for pseudovector Pauli-Lubanski $W_{\alpha\dot{\alpha}} = P_\alpha{}^{\dot{\beta}} \bar{M}_{\dot{\beta}\dot{\alpha}} - P^\beta{}_{\dot{\alpha}} M_{\beta\alpha}$ following value

$$W_{\alpha\dot{\alpha}} = (-\tfrac{i}{2}\bar{Z}^a Z_a) P_{\alpha\dot{\alpha}} .$$

Since the spinors λ_α and $\bar{\omega}^\alpha$ are canonically conjugated each other we use the convention $\lambda^\alpha = \varepsilon^{\alpha\beta}\lambda_\beta$, $\bar{\omega}_\alpha = \bar{\omega}_\beta\varepsilon^{\beta\alpha}$ so that $\lambda^\alpha\bar{\omega}_\alpha = \bar{\omega}^\alpha\lambda_\alpha$. Thus the classical consideration shows that model with twistor Lagrangian (5) describes massless particles of finite spin ($P^2 = 0, W^2 = 0$). Spin (helicity) is defined by the quantity $(-\frac{i}{2}\bar{Z}^a Z_a)$.

Relation of twistor formulation (5) to space–time formulation (1) is assigned twistor representation of light–like momentum

$$p_{\alpha\dot{\alpha}} = \lambda_\alpha\bar{\lambda}_{\dot{\alpha}} \tag{7}$$

and conditions

$$\omega^{\dot{\alpha}} = \tfrac{1}{2}x^{\dot{\alpha}\alpha}\lambda_\alpha\,, \qquad \bar{\omega}^\alpha = \tfrac{1}{2}\bar{\lambda}_{\dot{\alpha}}x^{\dot{\alpha}\alpha} \tag{8}$$

relating space–time coordinates and twistor variables. Fundamental relations of twistor transform (7), (8) are constructed thus that one of them, namely (7), resolves mass constraint $p^2 \approx 0$ in twistor variables whereas incidence conditions (8) resolve twistor constraint $\frac{i}{2}\bar{Z}^a Z_a + s \approx 0$ defining particle spin. But in traditional writing of incidence conditions (8) it should be isotropy condition for twistor

$$\bar{Z}^a Z_a = \bar{\omega}^\alpha\lambda_\alpha - \bar{\lambda}_{\dot{\alpha}}\omega^{\dot{\alpha}} = 0\,.$$

Since twistor Z_a of general form describes classical massless particle with helicity $-\frac{i}{2}\bar{Z}^a Z_a$ [4] the isotropic twistor describes massless particle of zero helicity $s = 0$. Thus basic conditions (7), (8) determine the correspondence between space–time formulation (1) and twistor one (5) of massless particle only at zero spin $s = 0$. Massless particles with nonzero helicity are obtained in quantum spectrum upon the transition from space–time formulation to twistor formulation either by introducing nonzero spin (helicity) constant s directly into the spin constraint $\bar{Z}Z - 2is \approx 0$ in the twistor action (5) 'by hand' or taking into account an ordering ambiguity of quantum variables in the spin constraint. In the standard case of ordering we obtain particles with helicities not more two [12]. It should be noted that twistor–like formulations [8], [15], [16], [18] using both space–time variables and twistor ones equivalents to systems (1) and (5) also only at $s = 0$.

In the Penrose twistor approach particles with nonzero helicity are described by nonisotropic twistors which are defined in complex space–time with coordinates $z^\mu = x^\mu + iy^\mu$ when matrix $z^{\dot{\alpha}\alpha} = z^\mu\sigma_\mu^{\dot{\alpha}\alpha}$ is in general nonHermitian. Corresponding basic relations $\omega^{\dot{\alpha}} = \frac{1}{2}z^{\dot{\alpha}\alpha}\lambda_\alpha$, $\bar{\omega}^\alpha = \frac{1}{2}\bar{\lambda}_{\dot{\alpha}}\bar{z}^{\dot{\alpha}\alpha}$ give $\frac{i}{2}\bar{Z}^a Z_a = y^\mu\lambda\sigma_\mu\bar{\lambda}$. Thus twistor is nonisotropic if $y^\mu \neq 0$ is nonparallel to vector $\lambda\sigma_\mu\bar{\lambda}$. In contrast to field theory where complex space–time is used effectively at definition of analyticity and spectral property of quantum fields, in mechanical model of point and extended objects it is more naturally and appropriately to work in usual real space–time.

Moreover taking into account basic ideology of twistor approach as alternative for space–time description [1]-[4], from set of formulations of spinning particle it must be space–time formulation corresponding to one (5) at fulfillment of basic relations for twistor transform which generalize the expressions (8). Below we establish the generalized relations and show that space–time formulation corresponding to twistor one (8) is the model of a spinning particle with index spinor [30]-[32].

3. SPINNING PARTICLE IN SPACE–TIME FORMULATION

As discussed above we will take index spinor formulation [30]-[32] of spinning particle as starting point for obtaining twistor formulation. Here we lead some details of this formulation.

In index spinor formalism spinning particle is described with space–time vector x^μ and commuting Weyl spinor ζ^α. In first order formalism its Lagrangian has the form [30]-[32]

$$L = -\tfrac{1}{2} p_{\alpha\dot\alpha} \dot\Pi^{\dot\alpha\alpha} + \tfrac{1}{2} V (p_{\alpha\dot\alpha} p^{\dot\alpha\alpha} - 2m^2) - \Lambda(\zeta^\alpha p_{\alpha\dot\alpha} \bar\zeta^{\dot\alpha} - s), \tag{9}$$

where the bosonic 'superform' is

$$\Pi^{\dot\alpha\alpha} \equiv \dot\Pi^{\dot\alpha\alpha} d\tau \equiv dx^{\dot\alpha\alpha} + 2i\bar\zeta^{\dot\alpha} d\zeta^\alpha - 2i d\bar\zeta^{\dot\alpha} \zeta^\alpha .$$

Here p_μ is momentum of particle. Real scalars V and Λ are Lagrange multipliers.

The spinning particle (9) is invariant under global conformal transformations:

- Poincare transformations

$$\delta x^{\dot\alpha\alpha} = a^{\dot\alpha\alpha} + x^{\dot\alpha\beta} l_\beta{}^\alpha + \bar l^{\dot\alpha}{}_{\dot\beta} x^{\dot\beta\alpha}, \quad \delta p_{\alpha\dot\alpha} = -l_\alpha{}^\beta p_{\beta\dot\alpha} - p_{\alpha\dot\beta} \bar l^{\dot\beta}{}_{\dot\alpha}, \quad \delta\zeta^\alpha = \zeta^\beta l_\beta{}^\alpha, \tag{10}$$

- dilatation transformations (only in massless case $m = 0$)

$$\delta x^{\dot\alpha\alpha} = x^{\dot\alpha\alpha} d, \quad \delta p_{\alpha\dot\alpha} = -p_{\alpha\dot\alpha} d, \quad \delta\zeta^\alpha = \tfrac{1}{2}\zeta^\alpha d, \quad \delta V = 2V d\,; \tag{11}$$

- conformal boosts (only in massless case $m = 0$)

$$\delta x^{\dot\alpha\alpha} = x^{\dot\alpha\beta} k_{\beta\dot\beta} x^{\dot\beta\alpha} - 4\zeta^\alpha \bar\zeta^{\dot\alpha} \zeta^\beta k_{\beta\dot\beta} \bar\zeta^{\dot\beta}, \quad \delta V = 2(x^{\dot\beta\beta} k_{\beta\dot\beta}) V,$$

$$\delta p_{\alpha\dot\alpha} = -(p_{\alpha\dot\beta} k_{\beta\dot\alpha} + k_{\alpha\dot\beta} p_{\beta\dot\alpha}) x^{\dot\beta\beta} - 2i(p_{\alpha\dot\beta} k_{\beta\dot\alpha} - k_{\alpha\dot\beta} p_{\beta\dot\alpha}) \bar\zeta^{\dot\beta} \zeta^\beta,$$

$$\delta\zeta^\alpha = \zeta^\beta k_{\beta\dot\beta} (x^{\dot\beta\alpha} - 2i\bar\zeta^{\dot\beta} \zeta^\alpha), \quad \delta\bar\zeta^{\dot\alpha} = (x^{\dot\alpha\beta} + 2i\bar\zeta^{\dot\alpha} \zeta^\beta) k_{\beta\dot\beta} \bar\zeta^{\dot\beta}; \tag{12}$$

- chiral (phase) transformations of spinor with real phase ϕ

$$\delta\zeta^\alpha = -\tfrac{5}{2} i\phi \zeta^\alpha, \quad \delta\bar\zeta^{\dot\alpha} = \tfrac{5}{2} i\phi \bar\zeta^{\dot\alpha}. \tag{13}$$

The first two terms $p_{\alpha\dot\alpha}\dot\Pi^{\dot\alpha\alpha} - V p_{\alpha\dot\alpha} p^{\dot\alpha\alpha}$ of the Lagrangian (9) are invariant under

- so called "even SUSY" transformations with commuting spinor parameter ε^α

$$\delta x^{\dot\alpha\alpha} = 2i(\bar\zeta^{\dot\alpha}\varepsilon^\alpha - \bar\varepsilon^{\dot\alpha}\zeta^\alpha), \quad \delta\zeta^\alpha = \varepsilon^\alpha, \quad \delta\bar\zeta^{\dot\alpha} = \bar\varepsilon^{\dot\alpha}; \tag{14}$$

- so called "bosonic superconformal boosts" (only in massless case $m = 0$)

$$\delta x^{\dot\alpha\alpha} = 2i(\bar\zeta^{\dot\alpha} \bar\psi_{\dot\beta} x^{\dot\beta\alpha} - x^{\dot\alpha\beta} \psi_\beta \zeta^\alpha) - 4\bar\zeta^{\dot\alpha} \zeta^\alpha (\zeta^\beta \psi_\beta + \bar\psi_{\dot\beta} \bar\zeta^{\dot\beta}),$$

$$\delta p_{\alpha\dot\alpha} = 4i(\psi_\alpha \zeta^\beta p_{\beta\dot\alpha} - p_{\alpha\dot\beta} \bar\zeta^{\dot\beta} \bar\psi_{\dot\alpha}),$$

$$\delta\zeta^{\alpha} = -4i\zeta^{\alpha}\zeta^{\beta}\psi_{\beta} + \bar{\psi}_{\dot\beta}(x^{\dot\beta\alpha} - 2i\bar{\zeta}^{\dot\beta}\zeta^{\alpha}), \quad \delta V = -4iV(\zeta^{\beta}\psi_{\beta} - \bar{\psi}_{\dot\beta}\bar{\zeta}^{\dot\beta}), \quad (15)$$

where ψ_{α}, $\bar{\psi}_{\dot\alpha} = \overline{(\psi_{\alpha})}$ is commuting spinor.
Below we show that in the twistor formulation the symmetry transformations (10)-(15) are linear in twistor variables and isomorphic to $U(3,2)$ group in massless case.

Apart from the mass and spin constraints

$$T \equiv \tfrac{1}{2}(p^2 + m^2) \approx 0, \quad (16)$$

$$\zeta^{\alpha} p_{\alpha\dot\alpha}\bar{\zeta}^{\dot\alpha} - s \approx 0 \quad (17)$$

it is presented also the even spinor constraints

$$d_{\zeta\alpha} \equiv ip_{\zeta\alpha} + p_{\alpha\dot\alpha}\bar{\zeta}^{\dot\alpha} \approx 0, \qquad \bar{d}_{\zeta\dot\alpha} \equiv -i\bar{p}_{\zeta\dot\alpha} + \zeta^{\alpha} p_{\alpha\dot\alpha} \approx 0. \quad (18)$$

On the constraints surface the spin constraint (17) is equivalent to the constraint

$$\boldsymbol{S} \equiv S - s \equiv \tfrac{i}{2}(\zeta^{\alpha} p_{\zeta\alpha} - \bar{p}_{\zeta\dot\alpha}\bar{\zeta}^{\dot\alpha}) - s \approx 0. \quad (19)$$

The constraint algebra is defined by the following non–zero brackets $\{d_{\zeta}, \bar{d}_{\zeta}\} = 2i\hat{p}$, $\{\boldsymbol{S}, d_{\zeta}\} = \frac{i}{2} d_{\zeta}$, $\{\boldsymbol{S}, \bar{d}_{\zeta}\} = -\frac{i}{2}\bar{d}_{\zeta}$. So, the constraints T and S belong to the first class. The spinor constraints $d_{\zeta\alpha}$ and $\bar{d}_{\zeta\dot\alpha}$ relate to the second class for particle with nonzero mass $\frac{1}{2}p_{\alpha\dot\alpha}p^{\alpha\dot\alpha} = m^2 > 0$. In case of massless particle $m = 0$ half of the spinor constraints (18) is first class constraints whereas other half is second class. The separation of the spinor constraints in classes is realized by projection of them on even spinors ζ^{α} and $\bar{\zeta}_{\dot\alpha}p^{\dot\alpha\alpha}$.

After quantization [30] the wave function $\Psi(x;\zeta,\bar{\zeta})$ of the spinning particle is expressed (in case $s > 0$) by holomorphic polynomial $\Phi(x;\zeta)$ on spinor ζ

$$\Psi(x;\zeta,\bar{\zeta}) = e^{-\zeta\hat{p}\bar{\zeta}}\Phi(x;\zeta), \qquad \Phi(x;\zeta) = \zeta^{\alpha_1}\ldots\zeta^{\alpha_{2J}}\phi_{\alpha_1\ldots\alpha_{2J}}(x)\,.$$

Component fields satisfy Klein–Gordon equation $(\partial^2 - m^2)\,\phi_{\alpha_1\ldots\alpha_{2J}} = 0$ and Fierz–Pauli equations $\partial^{\dot\beta\beta}\,\phi_{\beta\alpha_2\ldots\alpha_{2J}} = 0$ in massless case. Thus these fields are described particles with the spin (helicity) J which is equal to the constant s renormalized by ordering constants.

When potential term $-\Lambda(\zeta\hat{p}\bar{\zeta} - s)$ (the constraint (17)) in Lagrangian (9) is absent and the system is invariant under even SUSY transformation (14), the wave function is expressed by polynomial series on ζ, $\bar{\zeta}$ with non-fixed degree of homogeneity

$$\Phi(x;\zeta,\bar{\zeta}) = \sum_{n=0}^{\infty}\sum_{m=0}^{\infty}\zeta^{\alpha_1}\ldots\zeta^{\alpha_n}\bar{\zeta}^{\dot\alpha_1}\ldots\bar{\zeta}^{\dot\alpha_m}\phi_{\alpha_1\ldots\alpha_n\dot\alpha_1\ldots\dot\alpha_m}(x)\,.$$

In this case we have in spectrum the infinite tower of states with all arbitrary spins (helicities). Thus index spinor variables ζ^{α}, $\bar{\zeta}^{\dot\alpha}$ have the role analogous to role of the spinor variables in unfolded formulation of higher spin field theory [34]. Even variant of a supersymmetry transformation (14) was being arisen also in description of the infinite tower of the physical degrees of freedom of the critical open string with $N = 2$ superconformal symmetry on the worldsheet [35].

4. TWISTOR FORMULATION OF MASSIVE SPINNING PARTICLE

We shall now construct a twistor formulation of spinning particle which corresponds to the space–time formulation (9) considered in previous section. We begin with the consideration of the nonsingular case of the particle of nonzero mass $m \neq 0$. We introduce pure gauge variables in initial system (9) which are Lorentz harmonics. In what follows after canonical transformation we exclude space–time variables by means of gauge fixing and remain only with twistor variables. Canonical transformations and gauge fixing conditions have physical meaning. In particular they will produce the fundamental conditions of twistor transform relating the space–time formulation and twistor one.

4.1. Twistor formulation from Lorentz harmonic approach

We introduce two commuting spinors v^i_α, $\bar{v}_{\dot{\alpha} i} = \overline{(v^i_\alpha)}$, $i = 1,2$ which are considered as Lorentz harmonics [36]- [38]. They form 2×2 complex matrix with unit determinant. Thus the harmonic spinors v^i_α are subjected to the conditions

$$h \equiv v^{\alpha i} v_{\alpha i} + 2 \approx 0 , \quad \bar{h} \equiv \bar{v}_{\dot{\alpha} i} \bar{v}^{\dot{\alpha} i} + 2 \approx 0 , \tag{20}$$

where $v_{\alpha i} = \varepsilon_{ij} v^j_\alpha$, $\bar{v}^i_{\dot{\alpha}} = \varepsilon^{ij} \bar{v}_{\dot{\alpha} j}$ and ε^{ij} is the skew–symmetric tensor, $\varepsilon^{ij}\varepsilon_{jk} = \delta^i_k$.

Let us enlarge the system of the spinning particle (9) with a pure gauge sector of Lorentz harmonics. Namely we add to the Lagrangian (9) standard kinetic terms for harmonics v^i_α, $\bar{v}_{\dot{\alpha} i}$ and their canonically conjugate variables $p_{v}{}^{\alpha}_{i}$, $\bar{p}_v{}^{\dot{\alpha} i}$, $\{v^i_\alpha, p_{v}{}^{\beta}_{j}\} = \delta^\beta_\alpha \delta^i_j$, $\{\bar{v}_{\dot{\alpha} i}, \bar{p}_v{}^{\dot{\beta} j}\} = \delta^{\dot{\beta}}_{\dot{\alpha}} \delta^j_i$, and linear combination of the full set of the constraints, the coefficients of which are Lagrange multipliers. The number of constraints must be sufficient to exclude all harmonic variables. In addition to kinematic constraints (20) we impose the following natural set of constraints on harmonic variables

$$p_{v}{}^{\alpha}_{i} \approx 0 , \qquad \bar{p}_v{}^{\dot{\alpha} i} \approx 0 , \tag{21}$$

i.e. all conjugate variables for harmonics v^i_α, $\bar{v}_{\dot{\alpha} i}$ are zero in weak sense. Of course, already the constraints (21) are sufficient in order that the spinor variables v^i_α, p^α_{vi} and their complex conjugate to be pure gauge. But in order to use completeness conditions $\delta^\beta_\alpha = v^i_\alpha v^\beta_i$, $\delta^{\dot{\alpha}}_{\dot{\beta}} = \bar{v}^{\dot{\alpha}}_i \bar{v}^i_{\dot{\beta}}$, which inherent in harmonic approach, it is convenient to subject spinors v^i_α to the harmonic conditions (20).

The system of constraints (20) and (21) contains two pairs of second class constraints and six constraints of first class. The separation of constraints (21) on classes is realized by projection of them on spinors v^i_α, $\bar{v}_{\dot{\alpha} i}$. Because of nonsingularity of harmonic matrix v^i_α (20), the constraints (21) equivalent to Lorentz–invariant ones

$$p_{v}{}^{\alpha}_{i} v^j_\alpha \approx 0 , \qquad \bar{v}_{\dot{\alpha} i} \bar{p}_v{}^{\dot{\alpha} j} \approx 0 . \tag{22}$$

The trace parts of constraints (22) are conjugate ones for kinematic constraints (20). In real quantities the constraints

$$i(h-\bar{h}) = i(v^{\alpha i}v_{\alpha i} - \bar{v}_{\dot{\alpha}i}\bar{v}^{\dot{\alpha}i}) \approx 0\,, \quad h+\bar{h} = v^{\alpha i}v_{\alpha i} + \bar{v}_{\dot{\alpha}i}\bar{v}^{\dot{\alpha}i} + 4 \approx 0\,, \tag{23}$$

$$\mathcal{D}_0 \equiv i(p_{vi}{}^{\alpha}v^i_{\alpha} - \bar{v}_{\dot{\alpha}i}\bar{p}_v{}^{\dot{\alpha}i}) \approx 0\,, \quad \mathcal{B}_0 \equiv p_{vi}{}^{\alpha}v^i_{\alpha} + \bar{v}_{\dot{\alpha}i}\bar{p}_v{}^{\dot{\alpha}i} \approx 0\,, \tag{24}$$

form the pairs $(i(h-\bar{h}),\mathcal{D}_0)$ and $(h+\bar{h},\mathcal{B}_0)$ of mutually conjugate second class ones. Thus (20) can be considered as gauge fixing conditions for two constraints from (21).

The traceless parts of constraints (22), which commute with constraints (23) and (24), are first class constraints. It is convenient to represent these constraints in form real Lorentz invariant 3–vectors

$$\mathcal{D}_r \equiv \tfrac{i}{2}(\sigma_r)_i{}^j(p_{vj}{}^{\alpha}v^i_{\alpha} - \bar{v}_{\dot{\alpha}j}\bar{p}_v{}^{\dot{\alpha}i}) \approx 0\,, \quad \mathcal{B}_r \equiv \tfrac{1}{2}(\sigma_r)_i{}^j(p_{vj}{}^{\alpha}v^i_{\alpha} + \bar{v}_{\dot{\alpha}j}\bar{p}_v{}^{\dot{\alpha}i}) \approx 0\,. \tag{25}$$

Thus the spinning particle, enlarged with the pure gauge sector of the Lorentz harmonics, is described by phase space variables x^{μ}, p_{μ}; ζ^{α}, $\bar{\zeta}^{\dot{\alpha}}$, $p_{\zeta\alpha}$, $\bar{p}_{\zeta\dot{\alpha}}$; v^i_{α}, $\bar{v}_{\dot{\alpha}i}$, $p_{vi}{}^{\alpha}$, $\bar{p}_v{}^{\dot{\alpha}i}$ and constraints (16), (18), (19), (20), (24), (25). Using part of the constraints and corresponding gauge symmetries we can now eliminate all variables used in the space–time formulation, namely x, p, ζ, p_{ζ} and their complex conjugate.

To this end it is convenient using Lorentz harmonics to transform the initial variables x, p, ζ, p_{ζ} and their c.c. into Lorentz–invariant quantities

$$x^{(0)} = \tfrac{1}{2}\bar{v}_{\dot{\alpha}k}x^{\dot{\alpha}\alpha}v^k_{\alpha}\,, \quad x_{(r)} = \tfrac{1}{2}\bar{v}_{\dot{\alpha}j}x^{\dot{\alpha}\alpha}v^i_{\alpha}(\sigma_r)_i{}^j\,, \tag{26}$$

$$p_{(0)} = \tfrac{1}{2}v_k^{\alpha}p_{\alpha\dot{\alpha}}\bar{v}^{\dot{\alpha}k}\,, \quad p_{(r)} = \tfrac{1}{2}v_j^{\alpha}p_{\alpha\dot{\alpha}}\bar{v}^{\dot{\alpha}i}(\sigma_r)_i{}^j\,, \tag{27}$$

$$\xi^i = m^{1/2}\zeta^{\alpha}v^i_{\alpha}\,, \quad \bar{\xi}_i = m^{1/2}\bar{v}_{\dot{\alpha}i}\bar{\zeta}^{\dot{\alpha}}\,, \tag{28}$$

$$p_{\xi i} = m^{-1/2}v_i^{\alpha}p_{\zeta\alpha}\,, \quad \bar{p}^i_{\xi} = -m^{-1/2}\bar{p}_{\zeta\dot{\alpha}}\bar{v}^{\dot{\alpha}i}\,. \tag{29}$$

Since the twistor spinors have dimension equal half of mass dimension we will pass from the dimensionless harmonic variables v^i_{α}, $\bar{v}_{\dot{\alpha}i}$, p^{α}_{vi}, $\bar{p}^{\dot{\alpha}i}_v$ to the variables

$$\lambda^i_{\alpha} = m^{1/2}v^i_{\alpha}\,, \quad \bar{\lambda}_{\dot{\alpha}i} = m^{1/2}\bar{v}_{\dot{\alpha}i}\,. \tag{30}$$

The momenta $\bar{\omega}_i^{\alpha}$, $\omega^{\dot{\alpha}i}$ of the new variables λ^i_{α}, $\bar{\lambda}_{\dot{\alpha}i}$ are defined below.

The transformation (26)–(30) is canonical transformation. The generating function of the canonical transformation from system with phase variables x^{μ}, p_{μ}; ζ^{α}, $\bar{\zeta}^{\dot{\alpha}}$, $p_{\zeta\alpha}$, $\bar{p}_{\zeta\dot{\alpha}}$; v^i_{α}, $\bar{v}_{\dot{\alpha}i}$, $p_{vi}{}^{\alpha}$, $\bar{p}_v{}^{\dot{\alpha}i}$ to the system with phase variables $x^{(0)}$, $x_{(r)}$, $p_{(0)}$, $p_{(r)}$; ξ^i, $\bar{\xi}_i$, $p_{\xi i}$, $\bar{p}^i_{\xi}$; λ^i_{α}, $\bar{\lambda}_{\dot{\alpha}i}$, $\bar{\omega}_i^{\alpha}$, $\omega^{\dot{\alpha}i}$ has the form

$$F = p_{(0)}x^{(0)}(x,v,\bar{v}) + p_{(r)}x_{(r)}(x,v,\bar{v}) + p_{\xi i}\xi^i(\zeta,v) + \bar{p}^i_{\xi}\bar{\xi}_i(\bar{\zeta},\bar{v}) + \bar{\omega}_i^{\alpha}\lambda^i_{\alpha}(v) + \bar{\lambda}_{\dot{\alpha}i}(\bar{v})\omega^{\dot{\alpha}i}\,.$$

Here the expressions for new variables $x^{(0)}(x,v,\bar{v})$, $x_{(r)}(x,v,\bar{v})$, $\xi^i(\zeta,v)$, $\bar{\xi}_i(\bar{\zeta},\bar{v})$, $\lambda^i_{\alpha}(v)$, $\bar{\lambda}_{\dot{\alpha}i}(\bar{v})$ in term of old variables from the right hand side of the equations (26), (28), (30) have been used. This generating function reproduces the expressions (26)–(30) and gives the expressions of harmonic momenta $p_{v_i}{}^{\alpha}$, $\bar{p}_v{}^{\dot\alpha i}$ in the form

$$p_{v_i}{}^{\alpha} = m^{1/2}\left[\bar\omega_i^{\alpha} + \tfrac{1}{2m}p_{(0)}\bar\lambda_{\dot\alpha i}x^{\dot\alpha\alpha} + \tfrac{1}{2m}p_{(r)}\bar\lambda_{\dot\alpha j}x^{\dot\alpha\alpha}(\sigma_r)_i{}^j + p_{\xi i}\zeta^{\alpha}\right], \tag{31}$$

$$\bar{p}_v{}^{\dot\alpha i} = m^{1/2}\left[\omega^{\dot\alpha i} + \tfrac{1}{2m}p_{(0)}x^{\dot\alpha\alpha}\lambda_{\alpha}^{i} + \tfrac{1}{2m}p_{(r)}x^{\dot\alpha\alpha}\lambda_{\alpha}^{j}(\sigma_r)_j{}^i + \bar{p}_{\xi}^{i}\bar\zeta^{\dot\alpha}\right]. \tag{32}$$

Last three addends (31), (32) lead to additional terms in the constraints (24), (25) written in new variables. In new variables the constraints (24), (25) become

$$D_0 = \mathcal{D}_0 - i(\bar\xi_i\bar{p}_{\xi}^{i} - p_{\xi i}\xi^{i}) \approx 0, \tag{33}$$

$$B_0 = \mathcal{B}_0 + 2x^{(0)}p_{(0)} + 2x_{(r)}p_{(r)} + (\bar\xi_i\bar{p}_{\xi}^{i} + p_{\xi i}\xi^{i}) \approx 0, \tag{34}$$

$$D_r = \mathcal{D}_r - \tfrac{1}{2}\varepsilon_{rsp}p_{(s)}x_{(p)} - \tfrac{i}{2}(\sigma_r)_i{}^j(\bar\xi_j\bar{p}_{\xi}^{i} - p_{\xi j}\xi^{i}) \approx 0, \tag{35}$$

$$B_r = \mathcal{B}_r + x_{(r)}p_{(0)} + x^{(0)}p_{(r)} + \tfrac{1}{2}(\sigma_r)_i{}^j(\bar\xi_j\bar{p}_{\xi}^{i} + p_{\xi j}\xi^{i}) \approx 0. \tag{36}$$

Here in new constraints D_0, B_0, D_r, B_r the expressions for $\mathcal{D}_0$, $\mathcal{B}_0$, $\mathcal{D}_r$, $\mathcal{B}_r$ as in (24), (25) with new variables λ_{α}^{i}, $\bar\lambda_{\dot\alpha i}$, $\bar\omega_i^{\alpha}$, $\omega^{\dot\alpha i}$ in place of v_{α}^{i}, $\bar{v}_{\dot\alpha i}$, $p_{v_i}{}^{\alpha}$, $\bar{p}_v{}^{\dot\alpha i}$. In new variables the constraints (16)–(18) take the form

$$-(p_{(0)})^2 + p_{(r)}p_{(r)} + m^2 \approx 0, \tag{37}$$

$$-\tfrac{1}{m}\xi^{i}p_i{}^j\bar\xi_j - s \approx 0, \tag{38}$$

$$v_i^{\alpha}d_{\zeta\alpha} = m^{-1/2}\left(ip_{\xi i} + \tfrac{1}{m}p_i{}^j\bar\xi_j\right) \approx 0, \qquad \bar{d}_{\zeta\dot\alpha}\bar{v}^{\dot\alpha i} = m^{-1/2}\left(i\bar{p}_{\xi}^{i} - \tfrac{1}{m}\xi^{j}p_j{}^i\right) \approx 0, \tag{39}$$

where $p_i{}^j \equiv p_{(0)}\delta_i^j + p_{(r)}(\sigma_r)_i{}^j = v_i^{\alpha}p_{\alpha\dot\alpha}\bar{v}^{\dot\alpha j}$.

The transformations generated by the constraints (36) are Wigner transformations for massive particle which transform four–momentum to standard one. We impose gauge fixing conditions for constraints $B_r \approx 0$ in following form

$$p_{(r)} \approx 0, \quad r = 1,2,3. \tag{40}$$

Thus we take momentum of rest state as standard momentum in harmonic basis with coordinates $x^{(0)}$, $x_{(r)}$. Accounting of the constraints (36), (40) in strong sense exclude the variables $x_{(r)}$, $p_{(r)}$ from consideration. Dirac brackets of another variables coincide with Poisson ones because of resolved form of the gauge fixing conditions (40).

After that the mass–shell condition (37) takes the form

$$p_{(0)} \pm m \approx 0. \tag{41}$$

In this expression the sign $\pm$ defines the sign of energy $E \equiv p^{(0)} = -p_{(0)}$. The constraint (41) and the 'proper–time' condition

$$x^{(0)} \approx 0 \tag{42}$$

fixed gauge for it exclude the variables $x^{(0)}$, $p_{(0)}$. Again resolved form of gauge fixing condition (42) leads to the invariance of the commutative relations of other variables.

After fulfilled gauge fixing which gives $p_i{}^j = \mp m\delta_i^j$ the spin constraint (38) takes the form

$$\pm\bar{\xi}_i\xi^i - s \approx 0\,, \tag{43}$$

whereas the constraints (39) are written in a simple form

$$\psi_i \equiv ip_{\xi i} \mp \bar{\xi}_i \approx 0\,, \qquad \bar{\psi}^i \equiv i\bar{p}^i_{\xi} \pm \xi^i \approx 0\,. \tag{44}$$

The constraints (44) $\psi_i \approx 0$, $\bar{\psi}^i \approx 0$ are the pairs of second class constraints, $\{\psi_i, \bar{\psi}^j\} = \mp 2i\delta_i^j$. The kinetic term $\pm\frac{i}{2}(\dot{\bar{\xi}}_i\xi^i - \bar{\xi}_i\dot{\xi}^i)$ in the Lagrangian generates automatically such a structure of the constraints (44) and the self–conjugation of the variables ξ^i, $\bar{\xi}_i$.

Thus, after elimination of the space–time variables we arrive at the system described by the spinor variables λ^i_α, $\bar{\lambda}_{\dot{\alpha}i}$, $\bar{\omega}^\alpha_i$, $\omega^{\dot{\alpha}i}$ and two complex scalars ξ^i, $\bar{\xi}_i$ subjected the constraints

$$h \equiv \lambda^{\alpha i}\lambda_{\alpha i} + 2m \approx 0\,, \qquad \bar{h} \equiv \bar{\lambda}_{\dot{\alpha}i}\bar{\lambda}^{\dot{\alpha}i} + 2m \approx 0\,, \tag{45}$$

$$D_0 = \mathcal{D}_0 \pm 2\bar{\xi}_i\xi^i = i(\bar{\omega}^\alpha_i\lambda^i_\alpha - \bar{\lambda}_{\dot{\alpha}i}\omega^{\dot{\alpha}i}) \pm 2\bar{\xi}_i\xi^i \approx 0\,, \tag{46}$$

$$D_r = \mathcal{D}_r \pm (\sigma_r)_i{}^j\bar{\xi}_j\xi^i = (\sigma_r)_i{}^j\left[\tfrac{i}{2}(\bar{\omega}^\alpha_j\lambda^i_\alpha - \bar{\lambda}_{\dot{\alpha}j}\omega^{\dot{\alpha}i}) \pm \bar{\xi}_j\xi^i\right] \approx 0\,, \tag{47}$$

$$\mathbf{S} \equiv S - s \equiv \pm\bar{\xi}_i\xi^i - s \approx 0\,, \tag{48}$$

$$B_0 = \mathcal{B}_0 = \bar{\omega}^\alpha_i\lambda^i_\alpha + \bar{\lambda}_{\dot{\alpha}i}\omega^{\dot{\alpha}i} \approx 0\,. \tag{49}$$

Note that the constraints (46) and (47) can be presented in the form

$$D_i{}^j \equiv \tfrac{i}{2}\left(\bar{\omega}^\alpha_i\lambda^j_\alpha - \bar{\lambda}_{\dot{\alpha}i}\omega^{\dot{\alpha}j}\right) \pm \bar{\xi}_i\xi^j \approx 0\,. \tag{50}$$

Then $D_0 \approx 0$ is defined as the trace part of $D_i{}^j \approx 0$, $D_0 = D_i{}^i$, whereas $D_r \approx 0$ are traceless parts of $D_i{}^j - \frac{1}{2}\delta_i{}^j D_k{}^k \approx 0$, $D_r = (\sigma_r)_i{}^j D_j{}^i$.

Nonzero brackets of the constraints (45)-(49) are

$$\{h-\bar{h}, D_0\} = 2i(h+\bar{h}) - 8im\,, \quad \{h+\bar{h}, B_0\} = 2(h+\bar{h}) - 8m\,, \quad \{D_r, D_s\} = \varepsilon_{rsp}D_p\,.$$

Therefore the constraint $S - s \approx 0$ and traceless parts $D_r \approx 0$ are first class constraints, whereas pairs of the constraints ($h - \bar{h} \approx 0$, $D_0 \approx 0$) and ($h + \bar{h} \approx 0$, $B_0 \approx 0$) are conjugate second class ones. Thus in the obtained system we have 4 first class constraints and 4 of second class ones. These constraints eliminate 12 degrees of freedom. Since the phase space, which contains λ^i_α, $\omega^{\dot{\alpha}i}$ and ξ^i, is 20–dimensional, the number of the physical degrees of freedom in twistor model with constraints (45)-(49) is 8. This coincides with number of the physical degrees of freedom in space–time formulation of massive spinning particle with Lagrangian (9). Here we have 16 variables in x^μ, p_μ, ζ^α and $p_{\zeta\alpha}$ and 2 first class constraints (spin and mass constraints) and 4 second class constraints (spinor constraints). Thus the number of of physical degrees of freedom is also 8.

4.2. Equations of twistor transformation

Expressions of the canonical transformations obtained in previous section give us directly full set of equations which define the twistor transform.

Completeness conditions $\delta_\alpha^\beta = v_\alpha^i v_i^\beta = m^{-1}\lambda_\alpha^i \lambda_i^\beta$, $\delta_{\dot\beta}^{\dot\alpha} = \bar v_i^{\dot\alpha} \bar v_{\dot\beta}^i = m^{-1}\bar\lambda_i^{\dot\alpha}\bar\lambda_{\dot\beta}^i$ following from kinematic constraints (20) for harmonics v_α^i or from (45) for spinors λ_α^i give following expression of momentum

$$p_{\alpha\dot\alpha} = -\tfrac{1}{m}\lambda_\alpha^i p_i{}^j \bar\lambda_{\dot\alpha j} = -\tfrac{1}{m}\lambda_\alpha^i \left(p_{(0)}\delta_i^j + p_{(r)}(\sigma_r)_i{}^j\right)\bar\lambda_{\dot\alpha j}.$$

Then after gauge fixing (40) $p_{(r)} = 0$, $r = 1,2,3$ (adaptation of harmonic basis to space–time one) and mass–shall condition which are now written in the form (41) $p^{(0)} = -p_{(0)} = \pm m$ we obtain following representation for four–momentum $p_{\alpha\dot\alpha} = \pm\lambda_\alpha^i \bar\lambda_{\dot\alpha i}$. In what follows we will consider the case of positive energy $p^{(0)} = +m$. The consideration of $p^{(0)} = -m$ is analogous. Thus the expression of four–momentum is follows

$$p_{\alpha\dot\alpha} = \lambda_\alpha^i \bar\lambda_{\dot\alpha i}\,. \tag{51}$$

This representation of the momentum as a product of the spinors shows that spinors λ_α^i resolved momentum are twistor variables. Unlike the massless case (7) it is necessary to use two (or more) spinors for the twistor representation of the time–like momentum vector of the massive particle.

The relations (28), (30) give us the expressions for new variables ξ^i, $\bar\xi_i$ in term of the spinning variables ζ^α, $\bar\zeta^{\dot\alpha}$ of the space–time formulation and harmonic spinors λ_α^i, $\bar\lambda_{\dot\alpha i}$

$$\xi^i = \zeta^\alpha \lambda_\alpha^i\,, \qquad \bar\xi_i = \bar\lambda_{\dot\alpha i}\bar\zeta^{\dot\alpha}\,. \tag{52}$$

The variables ξ^i, $\bar\xi_i$ are intended for description of spin degrees of freedom in twistor formulation. In the first place, it follows from the expressions (52) defining the variables ξ^i, $\bar\xi_i$ as the spinning variables ζ^α, $\bar\zeta^{\dot\alpha}$ in harmonic (twistor) basis. In the second, the spin is defined by means the constraint (48) involving the variables ξ^i, $\bar\xi_i$.

After using gauge fixing conditions (40) fixed standard momentum in harmonic basis and constraint equations (41) and (21), the expressions of canonical transformation (31), (32) take the form of the incidence conditions

$$\bar\omega_i^\alpha = \tfrac{1}{2}\bar\lambda_{\dot\alpha i}x^{\dot\alpha\alpha} + i\bar\xi_i\zeta^\alpha\,, \qquad \omega^{\dot\alpha i} = \tfrac{1}{2}x^{\dot\alpha\alpha}\lambda_\alpha^i - i\xi^i\bar\zeta^{\dot\alpha}\,. \tag{53}$$

The expressions (51)-(53) are the equations of twistor transform which define full correspondence between the variables x^μ, p_μ; ζ^α, $\bar\zeta^{\dot\alpha}$ of the space–time formulation and the variables λ_α^i, $\bar\lambda_{\dot\alpha i}$, $\bar\omega_i^\alpha$, $\omega^{\dot\alpha i}$; ξ^i, $\bar\xi_i$ of the twistor formulation in case of the massive spinning particle.

4.3. Formulation of massive spinning particle in twistor formalism

In previous two sections we obtain twistor formulation of massive spinning particle by means of canonical transition from its space–time formulation. Let us summarize details of the model obtained.

The massive spinning particle in the twistor formulation is described by the variables

$$\lambda_\alpha^i\,,\ \bar\lambda_{\dot\alpha i} = \overline{(\lambda_\alpha^i)}\,; \qquad \omega^{\dot\alpha i}\,,\ \bar\omega_i^\alpha = \overline{(\omega^{\dot\alpha i})}\,; \qquad \xi^i\,,\ \bar\xi_i = \overline{(\xi^i)}\,; \qquad i=1,2 \tag{54}$$

for which canonical commutation relations are

$$\{\lambda_\alpha^i, \bar\omega_j^\beta\} = \delta_\alpha^\beta \delta_j^i\,, \qquad \{\bar\lambda_{\dot\alpha i}, \omega^{\dot\beta j}\} = \delta_{\dot\alpha}^{\dot\beta} \delta_i^j\,, \qquad \{\xi^i, \bar\xi_j\} = -\tfrac{i}{2}\delta_j^i\,.$$

The connection of twistor variables (54) with the space–time ones is defined by the conditions of twistor transform

$$p_{\alpha\dot\alpha} = \lambda_\alpha^i \bar\lambda_{\dot\alpha i}\,; \tag{55}$$

$$\bar\omega_i^\alpha = \tfrac{1}{2}\bar\lambda_{\dot\alpha i} x^{\dot\alpha\alpha} + i\bar\xi_i \zeta^\alpha\,, \qquad \omega^{\dot\alpha i} = \tfrac{1}{2} x^{\dot\alpha\alpha}\lambda_\alpha^i - i\xi^i\bar\zeta^{\dot\alpha}\,; \tag{56}$$

$$\xi^i = \zeta^\alpha \lambda_\alpha^i\,, \qquad \bar\xi_i = \bar\lambda_{\dot\alpha i}\bar\zeta^{\dot\alpha}\,. \tag{57}$$

Such form of twistor transformations give us desired form of kinetic terms in twistor variables. Precisely, using (55)-(57) the kinetic terms in twistor variables are $p_\mu \Pi^\mu = -\frac{1}{2} p_{\alpha\dot\alpha}\Pi^{\dot\alpha\alpha} = -\frac{1}{2}(d\bar\omega_i^\alpha\,\lambda_\alpha^i - d\bar\lambda_{\dot\alpha i}\,\omega^{\dot\alpha i} - \bar\omega_i^\alpha\, d\lambda_\alpha^i + \bar\lambda_{\dot\alpha i}\, d\omega^{\dot\alpha i}) + i(d\bar\xi_i\,\xi^i - \bar\xi_i\, d\xi^i)$.

The twistor phase space (54) of massive spinning particle is subjected to the set of the constraints (45)-(49)

$$h \equiv \lambda^{\alpha i}\lambda_{\alpha i} + 2m \approx 0\,, \qquad \bar h \equiv \bar\lambda_{\dot\alpha i}\bar\lambda^{\dot\alpha i} + 2m \approx 0\,, \tag{58}$$

$$D_i{}^j \equiv \tfrac{i}{2}\left(\bar\omega_i^\alpha \lambda_\alpha^j - \bar\lambda_{\dot\alpha i}\omega^{\dot\alpha j}\right) + \bar\xi_i \xi^j \approx 0\,. \tag{59}$$

$$\boldsymbol{S} \equiv S - s \equiv \bar\xi_i \xi^i - s \approx 0\,, \tag{60}$$

$$B_0 = \mathcal{B}_0 = \bar\omega_i^\alpha \lambda_\alpha^i + \bar\lambda_{\dot\alpha i}\omega^{\dot\alpha i} \approx 0\,. \tag{61}$$

In pairs of conjugated second class constraints $(h-\bar h, D_0)$ and $(h+\bar h, B_0)$ we can consider one of the constraints in pairs as gauge fixing condition for other constraint in this pair. The constraint $h+\bar h \approx 0$ containing mass parameter is analog of mass constraint in twistor formulation. Therefore in pair $(h+\bar h,\, B_0)$ of conjugated constraint it is naturally to leave only the constraint $h+\bar h\approx 0$ whereas the constraint $B_0 \approx 0$ will be considered as gauge fixing condition for it. Similarly, upon quantization the constraint $D_0 \approx 0$ defines the spin of the particle as we shall establish below. For this reason we can keep only the constraint $D_0 \approx 0$ from $(h-\bar h,\, D_0)$ and consider the second constraint $h - \bar h \approx 0$ as a corresponding gauge fixing condition. Thus we can consider the system described by only first class constraints

$$M \equiv h + \bar h \equiv \lambda^{\alpha i}\lambda_{\alpha i} + \bar\lambda_{\dot\alpha i}\bar\lambda^{\dot\alpha i} + 4m \approx 0\,, \tag{62}$$

$$D_i{}^j \equiv \tfrac{i}{2}\left(\bar{\omega}_i^{\alpha}\lambda_\alpha^j - \bar{\lambda}_{\dot{\alpha}i}\omega^{\dot{\alpha}j}\right) + \bar{\xi}_i\xi^j \approx 0\,. \qquad (63)$$

$$\boldsymbol{S} \equiv S - s \equiv \bar{\xi}_i\xi^i - s \approx 0\,. \qquad (64)$$

Here we have replaced the constraints D_0, D_r with the constraints $D_i{}^j$ as has been discussed above. Of course the system with the constraints (62)-(64) is equivalent to the system with the constraints (58)-(61). Note that contracting the incidence conditions (56) with $\bar{\lambda}_{\dot{\alpha}i}$ and λ_α^i we get the constraints (63) (or, equivalently, (46) and (47)).

Spinors λ^i and ω^i are components of the twistors Z^i

$$Z_a^i = \left(\lambda_\alpha^i,\, \omega^{\dot{\alpha}i}\right). \qquad (65)$$

Introducing in a standard way twistors conjugated to (65)

$$\bar{Z}_{\dot{a}i} = \overline{(Z_a^i)} = \left(\bar{\lambda}_{\dot{\alpha}i},\, \bar{\omega}_i^{\alpha}\right), \quad \bar{Z}_i^a = g^{a\dot{b}}\bar{Z}_{\dot{b}i} = \left(\bar{\omega}_i^{\alpha},\, -\bar{\lambda}_{\dot{\alpha}i}\right), \qquad (66)$$

where

$$g^{a\dot{b}} = \begin{pmatrix} 0 & \delta^\alpha{}_\beta \\ -\delta_{\dot{\alpha}}{}^{\dot{\beta}} & 0 \end{pmatrix}$$

is metric in twistor space, the kinetic terms are rewritten as

$$p_\mu\Pi^\mu = -\tfrac{1}{2}p_{\alpha\dot{\alpha}}\Pi^{\dot{\alpha}\alpha} = \tfrac{1}{2}(\bar{Z}_i^a\, dZ_a^i - d\bar{Z}_i^a\, Z_a^i) + i(d\bar{\xi}_i\,\xi^i - \bar{\xi}_i\, d\xi^i)\,. \qquad (67)$$

The twistor mass–shell constraints (62) take the form

$$h = Z_a^i I^{ab} Z_{bi} + 2m \approx 0\,, \quad \bar{h} = \bar{Z}_i^a I_{ab}\bar{Z}^{bi} + 2m \approx 0\,, \qquad (68)$$

where so–called asymptotic twistors [1]-[6]

$$I^{ab} = \begin{pmatrix} \varepsilon^{\alpha\beta} & 0 \\ 0 & 0 \end{pmatrix}, \quad I_{ab} = \begin{pmatrix} 0 & 0 \\ 0 & \varepsilon^{\dot{\alpha}\dot{\beta}} \end{pmatrix},$$

are used. Also the constraints (63) are represented as covariant contractions of twistors

$$D_i{}^j = \tfrac{i}{2}\bar{Z}_i^a Z_a^j + \bar{\xi}_i\xi^j \approx 0\,. \qquad (69)$$

Thus the twistor formulation of massive spinning particle is described by Lagrangian

$$L = \tfrac{1}{2}\left(\bar{Z}_i^a\dot{Z}_a^i - \dot{\bar{Z}}_i^a Z_a^i\right) - i\left(\dot{\bar{\xi}}_i\xi^i - \bar{\xi}_i\dot{\xi}^i\right) - \Lambda_m M - \Lambda_j{}^i D_i{}^j - \Lambda_s \boldsymbol{S}\,, \qquad (70)$$

where Λ_m, $\Lambda_j{}^i$, and Λ_s are Lagrange multipliers for constraints (62), (63) and (64).

5. TWISTORIAL FORMULATION OF MASSLESS PARTICLES

In level of zero mass $m \to 0$ the conditions (58) $\lambda^{\alpha 1}\lambda^2_\alpha = 0$ state that spinors λ^i_α are proportional each other, $\lambda^1_\alpha \sim \lambda^2_\alpha$. Therefore we can leave only one spinor λ_α which resolves the light–like vector of four–momentum p_μ. The equation of twistor transformation (55)-(57) for massless particle take the form

$$p_{\alpha\dot\alpha} = \lambda_\alpha \bar\lambda_{\dot\alpha}\,; \tag{71}$$

$$\bar\omega^\alpha = \tfrac{1}{2}\bar\lambda_{\dot\alpha} x^{\dot\alpha\alpha} + i\bar\xi\zeta^\alpha\,, \qquad \omega^{\dot\alpha} = \tfrac{1}{2}x^{\dot\alpha\alpha}\lambda_\alpha - i\xi\bar\zeta^{\dot\alpha}\,; \tag{72}$$

$$\xi = \zeta^\alpha\lambda_\alpha\,, \qquad \bar\xi = \bar\lambda_{\dot\alpha}\bar\zeta^{\dot\alpha}\,. \tag{73}$$

Here λ_α, $\omega^{\dot\alpha}$ and c. c. are spinor components of twistors $Z_a = (\lambda_\alpha\,, \omega^{\dot\alpha})$ and $\bar Z^a = (\bar\omega^\alpha\,, -\bar\lambda_{\dot\alpha})$ as in (65) and (66) whereas commuting complex scalar ξ, $\bar\xi = \overline{(\xi)}$ is corresponding spin variable in twistor formalism.

The phase space of the massless particle is subject to the first class constraints

$$D_0 \equiv i(\bar\omega^\alpha\lambda_\alpha - \bar\lambda_{\dot\alpha}\omega^{\dot\alpha}) + 2\bar\xi\xi = i\bar Z^a Z_a + 2\bar\xi\xi \approx 0\,, \tag{74}$$

$$\boldsymbol{S} \equiv S - s \equiv \bar\xi\xi - s \approx 0\,. \tag{75}$$

These constraints follow from the constraints (62)-(64) in case of using one twistor and zero mass. The constraint (74) is obtained also from definitions of the twistor variables i. e. from the conditions (72) and (73).

Thus in twistor variables the massless particles with nonzero helicity is described by Lagrangian

$$L = \tfrac{1}{2}(\bar Z^a \dot Z_a - \dot{\bar Z}{}^a Z_a) - i(\dot{\bar\xi}\,\xi - \bar\xi\,\dot\xi) - \Lambda D_0 - \Lambda_s \boldsymbol{S}\,. \tag{76}$$

This system is the space–time formulation (9) of the spinning particle reformulated in the twistor approach. Thus after using conditions (72) and (73) kinetic terms of the Lagrangian (9) transform in kinetic terms of the Lagrangian (76) $p_\mu \Pi^\mu = -\frac{1}{2}p_{\alpha\dot\alpha}\Pi^{\dot\alpha\alpha} = \frac{1}{2}(\bar Z^a\, dZ_a - d\bar Z^a Z_a) + i(d\bar\xi\,\xi - \bar\xi\, d\xi)$. Resolving the condition $p_{\alpha\dot\alpha} = \lambda_\alpha\bar\lambda_{\dot\alpha}$ and using the definitions (73) of scalars $\xi, \bar\xi$, the spin constraint $\zeta\hat p\bar\zeta - s \approx 0$ of the Lagrangian (9) takes the form (75) in twistor variables.

We can exclude the variables ξ and $\bar\xi$ with the help of first class constraint (75) $\bar\xi\xi - s \approx 0$. After that massless particle is described by Lagrangian (5)

$$L = \tfrac{1}{2}(\dot{\bar Z}{}^a Z_a - \bar Z^a \dot Z_a) - \Lambda(\tfrac{i}{2}\bar Z^a Z_a + s)\,. \tag{77}$$

Due to the spin constraint $\frac{i}{2}\bar Z^a Z_a + s \approx 0$ in Lagrangian (5) the twistor Z_a is nonisotropic (even on classical level) and describes massless particle with nonzero helicity. We obtain the twistor formulation with the nonisotropic twistor upon transition from the space–time formulation because of the presence of the second terms in the incidence conditions (72). The presence in the incidence conditions, written in real (not complex!) space–time, of the second terms for spinning case is well known [4]. Real rays which correspond to the twistor $Z = (\lambda, \omega)$ with the incidence conditions (72) (nonisotropic twistor) form the Robinson congruence.

Note that the conditions (72) for definition of spinors ω can be represented in form $\bar{\omega}^{\alpha} - i\bar{\xi}\zeta^{\alpha} = \frac{1}{2}\bar{\lambda}_{\dot{\alpha}}x^{\dot{\alpha}\alpha}$, $\omega^{\dot{\alpha}} + i\xi\bar{\zeta}^{\dot{\alpha}} = \frac{1}{2}x^{\dot{\alpha}\alpha}\lambda_{\alpha}$. Such forms show us that the transition from states with zero helicity to ones with nonzero helicity is realized by twistorial shift of spinor ω

$$\omega^{\dot{\alpha}} \to \omega^{\dot{\alpha}} - i\xi\bar{\zeta}^{\dot{\alpha}}, \qquad \bar{\omega}^{\alpha} \to \bar{\omega}^{\alpha} + i\bar{\xi}\zeta^{\alpha} \tag{78}$$

along spinor ζ. It is obvious since ζ and λ are orthogonal, $\zeta\lambda \neq 0$, and therefore we obtain variation of helicity

$$\tfrac{i}{2}(\bar{\lambda}_{\dot{\alpha}}\omega^{\dot{\alpha}} - \bar{\omega}^{\alpha}\lambda_{\alpha}) \to \tfrac{i}{2}(\bar{\lambda}_{\dot{\alpha}}\omega^{\dot{\alpha}} - \bar{\omega}^{\alpha}\lambda_{\alpha}) + \bar{\xi}\xi = \tfrac{i}{2}(\bar{\lambda}_{\dot{\alpha}}\omega^{\dot{\alpha}} - \bar{\omega}^{\alpha}\lambda_{\alpha}) + s\,.$$

That twistorial shift (78) is distinguished from twistor shift [19]

$$\omega^{\dot{\alpha}} \to \omega^{\dot{\alpha}} + l\bar{\lambda}^{\dot{\alpha}}, \qquad \bar{\omega}^{\alpha} \to \bar{\omega}^{\alpha} + l\lambda^{\alpha} \tag{79}$$

along spinor λ where l is a length constant. The shift (79) results in a modification of particle (or string) interactions with background fields [19] and does not produce any change of helicity since under this shift $\bar{\lambda}_{\dot{\alpha}}\omega^{\dot{\alpha}} - \bar{\omega}^{\alpha}\lambda_{\alpha} = inv$. Let us note that twistor shift similar to (79) has been used in [39] for description of AdS_5 particle.

The variables of the twistorial formulation (76), coordinates of twistor $Z_a = (\lambda_{\alpha}, \omega^{\dot{\alpha}})$ and complex scalar ξ, may be combined in quantity

$$\mathcal{Z}_{\mathcal{A}} = (Z_a;\xi) = (\lambda_{\alpha}, \omega^{\dot{\alpha}};\xi)\,, \tag{80}$$

which has five of complex components and can be called as 'bosonic supertwistor'. Introducing conjugate quantities $\bar{\mathcal{Z}}_{\dot{\mathcal{A}}} = (\bar{Z}_{\dot{a}};\bar{\xi}) = (\bar{\lambda}_{\dot{\alpha}}, \bar{\omega}^{\alpha};\bar{\xi})$, $\bar{\mathcal{Z}}^{\mathcal{A}} = g^{\mathcal{A}\dot{\mathcal{B}}}\bar{\mathcal{Z}}_{\dot{\mathcal{B}}} = (\bar{Z}^a;-2i\bar{\xi}) = (\bar{\omega}^{\alpha}, -\bar{\lambda}_{\dot{\alpha}};-2i\bar{\xi})$ where $g^{\mathcal{A}\dot{\mathcal{B}}} = \begin{pmatrix} g^{a\dot{b}} & 0 \\ 0 & -2i \end{pmatrix}$ we see that Lagrangian (76) without last term $-\Lambda_s(\bar{\xi}\xi - j)$ takes the form

$$\tfrac{1}{2}(\bar{\mathcal{Z}}^{\mathcal{A}}\dot{\mathcal{Z}}_{\mathcal{A}} - \dot{\bar{\mathcal{Z}}}^{\mathcal{A}}\mathcal{Z}_{\mathcal{A}}) - \Lambda\bar{\mathcal{Z}}^{\mathcal{A}}\mathcal{Z}_{\mathcal{A}}\,. \tag{81}$$

In the variables α, β, γ, δ introduced by $\lambda_1 = i(\alpha+\beta)$, $\lambda_2 = i(\gamma+\delta)$, $\omega^1 = \alpha - \beta$, $\omega^2 = \gamma - \delta$ the norm of 'bosonic supertwistor' $\frac{i}{2}\bar{\mathcal{Z}}^{\mathcal{A}}\mathcal{Z}_{\mathcal{A}} = \frac{i}{2}(\bar{\omega}^{\alpha}\lambda_{\alpha} - \bar{\lambda}_{\dot{\alpha}}\omega^{\dot{\alpha}}) + \bar{\xi}\xi$ takes the form

$$\tfrac{i}{2}\bar{\mathcal{Z}}^{\mathcal{A}}\mathcal{Z}_{\mathcal{A}} = \bar{\beta}\beta + \bar{\delta}\delta + \bar{\xi}\xi - \bar{\alpha}\alpha - \bar{\gamma}\gamma$$

and is quadratic Hermitian form of signature $(+++--)$. Thus the Lagrangian (81) is invariant under global $U(3,2)$ transformations. In the infinitesimal form these transformations are

$$\begin{pmatrix} \lambda_{\alpha} \\ \omega^{\dot{\alpha}} \\ \xi \end{pmatrix} = \begin{pmatrix} -l_{\alpha}{}^{\beta} - \frac{1}{2}(d - i\phi)\delta_{\alpha}{}^{\beta} & -2k_{\alpha\dot{\beta}} & 4i\psi_{\alpha} \\ \frac{1}{2}a^{\dot{\alpha}\beta} & \bar{l}^{\dot{\alpha}}{}_{\dot{\beta}} + \frac{1}{2}(d + i\phi)\delta^{\dot{\alpha}}{}_{\dot{\beta}} & -2i\bar{\varepsilon}^{\dot{\alpha}} \\ \varepsilon^{\beta} & 2\bar{\psi}_{\dot{\beta}} & -2i\phi \end{pmatrix} \begin{pmatrix} \lambda_{\beta} \\ \omega^{\dot{\beta}} \\ \xi \end{pmatrix}.$$

Using the conditions (72), (73) and (71), these transformations reduce to the transformations (10)-(15). Therefore the 'bosonic supersymmetric' transformations and 'bosonic superboosts' in the model of massless spinning particle are the parts of $U(3,2)$ group which play 'bosonic superconformal' group in twistor formalism.

6. TWISTOR TRANSFORM FOR MASSIVE FIELDS

Quantization of massive spinning particle in twistor formulation has been made in [26], [27]. After quantization twistor massive particle of spin J is described by the $(2J+1)$–component wave function

$$\Psi_M(\lambda,\bar{\lambda}), \qquad M=-J,-J+1,...,J-1,J$$

depending on the twistor spinors λ_α^i, $\bar{\lambda}_{\dot{\alpha}i}$. The numbers J and M are defined by $J=\frac{1}{2}(n_1+n_2)$ and $M=\frac{1}{2}(n_1-n_2)$ where non–negative integer numbers n_1 and n_2 are filling numbers of two–dimensional oscillator formed by the operators ξ^i, $\bar{\xi}_i$. The spin constraint (64) $\bar{\xi}_i\xi^i - J \approx 0$ fixed the value of J and then the number M takes $(2J+1)$ values $M=-J,-J+1,...,J-1,J$. In this constraint the constant J is classical constant s in (64) renormalized by ordering ambiguity constants. The wave function $\Psi_M(\lambda,\bar{\lambda})$ is subjected the first class constraints which take the form

$$\mathcal{D}_3\Psi_M = -M\Psi_M\,, \qquad \mathcal{D}_\pm\Psi_M = -\sqrt{(J\mp M)(J\pm M+1)}\Psi_{M\pm 1}\,, \tag{82}$$

where

$$\mathcal{D}_r \equiv \tfrac{1}{2}(\sigma_r)_i{}^j\left[\lambda_\alpha^i\frac{\partial}{\partial\lambda_\alpha^j} - \frac{\partial}{\partial\bar{\lambda}_{\dot{\alpha}i}}\bar{\lambda}_{\dot{\alpha}j}\right], \tag{83}$$

and $\mathcal{D}_\pm = \mathcal{D}_1 \pm i\mathcal{D}_2$. All $(2J+1)$ components of the wave function are obtained from one component, for example from component of highest weight Ψ_{+J} or lowest one Ψ_{-J}

$$\Psi_M = \sqrt{\frac{(J\pm M)!}{(J\mp M)!(2J)!}}(-1)^M(\mathcal{D}_\mp)^{J\mp M}\Psi_{\pm J}\,.$$

These components $\Psi_{\pm J}$ are defined by equations

$$\mathcal{D}_3\Psi_{\pm J} = \pm J\Psi_{\pm J}\,, \quad \mathcal{D}_\pm\Psi_{\pm J} = 0\,, \quad (\mathcal{D}_\mp)^{2J+1}\Psi_{\pm J} = 0\,. \tag{84}$$

The operators $\mathcal{D}_r$ are generators of $SU(2)$–transformations, acting on indices $i,j,k,...$ of $SL(2,C)$–matrix λ_α^i. The constraints (82) state that the wave function $\Psi_M(\lambda,\bar{\lambda})$ is defined up to local transformations acting on index M

$$\Psi'_M(\lambda') = \mathbf{D}^J_{MN}(h)\Psi_N(\lambda)\,, \tag{85}$$

where $h\in SU(2)$ and $\lambda_\alpha^{\prime i} = h^i_j\lambda_\alpha^j$. The $\mathbf{D}^J_{MN}$ is matrix of $SU(2)$–transformations of weight J. Thus the wave function is defined in fact on homogeneous space $\mathcal{M}=G/H=SL(2,C)/SU(2)$. In form of $SU(2)$-indices $i,j,...=1,2$ the index M is collective index $M=(i_1\dots i_{2J})$. Then the wave function (twistor field of massive spinning particle) is

$$\Psi_M(\lambda,\bar{\lambda}) = \Psi_{i_1...i_{2J}}(\lambda,\bar{\lambda})\,. \tag{86}$$

The wave function is completely symmetric $\Psi_{i_1...i_{2J}} = \Psi_{(i_1...i_{2J})}$.

The relation of the usual space–time spin–tensor fields $\Phi_{\alpha_1\ldots\alpha_{2J}}(x)$ with the twistor fields (86) is established by means of an integral transformation in the following way. One constructs $SU(2)$-invariant expressions contracting the twistor fields $\Psi_{i_1\ldots i_{2J}}(\lambda,\bar{\lambda})$ with twistor spinors $\lambda^{i_1}_{\alpha_1},\ldots,\lambda^{i_{2J}}_{\alpha_{2J}}$. Obtained expressions being Lorentz spin–tensors are defined on a homogeneous space $SL(2,C)/SU(2)$. After integration with an invariant measure $d^3\lambda$ of space $SL(2,C)/SU(2)$ with the standard Fourier exponent $e^{ix^\mu p_\mu}$ where $p_\mu = -\frac{1}{2}p_{\alpha\dot{\alpha}}\sigma^{\dot{\alpha}\alpha}_\mu = -\frac{1}{2}\lambda^i\sigma_\mu\bar{\lambda}_i$ we obtain usual space–time fields[1]

$$\Phi_{\alpha_1\ldots\alpha_{2J}}(x) = \int d^3\lambda e^{-\frac{i}{2}x^\mu\lambda^k\sigma_\mu\bar{\lambda}_k}\lambda^{i_1}_{\alpha_1}\ldots\lambda^{i_{2J}}_{\alpha_{2J}}\Psi_{i_1\ldots i_{2J}}(\lambda,\bar{\lambda}). \tag{87}$$

These fields are totally symmetric in spinor indices $\Phi_{\alpha_1\ldots\alpha_{2J}}(x) = \Phi_{(\alpha_1\ldots\alpha_{2J})}(x)$ and give us standard $(2J+1)$-component field description [42] of massive spin J. Due to the presence of the exponent in the integral representation for the fields (87) they satisfy automatically massive Klein–Gordon equation

$$(\partial^\mu\partial_\mu - m^2)\Phi_{\alpha_1\ldots\alpha_{2J}}(x) = 0. \tag{88}$$

We can obtain different description of free massive particles of spin J. The $SU(2)$-invariants can be constructed by contraction the twistor fields $\Psi_{i_1\ldots i_{2J}}(\lambda,\bar{\lambda})$ only with twistor spinors $\bar{\lambda}^i_{\dot{\alpha}}$. In this way we obtain the space–time fields

$$\Phi^{\dot{\alpha}_1\ldots\dot{\alpha}_{2J}}(x) = \int d^3\lambda e^{-\frac{i}{2}x^\mu\lambda^k\sigma_\mu\bar{\lambda}_k}\bar{\lambda}^{\dot{\alpha}_1 i_1}\ldots\bar{\lambda}^{\dot{\alpha}_{2J} i_{2J}}\Psi_{i_1\ldots i_{2J}}(\lambda,\bar{\lambda}) \tag{89}$$

only with primed spinor indices. These fields satisfy automatically massive Klein–Gordon equation

$$(\partial^\mu\partial_\mu - m^2)\Phi^{\dot{\alpha}_1\ldots\dot{\alpha}_{2J}}(x) = 0, \tag{90}$$

also and give us $(2J+1)$-component field description [42] of massive spin J in term of the primed spinors.

The fields (87)-(89) taken together give us $2(2J+1)$–component field description [42] of massive spin J. Using explicit expressions (87)-(89) for them we can easily show that they are related by equations

$$(i\partial_{\mu_1}\sigma^{\mu_1}_{\alpha_1\dot{\beta}_1})\cdots(i\partial_{\mu_{2J}}\sigma^{\mu_{2J}}_{\alpha_{2J}\dot{\beta}_{2J}})\Phi^{\dot{\beta}_1\ldots\dot{\beta}_{2J}} = m^{2J}\Phi_{\alpha_1\ldots\alpha_{2J}}, \tag{91}$$

$$(i\partial^{\mu_1}\sigma^{\dot{\alpha}_1\beta_1}_{\mu_1})\cdots(i\partial^{\mu_{2J}}\sigma^{\dot{\alpha}_{2J}\beta_{2J}}_{\mu_{2J}})\Phi_{\beta_1\ldots\beta_{2J}} = m^{2J}\Phi^{\dot{\alpha}_1\ldots\dot{\alpha}_{2J}}. \tag{92}$$

When constructing the $SU(2)$-invariants, the part of the $SU(2)$-indices in the twistor field $\Psi_{i_1\ldots i_{2J}}(\lambda,\bar{\lambda})$ can be contracted with spinors λ^i_α whereas the remaining ones are

[1] Analogous integral transformations for massless twistor fields have been considered in [37], [38], [40], [41]

contracted with $\bar{\lambda}^{i}_{\dot{\alpha}}$. As result we obtain the spin–tensor field

$$\begin{aligned}\Phi_{\alpha_1\dots\alpha_n}{}^{\dot{\alpha}_{n+1}\dots\dot{\alpha}_{2J}}(x) &= \int d^3\lambda e^{-\frac{i}{2}x^{\mu}\lambda^{k}\sigma_{\mu}\bar{\lambda}_{k}} \times \\ &\times \lambda^{i_1}_{\alpha_1}\dots\lambda^{i_n}_{\alpha_n}\bar{\lambda}^{\dot{\alpha}_{n+1}i_{n+1}}\dots\bar{\lambda}^{\dot{\alpha}_{2J}i_{2J}}\Psi_{i_1\dots i_{2J}}(\lambda,\bar{\lambda}), \quad (93)\end{aligned}$$

with n unprimed spinor indices and $2J-n$ primed ones. By construction the field (93) is symmetric with respect all unprimed and all primed indices, $\Phi_{\alpha_1\dots\alpha_n}{}^{\dot{\alpha}_{n+1}\dots\dot{\alpha}_{2J}} = \Phi_{(\alpha_1\dots\alpha_n)}{}^{(\dot{\alpha}_{n+1}\dots\dot{\alpha}_{2J})}$. Due to the presence of the exponent in the integral representation for the fields (93) they satisfy also massive Klein–Gordon equation

$$(\partial^{\mu}\partial_{\mu}-m^2)\Phi_{\alpha_1\dots\alpha_n}{}^{\dot{\alpha}_{n+1}\dots\dot{\alpha}_{2J}}(x)=0. \quad (94)$$

Using explicit expression (93) we can easily show that the fields (93) satisfy the condition

$$\partial^{\mu}\sigma_{\mu}{}^{\alpha_1}_{\dot{\alpha}_{n+1}}\Phi_{\alpha_1\dots\alpha_n}{}^{\dot{\alpha}_{n+1}\dots\dot{\alpha}_{2J}}(x)=0. \quad (95)$$

The fields (93) at fixed J, describing the massive spin J particle, are related to each other by Fierz–Pauli equations [43]

$$i\partial^{\mu}\sigma_{\mu}^{\dot{\beta}\alpha_n}\Phi_{\alpha_1\dots\alpha_n}{}^{\dot{\alpha}_1\dots\dot{\alpha}_{2J-n}} = m\Phi_{\alpha_1\dots\alpha_{n-1}}{}^{\dot{\alpha}_1\dots\dot{\alpha}_{2J-n}\dot{\beta}}, \quad (96)$$

$$i\partial^{\mu}\sigma_{\mu\beta\dot{\alpha}_n}\Phi_{\alpha_1\dots\alpha_{2J-n}}{}^{\dot{\alpha}_1\dots\dot{\alpha}_n} = m\Phi_{\alpha_1\dots\alpha_{2J-n}\beta}{}^{\dot{\alpha}_1\dots\dot{\alpha}_{n-1}}. \quad (97)$$

We can combine all the fields with fixed J (93) $\Phi_{\alpha_1\alpha_2\dots\alpha_{2J-1}\alpha_{2J}}$, $\Phi_{\alpha_1\alpha_2\dots\alpha_{2J-1}}{}^{\dot{\alpha}_{2J}}$, …, $\Phi_{\alpha_1}{}^{\dot{\alpha}_2\dots\dot{\alpha}_{2J-1}\dot{\alpha}_{2J}}$, $\Phi^{\dot{\alpha}_1\dot{\alpha}_2\dots\dot{\alpha}_{2J-1}\dot{\alpha}_{2J}}$ in a unique field $\Phi_{a_1\dots a_{2J}}$ in which Dirac indices $a=1,2,3,4$ are collective indices for lower undotted Weyl indices $\alpha=1,2$ and upper dotted ones $\dot{\alpha}=1,2$, i. e.

$$\Phi_{a_1a_2\dots a_{2J}} = \begin{pmatrix}\Phi_{\alpha_1a_2\dots a_{2J}}\\ \Phi^{\dot{\alpha}_1}{}_{a_2\dots a_{2J}}\end{pmatrix} \quad (98)$$

and similarly for the other indices $a_2,\dots,a_{2J}$. This field is total symmetric in all Dirac indices, $\Phi_{a_1\dots a_{2J}} = \Phi_{(a_1a_2\dots a_{2J})}$ and satisfies the equation

$$i\partial^{\mu}\gamma_{\mu a_1}{}^{b_1}\Phi_{b_1a_2\dots a_{2J}} = m\Phi_{a_1a_2\dots a_{2J}},$$

where $\gamma_{\mu} = \begin{pmatrix}0 & \sigma_{\mu}^{\dot{\alpha}\beta}\\ \sigma_{\mu\alpha\dot{\beta}} & 0\end{pmatrix}$ are Dirac matrices in Weyl representation. Thus the field (98) gives us Bargman–Wigner formalism [44] for description of massive spin J particle.

Using Pauli σ–matrices, undotted index $\alpha=1,2$ and dotted one $\dot{\alpha}=1,2$ are united in 4–component vector index μ. Therefore for half integer J the fields (93) can be cast into the fields

$$\Phi_{\mu_1\dots\mu_{[J]}\alpha} = \sigma_{\mu_1}{}^{\alpha_1}_{\dot{\alpha}_1}\dots\sigma_{\mu_{[J]}}{}^{\alpha_{[J]}}_{\dot{\alpha}_{[J]}}\Phi_{\alpha_1\dots\alpha_{[J]}}{}^{\dot{\alpha}_1\dots\dot{\alpha}_{[J]}\alpha}, \qquad \Phi_{\mu_1\dots\mu_{[J]}}{}^{\dot{\alpha}} = \sigma_{\mu_1}{}^{\alpha_1}_{\dot{\alpha}_1}\dots\sigma_{\mu_{[J]}}{}^{\alpha_{[J]}}_{\dot{\alpha}_{[J]}}\Phi_{\alpha_1\dots\alpha_{[J]}}{}^{\dot{\alpha}_1\dots\dot{\alpha}_{[J]}\dot{\alpha}}$$

where $[J] = J - \frac{1}{2}$ is integer part of J. These fields form a spin–tensor field

$$\Phi_{\mu_1 \dots \mu_{[J]}\, a} = \begin{pmatrix} \Phi_{\mu_1 \dots \mu_{[J]}\, \alpha} \\ \Phi_{\mu_1 \dots \mu_{[J]}}{}^{\dot{\alpha}} \end{pmatrix}, \qquad (99)$$

where $a = 1,2,3,4$ is Dirac index. Since the fields (93) are symmetric with respect the spinor indices, the field (99) is symmetric with respect all vector indices $\Phi_{\mu_1 \dots \mu_{[J]}\, a} = \Phi_{(\mu_1 \dots \mu_{[J]})\, a}$. Due to (95)-(97) and symmetry of the fields (93) with respect spinor indices the field (99) satisfies the equations

$$\partial^{\mu} \gamma_{\mu\, a}{}^{b} \Phi_{\mu_1 \dots \mu_{[J]}\, b} = m \Phi_{\mu_1 \dots \mu_{[J]}\, a}, \qquad \gamma^{\mu_1}{}_{a}{}^{b} \Phi_{\mu_1 \dots \mu_{[J]}\, b} = 0, \qquad \partial^{\mu_1} \Phi_{\mu_1 \dots \mu_{[J]}\, a} = 0.$$

Therefore the field (99) describes the spin J particle in Rarita–Schwinger formalism [45].

7. CONCLUSION

In this paper it is presented the twistor formulation of the particle of arbitrary spin. This formulation is obtained from the space–time formulation of the spinning particle enlarged with pure gauge harmonic variables. In the massive case, upon partial fixing of gauges we have obtained the model described with two twistors (bitwistor) and two complex scalars. As result of a canonical transformation we have obtained the conditions of the twistor transform. It is presented also the twistor transform relating the twistor formulation and the space–time one for the massless particle of nonzero helicity. Twistor formulation of massless particle is presented in term of one twistor and one complex scalar. It is obtained from twistor formulation of the massive particle in level of zero mass. Without term with spin constraint the model describes the infinite tower of states with all arbitrary spins (helicities). In massless case this system is invariant under $U(3,2)$ group which is 'even' analog superconformal group. Integral transformations (field twistor transform) relating the massive twistor fields with usual space–time spin–tensorial fields have been constructed.

ACKNOWLEDGMENTS

It is a pleasure to thank the organizers of the XIXth Max Born symposium "Fundamental Interactions and Twistor–like Methods" in Wrocław, 28.09-1.10.2004, for the kind invitation and the pleasant and stimulating atmosphere. The author are grateful to I.A. Bandos, E.A. Ivanov, S.O. Krivonos, J. Lukierski, and M.A. Vasiliev for interest to this work and useful discussions, and especially to A.Yu. Nurmagambetov and D.P. Sorokin for reading the manuscript and valuable comments which helped the authors to make some points more transparent. Thanks go to E.A. Ivanov for drawing my attention to the article [35]. This work was partially supported by INTAS Grant INTAS-2000-254 and by Ukrainian National Found of Fundamental Researches under the Project N 02.07/383.

REFERENCES

1. R. Penrose, J. Math. Phys. **8**, 345 (1967).
2. R. Penrose and M.A.H. MacCallum, Phys. Reports **6C**, 241 (1972).
3. R. Penrose, Rep. Math. Phys. **12**, 65 (1977).
4. R. Penrose and W. Rindler, Spinors and Space-Time, 1986 (Cambridge: Cambridge University Press)
5. L.P. Hughston, Twistor and Particles, Lecture Notes in Physics, Berlin, v. **97**, 1979, 153 pp.
6. S.A. Hugget and K.P. Tod, An Introduction to the Twistor Theory, Cambrige University Press, 1994, 178 pp.
7. A. Ferber, Nucl. Phys. **B132**, 55 (1978).
8. T. Shirafuji, Prog. Theor. Phys. **70**, 18 (1983).
9. E. Witten, Nucl. Phys. **B 266**, 245 (1986).
10. A.K.H. Bengtsson, I. Bengtsson, M. Cederwall and N. Linden, Phys. Rev. **D36**, 1766 (1987).
11. I. Bengtsson and M. Cederwall, Nucl. Phys. **B302**, 81 (1988).
12. Y. Eisenberg and S. Solomon, Nucl. Phys. **B309**, 709 (1988); Phys. Lett. **B220**, 562 (1989).
13. Y. Eisenberg, Phys. Lett. **B225**, 95 (1989).
14. M.S. Plyshchay, Mod. Phys. Lett. **A4**, 1827 (1989).
15. D.P. Sorokin, V.I. Tkach and D.V. Volkov, Mod. Phys. Lett. **A4**, 901 (1989).
16. D.P. Sorokin, V.I. Tkach, D.V. Volkov and A.A. Zheltukhin, Phys. Lett. **B216**, 302 (1989).
17. D.V. Volkov and A.A. Zheltukhin, Lett. Math. Phys. **17**, 141 (1989); Nucl. Phys. **B335**, 723 (1990).
18. D.P. Sorokin, Fortschr. Phys. **38**, 923 (1990);
    V.I. Gumenchuk and D.P. Sorokin, Sov. J. Nucl. Phys. **51**, 350 (1990).
19. V.A. Soroka, V.I. Tkach, D.V. Volkov and D.P. Sorokin, Int. J. Mod. Phys. **A7**, 5977 (1992); JETP Lett. **52**, 526 (1990).
20. P. Townsend, Phys. Lett. **B261**, 65 (1991).
21. D.P. Sorokin, Phys. Rept. **329**, 1 (2000) (and refs. therein)
22. Z. Perjés, Phys. Rev. **D11**, 2031 (1975); Reports Math. Phys. **12**, 193 (1977); Phys. Rev. **D20**, 1857 (1979).
23. P. Tod, Reports Math. Phys. **11**, 339 (1977).
24. A. Bette, J. Math. Phys. **25**(1984)2456; **37**, 1724 (1996); hep-th/0402150
25. S. Fedoruk and V.G. Zima, Nucl. Phys. (Proc. Suppl.) **B102 & 103**, 233 (2001).
26. S. Fedoruk and V.G. Zima, J. Kharkiv University **585**, 39 (2003). (hep-th/0308154)
27. S. Fedoruk and V.G. Zima, Bitwistor formulation of spinning particle, Intern. Workshop "Supersymmetry and Quantum Symmetry", Dubna, 24-29 July 3003 (hep-th/0401064)
28. A. Bette, A. de Azcarraga, J. Lukierski and C. Miquel-Espanya, Phys.Lett. **B595**, 491 (2004).
29. N. Berkovits and L. Motl, JHEP **0404**, 056 (2004);
    N. Berkovits and E. Witten, Conformal supergravity in twistor-string theory, hep-th/0406051;
    W. Siegel, Untwisting the twistor superstring, hep-th/0404255
30. V.G. Zima and S. Fedoruk, JETP Lett. **61**, 251 (1995).
31. V.G. Zima and S. Fedoruk, Class. Quantum Grav. **16**, 3653 (1999).
32. S. Fedoruk and V.G. Zima, Mod. Phys. Lett. **A15**, 2281 (2000).
33. J. Wess and J. Bagger, Supersymmetry and Supergravity, 1983 (Princeton: Princeton University Press)
34. M.A. Vasiliev, Phys. Rev. **D66**, 066006 (2002).
35. C. Devchand and O. Lechtenfeld, Nucl. Phys. **B516**, 255 (1998).
36. A. Galperin, E. Ivanov, E. Kalizin, V. Ogievetsky and E. Sokatchev, Class. Quantum Grav. **1**, 469 (1984); **2**, 155 (1985).
37. I.A. Bandos, Sov. J. Nucl. Phys. **51**, 906 (1990);
    I.A. Bandos and A.A. Zheltukhin, Class. Quantum Grav. **12**, 609 (1995).
38. V.G. Zima and S. Fedoruk, Theor. Math. Phys. **102**, 305 (1995).
39. P. Claus, J. Rahmfeld and Y. Zunger, Phys. Lett. **B466**, 181 (1999);
    P. Claus, R. Kallosh and J. Rahmfeld, Phys. Lett. **B462**, 285 (1999).
40. I. Bandos, J. Lukierski and D. Sorokin, Phys. Rev. **D61**, 040002 (2000).
41. M. Plyushchay, D. Sorokin and M. Tsulaia, JHEP **0304**, 013 (2003);
    Higher Spin from Tensorial Charges and $OSp(N|2n)$ *Symmetry*, hep-th/0301067.
42. S. Weinberg, Phys. Rev. **133**, B1318 (1964).

43. M. Fierz and W. Pauli, Proc. Roy. Soc. (London) **A173**, 211(1939).
44. V. Bargman and E. Wigner, Proc. Am. Acad. Sci. **34**, 211 (1948).
45. W. Rarita and J. Schwinger, Phys. Rev. **60**, 61 (1941).

Complex Minkowski Space as a Conformal Phase Space

G. Jakimowicz, A. Odzijewicz

Institute of Mathematics
University in Białystok
ul. Lipowa 41, PL-15-424 Białystok, Poland
e-mail: gjakim@alpha.uwb.edu.pl, aodzijew@labfiz.uwb.edu.pl

Abstract. The complex Minkowski phase space has the physical interpretation as the phase space of the scalar massive conformal particles. The aim of this presentation is the construction and investigation of the quantum complex Minkowski space.

1. INTRODUCTION

The twistor theory of Penrose provides the natural and effective formalism for the description of massless particles. This point of view can be extended to the case of massive elementary particles which opposite to the relativistic ones are allowed to change their mass during the evolution. According to [5] we shall call them conformal particles.

The conformally compactified complex Minkowski space $\mathbb{M}$ is defined as Grassmannian of 2-dimensional subspaces of the twistor space $\mathbf{T}$ and thus possesses canonically defined conformal structure invariant with respect to the group $SU(2,2)$. In [5, 6] it was shown that $\mathbb{M}$ can be decomposed into conformal compactification of real Minkowski space and the classical phase spaces of the scalar massless or massive conformal particles, antiparticles and tachyons.

The purpose of this presentation is the description of quantum version $\mathcal{M}^{++}$ of the phase space $\mathbb{M}^{++} \subset \mathbb{M}$ of the massive scalar conformal particle, which was introduced and studied in [2]. For this reason we will apply the method based on the notion of the coherent state map [9]. In our case it will be the map $\mathcal{K}_\lambda$ of $\mathbb{M}^{++}$ in the quantum phase space, i.e. in the complex projective Hilbert space $\mathbb{CP}(\mathcal{H})$, see (3.7). In the Section 3 we introduce the quantum Kähler polarization as the Banach algebra $\mathcal{P}^{++}$ of annihilation operators and quantum phase space $\mathcal{M}^{++}$ of the conformal scalar massive particle both defined by the coherent state map $\mathcal{K}_\lambda$. We show a few properties of $\mathcal{M}^{++}$, see Proposition 3.1, and present the physical interpretation.

CP767, *Fundamental Interactions and Twistor-Like Methods,* edited by J. Lukierski and D. Sorokin

2. PHASE SPACES FOR THE CONFORMAL SCALAR MASSIVE PARTICLES

Our starting point is the twistor description of Minkowski space-time. Let us recall that the twistors space $\mathbf{T}$ is $\mathbb{C}^4$ equipped with a Hermitian form η of signature $(++--)$. Hence, the conformal group $SU(2,2)$, realized by linear maps $g:\mathbb{C}^4\to\mathbb{C}^4$ satisfying the conditions $g^\dagger\eta g=\eta$ and $\det g=1$, is the group of automorphisms of the twistors space $\mathbf{T}$. One defines the compactified complex Minkowski space $\mathbb{M}$ as the Grassmannian of 2-dimensional complex vector subspaces $z\subset\mathbf{T}$ of the twistor space. The action

$$\sigma_g^{\mathbb{M}}: z\longrightarrow gz \tag{2.1}$$

of $SU(2,2)$ splits $\mathbb{M}$ into the orbits $\mathbb{M}^{kl}$ indexed by the signatures $sign\,\eta|_z=(k,l)$, where $k,l=+,-,0$, of the restricted Hermitian forms $\eta|_z$. The orbit $\mathbb{M}^{00}$ of isotropic subspaces is the conformally compactified Minkowski space and $\mathbb{M}$ is the complexification of $\mathbb{M}^{00}$ in the sense of the definition given in [10]. Fixing an element $\infty\in\mathbb{M}^{00}$, called point at infinity, one defines the Minkowski space $\mathbb{M}^{00}_\infty$ as the affine space of elements $z\in\mathbb{M}^{00}$ which are transversal to ∞, i.e. $z\oplus\infty=\mathbf{T}$. The elements $z\in\mathbb{M}^{00}$ which intersect with ∞ in more than one-dimension, i.e. $\dim_{\mathbb{C}}(z\cap\infty)\geq 1$, form a cone C^∞ at infinity, so

$$\mathbb{M}^{00}=\mathbb{M}^{00}_\infty\cup C^\infty\cong\mathbb{S}^1\times\mathbb{S}^3.$$

The cones $C_z=\{z'\in\mathbb{M}^{00}:\dim_{\mathbb{C}}(z\cap z')\geqq 1\}$ define a conformal structure on $\mathbb{M}^{00}$ invariant with respect to the conformal group action given by (2.1). The Poincaré group $\mathbf{P}_\infty$ extended by the dilatations is defined as the stabilizer $SU(2,2)_\infty$ of the element ∞. The intersections of the stabilizers $SU(2,2)_\infty\cap SU(2,2)_0$, where $0\in\mathbb{M}^{00}_\infty$ is the origin of the inertial coordinates system, is the Lorentz group extended by dilatations. One defines the Lorentz group $\mathbf{L}_{0,\infty}$ and the dilatations group $\mathbf{D}_{0,\infty}$ as commutator and centralizer of $SU(2,2)_\infty\cap\mathbf{S}U(2,2)_0$ respectively. Finally, the group of Minkowski space translations consists of the elements $\exp X$, where $X\in su(2,2)$ satisfies $\mathrm{im}X\subset\infty\subset\ker X$, while the elements $\exp X$ fulfilling $\mathrm{im}X\subset 0\subset\ker X$, define the commutative subgroup of four-accelerations.

Let us now explain what one means by the conformal phase space. In relativistic mechanics the elementary phase spaces are given by the coadjoint orbits of the Poincaré group, see [11], which are parametrized in this case by mass, spin, and signature of the energy of the relativistic particle. The conformal group is the extension of the Poincaré group by four-accelerations and dilatations which are transformations having clear physical interpretations. Therefore, the elementary conformal phase spaces are naturally identified with the coadjoint orbits of $SU(2,2)$. Since $su(2,2)$ is semisimple one has an isomorphism of the dual space $su(2,2)^*$ with the Lie algebra $su(2,2)$ given by the Cartan-Killing form:

$$\langle X,Y\rangle=\frac{1}{2}\mathrm{Tr}(XY), \tag{2.2}$$

where $X,Y\in$ su(2,2), and the coadjoint representation $\mathrm{Ad}^*_g: su(2,2)^*\to su(2,2)^*$ is identified with the adjoint one

$$\mathrm{Ad}_g X=gXg^{-1}, \tag{2.3}$$

where $g \in SU(2,2)$. For the complete description and physical interpretation of $\mathrm{Ad}^\star(SU(2,2))$-orbits see [6, 4]. In the following we will present the material based on those papers and we will restrict ourselves to the case of scalar conformal massive particle whose phase space is an eight-dimensional manifold.

Let us mention that if one assumes $\mathbb{M}^{00}$ as configuration space of the scalar conformal particle, then the cotangent bundle $T^\star\mathbb{M}^{00}$ is its canonically defined phase space. One has the $SU(2,2)$-vector bundle isomorphism:

$$\begin{array}{ccc} \mathbb{N} & \xrightarrow{L} & T^*\mathbb{M}^{00} \\ {\scriptstyle pr_1}\downarrow & & \downarrow{\scriptstyle \pi} \\ \mathbb{M}^{00} & \xrightarrow[\mathrm{id}]{} & \mathbb{M}^{00} \end{array} ,$$

where $\mathbb{N} := \{(x,X) \in \mathbb{M}^{00} \times su(2,2) : \mathrm{im}X \subset x \subset \ker X\}$ and $pr_1 : \mathbb{N} \mapsto \mathbb{M}^{00}$ is the projection on the first component of the product $\mathbb{M}^{00} \times su(2,2)$. The vector bundle map $\mathbf{L} : \mathbb{N} \longrightarrow T^*\mathbb{M}^{00}$ is defined by the following sequence $T_x^*\mathbb{M}^{00} \cong (su(2,2)/su(2,2)_x)^* \cong \{X \in su(2,2) : \mathrm{Tr}YX = 0 \ \forall Y \in su(2,2)_x\} \cong \{X \in su(2,2) : \mathrm{im}X \subset x \subset \ker X\} = pr_1^{-1}(x)$ of the vector space isomorphisms.

The conformal group action on $\mathbb{N}$ is defined by

$$\sigma_g^{\mathbb{N}} : (x,X) \mapsto (gx, gXg^{-1}) \tag{2.4}$$

for $g \in SU(2,2)$. The elements $X \in pr_1^{-1}(x)$ fulfilling the condition $\dim_{\mathbb{R}} \mathrm{im}X \leqslant 1$ form a cone C_x in $pr_1^{-1}(x) \cong T_x^*\mathbb{M}^{00}$. Thus one has a $\sigma^{\mathbb{N}}(SU(2,2))$-invariant conformal structure on the cotangent bundle $T^*\mathbb{M}^{00}$.

The 8-dimensional orbits of the $SU(2,2)$-action defined by (2.4) are:

(i) the bundle $\mathbb{N}^{++}$ of upper halves of the interiors of the cones;
(ii) the bundle $\mathbb{N}^{--}$ of bottom halves of the interiors of the cones;
(iii) the bundle $\mathbb{N}^{+-}$ of exteriors of the cones.

Similarly, the actions $\sigma^{\mathbb{M}}$ of $SU(2,2)$ on $\mathbb{M}$ defined by (2.1) generate three 8-dimensional orbits: $\mathbb{M}^{++}$, $\mathbb{M}^{--}$ and $\mathbb{M}^{+-}$.

On the union $\tilde{\mathbb{M}} := \mathbb{M}^{++} \cup \mathbb{M}^{--} \cup \mathbb{M}^{+-}$ of orbits one has the involution $\perp : \tilde{\mathbb{M}} \mapsto \tilde{\mathbb{M}}$, which maps $z \in \tilde{\mathbb{M}}$ on its orthogonal complement $z^\perp$ (with respect to the twistor forms η). Let us denote by $\pi_z : \mathbf{T} \mapsto \mathbf{T}$ and $\pi_{z^\perp} : \mathbf{T} \mapsto \mathbf{T}$ the projections defined by the decomposition: $\mathbf{T} = z \oplus z^\perp$. Now we define $SU(2,2)$-equivariant maps:

$$J_0(x,X) := X \tag{2.5}$$

$$J_\lambda(z) := i\lambda(\pi_z - \pi_{z^\perp}) \tag{2.6}$$

of $\mathbb{N}^{++}$, $\mathbb{N}^{+-}$, $\mathbb{N}^{--}$ on the 8-dimensional nilpotent $\mathrm{Ad}(SU(2,2))$-orbits and of $\mathbb{M}^{++}, \mathbb{M}^{+-}, \mathbb{M}^{--}$ on the 8-dimensional simple $\mathrm{Ad}(SU(2,2))$-orbits which consist of $X \in su(2,2)$ with eigenvalues $i\lambda$ and $-i\lambda$. Using the Kirillov construction [3] we obtain the $\sigma(SU(2,2))$-invariant symplectic form ω_0 on $T^*\mathbb{M}^{00} \cong \mathbb{N}^{++} \cup \mathbb{N}^{+-} \cup \mathbb{N}^{--} \cong \tilde{\mathbb{N}}$

which is identical with the canonical symplectic form of $T^*\mathbb{M}^{00}$. The maps (2.5),(2.6) are the momentum maps. Using $J_\lambda : \tilde{\mathbb{M}} \mapsto su(2,2)$ one obtains by the Kirillov construction the conformally invariant Kähler form ω_λ on $\tilde{\mathbb{M}}$. So $(\tilde{\mathbb{M}}, \omega_\lambda)$ and $(\tilde{\mathbb{N}}, \omega_0)$ are 8-dimensional conformal phase spaces.

In order to show that $\tilde{\mathbb{M}}$ and $\tilde{\mathbb{N}}$ are phase spaces of the conformal scalar particle let us take the following coordinate description of the presented models:

$$\eta = -i\begin{pmatrix} 0 & -\sigma_0 \\ \sigma_0 & 0 \end{pmatrix}, \infty = \left\{ \begin{pmatrix} \zeta \\ 0 \end{pmatrix} : \zeta \in \mathbb{C}^2 \right\}, \mathbf{0} = \left\{ \begin{pmatrix} 0 \\ \zeta \end{pmatrix} : \zeta \in \mathbb{C}^2 \right\}, \tag{2.7}$$

where we use the 2×2 matrix representation with Pauli basis:

$$\sigma_0 = \begin{pmatrix} 1 & 0 \\ 0 & 1 \end{pmatrix}, \quad \sigma_1 = \begin{pmatrix} 0 & 1 \\ 1 & 0 \end{pmatrix}, \quad \sigma_2 = \begin{pmatrix} 0 & i \\ -i & 0 \end{pmatrix}, \quad \sigma_3 = \begin{pmatrix} 1 & 0 \\ 0 & -1 \end{pmatrix}$$

in $Mat_{2\times 2}(\mathbb{C})$. This choice of $\eta, \infty, \mathbf{0}$ gives us the decomposition

$$su(2,2) = \mathcal{T}_\infty \oplus \mathcal{L}_{0,\infty} \oplus \mathcal{D}_{0,\infty} \oplus \mathcal{A}_0 \tag{2.8}$$

where the 4-translations, Lorentz, dilatations and 4-accelerations are given respectively by

$$\mathcal{T}_\infty = \{\begin{pmatrix} 0 & T \\ 0 & 0 \end{pmatrix} : T = T^\dagger \in Mat_{2\times 2}(\mathbb{C}) \text{ and } T = t^\mu \sigma_\mu\} \tag{2.9a}$$

$$\mathcal{L}_{o,\infty} = \{\begin{pmatrix} L & 0 \\ 0 & -L^\dagger \end{pmatrix} : \mathrm{Tr} L = 0 \text{ and } L \in Mat_{2\times 2}(\mathbb{C})\} \tag{2.9b}$$

$$\mathcal{D}_{0,\infty} = \{d\begin{pmatrix} \sigma_0 & 0 \\ 0 & -\sigma_0 \end{pmatrix} : d \in \mathbb{R}\} \tag{2.9c}$$

$$\mathcal{A}_0 = \{\begin{pmatrix} 0 & 0 \\ C & 0 \end{pmatrix} : C = C^\dagger \in Mat_{2\times 2}(\mathbb{C}) \text{ and } C = c^\mu \sigma_\mu\} \tag{2.9d}$$

The basis of $su(2,2)^* \cong su(2,2)$ dual to the one defined by Pauli matrices in the Lie subalgebras $\mathcal{T}_\infty$, $\mathcal{L}_{0,\infty}$, $\mathcal{D}_{0,\infty}$, $\mathcal{A}_0$ is

$$\mathcal{P}^*_\mu = \begin{pmatrix} 0 & 0 \\ \sigma_\mu & 0 \end{pmatrix} \tag{2.10a}$$

$$\mathcal{L}^*_{kl} = \frac{1}{2}\varepsilon_{klm}\begin{pmatrix} \sigma_m & 0 \\ 0 & \sigma_m \end{pmatrix} \tag{2.10b}$$

$$\mathcal{L}_{0k} = \frac{1}{2}\begin{pmatrix} \sigma_k & 0 \\ 0 & -\sigma_k \end{pmatrix} \tag{2.10c}$$

$$\mathcal{A}^*_\nu = \begin{pmatrix} 0 & \sigma_\nu \\ 0 & 0 \end{pmatrix} \tag{2.10d}$$

$$\mathcal{D}^* = \frac{1}{2}\begin{pmatrix} \sigma_0 & 0 \\ 0 & -\sigma_0 \end{pmatrix} \tag{2.10e}$$

One has the matrix coordinate map

$$\mathbb{M}_\infty \ni w \mapsto W \in Mat_{2\times 2}(\mathbb{C}) \tag{2.11}$$

defined by

$$w = \{\begin{pmatrix} W\zeta \\ \zeta \end{pmatrix} : \zeta \in \mathbb{C}^2\} \tag{2.12}$$

and $w \in \mathbb{M}^{00}_\infty$ iff $W = W^\dagger = X$. The element $(x, \mathfrak{X}) \in pr_1^{-1}(\mathbb{M}^{00}_\infty)$ is parametrized by

$$(x, \mathfrak{X}) \mapsto (X, \begin{bmatrix} XS & -XSX \\ S & -SX \end{bmatrix}), \tag{2.13}$$

where $X, S \in H(2)$ and $H(2)$ is the vector space of 2×2 Hermitian matrices.

The momentum maps (2.5) and (2.6) assume in the above defined coordinates the following forms

$$J_0(X,S) = \begin{bmatrix} XS & -XSX \\ S & -SX \end{bmatrix} \tag{2.14}$$

$$J_\lambda(W) = i\lambda \begin{bmatrix} (W+W^\dagger)(W-W^\dagger)^{-1} & -2W(W-W^\dagger)^{-1}W^\dagger \\ 2(W-W^\dagger)^{-1} & -\sigma_0 - 2(W-W^\dagger)^{-1}W^\dagger \end{bmatrix}. \tag{2.15}$$

Decomposing the value $J_0(X,S)$ of the momentum map J_0 in the basis given by (2.10) we obtain the standard expressions:

$$p_\mu = S_\mu \tag{2.16}$$

$$m_{\mu\nu} = x_\mu p_\nu - x_\nu p_\mu \tag{2.17}$$

$$d = x^\mu p_\mu \tag{2.18}$$

$$a_\mu = -2(x^\nu p_\nu)x_\mu + x^2 p_\mu \tag{2.19}$$

for the four-momentum p_μ, relativistic angular momentum $m_{\mu\nu}$ and four-acceleration a_ν respectively.

The symplectic form ω_0 in the coordinates $X = x^\mu \sigma_\mu$ and $P = p^\mu \sigma_\mu$ assumes the canonical form

$$\omega_0 = Tr(dX \wedge dP) = dx^\mu \wedge dp_\mu. \tag{2.20}$$

Concluding, one obtains that $(\mathbb{N}, \omega_o)$ is the phase space of the massive scalar conformal particle:

(i) an element $(x, \mathfrak{X})$ with canonical coordinates (x^μ, p_ν) describes the state of a conformal scalar particle iff $p^0 > 0$ and $(p^0)^2 - \vec{p}^2 > 0$, i.e.$(x, \mathfrak{X}) \in \mathbb{N}^{++}$;

(ii) it describes the state of a conformal massive scalar anti-particle iff $p^0 < 0$ and $(p^0)^2 - \vec{p}^2 > 0$, i.e.$(x, \mathfrak{X}) \in \mathbb{N}^{--}$;

(iii) it describes the state of a conformal scalar tachyon iff $(p^o)^2 - \vec{p}^2 < 0$, i.e. $(x, \mathfrak{X}) \in \mathbb{N}^{+-}$.

Let us now see how one interprets the physical sense of the other model. After decomposition of $J_\lambda(w)$ in the dual basis (2.10) we obtain the formulae for relativistic angular momentum and dilatations the same as in (2.17) and (2.18). However in this case the four-momentum depends on the complex coordinates $w^\nu = x^\nu + iy^\nu$ as follows

$$p^\nu = \lambda \frac{y^\nu}{y^2}, \tag{2.21}$$

and the four-accelerations assume also a different form

$$a_\mu = -2(x^\nu p_\nu)x_\mu + x^2 p_\mu - \frac{\lambda^2}{p^2} p_\mu \tag{2.22}$$

than in (2.22).

Similarly as in the nilpotent case, coordinate (x^μ, p_ν), where p_ν is given by (2.21), are the canonical coordinates on $\tilde{\mathbb{M}}$ since one has

$$\omega_\lambda = i\lambda \frac{\partial^2}{\partial w^\mu \partial \bar{w}^\nu} \log(w - \bar{w})^2 dw^\mu \wedge d\bar{w}^\nu = dx^\nu \wedge dp_\nu. \tag{2.23}$$

The state describes the conformal scalar massive particle, the conformal scalar massive antiparticle, and conformal scalar tachyon iff it belongs to $\mathbb{M}^{++}$, $\mathbb{M}^{--}$ and $\mathbb{M}^{+-}$ respectively.

Two above presented models do not differ if one considers them on the level of relativistic mechanics, since both of them behave towards Poincaré transformations in the same way. The different behavior appears after applying the acceleration transformations, which in canonical coordinate are

$$\tilde{X} = X(CX + \sigma_0)^{-1}, \tag{2.24}$$

$$\tilde{P} = (CX + \sigma_0)P(XC + \sigma_0) \tag{2.25}$$

for the standard model and

$$\tilde{X} = [XP + i\lambda\sigma_0 - i\lambda(XC - i\lambda P^{-1}C + \sigma_0)](CXP + i\lambda C + P)^{-1} \tag{2.26}$$

$$\tilde{P} = (CX + \sigma_0)P(XC + \sigma_0) + \lambda^2 CP^{-1}C \tag{2.27}$$

for the holomorphic model. We see from (2.27) that in the holomorphic model (opposite to the standard one) the four-momentum $P = p^\mu \sigma_\mu$ transforms in a non-linear way. This fact implies a lot of important physical consequences, e.g. the conformal scalar massive particle cannot be localized in space-time in conformal invariant way. Ending this section let us remark that the holomorphic model corresponds to the nilpotent one when $\lambda \to 0$.

3. QUANTUM COMPLEX MINKOWSKI SPACE

In this section we shall construct quantum conformal phase space. Since the case of the antiparticle can be transformed by the charge conjugation map to the particle one, see

[7], and the tachyon case is less interesting from physical point of view, we will work only with the phase space $\mathbb{M}^{++}$ of the conformal scalar massive particle.

The group $SU(2,2)$ acts on $\tilde{\mathbb{M}}$ by biholomorphism. For $g^{-1} = \begin{pmatrix} A & B \\ C & D \end{pmatrix} \in SU(2,2)$ and $w \in \mathbb{M}^{++}$ one has

$$\sigma_g^{\mathbb{M}} W = (AW + B)(CW + D)^{-1}, \tag{3.1}$$

where $W \in Mat_{2\times 2}(\mathbb{C})$ is the matrix holomorphic coordinate of $w \in \mathbb{M}^{++}$. Using complex matrix coordinates (2.12) one identifies $\mathbb{M}^{++}$ with the future tube

$$\mathbb{T} := \{W \in Mat_{2\times 2} : \mathrm{im} W > 0\}. \tag{3.2}$$

Applying the Caley transform

$$Z = (W - iE)(W + iE)^{-1}, W = i(Z + E)(Z - E)^{-1} \tag{3.3}$$

we map $\mathbb{T}$ on the generalized ball

$$\mathbb{D} := \{Z \in Mat_{2\times 2}(\mathbb{C}) : E - Z^* Z > 0\}. \tag{3.4}$$

Therefore we can also identify the conformal phase space $\mathbb{M}^{++}$ with the symmetric domain $\mathbb{D}$.

In order to quantize $\mathbb{M}^{++}$ we will use the method of coherent state map investigated in [8]. The essence of this method consists in replacing the classical state $m \in \mathbb{M}^{++}$ by the quantum pure state, what means, that one maps $\mathcal{K}_\lambda : \mathbb{M}^{++} \mapsto \mathbb{CP}(\mathcal{H})$ classical phase space $\mathbb{M}^{++}$ into the complex projective separable Hilbert space $\mathbb{CP}(\mathcal{H})$. We will call $\mathcal{K}_\lambda$ *coherent state map* if it fulfills the following conditions: $\mathcal{K}_\lambda^* \omega_{FS} = \omega_\lambda$ and $\mathcal{K}_\lambda(\mathbb{M}^{++})$ is linearly dense in $\mathcal{H}$.

Skipping the technical considerations we will now give the construction of $\mathcal{K}_\lambda$. Let

$$\left\{ \left| \begin{matrix} j & m \\ j_1 & j_2 \end{matrix} \right\rangle \right\}, \tag{3.5}$$

where $m, 2j \in \mathbb{N} \cup \{0\}$ and $-j \leqq j_1, j_2 \leqq j$, denote the orthonormal basis in $\mathcal{H}$, i.e.

$$\left\langle \begin{matrix} j & m \\ j_1 & j_2 \end{matrix} \middle| \begin{matrix} j' & m' \\ j_1' & j_2' \end{matrix} \right\rangle = \delta_{jj'} \delta_{mm'} \delta_{j_1 j_1'} \delta_{j_2 j_2'}. \tag{3.6}$$

Then the map $K_\lambda : \mathbb{M}^{++} \cong \mathbb{D} \mapsto \mathcal{H}$ given by

$$K_\lambda : Z \to |Z\rangle_\lambda := \sum_{j,m,j_1,j_2} \Delta^{jm}_{j_1 j_2}(Z) \left| \begin{matrix} j & m \\ j_1 & j_2 \end{matrix} \right\rangle, \tag{3.7}$$

where

$$\Delta^{jm}_{j_1 j_2}(Z) := (N^\lambda_{jm})^{-1} (\det Z)^m \sqrt{\frac{(j+j_1)!(j-j_1)!}{(j+j_2)!(j-j_2)!}} \times \tag{3.8}$$

$$\times \sum_{\substack{S\geqslant \max\{0,j_1+j_2\}\\ S\leqslant \min\{j+j_1,j+j_2\}}} \binom{j+j_2}{S}\binom{j-j_2}{S-j_1-j_2} z_{11}^S z_{12}^{j+j_1-S} z_{21}^{j+j_2-S} z_{22}^{S-j_1-j_2}$$

and

$$N_{jm}^{\lambda} := (\lambda-1)(\lambda-2)^2(\lambda-3)\frac{\Gamma(\lambda-2)\Gamma(\lambda-3)m!(m+2j+1)!}{(2j+1)!\Gamma(m+\lambda-1)\Gamma(m+2j+\lambda)} \tag{3.9}$$

for $\mathbb{N} \ni \lambda > 3$, defines a coherent state map

$$[K_\lambda] =: \mathcal{K}_\lambda : \mathbb{M}^{++} \mapsto \mathbb{CP}(\mathcal{H}) \tag{3.10}$$

with the properties mentioned above.

Let us take the space $H^\infty(\mathbb{D})$ of complex valued smooth functions $f : \mathbb{M}^{++} \to \mathbb{C}$ for which there exist bounded operators $a(f) \in L^\infty(\mathcal{H})$ such that

$$a(f)|Z\rangle_\lambda = f(Z)|Z\rangle_\lambda \tag{3.11}$$

for any $Z \in \mathbb{D} \cong \mathbb{M}^{++}$. Since the coherent states $|Z\rangle_\lambda$ form a linearly dense subset of $\mathcal{H}$ one has correctly defined isometric monomorphism $a : H^\infty(\mathbb{D}) \to L^\infty(\mathcal{H})$ of Banach algebra $H^\infty(\mathbb{D})$ with the norm given by

$$\|f\|_\infty := \sup_{Z\in\mathbb{D}} |f(Z)| \tag{3.12}$$

into the C^*-algebra $L^\infty(\mathcal{H})$ of bounded operators on H. By definition we will call $a(H^\infty(\mathbb{D})) =: \mathcal{P}^{++}$ the *quantum Kähler polarization* and elements $a(f) \in \mathcal{P}^{++}$ the *annihilation operators*.

The coordinate functions $f_{kl}(Z) := z_{kl}$, where $k,l = 1,2$ belong to $H^\infty(\mathbb{D})$. Therefore $a_{kl} := a(f_{kl}) \in \mathcal{P}^{++}$ and their action on the basis (3.5) is given by

$$\begin{aligned} a_{11}\left|\begin{matrix} j & m \\ j_1 & j_2 \end{matrix}\right\rangle &= \sqrt{\frac{(j-j_1+1)(j-j_2+1)m}{(2j+1)(2j+2)(m+\lambda-2)}}\left|\begin{matrix} j+\frac{1}{2} & m-1 \\ j_1-\frac{1}{2} & j_2-\frac{1}{2} \end{matrix}\right\rangle \\ &+\sqrt{\frac{(j+j_1)(j+j_2)(m+2j+1)}{m+2j+\lambda-1)2j(2j+1)}}\left|\begin{matrix} j-\frac{1}{2} & m \\ j_1-\frac{1}{2} & j_2-\frac{1}{2} \end{matrix}\right\rangle \end{aligned} \tag{3.13}$$

$$\begin{aligned} a_{12}\left|\begin{matrix} j & m \\ j_1 & j_2 \end{matrix}\right\rangle &= -\sqrt{\frac{(j-j_1+1)(j+j_2+1)m}{(2j+1)(2j+2)(m+\lambda-2)}}\left|\begin{matrix} j+\frac{1}{2} & m-1 \\ j_1-\frac{1}{2} & j_2+\frac{1}{2} \end{matrix}\right\rangle \\ &+\sqrt{\frac{(j+j_1)(j-j_2)(m+2j+1)}{m+2j+\lambda-1)2j(2j+1)}}\left|\begin{matrix} j-\frac{1}{2} & m \\ j_1-\frac{1}{2} & j_2+\frac{1}{2} \end{matrix}\right\rangle \end{aligned} \tag{3.14}$$

$$\begin{aligned} a_{21}\left|\begin{matrix} j & m \\ j_1 & j_2 \end{matrix}\right\rangle &= -\sqrt{\frac{(j+j_1+1)(j-j_2+1)m}{(2j+1)(2j+2)(m+\lambda-2)}}\left|\begin{matrix} j+\frac{1}{2} & m-1 \\ j_1+\frac{1}{2} & j_2-\frac{1}{2} \end{matrix}\right\rangle \\ &+\sqrt{\frac{(j-j_1)(j+j_2)(m+2j+1)}{m+2j+\lambda-1)2j(2j+1)}}\left|\begin{matrix} j-\frac{1}{2} & m \\ j_1+\frac{1}{2} & j_2-\frac{1}{2} \end{matrix}\right\rangle \end{aligned} \tag{3.15}$$

$$a_{22}\left|\begin{matrix} j & m \\ j_1 & j_2 \end{matrix}\right\rangle = \sqrt{\frac{(j+j_1+1)(j+j_2+1)m}{(2j+1)(2j+2)(m+\lambda-2)}}\left|\begin{matrix} j+\frac{1}{2} & m-1 \\ j_1+\frac{1}{2} & j_2+\frac{1}{2} \end{matrix}\right\rangle + \sqrt{\frac{(j-j_1)(j-j_2)(m+2j+1)}{m+2j+\lambda-1)2j(2j+1)}}\left|\begin{matrix} j-\frac{1}{2} & m \\ j_1+\frac{1}{2} & j_2+\frac{1}{2} \end{matrix}\right\rangle \tag{3.16}$$

In the formulas above we put by definition $\left|\begin{matrix} j & m \\ j_1 & j_2 \end{matrix}\right\rangle := 0$ if j, m, j_1, j_2 does not satisfy the condition $m, 2j \in \mathbb{N} \cup \{0\} \wedge -j \leq j_1, j_2 \leq j$.

The coordinate annihilation operators $a_{kl}, k, l = 1, 2$ generate Banach subalgebra $\mathcal{P}^{++}_{pol}$ of $\mathcal{P}^{++}$.

For the following considerations let us fix the matrix notation

$$\mathbb{A} := \begin{pmatrix} a_{11} & a_{12} \\ a_{21} & a_{22} \end{pmatrix} \in \mathcal{P}^{++}_{pol} \otimes Mat_{2\times 2}(\mathbb{C}), \tag{3.17}$$

$$\mathbb{A}^{+} := \begin{pmatrix} a^*_{11} & a^*_{21} \\ a^*_{12} & a^*_{22} \end{pmatrix} \in \overline{\mathcal{P}^{++}_{pol}} \otimes Mat_{2\times 2}(\mathbb{C}) \tag{3.18}$$

for the annihilation and creation operators. For example, in this notation the property (3.11) assumes the form

$$\mathbb{A}|Z\rangle_\lambda = Z|Z\rangle_\lambda. \tag{3.19}$$

We shall call the operator C^*-algebra $\mathcal{M}^{++} \subset L^\infty(\mathcal{H})$ generated by $\mathcal{P}^{++}$ the *quantum Kähler phase space* of the scalar conformal particle. Suitable for this by $\mathcal{M}^{++}_{pol}$ we shall denote the proper C^*-subalgebra of $\mathcal{M}^{++}$ generated by $\mathcal{P}^{++}_{pol} \subsetneq \mathcal{P}^{++}$.

Proposition 3.1.

(i) The autorepresentation of $\mathcal{M}^{++}_{pol}$ in $L^\infty(\mathcal{H})$ is irreducible and $\mathcal{P}^{++}_{pol} \cap \overline{\mathcal{P}^{++}_{pol}} = \mathbb{C}\mathbb{I}$.

(ii) $\mathcal{M}^{++}_{pol}$ is weakly (strongly) dense in $L^\infty(\mathcal{H})$.

(iii) $\mathcal{M}^{++}_{pol}$ contains the ideal $L^0(\mathcal{H})$ of compact operators.

(iv) $\mathcal{M}^{++}_{pol}$ is conformally invariant, i.e. $U_\lambda(g)\mathcal{M}^{++}_{pol}U_\lambda(g)^\dagger \subset \mathcal{M}^{++}_{pol}$, where U_λ is the discrete series representation of SU(2,2) *see [2].*

(v) The statements i), ii) and iii) are valid also for $\mathcal{M}^{++}$ and $\mathcal{P}^{++}$.

Let us define the vector space $\mathcal{A}^{++}$ of operators in $\mathcal{H}$ closed with respect to conjugation operation and all elements of which possess linear span of $\mathcal{K}_\lambda(\mathbb{M}^{++})$ as a common domain. Therefore for any operator $F \in \mathcal{A}^{++}$ the covariant Berezin covariant symbols

$$\langle F\rangle(Z^\dagger, Z) := \frac{\langle Z|FZ\rangle_\lambda}{\langle Z|Z\rangle_\lambda} \tag{3.20}$$

have a sense.

In the following we will use the coherent state weak topology, i.e. $A_n \xrightarrow{coh} A$ if $\langle Z|A_n|V\rangle_\lambda \to \langle Z|A|V\rangle_\lambda$ for all $Z,V \in \mathbb{D}$. Let us notice that it is weaker than weak topology.

The space $\mathcal{A}^{++}$ is closed with respect to coherent state weak topology. The quantum phase space $\mathcal{M}^{++}$ is contained in $\mathcal{A}^{++}$ as dense subset with respect to the coherent state topology. For any $F \in \mathcal{A}^{++}$ its Berezin symbol $f = \langle F\rangle$ belongs to the space $\mathcal{RO}(\mathbb{D})$ of real analytic functions

$$f(Z^\dagger, Z) = \sum f_{i_{11},i_{12},i_{21},i_{22},j_{11},j_{12},j_{21},j_{22}} \bar{Z}_{11}^{i_{11}} \bar{Z}_{12}^{i_{12}} \bar{Z}_{21}^{i_{21}} \bar{Z}_{22}^{i_{22}} Z_{11}^{j_{11}} Z_{12}^{j_{12}} Z_{21}^{j_{21}} Z_{22}^{j_{22}} \tag{3.21}$$

of the variables $(Z^\dagger, Z)$. One extends naturally the quantization $a : H^\infty(\mathbb{D}) \to L^\infty(\mathcal{H})$ to the space $\mathcal{RO}(\mathbb{D}) \supset H^\infty(D)$ in the following way

$$\begin{aligned} \mathcal{Q}_\lambda(f) :=& \sum f_{i_{11},i_{12},i_{21},i_{22},j_{11},j_{12},j_{21},j_{22}} a_{11}^{\dagger\, i_{11}} a_{12}^{\dagger\, i_{12}} a_{21}^{\dagger\, i_{21}} a_{22}^{\dagger\, i_{22}} a_{11}^{i_{11}} a_{12}^{i_{12}} a_{21}^{i_{21}} a_{22}^{i_{22}} = \\ =& : f(\mathbb{A}^\dagger, \mathbb{A}) :, \end{aligned} \tag{3.22}$$

where by the colon $:\cdot:$ one denotes as usually the normal ordering. The infinite sum in (3.22) is taken in the sense of coherent state weak topology. One defines the $*_\lambda$-product on the real analytic Berezin symbols $f, g \in \mathcal{RO}(\mathbb{D})$ by

$$(f *_\lambda g)(Z^\dagger, Z) := \frac{\langle Z^\dagger| : f(\mathbb{A}^\dagger, \mathbb{A}) :: g(\mathbb{A}^\dagger, \mathbb{A}) : |Z\rangle_\lambda}{\langle Z^\dagger|Z\rangle_\lambda}. \tag{3.23}$$

As an illustration let us quantize the 4-momentum p^ν, relativistic angular momentum $m_{\mu\nu}$, dilatation d and 4-acceleration a_μ given by respectively (2.21), (2.17), (2.18), (2.19) in terms of quantum coordinates $(\mathbb{A}^\dagger, \mathbb{A})$. Skipping the calculations we obtain

$$\mathcal{Q}_\lambda(p_\mu) = i\lambda : (\det(\mathbb{W} - \mathbb{W}^\dagger))^{-1} \operatorname{Tr}(\sigma_\mu(\mathbb{W} - \mathbb{W}^\dagger)) : \tag{3.24}$$

$$\begin{aligned} \mathcal{Q}_\lambda(m_{\mu\nu}) = i\lambda \bigg(& \frac{1}{2} \operatorname{Tr}(\sigma_\mu \mathbb{W}^\dagger) : (\det(\mathbb{W} - \mathbb{W}^\dagger))^{-1} \operatorname{Tr}(\sigma_\nu(\mathbb{W} - \mathbb{W}^\dagger)) : \\ & - \frac{1}{2} \operatorname{Tr}(\sigma_\nu \mathbb{W}^\dagger) : (\det(\mathbb{W} - \mathbb{W}^\dagger))^{-1} \operatorname{Tr}(\sigma_\mu(\mathbb{W} - \mathbb{W}^\dagger)) : \bigg) \end{aligned} \tag{3.25}$$

$$\mathcal{Q}_\lambda(d) = i\lambda \operatorname{Tr}(\sigma^\mu \mathbb{W}^\dagger) : (\det(\mathbb{W} - \mathbb{W}^\dagger))^{-1} \operatorname{Tr}(\sigma_\mu(\mathbb{W} - \mathbb{W}^\dagger)) : -2i\lambda \tag{3.26}$$

$$\begin{aligned} \mathcal{Q}_\lambda(a_\nu) = & \, i\lambda \det(\mathbb{W}^\dagger) : (\det(\mathbb{W} - \mathbb{W}^\dagger))^{-1} \operatorname{Tr}(\sigma_\nu(\mathbb{W} - \mathbb{W}^\dagger)) : \\ & - i\lambda \frac{1}{2} \operatorname{Tr}(\sigma_\nu \mathbb{W}^\dagger) \operatorname{Tr}(\sigma^\beta \mathbb{W}^\dagger) : (\det(\mathbb{W} - \mathbb{W}^\dagger))^{-1} \operatorname{Tr}(\sigma_\beta(\mathbb{W} - \mathbb{W}^\dagger)) : \\ & + i\lambda \operatorname{Tr}(\sigma_\nu \mathbb{W}^\dagger), \end{aligned} \tag{3.27}$$

where $(\mathbb{W}^\dagger, \mathbb{W})$ are matrix operator coordinates in $\mathcal{A}^{++}$ obtained from $(\mathbb{A}^\dagger, \mathbb{A})$ by the Caley transform

$$\mathbb{W} = i(\mathbb{A} + E)(\mathbb{A} - E)^{-1}, \tag{3.28}$$

which has sense in coherent state weak topology convergence.

It follows from (3.24)-(3.27) that

$$[Q_\lambda(p_\mu), Q_\lambda(p_\nu)] = 0, \tag{3.29}$$

$$[Q_\lambda(x^\mu), Q_\lambda(x^\nu)] = 0, \tag{3.30}$$

$$[Q_\lambda(x^\mu), Q_\lambda(p_\nu)] = -i\delta^\mu_\nu 1 \tag{3.31}$$

for the quantum canonical coordinates $(Q_\lambda(x^\mu), Q_\lambda(p_\nu))$.

Therefore we see that Heisenberg algebra generated by unbounded operators of 4-momenta $Q_\lambda(p_\nu)$ and 4-positions $Q_\lambda(x_\mu) = \frac{1}{2}\mathrm{Tr}(\sigma_\mu(\mathbb{W}+\mathbb{W}^\dagger))$ is included in $\mathcal{A}^{++}$. The creation and the annihilation operators

$$Q_\lambda(\bar{w}_\mu) = \frac{1}{2}\mathrm{Tr}(\sigma_\mu \mathbb{W}^\dagger) \qquad Q_\lambda(w_\nu) = \frac{1}{2}\mathrm{Tr}(\sigma_\nu \mathbb{W}) \tag{3.32}$$

generate the Caley transforms of quantum polarizations $\overline{\mathcal{P}^{++}_{pol}}$ and $\mathcal{P}^{++}_{pol}$ respectively. But their commutators $[Q_\lambda(\bar{w}^\mu), Q_\lambda(w_\nu)] \neq 0$ do not have so simple form as it has place in the case of quantum real polarization given by the canonical commutation relation (3.31).

Let us now discuss the physical sense of parameter $\mathbb{N} \ni \lambda > 3$. In the previous consideration for the technical reason we assumed that it is dimensionless. However, as one sees from (2.21) λ has dimension of action. Hence we assume the Planck constans h as the natural unit for λ. After this we obtain

$$w^\mu = x^\mu + i\lambda \frac{h}{mc}\frac{p^\mu}{mc}, \tag{3.33}$$

where $mc = \sqrt{p_0^2 - \vec{p}^2}$. The quantity $\frac{h}{mc}$ is the length of the Compton wave of the conformal particle. For example for the proton $\frac{h}{mc} \cong 10^{-13} cm$.

The $\frac{p^\mu}{mc}$ denotes relativistic 4-velocity measured with the speed of light as the unit. So the dimensional analysis shows that in the limit $\lambda \to \infty$ the theory describes physical phenomena characterized by the space-time scale much bigger than the Compton scale characteristic for the quantum phenomena. This physical argumentation is consistent with the following asymptotic behavior of of $*_\lambda$-product

$$f *_\lambda g \cong fg \tag{3.34}$$

$$f *_\lambda g - g *_\lambda f \cong i\lambda\{f, g\}_\lambda \tag{3.35}$$

for $\lambda \to \infty$, where right side of (3.34) is usual multiplication of functions and right side of (3.35) is the Poisson bracket defined by symplectic form (2.23). In order to show these asymptotic formulae for the case of general symmetric domain see [1]. They show the correspondences of the quantum description of the massive scalar conformal particle to its classical mechanical description in the large space-time scale limit.

The quantum effects are described by the transition amplitude between the coherent state w and v, which in the coordinates $(\bar{w}^\mu, w^\nu)$ is given by

$$a_\lambda(v^\dagger, w) = \left(\frac{((w-\bar{w})^2(v-\bar{v})^2)^{\frac{1}{2}}}{(w-\bar{v})^2}\right)^\lambda, \tag{3.36}$$

where $(w-\bar{v})^2 = \eta_{\mu\nu}(w^\mu - \bar{v}^\mu)(w^\nu - \bar{v}^\nu)$. One sees from (3.36) that transition probability $\left|a_\lambda(v^\dagger, w)\right|^2$ from w to v as a function of v forms a narrow peak around the coherent state $w \in \mathbb{T}$ if $\lambda \frac{h}{mc} \cong 0$. The detailed physical discussion can be found in [7].

ACKNOWLEDGMENTS

Authors thank T. Goliński for interest in the paper and corrections to the manuscript. Authors also thank prof. J. Lukierski for the possibility to present obtained results on the XIX Max Born Symposium 2004.

REFERENCES

1. F. A. Berezin, Commun. Math. Phys., **40**, (1975).
2. G. Jakimowicz and A. Odzijewicz, *to appear*, 2005.
3. A. Kirillov, *Elementy teorii predstawlenii*, Nauka, Moscow, 1972.
4. A. Karpio, A. Krzyszeń and A. Odzijewicz, Rep. Math. Phys., **24** (1), 65–80 (1986).
5. A. Odzijewicz, Inter. J. of Theor. Phys., **15** (8), 575–593 (1976).
6. A. Odzijewicz, Commun. Math. Phys., **107**, 561–575 (1986).
7. A. Odzijewicz, Commun. Math. Phys., **114**, 577–597 (1988).
8. A. Odzijewicz, Commun. Math. Phys., **150**, 385–413 (1992).
9. A. Odzijewicz, *Non-commutative Kähler-like structures in quantization*, to appear, 2005.
10. R. Penrose, Rep. Math. Phys., **12**, 65–76 (1977).
11. J. M. Souriau, *Structure des systemes dynamiques*, Paris: Dunod, 1970.

SUPERSYMMETRY, TWISTORS, AND HIGHER SPIN THEORY

BPS Preons in Supergravity and Higher Spin Theories. An Overview From the Hill of Twistor Approach

I.A. Bandos

Departamento de Física Teórica, Univ. de Valencia and IFIC (CSIC-UVEG), 46100-Burjassot (Valencia), Spain
Institute for Theoretical Physics, NSC "Kharkov Institute of Physics and Technology", UA61108, Kharkiv, Ukraine

Abstract. We review briefly the notion of BPS preons, first introduced in 11–dimensional context as hypothetical constituents of M–theory, in its generalization to arbitrary dimensions and emphasizing the relation with twistor approach. In particular, the use of a "twistor–like" definition of BPS preon (almost) allows us to remove supersymmetry arguments from the discussion of the relation of the preons with higher spin theories and also of the treatment of BPS preons as constituents. We turn to the supersymmetry in the second part of this contribution, where we complete the algebraic discussion with supersymmetric arguments based on the M–algebra (generalized Poincaré superalgebra), discuss the possible generalization of BPS preons related to the $osp(1|n)$ (generalized AdS) superalgebra, review a twistor–like κ–symmetric superparticle in tensorial superspace, which provides a point–like dynamical model for BPS preon, and the rôle of BPS preons in the analysis of supergravity solutions. Finally we describe resent results on the concise superfield description of the higher spin field equations and on superfield supergravity in tensorial superspaces.

INTRODUCTION

Twistor theory [1, 2] and twistor–like methods, which are the main subjects of this Max Born symposium, are becoming now increasingly popular in the light of the work of [3, 4] on the twistor string description of the Yang–Mills scattering amplitudes [5]. This can be considered as a significant progress towards a realization of the Penrose "twistor programme" [2] aimed to describe nature in terms of twistor space rather than spacetime.

The subject of this contribution is the notion of BPS preons, introduced in [6] in an M–theoretical context, but allowing for an easy 'generalization' to other dimensions (see [7, 8] and also [9]). In M-theory the BPS preons appeared as its (hypothetical) constituents [6]; a search 31/32 supersymmetric solutions of $D = 11$ supergravity, which would describe BPS preons, can be witnessed [10, 11, 12]. In some other dimensions, namely in $D = 4, 6$ and 10, the notion of BPS preons are related with higher spin theory (see [13, 14, 15, 8, 12, 16, 17, 18, 19]).

As it was noticed already in [6] the notion of BPS preons is related to the twistor approach [1] and its very simple orthosymplectic "generalization" [20] (hence the "twistorial constituents" name in the title of [6]). The discussion of this relation allows us to define the BPS preon in a simple and suggestive way, with a minimal use of supersymmetry. This observation suggests the following structure of this contribution.

CP767, *Fundamental Interactions and Twistor-Like Methods,* edited by J. Lukierski and D. Sorokin

We begin in Sec. I by a brief review of the known properties of twistor approach, massless particle mechanics and their supersymmetric generalization in the form which is useful to define and to discuss the properties of BPS preons. In Sec. II we present a purely bosonic definition of the of BPS preon [6] and discuss their properties (almost) without using (more precisely, with a minimal reference to) supersymmetry. In this framework we review, in particular, the rôle of BPS preons as constituents (of M–theory for $D = 11$) and the relation of BPS preon with higher spin theories. To establish this relation we use the point–like model for BPS preon provided by the twistor–like (super)particle model in tensorial (super)space; interestingly enough, this model had been proposed in [20] before the notion of BPS preons was introduced in [6]. The same can be said (at least up to some extent, see [13] and [14, 15, 16, 18, 19]) on the relation of this model with $D = 4, 6, 10$ higher spin theories. We use here the BPS preon notion to discuss these issues as it provides a universal framework allowing to discuss the higher spin theory in $D = 4, 6, 10$ and (some issues of) M–theory in the same term. The discussion of Sec. II is completed by supersymmetry arguments in Sec. III where we start form M–algebra, discuss the rôle of BPS preons in the classification of the BPS state [6] and in the analysis of the supergravity solitons [12], review the κ–symmetry of the "preonic superparticle" model [20] and its relation with preserved supersymmetry. We finish in Sec. IV by describing recent results on the superfield description of the tower of all possible conformal massless higher spin equations in $D = 4, 6, 10$ and on supergravity in tensorial superspaces, which might be relevant both in the search for a selfconsistent supersymmetric higher spin interaction and for M–theoretical applications.

I. PRELIMINARIES. TWISTOR APPROACH, MASSLESS PARTICLE AND SUPERPARTICLE IN $D = 4$.

This section contains a review of known issues on $D = 4$ twistors and supertwistors and their relation to massless particle and superparticle mechanics [21] (see also e.g. [22, 23] and a more recent [24, 25]) in a form convenient for the discussion on BPS preons.

I.1. Cartan–Penrose representation and Penrose correspondence

Let us begin by writing two basic relation of the original Penrose twistor approach in $D = 4$ [1]. One is the Cartan–Penrose representation for a real light–like vector, *e.g.* the momentum of massless particle,

$$p_{A\dot{A}} \;:=\; p_a\sigma^a_{A\dot{A}} = \lambda_A\bar{\lambda}_{\dot{A}}\,, \qquad \Leftrightarrow \qquad p_a p^a = 0 \tag{1}$$

($a = 0, 1, 2, 3$, $A = 1, 2$ and $\dot{A} = 1, 2$ are Weyl spinor indices and $\sigma^a_{A\dot{A}}$ are relativistic Pauli matrices). Another is the famous Penrose correspondence,

$$\mu^{\dot{A}} = x^{\dot{A}A}\lambda_A \;:=\; \frac{1}{2}x^a\sigma_a^{\dot{A}A}\lambda_A\,. \tag{2}$$

For a fixed x^a (real or complex), Eq. (2) is a homogeneous linear equation for the coordinates $Y_{0\alpha} = (\mu^{\dot{A}}, \lambda_A)$ of the complex space C^4. Imposing the topological restriction $Y_{0\alpha} \neq (0,0,0,0)$ (that is passing from C^4 to $C^4 - \{0\}$) and using the scaling symmetry $(\mu^{\dot{A}}, \lambda_A) \mapsto (z\mu^{\dot{A}}, z\lambda_A)$ as an identification relation, one can treat $Y_{0\alpha} = (\mu^{\dot{A}}, \lambda_A)$ as *homogeneous* coordinates of *the projective twistor space* CP^3 [1, 2]. Thus, as usually stated in twistor approach, Eq. (2) describes a correspondence the space of light–like lines in spacetime (which can be identified with the celestian sphere S^2) and the set of all surfaces in the projective twistor space $\mathbf{CP}^3$ ("curves of genus zero and degree one" [3]) which is isomorphic to $\mathbf{CP}^1$ (in this sense one can say that the Penrose correspondence illustrates the known identity $S^2 = \mathbf{CP}^1$).

To understand that the correspondence involves light–like lines rather then points x^a of the Minkowski spacetime M^4, one notices the symmetry of Eq. (2) under

$$\delta x^{\dot{A}A} = b\lambda^{\dot{A}}\lambda^{A} \, , \tag{3}$$

which is usually called *b*–symmetry. The presence of an arbitrary parameter b as a coefficient for light–like vector $\lambda^{\dot{A}}\lambda^{A}$ implies that the orbit of the b–symmetry transformations (3) is the light–like line $\hat{x}^{\dot{A}A}(b) = x^{\dot{A}A} + b\lambda^{A}\bar{\lambda}^{\dot{A}}$.

Let us notice that Eq. (2) with real x^a is the general solution of the single real equation for the twistor variables. This is usually called *helicity constraint*, and reads

$$\mathscr{S} = \bar{\mu}^{\dot{A}}\bar{\lambda}_{\dot{A}} - \mu^{A}\lambda_{A} = 0 \, . \tag{4}$$

If one substitute the complex vector x^a_L (the non–Hermitian $x_L^{\dot{A}A}$) for the real x^a in (2), thus studying the Penrose correspondence in the complexified Minkowski spacetime CM^4 (see *e.g.* [1]),

$$\mu^{\dot{A}} = x_L^{\dot{A}A}\lambda_A \, , \qquad x_L^{\dot{A}A} = x^{\dot{A}A} + iy^{\dot{A}A} \quad , \tag{5}$$

one finds, instead of (4), $\mathscr{S} = \bar{\mu}^{\dot{A}}\bar{\lambda}_{\dot{A}} - \mu^{A}\lambda_{A} = 2i\bar{\lambda}_{\dot{A}}y^{\dot{A}A}\lambda_A$, where $y^{\dot{A}A} := 1/2i(x_L^{\dot{A}A} - (x_L^{\dot{A}A})^*)$ is the imaginary part of $x_L^{\dot{A}A}$. The correspondence with the complexified Minkowski spacetime CM^4 can be used, in particular, to describe fields of nonzero helicity [21]: to this end one sets $\mathscr{S} = \bar{\mu}^{\dot{A}}\bar{\lambda}_{\dot{A}} - \mu^{A}\lambda_{A} = 2is$ with some half–integer s. [1] Notice also that the one–parametric *b*–symmetry (3) in the case of complex x^a_L is replaced by the complex–spinor–parametric symmetry $x^{\dot{A}A} + u^{\dot{A}}\lambda^{A}$ of (5). This allows to gauge away all the imaginary part y^a of x^a_L except for the one enclosed in the contraction $\lambda_A y^{\dot{A}A}\bar{\lambda}_{\dot{A}} \equiv 1/2i\mathscr{S}_c$. Thus the helicity constraint with nonvanishing *r.h.s.*, $\mathscr{S}_c = 2is$ may be used, together with gauge fixing, to define $y^a = \Im m\,(x^a_L)$ completely: $y^{\dot{A}A} = s w^{A}\bar{w}^{\dot{A}}$ where $w^A\lambda_A = 1$ and $\bar{w}^{\dot{A}} = (w^A)^*$.

An important observation is that, in distinction to (2), the equation (5) with an *arbitrary complex* x^a_L does not restrict the twistor $Y_{0\alpha} = (\mu^{\dot{A}}, \lambda_A)$ at all but rather provides

[1] Notice that nonzero helicities can appear as a result of the ordering ambiguity after quantization of the massless particle; the quantum consideration also indicates the quantization of helicity s in the units of $\hbar/2$, see [27, 28, 29] and refs. therein.

a possibility to change a set of basic variables form the set of two spinor $(\mu^{\dot{A}}, \lambda_A)$ to one spinor λ_A and a complex vector $x_L^{\dot{A}A} = x^{\dot{A}A} + iy^{\dot{A}A}$ defined modulo the (gauge) transformations $x_L^{\dot{A}A} \mapsto x_L^{\dot{A}A} + u^{\dot{A}}\lambda^A$. This can be treated as a reason for the existence of the formulation of bosonic higher spin theory with an auxiliary vector variable whose *AdS* version allows for a nontrivial interaction [17].

I.2.Twistors and massless particle in $D = 4$ Minkowski spacetime

The one parametric b symmetry (3), which is the invariance of Eq. (2), can be identified with the gauge symmetry of the massless particle action in its Ferber–Schirafuji form [21]

$$S_0 = \int_{W^1} d\hat{x}^{\dot{A}A}\hat{\lambda}_A\hat{\bar{\lambda}}_{\dot{A}} \equiv \int d\tau\, \partial_\tau \hat{x}^a(\tau)\, \tilde{\sigma}_a^{\dot{A}A}\, \hat{\lambda}_A\hat{\bar{\lambda}}_{\dot{A}}(\tau)\,. \tag{6}$$

A simple way to obtain this "twistor–like" action is to start with the first–order form of the Brink–Schwarz formulation of the massless particle action, $S_{0BS} = \frac{1}{2}\int_{W^1}(\hat{p}_{A\dot{A}}\, d\hat{x}^{\dot{A}A} + \frac{1}{2}d\tau\, e\, \hat{p}_{A\dot{A}}\hat{p}^{\dot{A}A})$, and to substitute the general solution

$$\hat{p}_{A\dot{A}} = \hat{\lambda}_A\hat{\bar{\lambda}}_{\dot{A}} \qquad (\Leftrightarrow \qquad \hat{p}_a\hat{p}^a \equiv 1/2\hat{p}_{A\dot{A}}\hat{p}^{\dot{A}A} = 0\,) \tag{7}$$

of the algebraic equation of motion $\frac{\delta S_{BS}}{\delta e} = \frac{1}{2}p_a p^a = 0$ for an arbitrary $\hat{p}_{A\dot{A}}$ in $S_{0\,BS}$. One can also obtain the b–symmetry (see (3) of the action (6),

$$\delta_b\hat{x}^{\dot{A}A} = b\hat{\lambda}^A\hat{\bar{\lambda}}^{\dot{A}}\,, \qquad \delta_b\hat{\lambda}^A = 0 \tag{8}$$

by substituting the Cartan–Penrose representation (1) for the light–like $\hat{p}_a$, given by Eq. (7), in the gauge symmetry $\delta\hat{x}^a = b\,\hat{p}^a$ of the action S_{BS}[2].

The massless particle action (6) provides a simple way to see the relation among the two basic formulae of the twistor approach, namely among the Cartan–Penrose representation of Eq. (1) and the Penrose correspondence relation (2). First one notices that the Hamiltonian formalism for the action (6) [21] reproduces the worldline version (7) of Eq. (1) as a primary constraint (see [65]). Secondly, the above observation that the action possesses the same b–symmetry as the Penrose correspondence (2) (see Eqs. (8) and (3)) suggests that (6) should also reproduce the worldline version of Eq. (2),

$$\hat{\mu}^{\dot{A}} = \hat{x}^{\dot{A}A}\hat{\lambda}_A\,. \tag{9}$$

This is indeed the case. Using the Leibnitz rule $(d\hat{x}\,\hat{\lambda}\hat{\bar{\lambda}} \equiv d(\hat{x}\hat{\lambda})\hat{\bar{\lambda}} - \hat{x}\hat{\lambda}\,d\hat{\bar{\lambda}})$ one can write the action (6) in the form (see [21])

$$S_0 = \int_{W^1}(d\bar{\mu}^{\dot{A}}\hat{\bar{\lambda}}_{\dot{A}} - \hat{\mu}^A d\hat{\lambda}_A)\,, \qquad \hat{\bar{\mu}}^{\dot{A}}\hat{\bar{\lambda}}_{\dot{A}} - \hat{\mu}^A\hat{\lambda}_A = 0\,, \tag{10}$$

---

[2] This symmetry also includes $\delta e = \partial_\tau b(\tau)$, $\delta\hat{p}_a = 0$ and constitutes a "variational version" of worldline reparametrization.

where $\hat{\mu}^{\dot{A}}$ is defined by Eq. (9); as (9) is the general solution of the helicity constraints (4), one can, alternatively, consider the twistor variables $\hat{Y}_{0\alpha} = (\hat{\mu}^{\dot{A}}, \hat{\lambda}_A)$ to be subject to the helicity constraint (4) [as it is written in (10)] and omit any reference on the spacetime coordinates. Taking the second point of view one finds that just the constraint (4) reduces the imaginary part of the action (10) to a total derivative. This constraint can be also incorporated into the action with a Lagrange multiplier $\Xi(\tau)$,

$$S = \int_{W^1} (d\hat{\bar{\mu}}^{\dot{A}} \hat{\bar{\lambda}}_{\dot{A}} - \hat{\mu}^A d\hat{\lambda}_A) + \int_{W^1} d\tau\, \Xi(\tau)\, (\hat{\bar{\mu}}^{\dot{A}} \hat{\bar{\lambda}}_{\dot{A}} - \hat{\mu}^A \hat{\lambda}_A)\,. \tag{11}$$

I.3. Supersymmetry: massless superparticle and supertwistors

The supersymmetric generalization of the action (6) can be obtained *e.g.* starting with the first order form of the Brink–Schwarz superparticle action and using there the general solution (7) of the mass shell constraints $p^2 = 0$ (see Sec.I.2). It reads [21]

$$S = \int_{W^1} \hat{\Pi}^{\dot{A}A} \hat{\lambda}_A \hat{\bar{\lambda}}_{\dot{A}} \equiv \int d\tau\, \hat{\lambda}_A \tilde{\sigma}_a^{\dot{A}A} \hat{\bar{\lambda}}_{\dot{A}}\, \hat{\Pi}_\tau^a(\tau), \tag{12}$$

where $\hat{\Pi}^{\dot{A}A} = d\tau \Pi_\tau^{\dot{A}A}$ is the pull–back to the particle worldline W^1 of the Volkov–Akulov one–form

$$\Pi^a = dx^a - id\theta_i \sigma^a \bar{\theta}^i + i\theta_i \sigma^a d\bar{\theta}^i \quad \Leftrightarrow \quad \Pi^{\dot{A}A} = dx^{\dot{A}A} - id\theta_i^A \bar{\theta}^{\dot{A}i} + i\theta_i^A d\bar{\theta}^{\dot{A}i} \tag{13}$$

for the $D = 4$ N–extended superspace $\Sigma^{(4|4N)}$ with the local coordinates

$$\Sigma^{(4|4N)} \ : \ z^M = (x^a\,,\, \theta_i^A\,,\, \bar{\theta}^{\dot{A}i});\quad a = 0,1,2,3\,,\ A = 1,2\,,\ \dot{A} = 1,2\,,\ i = 1,\ldots,N\,. \tag{14}$$

The (N–extended) global supersymmetry transformations which leave (13) invariant are

$$\delta x^a = -i\theta_i \sigma^a \bar{\varepsilon}^i + i\varepsilon_i \sigma^a \bar{\theta}^i\,, \qquad \delta\theta_i^A = \varepsilon_i\,, \qquad \delta\bar{\theta}^{\dot{A}i} = \bar{\varepsilon}^{\dot{A}i}\,. \tag{15}$$

As in the purely bosonic case, using the Leibnitz rule one can write the action (12) in the form (*cf.* (10))

$$S = \int_{W^1} (d\hat{\bar{\mu}}^{\dot{A}} \hat{\bar{\lambda}}_{\dot{A}} - \hat{\mu}^A d\hat{\lambda}_A - 2id\hat{\eta}_i \hat{\bar{\eta}}^i) = \int_{W^1} d\hat{\Upsilon}_\Lambda \hat{\bar{\Upsilon}}^\Lambda \equiv \int_{W^1} d\Upsilon_\Lambda \Omega^{\Lambda\Pi} (\Upsilon_\Pi)^*. \tag{16}$$

Here (the pull–backs of) the components of supertwistor

$$\Upsilon_\Lambda := (Y_{0\alpha}, \eta_i) = (\mu^{\dot{A}}\,, \lambda_A\,,\, \eta_i) \quad \left(\bar{\Upsilon}^\Lambda := \Omega^{\Lambda\Pi} (\Upsilon_\Lambda)^* \equiv (\bar{\lambda}_{\dot{A}},\quad \bar{\mu}^A\,, -2i\bar{\eta}^i)^T\right) \tag{17}$$

are related to the coordinates (14) (coordinate functions in (12)) by the following supersymmetric generalization of the Penrose correspondence relation (2) [21]

$$\begin{cases} \mu^{\dot{A}} = x_L^{\dot{A}A} \lambda_A := \frac{1}{2} x_L^a \sigma_a^{\dot{A}A} \lambda_A := (x^{\dot{A}A} + i\theta_i^A \bar{\theta}^{\dot{A}i})\, \lambda_A\,, \\ \quad \eta_i = \theta_i^A \lambda_A\,. \end{cases} \tag{18}$$

These expressions for the supertwistor gives the general solution of the superhelicity constraint

$$\mathscr{S} := \hat{\mu}^{\dot{A}}\hat{\bar{\lambda}}_{\dot{A}} - \hat{\mu}^{A}\hat{\lambda}_{A} - i\eta_i\bar{\eta}^i \equiv \Upsilon_\Lambda\bar{\Upsilon}^\Lambda \equiv \Upsilon_\Lambda\Omega^{\Lambda\Pi}(\Upsilon_\Pi)^* = 0\,, \tag{19}$$

in which (as well as in (16)) $\Omega^{\Lambda\Pi}$

$$\Omega^{\Lambda\Pi} := \begin{pmatrix} \Omega^{\alpha\beta} & 0 \\ 0 & -2i\delta_i{}^j \end{pmatrix} = \begin{pmatrix} 0 & \delta_{\dot{A}}{}^{\dot{B}} & 0 \\ -\delta^A{}_B & 0 & 0 \\ 0 & 0 & -2i\delta_i{}^j \end{pmatrix} \tag{20}$$

is the $SU(2,2|N)$ invariant matrix. Such an observation allows one to write the superparticle action in an equivalent form (*cf.* (11))

$$\begin{aligned} S \;&= \textstyle\int_{W^1}(d\hat{\mu}^{\dot{A}}\hat{\bar{\lambda}}_{\dot{A}} - \hat{\mu}^{A}d\hat{\lambda}_{A} - 2id\eta_i\bar{\eta}^i) + \int_{W^1} d\tau\,\Xi(\tau)\,(\hat{\mu}^{\dot{A}}\hat{\bar{\lambda}}_{\dot{A}} - \hat{\mu}^{A}\hat{\lambda}_{A} - 2i\eta_i\bar{\eta}^i) \equiv \\ &\equiv \textstyle\int_{W^1} d\Upsilon_\Lambda\Omega^{\Lambda\Pi}(\Upsilon_\Pi)^* + \int_{W^1} d\tau\,\Xi(\tau)\,\Upsilon_\Lambda\Omega^{\Lambda\Pi}(\Upsilon_\Pi)^*\,. \end{aligned} \tag{21}$$

In this form the $SU(2,2|N)$ symmetry of the superparticle action becomes manifest. The action (21), incorporating also the constraint (19) with the Lagrange multiplier Ξ, involves only one constant tensor $\Omega^{\Lambda\Pi} = -(-)^{\Lambda+\Pi}\Omega^{\Pi\Lambda}$, Eq. (20), and the invariance of such a tensor is the defining property of the $SU(2,2|N)$ supergroup.

In relation with the evident equivalence of the action (21) (or (16)) with (12) one can ask questions about degrees of freedom. In particular, the action (12) contains $4N$ fermionic fields (coordinate functions) θ^α while (21) (or (16)) involves $2N$ fermionic $\eta_i, \bar{\eta}^i$. This seeming mismatch indicates the presence of $2N$ local fermionic gauge symmetries in the action (12). These has the form

$$\delta_\kappa\hat{x}^{\dot{A}A}(\tau) = i\kappa\lambda^A\bar{\theta}^{\dot{A}} + i\bar{\kappa}\theta^A\bar{\lambda}^{\dot{A}} = i\delta_\kappa\theta^A\bar{\theta}^{\dot{A}}(\tau) - i\theta^A\delta_\kappa\bar{\theta}^{\dot{A}}(\tau) \tag{22}$$

$$\delta_\kappa\theta^A = \kappa(\tau)\lambda^A\,, \qquad \delta_\kappa\bar{\theta}^{\dot{A}}(\tau) = \bar{\kappa}(\tau)\bar{\lambda}^{\dot{A}}(\tau) \tag{23}$$

and provide an irreducible form (see [30, 31, 32, 23] and refs therein) of the seminal κ–symmetry [33] of the Brink–Schwarz superparticle which can be defined by $\delta_\kappa\bar{\theta}^{\dot{A}}\Pi_{\tau A\dot{A}} = 0$ and $i_\kappa\Pi^{\dot{A}A} := \delta_\kappa\hat{x}^{\dot{A}A} - i\delta_\kappa\theta^A\bar{\theta}^{\dot{A}} + i\theta^A\delta_\kappa\bar{\theta}^{\dot{A}} = 0$.

II. BPS PREONS WITHOUT SUPERSYMMETRY

II.1. BPS preons and generalized Cartan–Penrose representation

In this section we present the definition of the BPS preon from [6] (see also [8, 12]) in its bosonic form which makes transparent the relation with the twistor approach.

◇ The *BPS preon state* can be characterized by one bosonic spinor λ_α,

$$|BPS\ preon\rangle = |\lambda_\alpha\rangle\,, \qquad \alpha = 1,\ldots,n\,, \tag{24}$$

and is an eigenvector of the generalized momentum operator $P_{\alpha\beta} = P_{\beta\alpha}$ for the eigenvalue $\lambda_\alpha \lambda_\beta$ determined by the above mentioned spinor λ_α,

$$P_{\alpha\beta} | \lambda_\alpha \rangle = \lambda_\alpha \lambda_\beta | \lambda_\alpha \rangle \, . \tag{25}$$

◇ In the original M–theoretic context of [6] $\alpha = 1, \ldots, 32$ is the Majorana–Weyl spinor index of $SO(1,10)$ $(D = 11 = 1+10)$, but a generalization for $\alpha = 1, \ldots, n$ with other $n = 2^k$ allowing treatment as Majorana or pseudo–Majorana spinors of $SO(t, D-t)$ with other D is straightforward.[3]

In supersymmetric theory, where the generalized momentum is defined by the anticommutator of fermionic charges, $P_{\alpha\beta} = \{Q_\alpha Q_\beta\}$, the above definition implies (see [6, 8]) that the BPS preon state $|BPS\ preon\rangle = |\lambda_\alpha\rangle$ *preserves all but one supersymmetries generated by* Q_α with $\alpha = 1, \ldots, n$. Hence another notation for the preonic state is $|BPS\ preon\rangle = |BPS\ (n-1)\rangle$ reflecting the number of preserved supersymmetries; in the M–theoretic $n = 32$ case this is $|BPS\ preon\rangle = |BPS\ 31\rangle$. This notation, however, can be understood also without references on supersymmetry, as we will see in a moment.

II.2. BPS preons as fundamental constituents

The above definition of the BPS preon is based on the eigenvalue problem for the generalized momentum operator $P_{\alpha\beta}$ and, hence, assumes that different components of $P_{\alpha\beta}$ can be diagonalized simultaneously. This is the case when they are commuting,

$$[P_{\alpha\beta}, P_{\gamma\delta}] = 0 \, . \tag{26}$$

The general eigenvector $|p_{\alpha\beta}\rangle$ of the Abelian $P_{\alpha\beta}$,

$$P_{\alpha\beta} |p_{\alpha\beta}\rangle = p_{\alpha\beta} |p_{\alpha\beta}\rangle \, , \tag{27}$$

is characterized by an eigenvalue matrix $p_{\alpha\beta}$. One can rise the question how to classify these states. Such a classification problem looks much less academic in a supersymmetric context where (see Sec. III) it is equivalent to the search for a classification of the BPS states (M–theory BPS states for $n = 32$) [6].

The Abelian algebra of the generalized momenta, Eq. (26), possesses a manifest $GL(n)$ symmetry. The only property of $|p_{\alpha\beta}\rangle$ states which is invariant under this $GL(n)$ symmetry is the rank, rank$(p_{\alpha\beta})$, of the eigenvalue matrix $p_{\alpha\beta}$. Let us denote the matrix of rank $(n-k)$ by $p^{|k\rangle}_{\alpha\beta}$, rank$(p^{|k\rangle}_{\alpha\beta}) := (n-k)$ ($(32-k)$ in the M–theoretical case) and the state with the eigenvalue matrix $p^{|k\rangle}_{\alpha\beta}$ by $|BPS\ , p^{|k\rangle}_{\alpha\beta}\rangle$ or, shortly, $|BPS\ k\rangle = |\, k\,\rangle$,

$$P_{\alpha\beta} |k\rangle = p^{|k\rangle}_{\alpha\beta} |k\rangle \, , \qquad \text{rank}(p^{|k\rangle}_{\alpha\beta}) = n-k \qquad \text{((32-k) for n=32} \Leftarrow \text{D=11)} \, . \tag{28}$$

[3] The cases of $n \neq 2^k$ α allows for a treatment as multispinor index (a set of spinor indices); *e.g.* for the odd values of n, λ_α one can treat $\alpha = 1, \ldots, n$ as a set of n one–valued Majorana–Weyl spinors in $D = 2$.

The definition of BPS preon implies its identification with the state $|(n-1)\rangle$ with a generalized momentum eigenvalue matrix $p^{|(n-1)\rangle}{}_{\alpha\beta} = p^{(1)}{}_{\alpha\beta}$ of rank equal to one. Indeed, any matrix of rank one can be expressed by the direct product of two vectors,

$$p_{\alpha\beta} = \lambda_\alpha \lambda_\beta \quad \Leftrightarrow \quad p_{\alpha\beta} = p^{|(n-1)\rangle}_{\alpha\beta} := p^{(1)}_{\alpha\beta} = \lambda_\alpha \lambda_\beta \,, \qquad \alpha, \beta = 1, \ldots, n\,. \tag{29}$$

Clearly Eq. (29) provides [20] a generalization of the Cartan–Penrose representation (1) for the light–like four vector in the Minkowski spacetime $\mathbf{M}^4$.

In the supersymmetric theory, where $P_{\alpha\beta} = \{Q_\alpha\, Q_\beta\}$, (see Sec. III) the classification by the rank of the generalized momentum matrix $(32-k)$ provides the classification of the BPS states by the number of preserved supersymmetry (k) [6]. Here the states $|k\,\rangle$ *are the BPS states preserving k of the n supersymmetries generated by* Q_α*-s*. The BPS preons preserves *all but one supersymmetries*, $|BPS\ preon\rangle = |(n-1)\,\rangle$, which means 31 out of 32 supersymmetries in the M–theoretic ('$D=11$') case, $|BPS\ preon\rangle = |31\,\rangle$.

Now we are ready to discuss the rôle of BPS preons as possible constituents. Notice that a symmetric $n\times n$ matrix always can be diagonalized by $GL(n)$ transformations, *i.e.* there exists a matrix $g_\alpha{}^{(\gamma)} \in GL(n,\mathbf{R})$ such that

$$p^{|k\rangle}_{\alpha\beta} := p^{(32-k)}_{\alpha\beta} = g_\alpha{}^{(\gamma)} p_{(\gamma)(\delta)} g_\beta{}^{(\delta)} \tag{30}$$

with some diagonal matrix $p_{(\gamma)(\delta)} = diag(\ldots)$ holds. Moreover, this diagonal matrix can be put in the form $p_{(\gamma)(\delta)} = diag(1,\ldots,1,-1,\ldots,-1,0,\ldots,0)$, where the number of nonvanishing elements, all $+1$ or -1, is equal to $\tilde{n} = (n-k) = rank(p^{(k)}{}_{\alpha\beta})$.

Only at this stage we really need in a reference on supersymmetry. Indeed, the usual assumptions of unitary supersymmetric quantum mechanics do not allow for negative eigenvalues of $P_{\alpha\beta} = \{Q_\alpha, Q_\beta\}$. Thus, only positive eigenvalues are allowed and

$$p_{(\gamma)(\delta)} = diag(\underbrace{1,\ldots,1}_{\tilde{n}=32-k},\underbrace{0,\ldots,0}_{k})\,. \tag{31}$$

Substituting (31) into (30) and denoting $g_\alpha{}^1 = \lambda_\alpha{}^1$, ..., $g_\alpha{}^{\tilde{n}} = \lambda_\alpha{}^{\tilde{n}}$, one finds

$$P_{\alpha\beta}\ \ |BPS,k\rangle \ \ = \sum_{r=1}^{\tilde{n}:=32-k} \lambda_\alpha{}^r \lambda_\beta{}^r |BPS,k\rangle \equiv \left(\lambda_\alpha{}^1\lambda_\beta{}^1 + \ldots + \lambda_\alpha{}^{\tilde{n}}\lambda_\beta{}^{\tilde{n}}\right)|BPS,k\rangle\,. \tag{32}$$

Eq. (32) may be treated as a manifestation of the *composite structure* of any BPS state $|BPS,k\rangle$ with $k<(n-1)$. To this end one solves (32) by

$$|BPS,k\rangle = |\lambda^1\rangle \otimes \ldots \otimes |\lambda^{(32-k)}\rangle\,, \tag{33}$$

which implies that the BPS states $|BPS,k\rangle$ with $k<(n-1)$ are composed from $\tilde{n} = 32-k$ BPS preonic states $|\lambda^1\rangle$, ..., $|\lambda^{\tilde{n}}\rangle$ characterized by the spinors $\lambda_\alpha{}^1$, ..., $\lambda_\alpha{}^{\tilde{n}}$. Clearly for the vacuum states preserving all supersymmetries, $k=n$, Eq. (33) does not make sense; for $k=(n-1)$ it just identifies different notations for a BPS preon $|BPS,31\rangle = |\lambda^1\rangle$ *i.e.* it implies that BPS preons are fundamental.

In the light of a supersymmetric treatment this implies that [6] any BPS states preserving some (but not all) supersymmetries can be considered as a composite of BPS preons. In particular all the M–theory BPS states can be considered as composed of BPS preons, which allowed us to conjecture that the BPS preons may be considered as fundamental constituents of M–theory [6].

The supersymmetry is important in the following respect. *In non–supersymmetric* theory the generalized momentum matrix is not positive definite. This implies the possibility of minus signs in the diagonalized form of the generalized momentum matrix, *i.e.* $p_{(\gamma)(\delta)} = diag(1,\ldots,1,-1,\ldots,-1,0,\ldots,0)$ rather than (31). Then, to compose the state with such an eigenvalue of the generalized momentum one should introduce, in addition to BPS preons (25), their counterparts *with negative energy*, "antipreons" $|anti-BPSpreon\,,\,\lambda_\alpha\rangle \equiv |anti{-}\lambda_\alpha\rangle$ obeying $P_{\alpha\beta}|anti{-}\lambda_\alpha\rangle = -\lambda_\alpha\lambda_\beta|anti{-}\lambda_\alpha\rangle$.

II.3. BPS preon and generalized Penrose correspondence. Symplectic twistors and tensorial spaces

In the light of the discussion of the first sections, one may expect that some generalization of the Penrose correspondence (2) should be related with the generalization (29) of the Cartan–Penrose representation (1). As (2) incudes the coordinate x^a conjugate to the momentum p_a entering in (1), one may expect that the desired generalization of the (2) should include a spin–tensorial coordinate $X^{\alpha\beta} = X^{\beta\alpha}$ conjugate to $p_{\alpha\beta}$ of (29). In such a way one arrives at the $n(n+1)/2$ dimensional spacetime with coordinates $X^{\alpha\beta}$,

$$\Sigma^{(\frac{n(n+1)}{2}|0)}\;:\; X^{\alpha\beta} = X^{\beta\alpha}\,, \quad \alpha,\beta = 1,2,\ldots,n\;, \tag{34}$$

which is called "tensorial space" [49, 14, 18]. To justify this name one can notice that *e.g.* for $n=4$ one can decompose the symmetric spin–tensorial coordinate $X^{\alpha\beta} = X^{\beta\alpha}$ of the $\Sigma^{(\frac{4(4+1)}{2}|0)} = \Sigma^{(10|0)}$ space on the basis of $D=4$ Dirac matrices

$$X^{\alpha\beta} \;=\; X^{\beta\alpha} = \frac{1}{2}x^\mu\gamma_\mu^{\alpha\beta} + \frac{1}{4}y^{\mu\nu}\gamma_{\mu\nu}^{\alpha\beta}\,, \quad \mu\,,\nu = 0,1,2,3; \quad \alpha,\beta = 1,2,3,4, \tag{35}$$

arriving at the set of antisymmetric tensorial coordinates $y^{\mu\nu} = -y^{\nu\mu}$ in addition to the standard four–vector coordinates x^μ (see [49]). For $n=16$ one can use the decomposition on the basis of $D=10$ sigma–matrices,

$$\begin{aligned} X^{\alpha\beta} = X^{\beta\alpha} \;&=\; \tfrac{1}{16}x^\mu\tilde{\sigma}_\mu^{\alpha\beta} + \tfrac{1}{2\cdot 16\cdot 5!}y^{\mu_1\ldots\mu_5}\tilde{\sigma}^{\alpha\beta}_{\mu_1\ldots\mu_5}\,, \qquad \mu\,,\nu = 0,1,\ldots,9\,, \\ &\alpha\,,\beta = 1,\ldots,16, \quad y^{\mu_1\ldots\mu_5} = y^{[\mu_1\ldots\mu_5]} = (-)\tfrac{1}{5!}\varepsilon^{\mu_1\ldots\mu_5\nu_1\ldots\nu_5}y_{\nu_1\ldots\nu_5} \end{aligned} \tag{36}$$

one arrives at the parametrization of $\Sigma^{(\frac{10(10+1)}{2}|0)} = \Sigma^{(55|0)}$ space by 10 usual vector and 45 antisymmetric (anti-)selfdual 5–index tensorial coordinates $y^{\mu_1\ldots\mu_5} = y^{[\mu_1\ldots\mu_5]} = (-)\frac{1}{5!}\varepsilon^{\mu_1\ldots\mu_5\nu_1\ldots\nu_5}y_{\nu_1\ldots\nu_5}$ (see [28]). Finally, in $n=32$ one can use the set of $D=11$ gamma matrices to arrive at the parametrization of $\Sigma^{(528|0)}$ by the set of vectorial, x^μ, two–index tensorial $y^{\mu\nu} = -y^{\nu\mu}$ and five–index tensorial $y^{\mu_1...\mu_5} = y^{[\mu_1...\mu_5]}$ coordinates,

$$X^{\alpha\beta} = X^{\beta\alpha} \quad = \tfrac{1}{32} x^{\mu}\Gamma_{\mu}^{\alpha\beta} + \tfrac{i}{64\cdot 5!} y^{\mu\nu}\Gamma_{\mu\nu}^{\alpha\beta} + \tfrac{1}{32\cdot 5!} y^{\mu_1...\mu_5}\Gamma_{\mu_1...\mu_5}^{\alpha\beta} , \qquad (37)$$
$$\mu\,\nu = 0,1,\ldots,10; \quad \alpha,\beta = 1,\ldots,32.$$

These spaces appear as the bosonic body of the 'generalized' [20] or 'extended'/'enlarged' [36, 8] or 'tensorial' (see [18] and refs. therein) superspaces $\Sigma^{(10|4)}$, $\Sigma^{(55|16)}$ and $\Sigma^{(528|32)}$ which we will discuss in Secs III, IV.

The generalized Penrose correspondence has the simple form [20, 6]

$$\mu^{\alpha} = X^{\alpha\beta}\lambda_{\beta} , \qquad \alpha , \beta = 1,\ldots,n \qquad (38)$$

involving the $\Sigma^{(n(n+1)/2|n)}$ coordinates $X^{\alpha\beta}$ and $2n$ real components (n in μ^{α} and n in λ_{α}) of *symplectic twistor* $\Upsilon_{0\hat{\alpha}}$

$$\Upsilon_{0\hat{\alpha}} = (\mu^{\alpha} , \lambda_{\alpha}) \, . \qquad (39)$$

This parametrizes the space $\mathbf{R}^{2n} - \{0\}$ of the fundamental representation the $Sp(2n)$ group which leaves invariant the matrix

$$C_{\hat{\alpha}\hat{\beta}} = \begin{pmatrix} 0 & \delta_{\alpha}{}^{\beta} \\ -\delta^{\alpha}{}_{\beta} & 0 \end{pmatrix} \qquad \hat{\alpha},\hat{\beta} = 1,\ldots,2n \, , \qquad \alpha,\beta = 1,\ldots,n \, . \qquad (40)$$

The homogeneity of Eq. (38) allows one to treat it as an equation in the *projective symplectic twistor space* $\mathbf{RP}^{2n-1}$.

In distinction to (2) *with real* x^a (and in analogy with (5) *with complex* x_L^a), Eq. (38) with an arbitrary real $X^{\alpha\beta} = X^{\beta\alpha}$ does not set any restrictions on the bosonic spinors (or s–vectors [14])[4] μ and λ. Indeed, defining μ and λ in an arbitrary manner one always can find symmetric $X^{\alpha\beta}$ such that (38) holds. Moreover, such $X^{\alpha\beta}$ is not unique. The generalized Penrose correspondence (38) is invariant under the $n(n-1)/2$–parametric generalization of the b–symmetry (8) (see [20, 7])

$$\delta_b X^{\alpha\beta} \;=\; b^{IJ} u_I^{\alpha} u_J^{\beta} \quad (\quad \delta_b\lambda_{\alpha} = 0 \, , \qquad \delta_b\mu^{\alpha} = 0 \quad) \, , \qquad (41)$$

where u_I^{α}, spinors which are orthogonal to λ_{α},

$$u_I^{\alpha}\lambda_{\alpha} = 0 \, , \qquad I = 1,\ldots,(n-1) \, . \qquad (42)$$

Eq. (41) provides the general solution of the condition

$$\delta_b X^{\alpha\beta}\lambda_{\alpha} = 0 \qquad (43)$$

---

[4] One may find better to call λ_{α} and μ^{α} *s-vectors* [14] as the invariance of our basic equations is given by $GL(n)$ and, non-manifestly, by $Sp(2n)$ rather than by some thier $SO(1,D-1)$ subgroup.

which can be used as an alternative definition of the generalized b–symmetry. [5]

Thus Eq. (38) can be considered as a correspondence between the space of $(n-1)$ dimensional hyperplanes in $\mathbf{R}^{2n-1}$ (each parameterized by spinor λ modulo its scaling factor considered to be common with μ^{α}) and the space of $n(n-1)/2$ dimensional surfaces (given by $\hat{X}^{\alpha\beta}(b^{IJ}) = X^{\alpha\beta} + b^{IJ}u_I^{\alpha}u_J^{\beta}$) in $n(n+1)/2$ dimensional tensorial space $\Sigma^{(n(n+1)/2|0)}$. On the language of generalized particle–like mechanics this correspondence implies that the action [20]

$$S_0 = \int_{W^1} \lambda_\alpha \lambda_\beta d\hat{X}^{\alpha\beta} \equiv \frac{1}{2}\int d\tau \lambda_\alpha(\tau)\lambda_\beta(\tau)\partial_\tau \hat{X}^{\alpha\beta}(\tau)\,, \tag{44}$$

which possess a gauge b–symmetry given by pull–back of Eq. (43) on the worldline W^1, allows for the reformulation in terms of the symplectic twistor coordinate functions $\hat{\Upsilon}_{0\alpha}(\tau) = (\hat{\mu}^{\alpha}, \hat{\lambda}_{\alpha})$. Indeed, one can use the Leibniz rule ($\hat{\lambda}\partial\hat{X} \equiv \partial(\hat{\lambda}\hat{X}) - (\partial\lambda)\hat{X}$) to present the action (44) in the equivalent form

$$S \;=\; \int_{W^1} \left(d\hat{\mu}^{\alpha}\,\hat{\lambda}_\alpha - d\hat{\lambda}_\alpha\,\hat{\mu}^{\alpha}\right) \equiv \int_{W^1} d\hat{\Upsilon}_{0\hat{\alpha}} C^{\hat{\alpha}\hat{\beta}}\,\hat{\Upsilon}_{0\hat{\beta}}\,, \tag{45}$$

where $\hat{\mu}^{\alpha}$ is defined by the pull–back of the generalized Penrose correspondence relation (38), $\hat{\Upsilon}_{0\hat{\alpha}}$ by the pull–back of (39) and $C^{\hat{\alpha}\hat{\beta}}$ is the $Sp(2n)$ invariant of Eq. (40).

Notice that the action (44) produces the generalized Cartan–Penrose representation (29) as a primary constraint (see [65]) in the Hamiltonian formalism. This gives a reason (see the supersymmetric considerations in [8, 12] and Sec. III for more) to treat (44) as a *dynamical model for a point–like BPS preon.*

Hence a search for a generalized Penrose correspondence related to the generalized Cartan–Penrose representation (29) and the definition of the BPS preon (25) leads us, through Eq. (38), to the *(ortho)symplectic twistors* (39) and to the *generalized or tensorial (super)spaces* (34) which appeared also in different perspective, see [36, 37].

II.4. BPS preons and higher spin fields in $D = 4, 6, 10$

In distinction to (11), the "preonic" action (45) contains an *unconstrained twistors, i.e.* no counterpart of the helicity constraint (19) appears when one writes the equivalent twistor representation (45) of Eq. (44). After quantization of particle mechanics (11) one arrives at the wave function $\phi(\lambda_A, \bar{\lambda}_{\dot{A}})$ subject to the quantum counterpart of the constraint (4). Just the latter constraint makes $\phi(\lambda_\alpha) = \phi(\lambda_A, \bar{\lambda}_{\dot{A}})$ to describe massless

[5] To see that the transformations (41) provide the counterpart of the 'standard' b–symmetry (3), one notices that, allowing for the existence of a non–degenerate antisymmetric matrix $C_{\alpha\beta} = -C_{\beta\alpha}$ [clearly, for even n, see footnote 3; this also reduces $GL(n)$ symmetry down to $Sp(n)$], one can use its inverse $C^{\alpha\beta}$ to define $\lambda^{\alpha} = C^{\alpha\beta}\lambda_\beta$ which clearly obeys $\lambda^{\alpha}\lambda_{\alpha} = 0$. This allows us to identify this λ^{α} with one of the u_I^{α} [$u_I^{\alpha} = (\lambda^{\alpha}, u_{\bar{I}}^{\alpha})$, see [12]] and to find among (41) the counterpart $\delta_b X^{\alpha\beta} = b\lambda^{\alpha}\lambda^{\beta}$ of the $D=4$ transformations (3); for $D=3$ there is only one u_I^{α} which coincides with λ^{α} and thus all the b–symmetry is reproduced by the above formula.

particle of *a certain* helicity (see [21, 27, 28, 29], [13] and refs. therein), *e.g.* of helicity equal to zero. This explains the helicity constraint name. Now, the action (45) for the $n = 4$ case differs from that in Eq. (11) just by the absence of the helicity constraint (4) ($\lambda_\alpha = (\bar{\lambda}^{\dot{A}}, \lambda_A)$, $\mu^\alpha = (\bar{\mu}_{\dot{A}}, \mu^A)$). Hence, one may expect that the quantum state spectrum of this "preonic" particle mechanics would include a tower of massless field of all possible helicities. The analysis of [13] showed that this is indeed the case for $n = 4$ $D = 4$ and indicated an infinite tower of massless $D = 6$ and $D = 10$ 'higher spin' fields in the model with $n = 8$ and $n = 16$. As it is finally shown in [19] these are all the *conformal massless* higher spin fields in $D = 6$ and $D = 10$ dimensions, respectively.

Here we will try to create some image of the relation between preonic particle mechanics and higher spin fields with an emphasis on the rôle of twistor–like methods and notions. More technical details can be found in the original papers.

II.4.1. Higher spins from tensorial space

Interestingly enough, the tensorial space (34) with $n = 4$, Eq. (35), was proposed in [49] as a basis for the construction of D=4 higher–spin theories. It was known that a consistent interaction of higher spin fields requires i) *an infinite tower of all possible higher spin fields* and ii) a spacetime with a nonzero cosmological constant (see [50, 17, 26]). The assumption of [49] was that there may exist a theory in a ten–dimensional space $\Sigma^{(10|0)}$ whose (alternative–to–) Kaluza–Klein reduction may lead in $D = 4$ to an infinite tower of fields with increasing spins instead of the infinite tower of Kaluza–Klein particles of increasing mass. It was argued that the symmetry group of the theory should be $Sp(8) \supset SU(2,2)$, and $OSp(1|8)$ in supersymmetric case. The idea was that using a single representation of $OSp(1|8)$ (such that it contains each and every massless higher spin representation of the $D = 4$ superconformal group $SU(2,2) \subset OSp(1|8)$ only once) in the ten–dimensional tensorial space one could describe an infinite tower of massless higher spin fields in $D = 4$ space-time.

In this perspective the $\Sigma^{(n(n+1)/2|n)}$ superparticle action of [20], the *point–like model for BPS preon* in the light of [6] and [7, 8, 12], provided (rather accidentally; in its $n = 4$ $\Sigma^{(10|4)}$ version) a dynamical realization of the proposal from [49]. This *preonic superparticle action*, whose purely bosonic limit is given by Eq. (44), involves the auxiliary bosonic spinor variables $\lambda_\alpha(\tau)$. These provide a *twistorial dimensional reduction* (for $n > 2$) and, for $n = 4, 8, 10$, also a *twistorial compactification* mechanism [13] which results in the discreetness of the quantum state spectrum and its identification, for $n = 4$, with the spectrum of all the massless higher spin fields in $D = 4$ [13] and, for $n = 8, 16$, with the spectrum of all *conformal* massless higher spin fields in $D = 6$ and 10 [19]. The AdS generalization of the model [20] is provided by the superparticle on the $OSp(1|n)$ supergroup manifold [13, 51]. It was conjectured in [51, 14] and shown in [52, 23] that a field theory on $OSp(1|4)$ is classically equivalent to the $OSp(1|8)$–invariant free higher spin field theory in AdS_4.

The preonic particle model (44) possesses manifest $GL(n)$ and non–manifest $Sp(2n)$ symmetry ($OSp(1|2n)$ in supersymmetric case), thus showing the expected symmetry of higher spin theories. The latter becomes manifest symmetry after passing to an equivalent twistor form (45). The $Sp(2n)$ symmetry is also manifest in the following equivalent form of the action (44) [14]

$$S_0 = \int_{W^1} \left(\lambda_\alpha \lambda_\beta d\hat{X}^{\alpha\beta} + \tilde{\mu}^\alpha d\lambda_\alpha \right) . \tag{46}$$

One of the simple ways to show the equivalence of (44) and (46) is to notice that, moving the derivatives, one can rewrite (46) in the equivalent form (45) of the action (44). The only difference then will be a shift in definition of μ^α: Eq. (38) is replaced by $\mu^\alpha = X^{\alpha\beta}\lambda_\beta + \tilde{\mu}^\alpha$.

For Hamiltonian formalism the use of the action (46) instead of (44) looks like a simple method of *conversion* of the second class constraints, which are present for (44), into the first class ones (see [16] for further discussion and references; the conversion was also done in [13] but in a more complicated way). The quantization with such a conversion results in the preonic wave function $\Phi(X,\lambda)$ subject to the constraint [13]

$$\left(\partial_{\alpha\beta} - i\lambda_\alpha \lambda_\beta\right) \Phi(X,\lambda) \; = 0 \, , \qquad \partial_{\alpha\beta} := \partial/\partial X^{\alpha\beta} \, . \tag{47}$$

Eq. (47) is clearly the coordinate representation of the definition of the BPS preon (25) *provided the preon is considered as a point–like* object in tensorial space,

$$\Phi(X,\lambda) = < X^{\alpha\beta} \mid \lambda_\alpha\rangle \equiv < X^{\alpha\beta} \mid BPS \; preon \; \lambda_\alpha\rangle \, . \tag{48}$$

This gives one more reason to state that the generalized (super)particle model [20] with the bosonic limit (44) provides a model for a point–like BPS preon [6, 12].

The solution of Eq. (47) is given by the generalized plane wave, *preonic plane wave*,

$$\Phi(X,\lambda) = \phi(\lambda) \, \exp\{-i\lambda_\alpha X^{\alpha\beta} \lambda_\beta\} \, . \tag{49}$$

involving an arbitrary function $\phi(\lambda)$ of the bosonic spinor λ_α.[6] Its integration over λ

$$b(X) = \int d^n\lambda \; \Phi(X,\lambda) = \int d^n\lambda \; \phi(\lambda) \, \exp\{-i\lambda_\alpha X^{\alpha\beta} \lambda_\beta\} \tag{50}$$

provides the general solution of the following equations in tensorial space

$$\partial_{\alpha[\beta}\partial_{\gamma]\delta} b(X) \equiv 1/2(\partial_{\alpha\beta}\partial_{\gamma\delta} - \partial_{\alpha\gamma}\partial_{\beta\delta}) b(X) = 0 \, . \tag{51}$$

This was proposed in [14] as dynamical equations for massless higher spin fields. It was shown in [14] (see also [16]) that, for $n = 4$, decomposing the field $b(X) = b(x^\mu, y^{\mu\nu})$ (see (35)) in the series on $y^{\mu\nu} = -y^{\nu\mu}$ one finds all the field strengths of the massless *bosonic* $D = 4$ higher spin fields *i.e.* of the higher spin fields with integer spin; their equations of motion follow from (51). The details and further discussion on relation

---

[6] The choice of the class of functions where $\phi(\lambda)$ takes its values is not unique. One should fix it to provide the convergence of integrals used on the way to a spacetime treatment which, in its turn, is also not unique. This interesting issue is beyond the score of that contribution; see [15, 16] and also sec. IIA and footnote 5 in [13] for discussions.

between field theories in tensorial space and in the standard spacetime can be found in [15, 16] (mainly for $D=4$) and in [19] (also for $D=6,10$).

The field strengths of massless fields with all the possible half–integer spins are collected in the spinorial field $f_\alpha(X)$ obeying a tensorial space counterpart of the Dirac equation

$$\partial_{\alpha[\beta} f_{\gamma]}(X) \equiv 1/2(\partial_{\alpha\beta} f_\gamma(X) - \partial_{\alpha\gamma} f_\beta(X)) = 0\,. \tag{52}$$

The general solution of Eq. (52) is given by the integral of the preonic plane wave (49) with measure $d^n\lambda\,\lambda_\alpha$,

$$f_\alpha(X) = \int d^n\lambda\,\lambda_\alpha \Phi(X,\lambda) = \int d^n\lambda\,\lambda_\alpha\,\phi(\lambda)\exp\{-i\lambda_\alpha X^{\alpha\beta}\lambda_\beta\}\,. \tag{53}$$

Clearly, as for the even n (including $n=4$) the measure is symmetric under $\lambda \mapsto -\lambda$, the half integer fields collected in $f_\alpha(X)$ come from the odd part of $\phi(\lambda)$, $\phi_-(\lambda) = 1/2(\phi(\lambda)-\phi(-\lambda))$, while the integer fields collected in $b(X)$ come from the even part of $\phi(\lambda)$, $\phi_+(\lambda) = 1/2(\phi(\lambda)+\phi(-\lambda))$. However both integer and half–integer fields come from the same "twistorial wave function" $\phi(\lambda)$. [7] It appears directly as a result of quantization when one starts from the action (45).

II.4.2. Twistor wave function, Cartan–Penrose representation and the Hopf fibrations. D=3,4,6 and 10.

A simplest way to quantize the preonic particle model [20] is to use an equivalent twistor representation (45) of the preonic action (44) [13]. Indeed, the action (45) is i) written in terms of unconstrained symplectic twistor $(\mu^\alpha, \lambda_\alpha)$; ii) is the first order action, which allows to identify $(\mu^\alpha$ with the momentum conjugate to the coordinate λ_α (or *vice versa*); iii) the Hamiltonian of the system is identically zero, which implies that the system is free and that, after quantization, the Schrödinger equation just states an independence of the proper time parameter τ (reparametrization invariance). As a result, one sees that the wave function of the preonic particle is an arbitrary function $\phi(\lambda_\alpha)$ of the bosonic spinor λ_α (see footnote 6).

To provide the spacetime treatment of such a wavefunction one applies [13] the generalized Penrose correspondence (29) and extracts the D dimensional momentum from the generalized momentum using the $n\times n$ real (or pseudo–real) representation of D–dimensional gamma matrices ($n=2^{[D/2]}$ for $D\neq 10$ and $n=2^{[D/2]}$ for other D)

$$p_{\alpha\beta} = \lambda_\alpha\lambda_\beta\,, \qquad p_\mu \equiv \frac{1}{n} p_{\alpha\beta}\Gamma_\mu^{\alpha\beta} = \lambda\Gamma_\mu\lambda \qquad \begin{cases} \alpha,\beta = 1,\dots,n \\ \mu = 0,1,\dots,D-1 \end{cases}. \tag{54}$$

Thus the wave function $\phi(\lambda)$ can be treated as dependent on the spacetime momentum p_μ and the additional variables parameterizing the fibration $\Im_n^D$ of the space $\pounds := \mathbf{R}^n -$

---

[7] In this respect one notices (see [13] and [19]) that the quantum state spectrum of the preonic particle is already supersymmetric (contains all integer and all half integer fields) while the spectrum of its supersymmetric generalization [13] is doubly degenerate. To resolve the degeneracy and to provide the physical spin–statistics correspondence one uses a projection relating the Grassmann parity and the parity with respect to $\lambda \mapsto -\lambda$; see [13], [18] and refs. therein for further discussion.

$\{0\} = \mathbf{S}^{n-1} \otimes \mathbf{R}_+$ of nonvanishing spinors $\lambda_\alpha \neq (0,\dots,0)$ over the (base) space $\wp :=$ $\{p_\mu : p_\mu = \lambda\Gamma_\mu\lambda\}$ of momentums determined by the Cartan–Penrose representation $p_\mu = \lambda\Gamma_\mu\lambda$,

$$\phi(\lambda) = \phi(p; \Im_n^D)|_{p=\lambda\Gamma\lambda} \quad \Leftrightarrow \quad \phi(\pounds) = \phi(\wp, \Im), \quad \Im \approx \tfrac{\pounds}{\wp}\,. \tag{55}$$

The properties of the base space $\wp = \{p_\mu : p_\mu = \lambda\Gamma_\mu\lambda\}$ depends strongly on D and n.

For $n = 2,4,8,16$ corresponding to $D = 3,4,6,10$ the famous identity $\Gamma^{\mu(\alpha\beta}\Gamma_\mu{}^{\gamma)\delta} \equiv 0$ holds. It results in a light–like momentum p_μ

$$D = 3,4,6,10 \; : \quad p_\mu = \lambda\Gamma_\mu\lambda \quad \Rightarrow p_\mu p^\mu = 0\,. \tag{56}$$

Hence the space $\wp$ spanned by momenta $p_\mu = \lambda\Gamma_\mu\lambda$ is $D-1$ dimensional (rather than D–dimensional), $\wp = \mathbf{R}^{(D-1)} - \{0\} = \mathbf{S}^{(D-2)} \otimes \mathbf{R}_+$. Hence the space $\Im_D^n$ of additional variables in (55) is the fibration $\pounds^n/\wp$ of $\pounds = \mathbf{S}^{(n-1)} \otimes \mathbf{R}_+$ over $\wp = \mathbf{S}^{(D-2)} \otimes \mathbf{R}_+$ which, in the light of identification of scales by $p_\mu = \lambda\Gamma_\mu\lambda$, is the fibrations of spheres over spheres, $\Im_D^n := \frac{\pounds^n}{\wp^D} = \frac{\mathbf{S}^{(n-1)}\otimes\mathbf{R}_+}{\mathbf{S}^{(D-2)}\otimes\mathbf{R}_+} = \frac{\mathbf{S}^{(n-1)}}{\mathbf{S}^{(D-2)}}$. Furthermore, as in the dimensions $D =$ $3,4,6$ and 10 the number of components of minimal real (or, in $D = 6$, pseudoreal) representation can be written as $n = 2(D-2)$ (see [23] and refs. therein), the spaces of nonvanishing bosonic spinors are $\pounds^n = \pounds^{2(D-2)} = \mathbf{S}^{(n-1)} \otimes \mathbf{R}_+ = \mathbf{S}^{(2D-5)} \otimes \mathbf{R}_+$ and the fibrations parametrized by additional variables can be presented in a more transparent form $\Im_D^n = \Im_D^{2(D-2)} = \mathbf{S}^{(2D-5)}/\mathbf{S}^{(D-2)}$. This form makes evident that for $D = 3,4,6$ and 10 the spaces $\Im$ of auxiliary variables in (55) are given by the Hopf fibration of spheres over spheres which are isomorphic to spheres $S^{(D-3)}$,

$$D = 3,4,6,10 \quad : \qquad \Im_D^n \equiv \frac{\mathbf{S}^{(n-1)}}{\mathbf{S}^{(D-2)}} = \frac{\mathbf{S}^{(2D-5)}}{\mathbf{S}^{(D-2)}} = \mathbf{S}^{(D-3)} \quad (= (\mathbf{Z}_2, \mathbf{S}^1, \mathbf{S}^3, \mathbf{S}^7))\;. \tag{57}$$

For $D = 3$ Eq. (57) gives $\Im_3^2 = \mathbf{Z}_2$. Thus the wave function $\phi(\lambda)$, Eq. (55), can be treated as function of a light–like momentum (56) and a sign variable $\phi(\lambda) = \phi(p_\mu :$ $p^2 = 0; \pm)$. In the cases of $D = 4,6,10$ the wave function $\phi(\lambda)$ of Eq. (55) depends, in addition to the light–like momentum $p_\mu = \lambda\Gamma_\mu\lambda$, on a number of angular variables which parameterize the compact spaces $S^{(D-3)}$

$$\phi(\lambda_\alpha) = \Phi(p_\mu\,,\, S^{(D-3)})|_{p_\mu p^\mu=0} \;\; \equiv \Phi(p_\mu\,, \alpha_1\,,\dots,\alpha_{D-3})|_{p_\mu p^\mu=0}\;,\; D = 4,6,10\,. \tag{58}$$

Thus *for D=4,6,10 the space of additional variable* $\Im_D^{2(D-2)}$ *in (55) is compact and isomorphic to the spheres* $S^{(D-3)}$. This phenomenon was called *twistor compactification* in [13]. This twistorial compactification is alternative to the Kaluza–Klein one in particular as it occurs in momentum space and hence makes the coordinates discrete. These discrete "coordinates" can be identified with quantum numbers enumerating the possible helicity states of all massless higher spin fields in $D = 4$ [13] and all *conformal* massless higher spin fields in $D = 6,10$ [19].

In the quantization of preonic mechanics [13] the twistor compactification occurs due to the treatment of the gamma–trace of generalized momentum $p_{\alpha\beta}$ as spacetime momentum (see (54)) and due to the generalized Cartan–Penrose representation for $p_{\alpha\beta}$. It was also noticed [13] that the generalized Cartan–Penrose representation (54), $p_{\alpha\beta} = \lambda_\alpha \lambda_\beta$ provides, for $n > 2$, a mechanism of *twistorial dimensional reduction* which also occur in momentum space and reduces the number $n(n+1)/2$ of degrees of freedom in $p_{\alpha\beta}$ to the smaller number n of degrees of freedom in the bosonic spinor λ_α .

II.4.3. Problems in M–theoretical $D = 11$ case $n = 32$.

What turns out to be different in the M–theoretical $D = 11$ case is that the momentum $p_\mu = 1/32\lambda\Gamma_\mu\lambda$ *is not* light–like, $p_\mu p^\mu \neq 0$. Its square (the $D = 11$ mass operator) can be rather expressed through the values of tensorial charges, $p_\mu p^\mu = -2Z^{\mu\nu}Z_{\mu\nu} - 5!Z_{\mu_1\ldots\mu_5}Z^{\mu_1\ldots\mu_5}$ [48], also constructed from the bosonic spinor: $Z_{\mu\nu} = \frac{i}{64}\lambda\Gamma_{\mu\nu}\lambda$ and $Z_{\mu_1\ldots\mu_5} = \frac{1}{32\cdot 5!}\lambda\Gamma_{\mu_1\ldots\mu_5}\lambda$. Thus, if one identifies $p_\mu = 1/32\lambda\Gamma_\mu\lambda$ with the eleven–dimensional momentum, this is not restricted by a mass shell condition and parametrizes $\mathbf{R}^D - \{0\} = \mathbf{S}^{(D-1)} \times \mathbf{R}_+ = \mathbf{S}^{10} \times \mathbf{R}_+$ (instead of $\mathbf{S}^{(D-2)} \times \mathbf{R}_+ = \mathbf{S}^9 \times \mathbf{R}_+$ as it would be if the momentum were light–like). Then the additional variables in (55) parametrize the fibration $\frac{\mathbf{S}^{31}}{\mathbf{S}^{10}}$. Such a 21 dimensional space is not (is not known to be) isomorphic to a sphere or a well–studied manifold. The indefiniteness of the $p_\mu p^\mu$ for $p_\mu = 1/32\lambda\Gamma_\mu\lambda$ in $D = 11$ can be treated as the continuity of the mass spectrum (see [48]), which is another possibility left by conformal invariance of the particle or particle–like mechanics (the latter case includes null–(super)–*p*–branes, see [31] and refs. therein).

Notice that a similar problem appeared for eleven—dimensional supermembrane (now M2–brane) [60] and that now this is treated [61] as an indication that the supermembrane is a composite state, a system of ten–dimensional *D*–branes, in the spirit of Matrix model [62]. One might try to understand the relation between the D=11 BPS preons and spacetime fields ("higher spin" $D = 11$ or $D = 10$ fields) in such a perspective. However, for a moment we do not have much to say on this direction.

III. BPS PREONS AND SUPERSYMMETRY

We discussed the definition and some properties of BPS preons, including their relation with massless higher spin theories in $D = 4, 6, 10$ (almost) without the use of supersymmetry. Now we move to the supersymmetric aspects of the BPS preon conjecture. The pure bosonic definition of the BPS preon (25) uses the generalized momentum. As we have noticed, this appears to be related with the most general supersymmetry algebra.

III.1. Generalized momentum, M–algebra and BPS states

For any $n = 2^m$ the generalized momentum operator $P_{\alpha\beta} = P_{\beta\alpha}$ is associated with the bosonic central generator of most general supersymmetry algebra characterized by the most general form of the commutator of two fermionic supercharges Q_α,

$$\{Q_\alpha, Q_\beta\} = P_{\alpha\beta}\,, \quad P_{\alpha\beta} = P_{\beta\alpha}\,, \quad \alpha = 1, 2, \ldots n\,; \tag{59}$$

the property of $P_{\alpha\beta}$ to be central is expressed by

$$[Q_\alpha, P_{\beta\gamma}] = 0 , \qquad [P_{\alpha\beta}, P_{\gamma\delta}] = 0 . \tag{60}$$

The algebra (59) with $n = 32$ ($\alpha = 1,2,\ldots 32$) [34, 35] is usually called M–theory superalgebra or M–algebra [35] [8]. It encodes a full information about the nonperturbative BPS states of the hypothetical underlying M–theory and also its duality symmetries [42, 35]. Indeed, for instance, treating α, β as eleven dimensional ($SO(1,10)$) spinor indices, one may decompose the symmetric spin–tensor generator $P_{\alpha\beta}$ on the basis provided by antisymmetric products of $D = 11$ gamma–matrices

$$\alpha, \beta = 1,2,\ldots 32 \quad , \qquad P_{\alpha\beta} = P_\mu \Gamma^\mu_{\alpha\beta} + Z_{\mu\nu} i\Gamma^{\mu\nu}_{\alpha\beta} + Z_{\mu_1\ldots\mu_5}\Gamma^{\mu_1\ldots\mu_5}_{\alpha\beta} . \tag{61}$$

Here P_μ is treated as the $D = 11$ spacetime momentum, while the additional central bosonic generators $Z_{\mu\nu} = -Z_{\nu\mu} = Z_{[\mu\nu]}$, $Z_{\mu_1\ldots\mu_5} = Z_{[\mu_1\ldots\mu_5]}$ can be treated [43] as topological charges of the BPS states corresponding to supersymmetric extended objects, branes, living in the eleven–dimensional world. For instance, the spacial components Z_{IJ} and $Z_{J_1\ldots J_5}$ of $Z_{\mu\nu}$ and $Z_{\mu_1\ldots\mu_5}$ are related to the topological charges [43] of the supermembrane and the super–M5–brane [43, 44].

A simple but important observation is that, *e.g.* for $n = 32$, the M–algebra (59), (60) possesses the $GL(32)$ ($GL(n)$) automorphism symmetry, which is broken down to $SO(1,10)$ only upon the use of decomposition (61). This implies the "brane rotating" nature of the $GL(32)/SO(1,10)$ symmetry [64]. On the other hand it indicates that the above eleven–dimensional treatment based on the decomposition (61) is, certainly, not a unique one. Treating α, β in (59), (60) as multiindices and using the set of gamma–matrices for other D in direct product with the internal space gamma–matrices one may provide the D=10 type IIA, D=10 type IIB, D=4 N=8 [35], as well as a more exotic D=2+10 treatment [63]. As a result, the information about nonperturbative BPS states of, say, $D = 10$ superstring theories (including Dirichlet superbranes) can also be extracted from (59). This also explains why the M–algebra (59) encodes as well all the duality relations between different $D = 10$ and $D = 11$ superbranes.

III.2. BPS preons states preserving all but one supersymmetry. A classification of M-theory BPS states

Associating the generalized momentum matrix with the right hand side of the general supersymmetry algebra (59) one finds (see [6, 12, 8]) that the definition of the BPS preon in Eqs. (24), (25) is equivalent to defining the BPS preon as a state preserving all supersymmetries but one (hence the notation $|BPS,(n-1)\rangle$; $|BPS,31\rangle$ in $D = 11$) [6].

---

[8] See [36] for a treatment of (59)–(60) as central extension of the abelian fermionic translation algebra, [38, 36] and refs. therein for further generalizations of the M–theory superalgebra and for their structure. For $n \neq 2^l$, including odd values of n one can treat the algebra as (59) as $d = 2$ extended supersymmetry algebra with central charges.

The bosonic spinor parameters $\in_I{}^{\alpha}$ corresponding to the supersymmetries preserved by a BPS preon $|\lambda\rangle$,

$$\in_I^{\alpha} Q_\alpha|\lambda\rangle = 0\,, \qquad I = 1,\ldots,(n-1) \qquad (I = 1,\ldots,31 \;\; in \;\; D = 11) \tag{62}$$

are 'orthogonal' to the bosonic spinor λ_α that labels it,

$$\in_I{}^{\alpha}\lambda_\alpha = 0\,, \qquad I = 1,\ldots,(n-1) \qquad (I = 1,\ldots,31 \;\; in \;\; D = 11). \tag{63}$$

Notice that these are the same bosonic spinors as in Eq. (42), which completes the definition (41) of the b–symmetry transformations,

$$\in_I{}^{\alpha} = u_I{}^{\alpha}\,. \tag{64}$$

In general, the number $n-k$ ($32-k$ in D=11) of supersymmetries preserved by a BPS state $|BPS,k\rangle$ coincides [6, 8, 12] with the rank of the eigenvalue matrix $p^{(k)}{}_{\alpha\beta}$ (28) of the generalized momentum $P_{\alpha\beta}$

$$\in_{\check{I}}{}^{\alpha}Q_\alpha|BPS\,k\rangle = 0\,, \quad \check{I} = 1,\ldots,k \quad \Rightarrow \quad \begin{cases} P_{\alpha\beta}|BPS\,k\rangle = p^{|k\rangle}_{\alpha\beta}|BPS\,k\rangle\,, \\ \mathrm{rank}(p^{|k\rangle}_{\alpha\beta}) = 32-k\,. \end{cases} \tag{65}$$

In this respect *all the BPS states* related to a general supersymmetry algebra (59), including the M-theory BPS states for $n = 32$, *may be classified by the number of preserved supersymmetries* [6]. This is the same as the classification by rank of the generalized momentum matrix considered in Sec. II.1. Then the discussion of section II.2. implies that *a BPS state preserving k supersymmetry can be treated as a composite of $\tilde{n} = n-k$ BPS preons*

$$\begin{aligned} \in_{\check{I}}{}^{\alpha}Q_\alpha|BPS\,k\rangle = 0\,, \quad \check{I} = 1,\ldots,k \quad &\Longrightarrow \quad |BPS,k\rangle = |\lambda^1\rangle\otimes\ldots\otimes|\lambda^{(32-k)}\rangle\,, \\ \in_{\check{I}}{}^{\alpha}\lambda^{(1)}_\alpha = 0 \quad &, \ldots, \quad \in_{\check{I}}{}^{\alpha}\lambda^{(n-k)}_\alpha = 0\,. \end{aligned} \tag{66}$$

III.3. On "AdS generalizations" of the M–algebra and of the BPS preon definition

Eqs. (59)–(60) give the generalization of the super–Poincaré algebra. The corresponding generalization of the superconformal algebra is suggested to be $OSp(1|2n)$ [see [39, 34] as well as [20] and [8] for more references]. One can ask how the analogous generalization of the AdS superalgebra look like. The study of [13, 14, 52, 16] suggests that this is given by the Lie superalgebra of the $OSp(1|n)$ supergroup,

$$[P_{\alpha\beta}, P_{\gamma\delta}] = i\varsigma\,(C_{\alpha(\gamma}P_{\delta)\beta} + C_{\beta(\gamma}P_{\delta)\alpha})\,, \tag{67}$$

$$[P_{\alpha\beta}\,, Q_\gamma] = i\varsigma\, C_{\gamma(\alpha}Q_{\beta)}\,, \tag{68}$$

$$\{Q_\alpha, Q_\beta\} \;=\; P_{\alpha\beta}\,, \tag{69}$$

where the dimensional parameter ς, the inverse AdS radius, is introduced to make transparent the contraction of $OSp(1|n)$ down to the generalized superPoincaré supergroup $\Sigma^{(n(n+1)/2|n)}$ with the algebra (59)–(60); this occurs in $\varsigma \mapsto 0$ limit.

Clearly, the noncommutative $P_{\alpha\beta}$ cannot be diagonalized and the above definition of the BPS preon should be modified for that case. The study of [52, 16] suggests the following definition of the BPS preons for that case (*cf.* [18])

$$(P_{\alpha\beta} - Y_{(\alpha}Y_{\beta)})|BPS\; preon\,;\lambda_\alpha\rangle = 0\,, \qquad Y_\alpha := \lambda_\alpha - \varsigma P^{(\lambda)}_\alpha\,. \tag{70}$$

Here $P^{(\lambda)}_\alpha$ is the operator of momentum conjugate to λ_α and, hence, the spinorial operators Y_α do not commute for $\varsigma \neq 0$,

$$[Y_\alpha\,, Y_\beta] = 2i\varsigma C_{\alpha\beta}\,. \tag{71}$$

Thus, in distinction to generalized super–Poicaré ($\varsigma = 0$) case, the definition of BPS preons for the generalized AdS or $OSp(1|n)$ superalgebra refers to a factorization of the noncommutative spin–tensorial operator rather than to an eigenvalue problem.

Here, however, we mainly consider the case of the generalized super–Poincaré algebra with central $P_{\alpha\beta}$ which allows for the above simple definition (25) of the BPS preon [6].

III.4. BPS preons and BPS states in supergravity

As discussed above the BPS states $|BPS\; k\rangle$, $k \neq 0$, preserve a fraction (k/n) of the supersymmetries; due to this fact they saturate the Bogomolny–Parasad–Sommerfield or BPS bound (hence the "BPS state" name) and, as a result, are stable.

In supergravity the algebraic notion of BPS state is realized as a supersymmetric solution of the supergravity equations *i.e.* the solitonic solution preserving a fraction k/n ($k/32$ in the M–theoretic $n = 32$ case) of the *local supersymmetries* characteristic of the supergravity theory. The k supersymmetries (65) preserved by the BPS state $|BPS\; k\rangle$ are represented in this "solitonic" picture by a set of k linearly independent *Killing spinors* $\epsilon_{J}{}^{\alpha}(x)$ obeying the *Killing spinors equation*

$$\mathscr{D} \epsilon_{J}{}^{\alpha} := d\epsilon_{J}{}^{\alpha} - \epsilon_{J}{}^{\beta}\omega_\beta{}^\alpha \equiv D\epsilon_{J}{}^{\alpha} - \epsilon_{J}{}^{\beta}\omega_\beta{}^\alpha = 0\,, \qquad J = 1,\ldots,k\,. \tag{72}$$

In many cases, including higher dimensional supergravity (*e.g.* $D = 10, 11$) and extended supergravity in $D = 4$ (*e.g.* $N = 4, 8$) the generalized covariant derivative $\mathscr{D}$ is constructed with the use of *generalized connection* $\omega_\beta{}^\alpha$ which includes, besides the Lorentz (spin) connection $\omega_{L\beta}{}^\alpha = 1/4\omega_L^{ab}\Gamma_{ab\beta}{}^\alpha$, a tensorial part $t_{1\beta}{}^\alpha = \omega_\beta{}^\alpha - \omega_{L\beta}{}^\alpha$ constructed from the fields of supergravity multiplet or their field strengths. Among the cases where the generalized connection is reduced to the Lorentz connection ($\mathscr{D} = D$, $t_{1\alpha}{}^\beta = 0$) is the simple, $D = 4$ $N = 1$, supergravity. In $D = 11$ supergravity the Lorentz covariant part $t_{1\alpha}{}^\beta$ of the generalized connection $\omega_\beta{}^\alpha$ is constructed in terms of the field strength F_{abcd} of the three–form gauge field A_3,

$$\omega_\beta{}^\alpha \;=\; \frac{1}{4}\omega_L^{ab}\Gamma_{ab\beta}{}^\alpha + \frac{i}{18}e^a\left(F_{ab_1b_2b_3}\Gamma^{b_1b_2b_3} + \frac{1}{8}F^{b_1b_2b_3b_4}\Gamma_{ab_1b_2b_3b_4}\right)_\beta{}^\alpha\,. \tag{73}$$

The *necessary condition* for the existence of Killing spinors is given by an algebraic equation coming from the integrability condition for (72) $\mathscr{D}\mathscr{D} \in_J{}^{\alpha} = 0$. It has the suggestive form [10],

$$\in_{\check{J}}{}^{\beta} \mathscr{R}_{\beta}{}^{\alpha} = 0 \qquad \Leftrightarrow \qquad \mathscr{D}\mathscr{D} \in_J{}^{\alpha} = 0 \, , \tag{74}$$

in terms of the *generalized curvature*

$$\mathscr{R}_{\beta}{}^{\alpha} = d\omega_{\beta}{}^{\alpha} - \omega_{\beta}{}^{\gamma} \wedge \omega_{\gamma}{}^{\alpha} \tag{75}$$

or curvature of generalized connection taking values in the Lie algebra of the so–called *generalized holonomy group* [10, 11]. For $D = 11$ supergravity the generalized holonomy group has to be a subgroup of $SL(32, \mathbf{C})$ [11] (see [56] for further discussion with concrete solutions). The same is true for type IIB supergravity [59].

The rôle of BPS preons in the analysis of supersymmetric supergravity solutions was discussed in [12]. The fact that a BPS state $|BPS\ k\rangle$ preserving k of 32 (in general k of n) supersymmetry can be considered as a composite of $(32-k)$ (in general $(n-k)$) BPS preons, Eq. (66), is reflected, in the language of supergravity solutions, by the possibility of finding $\tilde{n} := (32-k)$ spinors (spinor fields) $\lambda^r_{\alpha}(x)$, $r = 1, \ldots, \tilde{n}$ which are orthogonal to the Killing spinors, $\in_{\check{I}}{}^{\alpha}(x)$, $\check{I} = 1, \ldots, k$, Eq. (72), characterizing the solution,

$$\in_{\check{J}}{}^{\alpha}(x)\, \lambda_{\alpha}{}^{r}(x) \;\; = \;\; 0, \qquad \check{J} = 1, \ldots, k, \quad r = 1, \ldots, (32-k) \, . \tag{76}$$

Thus, BPS preonic spinors and Killing spinors provide an alternative (dual) characterization of a $\nu = k/32$–supersymmetric solution ($\nu = k/n$ in general); either one can be used and, for solutions with supernumerary supersymmetries [55], the description provided by BPS preons is clearly a more economic one. Moreover, the use of both BPS preonic ($\lambda_{\alpha}{}^{r}$) and Killing ($\varepsilon_J{}^{\alpha}$) spinors allowed us to develop [12] a *moving G–frame* method, which may be useful in the search for new supersymmetric solutions of supergravity.

As a simplest application of the moving G–frame method let us present the general expression for the generalized curvature (75) of the $k/32$–supersymmetric solution of $D = 11$ Cremmer–Julia–Scherk supergravity [12]. It is given by

$$\mathscr{R}_{\alpha}{}^{\beta} = G_r{}^{s} \lambda_{\alpha}{}^{r} w_s{}^{\beta} + \nabla B^{\check{I}}_r \lambda_{\alpha}{}^{r} \in_{\check{I}}{}^{\beta} \, , \tag{77}$$

where $w_s{}^{\beta}$ is a set of $(32-k)$ spinors obeying $w_s{}^{\beta} \lambda_{\beta}{}^{r} = \delta_s{}^{r}$ and forming, together with the Killing spinor $\in_{\check{J}}{}^{\alpha}$, the nondegenerate matrix

$$g^{-1}{}_{(\beta)}{}^{\alpha} = \begin{pmatrix} w_s{}^{\alpha} \\ \in_{\check{J}}{}^{\alpha} \end{pmatrix} \qquad \left(g_{\alpha}{}^{(\beta)} = \left(\lambda_{\alpha}{}^{s} \, , \, w_{\alpha}{}^{\check{J}} \right) \right) \tag{78}$$

the moving G–frame matrix. Finally,

$$G_r{}^{s} := (dA - A \wedge A)_r{}^{s} \, , \qquad \nabla B^{\check{I}}_r := dB^{\check{I}}_r - A_r{}^{s} \wedge B^{\check{I}}_s \, , \tag{79}$$

where $A_s{}^{r}$ and $B_r{}^{\check{I}}$ are $\tilde{n} \times \tilde{n} \equiv (32-k) \times (32-k)$ and $\tilde{n} \times k \equiv (32-k) \times k$ matrix valued one–forms which have to be fixed by the concrete $k/32$–supersymmetric solution. The condition that the generalized holonomy group H (as well as the generalized structure group G) should be inside $SL(32)$, $\mathscr{R}_\alpha{}^\alpha = 0$, [11] implies that $G_r{}^r = 0$ ($A_r{}^r = 0$).

On one hand, Eq. (77) with $G_r{}^r = 0$ provides an explicit expression for the results of [11, 59] that the generalized holonomies of $k/32$–supersymmetric solutions of $D = 11$ and of $D = 10$ type IIB supergravity should be $H \subset SL(32-k, \mathbf{R}) \otimes \mathbf{R}^{k(32-k)}$. In the light of the fact that, when fermions vanish, $\psi_\mu^\alpha = 0$, all the free bosonic equations of the 11–dimensional supergravity as well as the Bianchi identities for the Riemann tensor and for the three–form gauge field strengths can be collected in the following simple equation for generalized curvature (75) [58, 12]

$$\mathscr{R}_{ab\,\alpha}{}^\gamma \, \Gamma^b{}_\gamma{}^\beta = 0\,, \tag{80}$$

we expect that the explicit form (77) of the generalized curvature $\mathscr{R}$ to be useful in the search for new supergravity solutions [9].

In particular, the moving G–frame formalism might be useful to settle the question whether a BPS preonic solution preserving 31 out of the 32 supersymmetries exists in $D = 11$ and/or $D = 10$ type IIB supergravities. Although this problem was addressed in [10, 11, 12], neither a solution with such a property has been found nor a statement forbidding an existence of such a solution has been proved yet. However in [12] it was observed that BPS preonic configurations do solve the equations of a Chern–Simons like supergravity. This follows from the fact that the generalized curvature of a BPS preon is nilpotent, $\mathscr{R}_\alpha{}^\gamma \wedge \mathscr{R}_\gamma{}^\beta = 0$. This actually follows from the statement of [11] that the generalized holonomy of (a hypothetical) $\nu = 31/32$ supersymmetric solution is a subgroup of $\mathbf{R}^{31}$. More explicitly, according to [12] the generalized curvature for the preonic ($\nu = 31/32$) solution has the form

$$\mathscr{R}_\alpha{}^\beta = dB^I \, \lambda_\alpha \, \in_I{}^\beta \tag{81}$$

which, in the light of (76), implies the nilpotency not only for the two form $\mathscr{R}_\alpha{}^\beta$ but also for the tensor $\mathscr{R}_{ab\,\alpha}{}^\beta$. See [6] for a further discussion on a hypothetical preonic solution in $D = 11$ Cremmer–Julia–Scherk supergravity.

III.5. Superparticle model for BPS preon

Interestingly enough, the point–like model for BPS preon [6] is provided by the action that had been proposed in [20] before the notion of BPS preons was introduced. This describes a superparticle in *tensorial superspace* (which was called "generalized superspace" in [20], "extended superspace" in [36] and enlarged superspace in [8]) with the bosonic body (34),

$$\Sigma^{(\frac{n(n+1)}{2}|n)} \;\; : \;\; \mathscr{Z}^{\mathscr{M}} = (X^{\alpha\beta}, \theta^\alpha)\,, \quad X^{\alpha\beta} = X^{\beta\alpha}\,, \qquad \alpha,\beta = 1,2,\ldots,n\,. \tag{82}$$

[9] The concise form of all the bosonic equations for $D = 11$ supergravity generalizing (80) for the case of nonvanishing fermions can be found in [57].

The action of ref. [20] is the straightforward supersymmetric generalization of the bosonic functional (44), which can be obtained by substituting the pull–back $\hat{\Pi}^{\alpha\beta} \equiv d\tau\hat{\Pi}^{\alpha\beta}_{\tau} = d\hat{X}^{\alpha\beta}(\tau) - id\hat{\theta}^{(\alpha}\hat{\theta}^{\beta)}(\tau)$ of the supersymmetric Volkov–Akulov one–form

$$\Pi^{\alpha\beta} := dX^{\alpha\beta} - id\theta^{(\alpha}\theta^{\beta)} \tag{83}$$

of the $\Sigma^{(\frac{n(n+1)}{2}|n)}$ superspace for $d\hat{X}^{\alpha\beta}$ in (44),

$$S = \frac{1}{2}\int_{W^1} \lambda_\alpha\lambda_\beta\hat{\Pi}^{\alpha\beta} \equiv \frac{1}{2}\int d\tau\, \hat{\Pi}^{\alpha\beta}_{\tau}(\tau)\lambda_\alpha(\tau)\lambda_\beta(\tau)\,. \tag{84}$$

The action (84) is invariant under the global supersymmetry transformations,

$$\delta_\varepsilon\hat{X}^{\alpha\beta}(\tau) = -i\varepsilon^{(\alpha}\hat{\theta}^{\beta)}(\tau), \quad \delta_\varepsilon\hat{\theta}^{\alpha}(\tau) = \varepsilon^{\alpha}, \qquad \delta_\varepsilon\lambda_\alpha(\tau) = 0\,, \tag{85}$$

and under $(n-1)$ local fermionic κ–symmetries

$$i_\kappa\hat{\Pi}^{\alpha\beta} = 0 \quad \Leftrightarrow \quad \delta_\kappa\hat{X}^{\alpha\beta} = 2i\delta_\kappa\hat{\theta}^{(\alpha}\hat{\theta}^{\beta)}\,, \qquad \delta_\kappa\lambda_\alpha = 0\,, \tag{86}$$

$$\delta_\kappa\hat{\theta}^{\alpha}\lambda_\alpha(\tau) = 0 \quad \Leftrightarrow \quad \delta_\kappa\hat{\theta}^{\alpha} = \hat{u}_I{}^{\beta}(\tau)\kappa^{I}(\tau)\,, \quad I = 1,\ldots,(n-1)\,. \tag{87}$$

In (86) $\kappa^I(\tau)$ are the $(n-1)$ (31 for $D=11$) Grassmann parameters of the κ–symmetry and $\hat{u}_I{}^{\beta}(\tau)$ are $(n-1)$ (31) auxiliary spinor fields which are orthogonal to $\lambda_\alpha(\tau)$, Eq. (42) or Eq. (63) with (64). These can be omitted from the consideration as one can use the first equation in (87), $\delta_\kappa\hat{\theta}^{\alpha}\lambda_\alpha(\tau) = 0$, as the definition of the κ–symmetry.

Just the presence of 31–parametric (n–parametric) κ–symmetry, allows one to treat the action (84) as a model for BPS preons: *the κ symmetry of the worldvolume action reflects the supersymmetry preserved by the ground state of the point–like or extended object [66, 44]*. Indeed the requirement of the Lorentz invariance (or more powerful $GL(n)$ invariance) of the ground state leads to the conclusion that in this state all fermions vanish, $\hat{\theta}^{\alpha}(\tau) = 0$. Then, as the fermionic coordinate function transforms both under the supersymmetry and under the κ–symmetry, $\delta\hat{\theta}^{\alpha} = \varepsilon^{\alpha} + \delta_\kappa\hat{\theta}^{\alpha}$, the invariance of the ground state of the superparticle (84) is defined by the equation

$$\hat{\theta}^{\alpha} = 0 \Rightarrow \qquad 0 = \delta\hat{\theta}^{\alpha} = \varepsilon^{\alpha} + \delta_\kappa\tilde{\theta}^{\alpha} = \varepsilon^{\alpha} + \hat{u}^{\alpha}_{I}\kappa^{I}(\tau)\,. \tag{88}$$

Thus the parameters of the symmetries preserving ground state solution should obey

$$\text{susy preserved by ground state with } \hat{\theta}^{\alpha} = 0\text{:} \qquad \varepsilon^{\alpha} = -\hat{u}^{\alpha}_{I}\kappa^{I}\,. \tag{89}$$

The extended object models for BPS preons are provided by tensionless superbranes in $\Sigma^{(528|32)}$ ($\Sigma^{(n(n+1)/2|n)}$) superspace [53, 7].

IV. SUPERFIELDS AND SUPERGRAVITY IN TENSORIAL SUPERSPACE

IV.1. Superfield generalization of the massless higher spin equations

The superparticle models [20] with the the properties of BPS preon [6] were studied in the flat tensorial superspace $\Sigma^{(n(n+1)/2|n)}$ [13, 14, 16] and on the $OSp(1|2n)$ supergroup manifold [51, 52, 16]. The latter are the "AdS–like" version of tensorial superspace (see [51, 14, 52, 16], Sec. II.4 for a brief discussion and [19] for $D = 6, 10$ generalization of this statement). The quantum state spectrum of the preonic superparticle in $D = 4$ contains a tower of conformal massless fields of all possible 'helicities'; which can be described all together (see [13, 19] and Sec. II.4) by the scalar bosonic and spinor (s-vector) fermionic fields obeying Eqs. (51) and (52).

This spectrum is manifestly supersymmetric. Then the question arises: is there any superfield generalization of these equations, *i.e.* is there a superfield equation which collects the scalar and spinor field in tensorial space and implies Eqs. (51) and (52) on these fields? The answer on this question is affirmative. As it was shown in [18], such a superfield equation does exist and has the form

$$D_{[\alpha}D_{\beta]}\Phi(X,\theta)=0\,, \qquad (90)$$

where $D_\alpha = \partial/\partial\theta^\alpha + i\theta^\beta\partial_{\beta\alpha}$ is the flat Grassmann covariant derivative in the flat tensorial superspace $\Sigma^{(n(n+1)/2|n)}$, (82), ($\{D_\alpha, D_\beta\} = 2i\partial_{\alpha\beta}$). Eq. (90) sets to zero all the higher components $\phi_{\alpha_1\cdots\alpha_i}(X)$, $i \geq 2$, of the scalar superfield $\Phi(X,\theta) = b(X) + f_\alpha(X)\,\theta^\alpha + \Sigma_{i=2}^n\,\phi_{\alpha_1\cdots\alpha_i}(X)\,\theta^{\alpha_1}\cdots\theta^{\alpha_i}$, thus reducing it to the form

$$\Phi(X^{\alpha\beta},\theta^\gamma) = b(X) + f_\alpha(X)\,\theta^\alpha\;; \qquad (91)$$

it also imposes on the surviving components the dynamical equations (51) and (52). [10] The generalization of the "preonic equation" (47) has the form [18]

$$(D_\alpha D_\beta + \lambda_\alpha\lambda_\beta)\,\Phi(X\,,\,\theta\,,\,\lambda) = 0\,. \qquad (92)$$

Its antisymmetric part gives Eq. (90) while the symmetric part produces Eq. (47).

The AdS generalization of Eq. (90) reads

$$\left(\nabla_{[\alpha}\nabla_{\beta]} + i\frac{\varsigma}{4}C_{\alpha\beta}\right)\Phi(X,\theta) = 0\,, \qquad (93)$$

where ∇_α are spinorial covariant derivatives on the $OSp(1|n)$ supergroup manifold obeying the superalgebra (69), ((68), (67). The equation generalizing (92) for the case of $OSp(1|n)$ supergroup manifold reads can be found in [18] where it was also discussed the way of derivation of (92) from the equation for Clifford superfield wave function which appeared in the quantization [13] of the preonic superparticle [20], Eq. (84) with the conversion method.

---

[10] One can also collect the same field content inside a spinor superfield Ψ_α, but this should be subject to a set of two equations, $D_{[\alpha}\Psi_{\beta]}(X,\theta) = 0$ and $\partial_{\alpha[\beta}\Psi_{\gamma]}(X,\theta) = 0$ [18].

IV.2. Superfield supergravity in tensorial superspace

When one considers the standard superparticles and superbranes, the natural starting point is also an action in flat superspace. Then one finds [46, 45] that considering the natural generalization of the model in curved superspace and assuming the existence of a smooth flat superspace limit one arrives at the requirement that curved superspace has to satisfy the supergravity constraints. For higher dimensional $D > 6$ superspaces (and also for the extended, $N > 2$, superspaces in $D = 4$) these are the *on–shell* supergravity constraints which contain all the dynamical equations of motion among their consequences. This is not the case in $D = 4$ $N = 1$ superspace where one arrives at off-shell constraints which do not imply dynamical equations of motion (see refs. in [47] which is devoted to a complete Lagrangian description of the supergravity—superstring interaction); in a simpler $D = 3$ and $D = 2$ cases the supergravity is not dynamical.

It is natural to ask what are the generalized supergravity constraints which might appear from the consistency requirement for a preonic model in a curved tensorial superspace. Such a supergravity in a curved $\Sigma^{(528|32)}$ superspace may be interesting in an M–theoretical perspective, while the models in $\Sigma^{(n(n+1)/2|n)}$ with $n = 4,6$ and 10 [$\Sigma^{(10|4)}$, $\Sigma^{(36|8)}$ and $\Sigma^{(136|16)}$] could provide a basis for interacting higher spin theories. One might even hope that such a tensorial supergravity could itself provide an interacting higher spin theory; however, as shown in [18], this is not the case, at least when the supergravity with $SL(n)$ or $GL(n)$ holonomy groups are considered.

The natural generalization of the point like preonic action (84) for the case of curved tensorial superspace $\Sigma^{(n(n+1)/2|n)}$ reads [18]

$$S = \frac{1}{2}\int_{W^1} \lambda_\alpha \lambda_\beta \hat{E}^{\alpha\beta} \equiv \frac{1}{2}\int d\tau \, \hat{E}_\tau^{\alpha\beta}(\tau)\, \lambda_\alpha(\tau)\lambda_\beta(\tau)\,, \tag{94}$$

where $\hat{E}^{\alpha\beta} := d\tau \hat{E}_\tau^{\alpha\beta} = d\tau \partial_\tau \hat{\mathscr{Z}}^{\mathscr{M}}\, E_{\mathscr{M}}{}^{\alpha\beta}(\hat{\mathscr{Z}})$ is the pull–back to the worldline W^1 of the bosonic supervielbein form $E^{\alpha\beta} := d\mathscr{Z}^{\mathscr{M}}\, E_{\mathscr{M}}{}^{\alpha\beta}(\mathscr{Z})$ of the curved tensorial superspace $\Sigma^{(n(n+1)/2|n)}$ (82) with supervielbein

$$E^{\mathscr{A}} = (E^{\alpha\beta}, E^\alpha) = d\mathscr{Z}^{\mathscr{M}} E_{\mathscr{M}}{}^{\mathscr{A}}(\mathscr{Z})\,, \qquad \alpha, \beta = 1, \ldots, n \tag{95}$$

including also n fermionic one forms E^α whose pull–backs $\hat{E}^\alpha$ do not enter Eq. (94).

The supergravity in tensorial superspace should be the theory of the supervielbein superfields in (95). However, to make the formalism covariant one also introduces in superfield supergravity the connection taking values in a structure group of the superspace. For the usual superspace the structure group is the Lorentz group which in the flat superspace appears as a global symmetry of the superparticle action. With this in mind, and taking into attention that the flat superspace preonic superparticle action (84) is invariant under $GL(n)$ group, one finds natural to consider $GL(n)$ as the structure group of tensorial superspace. Hence, by analogy with the conventional spin connection of general relativity and the standard supergravity, the $GL(n)$ connection was introduced in [18],

$$\Omega_\beta{}^\alpha := d\mathscr{Z}^{\mathscr{M}} \Omega_{\mathscr{M}\beta}{}^\alpha \equiv E^{\mathscr{A}} \Omega_{\mathscr{A}\beta}{}^\alpha\,. \tag{96}$$

The torsion 2-forms and the curvature of the $GL(n)$ connection were defined by

$$T^{\alpha\beta} := \mathscr{D}E^{\alpha\beta} \equiv dE^{\alpha\beta} - E^{\alpha\gamma}\wedge\Omega_\gamma{}^\beta - E^{\beta\gamma}\wedge\Omega_\gamma{}^\alpha , \qquad (97)$$

$$T^{\alpha} := \mathscr{D}E^{\alpha} \equiv dE^{\alpha} - E^{\beta}\wedge\Omega_\beta{}^\alpha , \qquad (98)$$

$$\mathscr{R}_\beta{}^\alpha := d\Omega_\beta{}^\alpha - \Omega_\beta{}^\gamma\wedge\Omega_\gamma{}^\alpha . \qquad (99)$$

The requirement of preservation of the κ–symmetry of the superparticle (94) imposes the constraints $T_{\gamma\,\delta}{}^{\alpha\beta} \propto \delta_\gamma{}^{(\alpha}\,\delta_\delta{}^{\beta)}$, $T_{\gamma\gamma'\,\delta}{}^{\alpha\beta} \propto \delta_{(\gamma}{}^{\alpha}\,t_{\gamma')\delta}{}^{\beta}$, on the bosonic torsion (97), $T^{\alpha\beta} := \frac{1}{2}E^{\mathscr{D}}\wedge E^{\mathscr{C}}T_{\mathscr{C}\mathscr{D}}{}^{\alpha\beta}$. Then imposing the conventional constraints, which fix the freedom in redefinition of the basic superfields, and studying the Bianchi identities one finds the following complete expressions for the torsion and curvature two–forms [18]

$$T^{\alpha\beta} = -iE^\alpha\wedge E^\beta + 2E^{\gamma(\alpha}\wedge E^{\beta)\delta}R_{\gamma\delta} , \qquad (100)$$

$$T^{\alpha} = 2E^{\alpha\beta}\wedge E^{\gamma}R_{\beta\gamma} + E^{\alpha\beta}\wedge E^{\gamma\delta}U_{\beta\gamma\delta} , \qquad (101)$$

$$\begin{aligned}\mathscr{R}_\beta{}^\alpha = {}& iE^{\gamma\delta}\wedge E^{\alpha}U_{\beta\gamma\delta} - E^{\alpha\gamma}\wedge E^{\delta}(F_{\delta\beta\gamma} + \mathscr{D}_\delta R_{\beta\gamma}) - \\ & - E^{\alpha\gamma}\wedge E^{\delta\varepsilon}(\mathscr{D}_{(\beta}U_{\gamma)\delta\varepsilon} + \mathscr{D}_{\delta\varepsilon}R_{\beta\gamma}) . \end{aligned} \qquad (102)$$

Here $R_{\gamma\delta}(\mathscr{Z}) = -R_{\delta\gamma}(\mathscr{Z})$ and $U_{\alpha\beta\gamma}(\mathscr{Z}) = U_{\alpha\gamma\beta}(\mathscr{Z})$ are 'main' superfields which are related by the equations

$$\mathscr{D}_{[\alpha}U_{\beta]\gamma\delta} = -\mathscr{D}_{\gamma\delta}R_{\alpha\beta} , \qquad (103)$$

$$\mathscr{D}_{(\alpha}U_{\beta)\gamma\delta} = -i\mathscr{D}_{(\gamma}F_{\delta)\,\alpha\beta} , \qquad F_{\alpha\beta\gamma} = 2iU_{(\beta\gamma)\alpha} - iU_{\alpha\beta\gamma} - 2\mathscr{D}_{(\beta}R_{\gamma)\alpha} , \qquad (104)$$

$$\mathscr{D}_{\alpha\beta}U_{\gamma\delta\sigma} - \mathscr{D}_{\delta\sigma}U_{\gamma\alpha\beta} + 2U_{\gamma\alpha(\sigma}R_{\delta)\beta} + 2U_{\gamma\beta(\sigma}R_{\delta)\alpha} = 0 . \qquad (105)$$

Setting the main superfields to zero, $R_{\gamma\delta}(\mathscr{Z}) = 0$, $U_{\alpha\beta\gamma}(\mathscr{Z}) = 0$ and ignoring the trivial $GL(n)$ connection (setting $\Omega_\alpha{}^\beta = 0$) one reduces the constraints to the Maurer–Cartan equations of flat tensorial superspace $\Sigma^{(n(n+1)/2|n)}$ with the solution

$$R_{\gamma\delta}(Z) = 0 , \quad U_{\alpha\beta\gamma}(Z) = 0 \quad \Rightarrow \quad E^{\alpha\beta} = \Pi^{\alpha\beta} , \quad E^\alpha = d\theta^\alpha . \qquad (106)$$

On the other hand, setting $R_{\gamma\delta} = \varsigma C_{\gamma\delta}$ and $U_{\alpha\beta\gamma}(Z) = 0$ one arrives in the Maurer–Cartan equations of the $OSp(1|2n)$ supergroup

$$\begin{aligned} d\mathscr{E}^{\alpha\beta} &= -i\mathscr{E}^\alpha\wedge\mathscr{E}^\beta - \zeta\mathscr{E}^{\alpha\gamma}\wedge\mathscr{E}^{\delta\beta}C_{\gamma\delta} , \\ d\mathscr{E}^{\alpha} &= -\zeta\mathscr{E}^{\alpha\gamma}\wedge\mathscr{E}^{\delta}C_{\gamma\delta} . \end{aligned} \qquad (107)$$

In both cases the curvature is equal to zero which allows one to gauge away the trivial $GL(n)$ connections $\Omega_\alpha{}^\beta = 0$.

In the superspace subject to the constraints (100) the preonic superparticle action possesses the gauge invariance under the local fermionic κ–symmetry (*cf.* (86), (87))

$$i_\kappa E^{\alpha\alpha'} := \delta_\kappa Z^M E_M^{\alpha\alpha'} = 0 , \quad i_\kappa E^\alpha := \delta_\kappa Z^M E_M^\alpha = u_I^\alpha\kappa^J(\tau) , \qquad (108)$$

where u_I^α is defined by Eq. (42), and under the b–symmetry transformations (*cf.* (41))

$$i_b E^{\alpha\alpha'} := \delta_b Z^M E_M^{\alpha\alpha'} = u_I^\alpha u_J^{\alpha'} b^{IJ}(\tau)\,, \quad i_b E^\alpha := \delta_b Z^M E_M^\alpha = 0\,. \tag{109}$$

One can ask whether the scalar superfield equation (90) allows for a consistent generalization to the curved tensorial superspace. It does and the desired generalization reads

$$\mathscr{D}_{[\alpha}\mathscr{D}_{\beta]}\Phi = \frac{i}{2}R_{\alpha\beta}\Phi \tag{110}$$

and is consistent when the holonomy group of the tensorial superspace is restricted to be $SL(n)$ (which means that the curvature tensor is traceless $\mathscr{R}_\alpha{}^\alpha = 0$) [18].

The first impression might be that the tensorial supergravity defined by the constraints (100)–(102) should contain a huge number of extra nonphysical fields. However this is not the case. As it is shown in [18] the general solution of the tensorial supergravity constraints contains only two classes of superspaces: the *superconformally flat superspaces* and the superspaces superconformally related to the $OSp(1|n)$ supergroup manifold. The superconformally flat superspaces are described by

$$E^{\alpha\beta} = e^{\frac{2W(Z)}{n}}\Pi^{\alpha'\beta'}L_{\alpha'}^{\alpha}(Z)\,L_{\beta'}^{\beta}(Z)\,, \qquad E^\alpha = e^{\frac{W(Z)}{n}}(d\theta^{\alpha'} - i\Pi^{\alpha'\beta'}D_{\beta'}W)L_{\alpha'}^{\alpha}(Z)\,,$$
$$\Omega_\beta^{\ \alpha} = \tfrac{1}{n}dW\,\delta_\beta{}^\alpha - L^{-1}{}_\beta{}^{\beta'}\left[d\theta^{\alpha'}D_{\beta'}W + \Pi^{\alpha'\gamma}(D_{\gamma\beta'}W + \tfrac{i}{2}D_\gamma W\,D_{\beta'}W)\right]L_{\alpha'}^{\alpha}\,, \tag{111}$$

where $L_\beta^{\ \alpha}(Z)$ is a matrix of local $SL(n)$ transformations which together with $exp\{W(Z)/n\}$ form a $GL(n)$ matrix $G_\beta^{\ \alpha} = L_\beta^{\ \alpha}exp\{W(Z)/n\}$. The extraction of the scaling factor allows to apply Eqs. (111) to supergravity with $SL(n)$ structure group.

Working with the structure group $GL(n)$ (which does not forbid reduction of the *holonomy* group down to its subgroup $SL(n)$), one can obtain all superspaces superconformally–related to the $OSp(1|n)$ supergroup manifold by making first the following "generalized super–Weyl transformations" [18]

$$\begin{aligned} E^{\alpha\beta} &= \mathscr{E}^{\alpha\beta}\,, \qquad E^\alpha = \mathscr{E}^\alpha + \mathscr{E}^{\alpha\beta}W_\beta\,, \\ \Omega_\beta^{\ \alpha} &= -i\mathscr{E}^\alpha W_\beta - \mathscr{E}^{\alpha\gamma}(\nabla_\gamma W_\beta + iW_\gamma W_\beta)\,, \end{aligned} \tag{112}$$

of the $OSp(1|n)$ supervielbein ($\mathscr{E}^{\alpha\beta}$, $\mathscr{E}^\alpha$), Eq. (107), with $W_\gamma = -i\nabla_\gamma W$ and then performing a $GL(n)$ "rotation", if needed. In (112) ∇_γ is the $OSp(1|n)$ covariant derivative, $d = \mathscr{E}^{\alpha\beta}\nabla_{\alpha\beta} + \mathscr{E}^\alpha\nabla_\alpha$. The flat tensorial superspaces $\Sigma^{(n(n+1)/2|n)}$ can be recovered in $\varsigma = 0$ limit.

The fact that superconformally flat and $OSp(1|n)$ related superspaces provide the general (modulo topological subtleties) solution of the tensorial supergravity constraints (100)–(102) implies that the main superfields can always be expressed by

$$R_{\alpha\beta} = -\frac{\varsigma}{2}C_{\alpha\beta} + \mathscr{D}_{[\alpha}W_{\beta]} + \frac{i}{2}W_\alpha W_\beta\,, \qquad U_{\alpha\beta\gamma} = -\mathscr{D}_{\beta\gamma}W_\alpha + W_{(\gamma}\mathscr{D}_{\beta)}W_\alpha\,. \tag{113}$$

One can check that the holonomy group of the superspace reduces to $SL(n)$ (*i.e.* that $\mathscr{R}_\alpha{}^\alpha = 0$) when $W_\alpha = -i\mathscr{D}_\alpha W$. The 'super–Weyl transformations' (107) with $W_\alpha \neq$

$-i\mathscr{D}_\alpha W$ result in the connection with $GL(n)$ holonomy. In the $OSp(1|n)$ covariant derivatives the main superfields of the superspace with $SL(n)$ holonomy group read

$$R_{\alpha\beta} = ie^{-\frac{2W}{n}}\left[i\tfrac{\varsigma}{2}C_{\alpha\beta} + \nabla_{[\alpha}\nabla_{\beta]}W + \tfrac{1}{2}\nabla_\alpha W\,\nabla_\beta W\right], \tag{114}$$

$$U_{\beta\gamma\delta} = e^{-\frac{3W}{n}}\left[-i\nabla_{\gamma\delta}\nabla_\beta W + \nabla_{(\gamma}W\,\nabla_{\delta)}\nabla_\beta W\right]. \tag{115}$$

One can make the (seemingly important) observation that, formally, putting in (114) $R_{\alpha\beta} = -\frac{\varsigma}{2}C_{\alpha\beta}\,e^{(1+4/n)W/2}$ one finds an equation

$$\nabla_{[\alpha}\nabla_{\beta]}W + \tfrac{1}{2}\nabla_\alpha W\,\nabla_\beta W = -\tfrac{i\varsigma}{2}C_{\alpha\beta}\left(1 - e^{-\frac{W}{2}}\right) \tag{116}$$

for the scalar superfield W. However, first one observes that, after the field redefinition $W = 2\ln\left(\frac{\Phi+a}{a}\right)$ (with $a > 0$) this reduces to the scalar superfield equation (93) on the $OSp(1|n)$ supergroup manifold. Moreover, this does *not* imply a nontrivial embedding of even the free scalar superfield (and of the higher spin theories) in tensorial supergravity. The reason can be traced to the super–Weyl invariance of both the constraints, Eqs. (100)–(102), and the scalar superfield equation in supergravity background, Eq. (110). Thus one may use (112) as a field redefinition (leaving the constraint invariant) pass form the superconformally–$OSp(1|n)$–related geometry to the rigid $OSp(1|n)$ supergroup manifold.

In other words, like the $D = 3$ $N = 1$ Poincaré and AdS supergravities, the supergravity in tensorial superspace is shown to be nondynamical: the general solution of its constraints is given by superconformally flat and $OSp(1|n)$ related superspaces which may be reduced to the rigid $\Sigma^{(n(n+1)/2|n)}$ or $OSp(1|n)$ superspaces by the (super)field redefinition (Eq. (112) plus $GL(n)$ transformations or, equivalently, Eqs. (111)) [18].

This implies, in particular, that to proceed with the search for $D = 4, 6, 10$ interacting higher spin theories on the basis of curved tensorial superspaces $\Sigma^{(n(n+1)/2|n)}$ with $n = 4, 8, 16$ one has *to extend* the tensorial superspace rather than to restrict it, as it might be expected. On the other hand, such an extension looks natural in the light of the existing results on interacting higher spin theories [17, 50], as these imply the necessity of doubling of the auxiliary variables. Such auxiliary variables responsible for the spin degrees of freedom can be chosen to be spinors or antisymmetric tensors ($y^{[\mu\nu]}$ for $D = 4$). Hence a natural candidate for the variables to use for the extension of the tensorial superspace $X^{(\alpha\beta)}, \theta^\alpha)$ ($= (x^\mu, y^{\mu\nu}, , \theta^\alpha)$ for $D = 4$) in a search for consistent interacting higher spin theories are the bosonic spinors λ_α which are used to define the notion of BPS preon and are present in the action [20] for the "preonic superparticle".

The study of supergravity and super–Yang–Mills theories in tensorial superspace enlarged by additional bosonic spinors is an interesting subject for future study. Another interesting direction is an M–theoretic use of the superconformally flat and $OSp(1|32)$–related superspaces in an M–theoretical context. This might be related with the direction described in the contribution of José A. de Azcárraga to this volume [37].

CONCLUDING REMARKS

In this contribution we made a brief review of the notion of BPS preon, both in its original $D = 11$ image as hypothetical constituents of M-theory [6] and in its natural generalization to arbitrary dimension D [8, 12]. Actually the definition of BPS preon possesses a wider $GL(n)$ symmetry and, thus is rather characterized by the number n of possible values of the spinor (or 's–vector' [14]) index α than by the number D related to an invariance under a subgroup $SO(t, D-t) \subset GL(n)$. For $n = 4, 8, 16$ cases the BPS preon may be identified with the tower of massless higher spin fields in $D = 4, 6$ and 10. This can be established by quantization [13] of the "preonic superparticle model" [20] which, interestingly enough, had been carried out some times before the notion of BPS preon was introduced in [6]. The present treatment [8, 12, 7] of the results of [13, 20] in terms of BPS preon notion [6] is justified by a search for a universal language which might provide a bridge between (essentially) eleven–dimensional M–theory and the higher spin theories in $D = 4, 6, 10$.

We have also reviewed the "preonic superparticle" action of [20] bringing us to the tensorial superspace, as well as the rôle of BPS preons in the classification and study of supergravity solitons [12], the concise superfield description of the higher spin theories [18] and the results of the study of supergravity in tensorial superspace [18].

Actually, in the light of the algebraic classification of the M–theory BPS states proposed in [6], the possibility of treating BPS preons as constituents of M–theory is a bit more than conjecture; a conjecture concerns rather a usefulness of such a treatment. One might express doubts on such a usefulness arguing that the symmetry of the BPS preon is too high to describe the M-theory physics. However, its identification with higher spin theory in lower dimensions suggests an answer. Higher spin theories were (and are) conjectured (see [50] and *e.g.* [26]) to be related to the "symmetric" phase of string theory characterized by an enhanced symmetry whose spontaneous breaking should reproduce the complete string theory. In the same way one may conjecture that the $GL(32)$–invariant (actually $OSp(1|64)$–invariant) description provided by the BPS preons corresponds to a symmetric phase of M–theory, while the complete description of M–theory might require the spontaneous breaking of these $OSp(1|64)$ symmetry.

ACKNOWLEDGMENTS

The author is grateful to José A. de Azcárraga, Jurek Lukierski, Dmitri Sorokin for discussions and useful comments. He thanks M. Roček, W. Siegel, C. Vafa and other organizers of the Simons Workshop in Mathematics and Physics "Superstrings and Topological Strings" for the conversations and hospitality in Stony Brook. The contribution is based on work done in collaboration with J. A. de Azcárraga, J. Lukierski, D. Sorokin, J. Izqierdo, M. Picón, O. Varela and, more recently, P. Pasti and M. Tonin, which is acknowledged with great pleasure. This work has been partially supported by the research grants BFM2002-03681 from the Ministerio de Educación y Ciencia and from EU FEDER funds, N383 from the Ukrainian State Fund for Fundamental Research and by the Grupos 03-124 grant from the Generalidad Valenciana.

REFERENCES

1. R. Penrose and M. A. H. MacCallum, Phys. Rep. **6**, 241 (1972) and refs. therein.
2. R. Penrose, Rep. Math. Phys. **12**, 65 (1977).
3. E. Witten, Commun. Math. Phys. **252**, 189 (2004), [hep-th/0312171];
   N. Berkovits, Phys. Rev. Lett. **93**, 011601 (2004), [hep-th/0402045];
   W. Siegel, *Untwisting the twistor superstring*, hep-th/0404255.
4. R. Roiban, M. Spradlin and A. Volovich, JHEP **0404**, 012 (2004), [hep-th/0402016];
   Dissolving N = 4 loop amplitudes into QCD tree amplitudes, hep-th/0412265;
   F. Cachazo, P. Svrcek and E. Witten, JHEP **0410**, 074 (2004), [arXiv:hep-th/0406177];
   JHEP **0410**, 077 (2004), [hep-th/0409245]; and refs therein.
5. V. P. Nair, Phys. Lett. **B214**, 215 (1988).
6. I.A. Bandos, J.A. de Azcárraga, J.M. Izquierdo and J. Lukierski, Phys. Rev. Lett. **86**, 4451 (2001), [hep-th/0101113].
7. I.A. Bandos, Phys. Lett. **B558**, 197 (2003), [hep-th/0208110].
8. I.A. Bandos, J.A. de Azcárraga, M. Picón and O. Varela, Phys. Rev. **D69**, 085007 (2004), [hep-th/0307106].
9. R. Mkrtchyan, Phys. Lett. B **558**, 205 (2003), [arXiv:hep-th/0212174]; hep-th/0209175;
   H. Mkrtchyan and R. Mkrtchyan, Mod. Phys. Lett. **A18**, 2665 (2003), [arXiv:hep-th/0308065];
   H. Mkrtchyan and R. Mkrtchyan, Mod. Phys. Lett. **A19**, 931 (2004), [arXiv:hep-th/0312281] and refs. therein.
10. M. J. Duff and J. T. Liu, Nucl. Phys. **B674**, 217 (2003), [hep-th/0303140];
    M. J. Duff, *Erice lectures on 'The status of local supersymmetry*, 29 Aug - 7 Sep 2003, hep-th/0403160.
11. C. Hull, *Holonomy and Symmetry in M-theory*, hep-th/0305039.
12. I.A. Bandos, J.A de Azcárraga, J.M. Izquierdo, M. Picón and O. Varela, Phys. Rev. **D69**, 105010 (2004), [hep-th/0312266].
13. I. Bandos, J. Lukierski and D. Sorokin, Phys. Rev. **D61**, 045002 (2000), [hep-th/9904109].
14. M.A. Vasiliev, Phys. Rev. **D66**, 066006 (2002), [hep-th/0106149].
15. M.A. Vasiliev, *Relativity, causality, locality, quantization and duality in the* $Sp(2M)$ *invariant generalized space-time*, hep-th/0111119.
16. M. Plyushchay, D. Sorokin and M. Tsulaia, JHEP **0304**, 013 (2003), [hep-th/0301067];
    GL flatness of $OSp(1|2n)$ *and higher spin field theory from dynamics in tensorial spaces*, hep-th/0310297.
17. M. A. Vasiliev, Phys. Lett. B **567**, 139 (2003), [hep-th/0304049];
    Fortsch. Phys. **52**, 702 (2004), [hep-th/0401177].
18. I. Bandos, P. Pasti, D. Sorokin and M. Tonin, JHEP **0411**, 023 (2004), [hep-th/0407180].
19. I.A. Bandos, X. Bekaert, J. A. de Azcárraga, D. Sorokin and M. Tsulaia, *Tensorial Space and the Dynamics of Conformal Higher Spin Fields*, hep-th/0501113.
20. I. Bandos and J. Lukierski, Mod. Phys. Lett. **14**, 1257 (1999), [hep-th/9811022].
21. A. Ferber, Nucl. Phys. **B132**, 55 (1978);
    T. Shirafuji, Prog. Theor. Phys. **70**, 18 (1983).
22. D. P. Sorokin, Fortsch. Phys. **38**, 923 (1990).
23. D. P. Sorokin, Phys. Rep. **329**, 1 (2000), [hep-th/9906142].
24. S. Fedoruk and V. G. Zima, Journal of Kharkiv Univ. **585**, 39-48, (2003), hep-th/0308154;
    Bitwistor formulation of the spinning particle, hep-th/0401064.
25. A. Bette, J. A. de Azcarraga, C. Miquel-Espanya and J. Lukierski, Phys. Lett. **B595**, 491 (2004), [hep-th/0405166].
26. D. Sorokin, *Introduction to the classical theory of higher spins*, hep-th/0405069, contribution to the present volume.
27. A. P. Balachandran, G. Marmo, B. S. Skagerstam and A. Stern, Lect. Notes Phys. **188**, 1 (1983).
28. Y. Eisenberg and S. Solomon, Nucl. Phys. **B309**, 709 (1988);
    Phys. Lett. **B220**, 562 (1989).
29. I. A. Bandos, Sov. J. Nucl. Phys. **51**, 906–914 (1990);
    JETP Lett. **52**, 205 (1990).
30. D. P. Sorokin, V. I. Tkach and D. V. Volkov, Mod. Phys. Lett. **A4**, 901 (1989).

31. I.A. Bandos and A.A. Zheltukhin, Quantum Fortschr. Phys. **41**, 619 (1993).
32. I.A. Bandos and A. A. Zheltukhin, JETP Lett. **54**, 421 (1991);
Phys. Lett. **B288**, 77 (1992).
33. J.A. de Azcárraga and J. Lukierski, Phys. Lett. **113B**, 170 (1982);
W. Siegel, Phys.Lett. **B128**, 397 (1983).
34. J.W. van Holten and A. van Proeyen, J. Phys. **A15**, 3763–3783 (1982).
35. P.K. Townsend, *p-brane democracy*, hep-th/9507048, *M-theory from its superalgebra*, hep-th/9712004.
36. C. Chryssomalakos, J.A. de Azcárraga, J.M. Izquierdo and J.C. Pérez Bueno, Nucl. Phys. **B567**, 293 (2000), [hep-th/9904137];
J.A. de Azcárraga and J.M. Izquierdo, Am. Inst. of Phys. Conf. Proc. **589**, 3 (2001), [hep-th/0105125] and refs. therein.
37. José A. de Azcárraga, *Superbranes, D=11 CJS supergravity and enlarged superspace coordinates/fields correspondence*, hep-th/0501198, contribution to the presnet volume.
38. E. Sezgin, Phys. Lett. **B392**, 323-331 (1997), [hep-th/9609086].
39. R. D'Auria and P. Fré, Nucl. Phys. **B201**, 101 (1982); *erratum:ibid* **B206**, 496 (1982).
40. I. A. Bandos, J. A. de Azcarraga, J. M. Izquierdo, M. Picon and O. Varela, Phys. Lett. **B596**, 145-155 (2004), [hep-th/0406020].
41. I.A. Bandos, J.A. de Azcarraga, M. Picon and O. Varela, *On the formulation of D = 11 supergravity and the composite nature of its three-form field*, to appear in Ann. Phys., hep-th/0409100.
42. C.M. Hull and P. K. Townsend, Nucl. Phys. **B438**, 109 (1995), [hep-th/9410167];
E. Witten, Nucl. Phys. **B443**, 85 (1995), [hep-th/9503124];
J.H. Schwarz, Nucl. Phys. Proc. Suppl. **55B**, 1 (1997).
43. J.A. de Azcárraga, J.P. Gauntlett, J.M. Izquierdo and P.K. Townsend, Phys. Rev. Lett. **63**, 2443 (1989).
44. D.P. Sorokin and P.K.Townsend, Phys. Lett. **B412**, 265 (1997), [hep-th/9708003].
45. M.T. Grisaru, P.S. Howe, L. Mezincescu, B. Nilsson and P.K. Townsend, Phys. Lett. **B162**, 116 (1985);
J.A. Shapiro and C.C. Taylor, Phys. Rep. **191**, 221-287 (1990) and refs. therein.
46. E. Bergshoeff, E. Sezgin and P.K. Townsend, Phys. Lett. **B189**, 75 (1987).
47. I. A. Bandos and J. M. Isidro, Phys. Rev. **D69**, 085009 (2004), [hep-th/0308102].
48. I. Bandos and J. Lukierski, Lect. Not. Phys. **539**, 195 (2000), [hep-th/9812074].
49. C. Fronsdal, *Massless particles, orthosymplectic symmetry and another type of Kaluza–Klein theory*, in *Essays on Supersymmetry* (C. Fronsdal Ed.), Math. Phys. Stud. **8**, D. Reidel Pub. Co. (1986), p. 163.
50. M.A. Vasiliev, Ann. Phys. (NY) **190**, 59 (1989).
51. I. A. Bandos, J. Lukierski, C. Preitschopf and D. P. Sorokin, Phys. Rev. **D61**, 065009 (2000), [hep-th/9907113].
52. V.E. Didenko and M.A. Vasiliev, J. Math. Phys. **45**, 197 (2004), [hep-th/0301054].
53. A.A. Zheltukhin and D.V. Uvarov, JHEP **0208**, 008 (2002), [hep-th/0206214].
54. D. V. Uvarov and A. A. Zheltukhin, JHEP **0403**, 063 (2004), [hep-th/0310284];
Whether conformal supersymmetry is broken by quantum p-branes with exotic supersymmetry?, hep-th/0401059.
55. M. Cvetič, H. Lu and C.N. Pope, Nucl. Phys. **B644**, 65 (2002), [hep-th/0203229];
J.P. Gauntlett and C.M. Hull, JHEP **0206**, 013 (2002), [hep-th/0203255];
J. Michelson, Phys. Rev. **D66**, 066002 (2002), [hep-th/0203140]; Class. Quant. Grav. **19**, 5935 (2002), [hep-th/0206204];
I. Bena and R. Roiban, Phys. Rev. **D67**, 125014 (2003), [hep-th/0206195].
56. A. Batrachenko, M. J. Duff, J. T. Liu and W. Y. Wen, *Generalized holonomy of M-theory vacua*, hep-th/0312165;
A. Batrachenko, J. T. Liu, O. Varela and W. Y. Wen, *Higher order integrability in generalized holonomy*, hep-th/0412154.
57. I. A. Bandos, J. A. de Azcarraga, M. Picon and O. Varela, *Generalized curvature and equations of D=11 supergravity*, hep-th/0501007.
58. J. P. Gauntlett and S. Pakis, JHEP **0304**, 039 (2003), [hep-th/0212008].
59. G. Papadopoulos and D. Tsimpis, Class. Quant. Grav. **20**, L253 (2003), [hep-th/0307127].

60. B. de Wit, J. Hoppe and H. Nicolai, Nucl. Phys. **B305**, 545 (1988).
61. H. Nicolai and J. Plefka, Phys. Lett. **B477**, 309 (2000), [hep-th/0001106];
A. Dasgupta, H. Nicolai and J. Plefka, Grav. Cosmol. **8**, 1 (2002), [hep-th/0201182] and refs. therein.
62. T. Banks, W. Fischler, S.H. Shenker and L. Susskind, Phys. Rev. **D55**, 5112 (1997), [hep-th/9610043].
63. I. Bars, *S-theory*, Phys. Rev. **D55**, 2373 (1997), [hep-th/9607112]; *2T physics 2001*, hep-th/0106021; Phys. Rev. **D70**, 104022 (2004), [hep-th/0407239] and refs. therein.
64. O. Baerwald and P. West, Phys. Lett. **B476**, 157 (2000), [arXiv:hep-th/9912226];
P. West, *Brane dynamics, central charges and* E_{11}, hep-th/0412336 and refs. therein.
65. P.A.M. Dirac, *Lectures on quantum mechanics*, Academic Press, NY (1967).
66. E. Bergshoeff, R. Kallosh, T. Ortin and G. Papadopoulos, Nucl. Phys. **B502**, 149 (1997), [hep-th/9705040].

Introduction to the Classical Theory of Higher Spins

D. Sorokin

Istituto Nazionale di Fisica Nucleare, Sezione di Padova & Dipartimento di Fisica, Università degli Studi di Padova, Via F. Marzolo 8, 35131 Padova, Italia

Abstract. We review main features and problems of higher spin field theory and flash some ways along which it has been developed over last decades.

INTRODUCTION

Several dozens of years of intensive study, which involved enormous amount of theoretical brain power, have resulted in a deep insight into various fundamental features of String Theory, and every new year of research brings us new and new aspects of its immense structure. By now, for example, we know in detail low energy field–theoretical limits of String Theory which correspond to massless excitations over different string vacua. These are described by ten–dimensional supergravities whose supermultiplets consist of fields of spin not higher than two and higher order corrections thereof. This region of String Theory also contains various types of branes which reflect dualities between different string vacua. The classical dynamics of the supergravity fields is well known from the analysis of the classical supergravity actions and equations of motion which possess an interesting geometrical and symmetry structure based on supersymmetry. However, from the perspective of quantization the ten–dimensional supergravities look not so promising since they are non–renormalizable as field theories containing gravitation usually are. At the same time (Super)String Theory is believed to be a renormalizable and even finite quantum theory in the ultraviolet limit, and therefore it consistently describes quantum gravity. A field theoretical reason behind this consistent quantum behaviour is the contribution to quantum corrections of an infinite tower of massive higher spin excitations of the string, whose mass squared is proportional to string tension and spin (e.g. in open string theory $M_s^2 \sim T\,(s-1) \sim \frac{1}{\alpha'}\,(s-1)$, where s is the maximum spin value of a state). Therefore, a better understanding of the dynamics of higher spin states is important for the analysis of quantum properties of String Theory.

Until recently the field of higher spins has remained a virgin land cultivated by only a few enthusiasts. But higher spin field theory may become a fashionable topic if a breakthrough happens in understanding its basic problems.

Our experience in quantum field theory teaches us that massive fields with spin 1 and higher are non–renormalizable unless their mass was generated as a result of spontaneous breaking of a gauge symmetry associated with corresponding massless gauge fields. So first of all we should understand the structure of the theory of massless

CP767, *Fundamental Interactions and Twistor-Like Methods,* edited by J. Lukierski and D. Sorokin

higher spin fields.

In String Theory higher spin excitations become massless in the limit of zero string tension. Thus in this limit one should observe an enhancement of String Theory symmetry by that of the massless higher spin fields, and one can regard string tension generation as a mechanism of spontaneous breaking of the higher spin symmetry. If the conjecture that String Theory is a spontaneously broken phase of an underlying gauge theory of higher spin fields is realized, it can be useful for better understanding of string/M theory and of the (A)dS/CFT correspondence (see e.g. [1, 3, 4, 5, 6, 7] and references therein). This is one of the motivations of the development of the theory of interacting higher spin fields.

A direct but, perhaps, too involved way of studying the higher spin string states would be the one in the framework of String Field Theory, which itself is still under construction as far as supersymmetric and closed strings are concerned. Another possibile way is, as in the case of lower spin excitations, to derive an effective field theory of higher spins and to study its properties using conventional field theoretical methods.

In fact, higher spin field theory, both for massive and massless fields, has been developed quite independently of String Theory for a long period of time starting from papers by Dirac [8], Wigner [9], Fierz and Pauli [10], Rarita and Schwinger [11], Bargmann and Wigner [12], Fronsdal [13], Weinberg [14] and others. In last decades a particular attention has been paid to massless higher spin fields whose study revealed a profound and rich geometrical and group–theoretical (conformal) structure underlying their dynamics.

Understanding the interactions of higher spin fields is a main long standing problem of the construction of the higher spin field theory. The interaction problem already reveals itself when one tries to couple higher spin fields to an electromagnetic field [16] or to gravity [17, 18], or to construct (three–vertex) self–interactions [19, 20, 21]. In the case of massless higher spin fields the problem is in introducing interactions in such a way that they do not break (but may only properly modify) gauge symmetries of the free higher spin field theory. Otherwise the number of degrees of freedom in the interacting theory would differ from that of the free theory, which apparently would result in inconsistencies.

One should also note that the general (Coleman–Mandula and Haag–Lopuszanski–Sohnius) theorem of the possible symmetries of the unitary S–matrix of the quantum field theory in $D = 4$ Minkowski space [23] does not allow conserved currents associated with symmetries of fields with spin greater than two to contribute to the S–matrix. This no–go theorem might be overcome if the higher spin symmetries would be spontaneously broken, as probably happens in String Theory.

Another way out is that one should construct the interacting higher spin field theory in a vacuum background with a non–zero cosmological constant, such as the Anti de Sitter space, in which case the S–matrix theorem does not apply. This has been realized in [24], where consistent interactions of massless higher spin fields with gravity were constructed in the first non–trivial (cubic) order. Until now the extension of these results to higher orders in the coupling constant at the level of the action has encountered difficulties of a group–theoretical and technical nature related to the problem of finding the full algebraic structure of interacting higher spin symmetries. As has been noted in [20, 21, 24, 25], such an algebraic structure and consistent interactions should involve higher derivative terms and infinite tower of fields with increasing spins, and this again resembles the situation which we have in String Theory. At the level of so called unfolded equations of motion non–linear gauge field models of interacting massless higher spin fields have been constructed in [26, 2, 27].

To study the relation of higher spin field theory to superstring theory one should work in ten–dimensional space–time. Here we encounter a "technical" problem. In $D=4$ all states of higher spin can be described either by the higher rank symmetric tensors or spin tensors, since all tensor fields with mixed, symmetric and antisymmetric, components can be related via Poincaré duality to the symmetric tensors. This is not the case, for instance, in $D=10$ where mixed symmetry tensor fields describe independent higher spin modes and should be studied separately [28]. From the group–theoretical point of view this is related to the fact that in $D=4$ the compact subgroup of the Wigner little group, which is used to classify all the massless irreducible representations of the Poincaré group, is $SO(2)$ whose Young tableaux are single symmetric rows [1], while in $D=10$ the compact subgroup of the little group is $SO(8)$ whose representations are described by Young tableaux with both (symmetric) rows and (antisymmetric) columns. An essential progress in studying the mixed symmetry fields has been made only quite recently [29]–[38].

In these lectures, with the purpose of simplifying a bit the comprehension of the material, we shall mainly deal with higher spin fields described by symmetric tensors and spin tensors.

These notes are not a comprehensive review but rather an attempt to write an elementary introduction to only few aspects of higher spin field theory and its history. I apologize to the authors whose work has not been reflected in what follows.

FREE HIGHER SPIN FIELD THEORY

The choice of Lorentz representations for describing higher spin fields. Symmetric tensors and spin–tensors

One of the possible choices is to associate potentials of integer higher spin fields with symmetric tensors. In $D=4$ the symmetric tensors describe all possible higher spin representations of the Poincaré group because the antisymmetric second rank potentials are dual to scalar fields and the three and four form potentials do not carry physical degrees of freedom. As we have already mentioned, this can also be understood using the fact that the compact subgroup of the little group of the $D=4$ Lorentz group is one–dimensional $SO(2)$. In higher dimensions, for instance in $D=10$, the symmetric tensors do not embrace all the integer higher spins, and one should also consider tensors with the indices of mixed symmetry (symmetric and antisymmetric).

---

[1] In the case of the massive higher spin fields in $D=4$ the Wigner little group is $SU(2)\sim SO(3)$, whose irreducible representations are also described by only single row Young tableaux because of the degeneracy of the antisymmetric three–dimensional matrix.

We shall restrict ourselves to the consideration of the symmetric tensor fields $\phi_{m_1\cdots m_s}(x)$ which, under some conditions to be discussed below, describe higher spin states of an integer spin s. To describe the physical states of half integer spin s one should consider spinor tensor fields $\psi^{\alpha}_{m_1\cdots m_{s-\frac{1}{2}}}(x)$.

In string theory symmetric tensor fields arise, for examples, as string states obtained by acting on the vacuum by a single string oscillator a_m^{-n} with fixed integer n (e.g. n=1)

$$\phi_{m_1\cdots m_s} = a_{m_1}^{-1}\cdots a_{m_s}^{-1}|0> .$$

Alternatively, the field strengths of half integer and integer spin can be described by symmetric spin–tensors $\varphi_{\alpha_1\cdots\alpha_{2s}}(x)$ depending on whether s is half integer or integer [14]. The advantage of this formulation is that all the spins (integer and half integer) are treated on an equal footing.

In the Green–Schwarz formulation of the superstring such fields arise as string states obtained by acting on the string vacuum with an antisymmetrized product of different fermionic oscillators θ_α^{-n}

$$\varphi_{\alpha_1\cdots\alpha_{2s}} = \theta_{\alpha_1}^{[-1}\cdots\theta_{\alpha_{2s}}^{-2s]}|0>,$$

where $n = 1,\cdots 2s$ labels fermionic oscillator modes.

We shall briefly consider the spin–tensor formulation in Subsection 3.4.

Symmetric tensor description of massless higher spin fields

As it has already been mentioned, in quantum field theory massive fields with spin 1 and higher are not renormalizable unless their mass is generated as a result of spontaneous breaking of a gauge symmetry associated with corresponding massless gauge fields. So first of all we should understand the structure of the theory of massless higher spin fields and I shall concentrate on this problem. The theory of massive higher spin fields and their interactions (in particular with electromagnetic fields and gravity) was discussed e.g. in [9, 10, 11, 13, 14, 15, 16, 22, 34, 35, 36] .

Note that in $D=4$ the physical fields of spin $s\leq 2$ are part of the family of the symmetric (spin) tensors. Their well known equations of motion and gauge transformations are reproduced below in a form suitable for the generalization to the case of the higher spin fields

s=0 $\phi(x)$ – scalar field, $\partial_m\partial^m\phi \equiv \partial^2\phi = 0$, matter field, no gauge symmetry;

s $=\frac{1}{2}$ $\psi^\alpha(x)$ – spinor field, $\gamma^{m\alpha}{}_{\beta}\partial_m\psi^\beta = (\partial\!\!\!/\psi)^\alpha = 0$, matter field, no gauge symmetry;

s=1 $\phi_m(x) = A_m(x)$ – Maxwell field, $\partial^m F_{mn} = \partial^2 A_n - \partial_n\partial_m A^m = 0$, $\delta A_m = \partial_m\xi(x)$;

s $=\frac{3}{2}$ $\psi^\alpha_m(x)$ – Rarita–Schwinger field, $\gamma_{mnp}\partial^n\psi^p = \partial\!\!\!/\psi_m - \partial_m\gamma^n\psi_n = 0$, $\delta\psi^\alpha_m = \partial_m\xi^\alpha(x)$;

s=2 $\phi_{m_1m_2}(x) = g_{m_1m_2}(x)$ – graviton, $R_{m_1m_2} = 0$, $\delta g_{m_1m_2} = D_{m_1}\xi_{m_2} + D_{m_2}\xi_{m_1}$, where $D_m = \partial_m + \Gamma_{mn}^{p}$ is the covariant derivative and $\Gamma_{mn,p} = \frac{1}{2}(\partial_p g_{mn} - \partial_m g_{np} - \partial_n g_{mp})$ is the Christoffel connection;

in the linearized limit where the deviation of $g_{m_1m_2}(x)$ from the Minkowski metric $\eta_{m_1m_2}$ is infinitesimal the Einstein equation and the diffeomorphisms reduce to $\partial^2 g_{m_1m_2} - \partial_{m_1}\partial_n g^n{}_{m_2} - \partial_{m_2}\partial_n g^n{}_{m_1} + \partial_{m_1}\partial_{m_2} g^n{}_n = 0$, $\delta g_{m_1m_2} = \partial_{m_1}\xi_{m_2} + \partial_{m_2}\xi_{m_1}$.

Except for the scalar and the spinor field, all other massless fields are gauge fields. The associated gauge symmetry eliminates (unphysical) lower spin components of these fields and thus ensures that they have a positive norm. So it is natural to assume that all massless higher spin fields are also the gauge fields with the gauge transformations being an appropriate generalization of those of the Maxwell, Rarita–Schwinger and Einstein field. In the linear (free field) approximation the higher spin gauge transformations (for the integer and half integer spins) have the form

$$\begin{aligned}\delta\phi_{m_1\cdots m_s}(x) = \partial_{m_1}\xi_{m_2\cdots m_s} + \partial_{m_2}\xi_{m_1\cdots m_s} + \cdots \equiv \sum \partial_{m_1}\xi_{m_2\cdots m_s}\,, \\ \delta\psi^{\alpha}_{m_1\cdots m_{s-\frac{1}{2}}}(x) = \sum \partial_{m_1}\xi^{\alpha}_{m_2\cdots m_{s-\frac{1}{2}}}\,,\end{aligned} \tag{1}$$

where $\sum$ will denote (almost everywhere) the symmetrized sum with respect to all non-contracted vector indices.

Free equations of motion

We assume that the free equations of motion of the higher spin fields are second order linear differential equations in the case of the integer spins and the first order differential equations in the case of the half integer spins. This is required by the unitary and ensures that the fields have a positive–definite norm. The massless higher spin equations have been derived from the massive higher spin equations [16] by Fronsdal [39] and studied in more detail in [41].

The bosonic equations, which I shall denote by $G_{m_1\cdots m_s}(x)$ are a natural generalization of the Klein-Gordon, Maxwell and linearized Einstein equations

$$G_{m_1\cdots m_s}(x) \equiv \partial^2\phi_{m_1\cdots m_s}(x) - \sum \partial_{m_1}\partial_n\phi^n{}_{m_2\cdots m_s}(x) + \sum \partial_{m_1}\partial_{m_2}\phi^n{}_{nm_3\cdots m_s}(x) = 0\,. \tag{2}$$

The first order fermionic equations are a natural generalization of the Dirac and Rarita–Schwinger equation

$$G^{\alpha}_{m_1\cdots m_{s-\frac{1}{2}}}(x) \equiv (\partial\!\!\!/\psi)^{\alpha}_{m_1\cdots m_{s-\frac{1}{2}}} - \sum \partial_{m_1}(\gamma^n\psi)^{\alpha}_{nm_2\cdots m_{s-\frac{1}{2}}} = 0\,. \tag{3}$$

Constraints on higher spin symmetry parameters and on higher spin fields

We should now verify that the equations of motion (2) and (3) are invariant under gauge transformations (1). The direct computations give

$$\delta G_{m_1\cdots m_s} = 3\sum \partial^3_{m_1m_2m_3}\xi^n{}_{nm_4\cdots m_s}, \qquad \delta G^{\alpha}_{m_1\cdots m_{s-\frac{1}{2}}} = -2\sum \partial^2_{m_1m_2}\gamma^{n\alpha}{}_{\beta}\,\xi^{\beta}{}_{nm_3\cdots m_{s-\frac{1}{2}}}\,, \tag{4}$$

where $\partial^2_{m_1m_2} = \partial_{m_1}\partial_{m_2}$ and $\partial^3_{m_1m_2m_3} = \partial_{m_1}\partial_{m_2}\partial_{m_3}$.

We see that these variations vanish if the parameters of the transformations of the bosonic higher spin fields (for $s \geq 3$) are traceless

$$\xi^n{}_{nm_4\cdots m_s} = 0 \tag{5}$$

and the parameters of the transformations of the fermionic higher spin fields for ($s \geq 5/2$) are γ–traceless

$$(\gamma^n \xi)^\alpha_{nm_3\cdots m_{s-\frac{1}{2}}} = 0 \,. \tag{6}$$

Other constraints in the theory of higher spins appear for bosons with $s \geq 4$ and for fermions with $s \geq \frac{7}{2}$. Since the theory is gauge invariant there should exist Bianchi identities analogous to those in Maxwell and Einstein theory which are identically satisfied. The Bianchi identities (or, equivalently, integrability conditions) imply that the traceless divergence of the left–hand–side of equations of motion must vanish identically. This also implies that the currents of the matter fields which could couple to the gauge fields are conserved.

For instance in Maxwell theory we have

$$\partial_n(\partial_m F^{mn}) \equiv 0, \tag{7}$$

and, hence the electric current which enters the r.h.s. of the Maxwell equations $\partial_m F^{mn} = J^m$ is conserved $\partial_m J^m = 0$.

In the theory of gravity coupled to matter fields and described by the Einstein equation

$$R_{mn} - \frac{1}{2} g_{mn} R = T_{mn}$$

the energy–momentum conservation $D_m T^{mn} = 0$ is related to the Bianchi identity

$$D_m R^m_{\ n} - \frac{1}{2} D_n R^m_{\ m} \equiv 0 \,. \tag{8}$$

The linearized form of (8) generalized to the case of the bosonic higher spin fields results in the following Bianchi identity (or the integrability condition)

$$\partial_n G^n{}_{m_2\cdots m_s} - \frac{1}{2} \sum \partial_{m_2} G^n{}_{nm_3\cdots m_s} = -\frac{3}{2} \sum \partial^3_{m_2 m_3 m_4} \phi^{np}{}_{npm_5\cdots m_s} \,, \tag{9}$$

and in the case of the fermionic fields we have

$$\partial_n G^{\alpha n}{}_{m_2\cdots m_{s-\frac{1}{2}}} - \frac{1}{2} \sum \partial_{m_2} G^{\alpha n}{}_{nm_3\cdots m_{s-\frac{1}{2}}} - \frac{1}{2} (\partial\!\!\!/ \gamma^n G)^\alpha_{nm_2\cdots m_{s-\frac{1}{2}}} = \sum \partial^2_{m_2 m_3} (\gamma^n \psi)^{\alpha p}{}_{npm_4\cdots m_{s-\frac{1}{2}}} \,. \tag{10}$$

We see that the right–hand–sides of (9) and (10) do not vanish identically and require that the bosonic fields with $s \geq 4$ are double–traceless

$$\phi^{np}{}_{npm_5\cdots m_s} = 0 \tag{11}$$

and the fermionic fields with $s \geq \frac{7}{2}$ are triple–gamma–traceless

$$(\gamma^n \gamma^p \gamma^r \psi)^\alpha{}_{nprm_4\cdots m_{s-\frac{1}{2}}} \equiv (\gamma^n \psi)^{\alpha p}{}_{npm_4\cdots m_{s-\frac{1}{2}}} = 0 \,. \tag{12}$$

It turns out that for the consistency of the theory the fields should satisfy the double–triple traceless conditions identically, i.e. off the mass shell. Note that the double– and triple–traceless conditions are the strongest possible gauge invariant algebraic constraints on the fields, provided that the gauge parameters are traceless.

Physically the requirement of the double tracelessness, together with the gauge fixing of higher spin symmetry, ensures that the lower spin components contained in the symmetric tensor fields are eliminated, so that only the massless states with helicities $\pm s$ propagate. And as we know very well for lower spin fields $\frac{1}{2} \leq s \leq 2$, in $D=4$ each massless field has only two physical degrees of freedom which are characterized by the helicities $\pm s$.

Pure gauge degrees of freedom of the integer higher spin fields can be eliminated by imposing gauge fixing conditions analogous to the Lorentz gauge of the vector field and the de Donder gauge in the case of gravity

$$\partial_p \phi^p{}_{m_2 \cdots m_s} - \frac{1}{2} \sum \partial_{m_2} \phi^p{}_{p m_3 \cdots m_s} = 0. \tag{13}$$

Then the higher spin equations of motion (2) reduce to the Klein–Gordon equation $\partial^2 \phi_{m_1 \cdots m_s} = 0$, which implies that we indeed deal with *massless* fields.

Covariant gauge fixing condition for fermion fields are [41]

$$\gamma^n \psi_{n m_2 \cdots m_{s-\frac{1}{2}}} = 0, \quad \psi^n{}_{n m_3 \cdots m_{s-\frac{1}{2}}} = 0. \tag{14}$$

They reduce the field equation (3) down to the massless Dirac equation.

Thus, the double–and triple–traceless constraints along with the gauge fixing conditions single out physical components of the massless higher spin fields. Another role of the double– and triple–traceless constraints (11), (12) is that only when the fields identically satisfy (11) and (12), the higher spin field equations (2) and (3) can be obtained from appropriate actions [39, 41].

Unconstrained formulations of higher spin field dynamics will be considered in the next Section.

The free higher spin field actions

For the bosonic fields the action in D–dimensional space–time is

$$S_B = \int d^D x \left(\frac{1}{2} \phi^{m_1 \cdots m_s} G_{m_1 \cdots m_s} - \frac{1}{8} s(s-1) \phi_n{}^{n m_3 \cdots m_s} G^p{}_{p m_3 \cdots m_s} \right) \tag{15}$$

and for fermions

$$S_F = \int d^D x \Big(-\tfrac{1}{2} \bar{\psi}^{m_1 \cdots m_{s-\frac{1}{2}}} G_{m_1 \cdots m_{s-\frac{1}{2}}} + \tfrac{1}{4} s \, \bar{\psi}^{m_2 \cdots m_{s-\frac{1}{2}} n} \gamma_n \gamma^p G_{p m_2 \cdots m_{s-\frac{1}{2}}}$$
$$+ \tfrac{1}{8} s(s-1) \bar{\psi}_n{}^{n m_3 \cdots m_{s-\frac{1}{2}}} G^p{}_{p m_3 \cdots m_{s-\frac{1}{2}}} \Big), \tag{16}$$

where $G_{m_1\cdots m_s}$ and $G^{\alpha}_{m_1\cdots m_{s-\frac{1}{2}}}$ stand for the left hand sides of the equations (2) and (3).

The actions are invariant under the gauge transformations (1) with the traceless parameters (5) and (6), and the higher spin fields are supposed to be double or gamma–triple traceless (11), (12).

GEOMETRIC FORMULATIONS OF FREE HIGHER SPIN FIELD THEORY

The presence of the constraints on the gauge parameters and higher spin fields in the formulation of Fronsdal and of Fang and Fronsdal may look as an odd feature of the theory and point out that such a formulation is incomplete. A modification of the equations of motion (2) and (3) which would remove the constraints (5) and (6) on the gauge parameters and the double–traceless conditions (11), (12) can be achieved in three different though related ways.

One of the ways to remove the tracelessness constraints is to use, in addition to the physical higher spin field, an appropriate number of auxiliary tensorial fields satisfying certain equations of motion, as was shown in [29] [2]. The higher spin field equations remain lagrangian.

Another way was proposed by Francia and Sagnotti [42]. Its key point is to renounce locality of the theory. It was shown that the equations of motion of the unconstrained higher spin fields and corresponding actions can be made invariant under the unconstrained gauge transformations if they are enlarged with *non–local* terms. A motivation of Francia and Sagnotti for removing constraints on gauge parameters has been based on the observation that symmetries of String Field Theory do not have such restrictions. Another motivation was to find a more conventional *geometric* form of the higher–spin field equations in terms of conditions on generalized curvatures introduced in [41]. This would be a generalization of the Maxwell and Einstein equations written in terms of F_{mn} and R_{mn}, respectively. The choice of non–local terms in the higher spin field equations is not unique. Choosing a suitable form of non–local equations one manages to keep their Lagrangian nature. We will not go into details of this formulation and address the interested reader to [42, 43].

A third possibility of removing the constraints is to allow the higher spin field *potentials* to satisfy *higher order* differential equations, which can be constructed in a manifestly gauge invariant way as conditions imposed on the higher spin field curvatures. Let us consider this geometric formulation of the free higher spin field theory in more detail. Actually, in $D = 4$ space–time it has been constructed many years ago by Bargmann and Wigner [12]. As we shall see, the higher order derivative structure of the higher spin curvature equations does not spoil the unitarity of the theory. These equations are physically equivalent to the Fang–Fronsdal and Francia–Sagnotti equations.

Generalized curvatures for the higher spin fields $\phi_{m_1\cdots m_s}(x)$ and $\psi^{\alpha}_{m_1\cdots m_{s-\frac{1}{2}}}(x)$ which

---

[2] In the case of massive higher spin fields, auxiliary fields to construct higher spin field actions were introduced by Fierz and Pauli [10].

are invariant under the *unconstrained* gauge transformations (1) can be constructed as a direct generalization of the spin 1 Maxwell field strength

$$F_{mn} = \partial_m A_n - \partial_n A_m$$

and of the linearized Riemann tensor in the case of spin 2

$$R_{m_1n_1,m_2n_2} = \partial_{m_1}\partial_{m_2} g_{n_1n_2} - \partial_{n_1}\partial_{m_2} g_{m_1n_2} - \partial_{n_1}\partial_{m_2} g_{n_1m_2} + \partial_{n_1}\partial_{n_2} g_{m_1m_2}. \tag{17}$$

Thus, for an arbitrary integer spin s the gauge invariant curvature is obtained by takin s derivatives of the field potential $\phi_{n_1\cdots n_s}(x)$

$$\begin{aligned} R_{m_1n_1,m_2n_2,\cdots,m_sn_s} &= \partial_{m_1}\partial_{m_2}\cdots\partial_{m_s}\phi_{n_1n_2\cdots n_s} - \partial_{n_1}\partial_{m_2}\cdots\partial_{m_s}\phi_{m_1n_2\cdots n_s} \\ &\quad -\partial_{m_1}\partial_{n_2}\cdots\partial_{m_s}\phi_{n_1m_2\cdots n_s} + \partial_{n_1}\partial_{n_2}\cdots\partial_{m_s}\phi_{m_1m_2\cdots n_s} + \cdots \\ &\equiv \partial^s_{m_1\cdots m_s}\phi_{n_1\cdots n_s} - \sum(m_i \leftrightarrow n_i). \end{aligned} \tag{18}$$

Analogously, for an arbitrary half integer spin s the curvature is obtained by taking $(s-\frac{1}{2})$ derivatives of the field potential $\psi^\alpha_{n_1\cdots n_{s-\frac{1}{2}}}(x)$

$$\begin{aligned} \mathscr{R}^\alpha_{m_1n_1,m_2n_2,\cdots,m_{s-\frac{1}{2}}n_{s-\frac{1}{2}}} &= \partial_{m_1}\partial_{m_2}\cdots\partial_{m_{s-\frac{1}{2}}}\psi^\alpha_{n_1n_2\cdots n_{s-\frac{1}{2}}} - \partial_{n_1}\partial_{m_2}\cdots\partial_{m_{s-\frac{1}{2}}}\psi^\alpha_{m_1n_2\cdots n_{s-\frac{1}{2}}} \\ &\quad -\partial_{m_1}\partial_{n_2}\cdots\partial_{m_{s-\frac{1}{2}}}\psi^\alpha_{n_1m_2\cdots n_{s-\frac{1}{2}}} + \partial_{n_1}\partial_{n_2}\cdots\partial_{m_{s-\frac{1}{2}}}\psi^\alpha_{m_1m_2\cdots n_{s-\frac{1}{2}}} + \cdots \\ &\equiv \partial^{s-\frac{1}{2}}_{m_1\cdots m_{s-\frac{1}{2}}}\psi^\alpha_{n_1\cdots n_{s-\frac{1}{2}}} - \sum(m_i \leftrightarrow n_i). \end{aligned} \tag{19}$$

In the right hand side of (18) and (19) it is implied that the sum is taken over all the terms in which the indices within the pairs of $[m_i, n_i]$ with the same label $i = 1, \cdots, s$ are antisymmetrized.

By construction similar to the Riemann tensor (17), the higher spin curvatures (18) and (19) are completely symmetric under the exchange of any two pairs of their antisymmetric indices and they obey for any pair of the antisymmetric indices $[m_i n_i]$ the same Bianchi identities as the Riemann tensor, *e.g.* for the bosonic spin s field

$$R_{m_1n_1,m_2n_2,\cdots,m_sn_s} = -R_{n_1m_1,m_2n_2,\cdots,m_sn_s} = R_{m_2n_2,m_1n_1,\cdots,m_sn_s}, \tag{20}$$

$$R_{[m_1n_1,m_2]n_2,\cdots,m_sn_s} = 0, \tag{21}$$

$$\partial_{[l_1}R_{m_1n_1],m_2n_2,\cdots,m_sn_s} = 0. \tag{22}$$

On the other hand, if a rank $2[s]$ (spinor)–tensor (where $[s]$ is the integer part of s) possesses the properties (20)–(22), in virtue of the generalized Poincaré lemma of [44, 45, 46] this tensor can be expressed as an 'antisymmetrized' $[s]$–th derivative of a symmetric rank $[s]$ field potential, as in eqs. (18) and (19).

Let us note that de Wit and Freedman [41] constructed curvature tensors out of the $[s]$ derivatives of the symmetric higher spin gauge fields $\phi_{m_1\cdots m_s}(x)$ and $\psi^{\alpha}_{m_1\cdots m_{s-\frac{1}{2}}}(x)$ in an alternative way. Their curvatures have two pairs of the groups of $[s]$ symmetric indices and they are symmetric or antisymmetric under the exchange of these groups of indices depending on whether $[s]$ is even or odd

$$\tilde{R}_{m_1\cdots m_{[s]},n_1\cdots n_{[s]}} = (-1)^{[s]}\tilde{R}_{n_1\cdots n_{[s]},m_1\cdots m_{[s]}}\,. \tag{23}$$

The de Wit–Freedman curvatures satisfy the cyclic identity, which is a symmetric analog of (21),

$$\tilde{R}_{m_1\cdots m_{[s]},n_1\cdots n_{[s]}} + \sum_{n_i}\tilde{R}_{n_1 m_2\cdots m_{[s]},m_1 n_2\cdots n_{[s]}} = 0\,, \tag{24}$$

where $\sum_{n_i}$ denotes the symmetrized sum with respect to the indices n_i. The form of the analogs of the differential Bianchi identities (22) for $\tilde{R}_{m_1\cdots m_{[s]},n_1\cdots n_{[s]}}$ is less transparent. They can be obtained from (22) using the fact that the de Wit–Freedman tensors (23) are related to the generalized 'Riemann' tensors (18) and (19) by the antisymmetrization in the former of the indices of each pair $[m_i, n_i]$. In what follows we shall work with the generalized Riemann curvatures.

To produce the dynamical equations of motion of the higher spin fields one should now impose additional conditions on the higher spin curvatures. Since the curvatures (18) and (19) have the same properties (20)–(22) as the Riemann tensor (17), and the linearized Einstein equation amounts to putting to zero the trace of the Riemann tensor

$$R^{m}{}_{n_1,mn_2} = R_{n_1n_2} = \partial^2\partial^2 g_{n_1n_2} - \partial_{n_1}\partial_m g^{m}{}_{n_2} - \partial_{n_2}\partial_m g^{m}{}_{n_1} + \partial_{n_1}\partial_{n_2} g^{m}{}_{m} = 0\,, \tag{25}$$

one can naturally assume that in the case of the integer spins the equations of motion are also obtained by requiring that the trace of the generalized Riemann tensor (18) with respect to any pair of indices is zero, e.g.

$$R^{m}{}_{n_1,mn_2,m_3n_3,\cdots,m_sn_s} = 0\,. \tag{26}$$

In the case of the half integer spins we can assume that the fermionic equations of motion are a generalization of the Dirac and Rarita–Schwinger equations and that they are obtained by putting to zero the gamma–trace of the fermionic curvature (19)

$$(\gamma^{m_1}\mathscr{R})^{\alpha}_{m_1n_1,m_2n_2,\cdots,m_{s-\frac{1}{2}}n_{s-\frac{1}{2}}} = 0\,. \tag{27}$$

The equations (26) and (27) are non–lagrangian for $s > 2$.

Let us remind the reader that, because of the gauge invariance of the higher spin curvatures, eqs. (26) and (27) are invariant under the higher spin gauge transformations (1) with *unconstrained* parameters, and the higher spin field potentials are also *unconstrained* in contrast to the Fronsdal and Fang–Fronsdal formulation considered in the previous Section .

The reason and the price for this is that eqs. (26) and (27) are higher order differential equations, which might cause a problem with unitarity of the quantum theory. However, as we shall now show the equations (26) and (27) reduce to, respectively, the second and first order differential equations for the higher spin field potentials related to those of Fronsdal (2) and of Fang and Fronsdal (3).

Integer spin fields

In the integer spin case, analyzing the form of the left hand side of Eq. (26) in terms of the gauge field potential (18) one gets the higher spin generalization of the spin 3 Damour–Deser identity [47] which relates the trace of the higher spin curvature to the "Fronsdal kinetic operator", namely

$$tr R_{m_1n_1,\cdots,m_sn_s} = R_{m_1n_1,m_2n_2,\cdots,n_{s-1}m,n_s}{}^{m}$$
$$= \partial^{s-2}_{m_1m_2\cdots m_{s-2}} G_{n_1\cdots n_{s-2}n_{s-1}n_s} - \sum_{i=1}^{i=s-2} (m_i \leftrightarrow n_i) \qquad (28)$$

where the symmetric tensor $G(x)$ stands for the left hand side of the Fronsdal equations (2) (sometimes called the "Fronsdal kinetic operator", and the indices $[m_i, n_i]$ with $(i = 1,\cdots, s-2)$ are anti–symmetrized.

When the curvature tensor satisfies the tracelessness condition (26) the left hand side of Eq. (28) vanishes, which implies that the tensor G is ∂^{s-2}–closed. In virtue of the generalized Poincaré lemma [44, 45, 46] this means that (at least locally) G is ∂^3–exact, *i.e.* has the form [48]

$$G_{n_1\cdots n_s} = \sum \partial^3_{n_1n_2n_3} \rho_{n_4\cdots n_s}, \qquad (29)$$

where the sum implies the symmetrization of all the indices n_i and $\rho(x)$ is a symmetric tensor field of rank $(s-3)$ called 'compensator' since its gauge transformation

$$\delta \rho_{n_1\cdots n_{s-3}} = 3\,\xi^{m}{}_{mn_1\cdots n_{s-3}} \qquad (30)$$

compensates the non–invariance (4) of the kinetic operator $G(x)$ under the unconstrained local variations (1) of the gauge field potential $\phi(x)$. Eq. (29) was discussed in [42, 67]. Here we have obtained it from the geometric equation on the higher spin field curvature (26) following ref. [48] where such a derivation was carried out for a generic mixed symmetry field [3].

The trace of the gauge parameter (30) can be used to eliminate the compensator field. Then Eq. (29) reduces to the Fronsdal equation (2) which is invariant under residual gauge transformations with traceless parameters.

A non–local form of the higher spin equations

We shall now demonstrate how the higher spin field equations with the compensator (29) are related to the non–local equations of [42]. We shall consider the simple (standard) example of a gauge field of spin 3. The case of a generic spin s can be treated in a similar but more tedious way. In a somewhat different way the relation of the compensator equations (29) to non–local higher spin equations was discussed in the second paper of [42].

---

[3] For a relevant earlier discussion of the relationship of the equations on the Riemann and de Wit–Freedman curvatures to the equations of motion of symmetric (spinor)–tensor field potentials see [49].

For the *spin* 3 field the compensator equation takes the form

$$G_{mnp} := \partial^2 \phi_{mnp} - 3\partial_q \partial_{(m} \phi_{np)}{}^q + 3\partial_{(m} \partial_n \phi_{p)q}{}^q = \partial^3_{mnp} \rho(x)\,, \tag{31}$$

where () stand for the symmetrization of the indices with weight one and $\rho(x)$ is the compensator, which is a scalar field in the case of spin 3.

We now take the derivative and then the double trace of the left and the right hand side of this equation and get

$$\partial_m G^{mn}{}_n = \partial^2 \partial^2 \rho(x)\,. \tag{32}$$

Modulo the *doubly harmonic* zero modes $\rho_0(x)$, satisfying $\partial^2 \partial^2 \rho_0(x) = 0$, one can solve Eq. (32) for $\rho(x)$ in a non–local form

$$\rho(x) = \frac{1}{\partial^2 \partial^2} \partial_m G^{mn}{}_n\,. \tag{33}$$

Substituting this solution into the spin 3 field equation (31) we get one of the non–local forms of the spin 3 field equation constructed in [42]

$$G_{mnp} := \partial^2 \phi_{mnp} - 3\partial_q \partial_{(m} \phi_{np)}{}^q + 3\partial_{(m} \partial_n \phi_{p)q}{}^q = \frac{1}{\partial^2 \partial^2} \partial^3_{mnp} (\partial_q G^{qr}{}_r)\,. \tag{34}$$

Let us now consider a more complicated example of *spin* 4. In the Fronsdal formulation, the fields of spin 4 and higher feature one more restriction: they are double traceless. We shall show how this constraint appears upon gauge fixing the compensator equation, which for the spin 4 field has the form

$$G_{mnpq} := \partial^2 \varphi_{mnpq} - 4\partial_r \partial_{(m} \varphi_{npq)}{}^r + 6\partial_{(m} \partial_n \varphi_{pq)r}{}^r = 4\partial^3_{(mnp} \rho_{q)}(x)\,. \tag{35}$$

Taking the double trace of (35) we have

$$G^{mn}{}_{mn} = 3\partial^2 \varphi^{mn}{}_{mn} = 4\partial^2 \partial_m \rho^m\,. \tag{36}$$

Taking the derivative of (35) and the double trace we get

$$\partial_m G^{mn}{}_{np} = \partial^2 \partial^2 \rho_p + 3\partial_p \partial^2 \partial_m \rho^m = \partial^2 \partial^2 \rho_p + \frac{3}{4} \partial_p G^{mn}{}_{mn}\,, \tag{37}$$

where we have used (36) to arrive at the right hand side of (37).

From (37) we find that, modulo the zero modes ρ_0^p of $\partial^2 \partial^2 \rho_0^p = 0$, the compensator field is non–locally expressed in terms of the (double) trace of the Fronsdal kinetic term

$$\rho_p = \frac{1}{\partial^2 \partial^2} (\partial_m G^{mn}{}_{np} - \frac{3}{4} \partial_p G^{mn}{}_{mn})\,. \tag{38}$$

Substituting (38) into (35) we get one of the non–local forms [42] of the spin 4 field equation [42].

Consider now the following identity

$$\partial_q G^q{}_{mnp} - \partial_{(m} G^q{}_{np)q} = -\frac{3}{2}\partial_m \partial_n \partial_p \phi^{qr}{}_{qr} = -2\,\partial_m \partial_n \partial_p \left(\partial_q \rho^q\right). \tag{39}$$

On the other hand, from (36) and (39) it follows that modulo a linear and quadratic term in x^m (which can be put to zero by requiring an appropriate asymptotic behaviour of the wave function at infinity) the double trace of the gauge field $\phi(x)$ is proportional to the divergence of $\rho_q(x)$

$$\phi^{mn}{}_{mn} = \frac{4}{3}\partial_q \rho^q\,. \tag{40}$$

Therefore, when we partially fix the gauge symmetry by putting $\rho_q(x) = 0$, the double trace of the gauge field also vanishes and we recover the Fronsdal formulation with the traceless gauge parameter and the double traceless gauge field.

Half integer spin fields

Let us generalize the previous consideration to the case of fermions. The fermionic spin–s field strength $\mathscr{R}^\alpha$ is the spinor–tensor

$$\mathscr{R}^\alpha_{m_1 n_1,\cdots,m_{s-\frac{1}{2}} n_{s-\frac{1}{2}}}(x)\,. \tag{41}$$

It satisfies the Bianchi identities analogous to (21), (22) and thus can be expressed in terms of $s-\frac{1}{2}$ derivatives of a fermionic field potential (19).

The fermionic generalization of the Damour–Deser identity is

$$(\gamma^m \mathscr{R})^\alpha_{mn_1,m_2n_2,\cdots m_{s-\frac{1}{2}} n_{s-\frac{1}{2}}} = \partial^{s-\frac{3}{2}}_{m_1 \cdots m_{s-\frac{3}{2}}} G^\alpha{}_{n_1 \cdots n_{s-\frac{3}{2}} n_{s-\frac{1}{2}}} - \sum_{i=1}^{i=s-\frac{3}{2}} (m_i \leftrightarrow n_i)\,, \tag{42}$$

where the some is taken over terms with the indices $[m_i, n_i]$ anti–symmetrized, and the Fang–Fronsdal fermionic kinetic operator G^α acting on the gauge field ψ^α is defined in (3). The field strength (19) is invariant under the *unconstrained* gauge transformations (1).

When the fermionic field strength satisfies the γ–tracelessness condition (27)

$$(\gamma^m \mathscr{R})^\alpha_{mn_1,m_2n_2,\cdots m_{s-\frac{1}{2}} n_{s-\frac{1}{2}}} = 0\,,$$

Eq. (42) implies that G^α is $\partial^{s-\frac{3}{2}}$–closed. Since $\partial^{s+\frac{1}{2}} \equiv 0$, by virtue of the generalized Poincaré lemma G^α is ∂^2–exact

$$G^\alpha{}_{n_1 \cdots n_{s-\frac{1}{2}}} = \sum \partial_{n_1} \partial_{n_2} \rho^\alpha_{n_3 \cdots n_{s-\frac{1}{2}}}\,, \tag{43}$$

where $\sum$ implies the symmetrization of all the indices n_i.

Equation (43) is the compensator equation given in [42, 67]. The demonstration of its relation to the gamma–traceless part of the fermionic higher spin field strength has been given in [50].

The gauge variation of $G^{\alpha}(x)$ has been presented in (4). It is compensated by a gauge shift of the field $\rho^{\alpha}(x)$ given by the γ–trace of the gauge parameter

$$\delta\rho^{\alpha}_{n_3\cdots n_{s-\frac{1}{2}}} = -2\sum\gamma^{m\alpha}{}_{\beta}\,\xi^{\beta}_{mn_3\cdots n_{s-\frac{1}{2}}}\,. \tag{44}$$

Thus, the compensator can be gauged away by choosing a gauge parameter $\xi^{\alpha}(x)$ with the appropriate γ–trace. Then, the equations of motion of the gauge field $\psi^{\alpha}(x)$ become the first order differential equations (3) which are invariant under the gauge transformations (1) with γ–traceless parameters.

Alternatively, one can get non–local Francia–Sagnotti equations for fermions by taking a particular non–local solution for the compensator field in terms of the fermionic kinetic operator G^{α}. As a simple example consider the $s = 5/2$ case. Eq. (43) takes the form

$$G^{\alpha}_{mn} := \partial\!\!\!/\,\psi^{\alpha}_{mn} - 2\partial_{(m}(\gamma^{q}\psi)^{\alpha}_{n)q} = \partial_m\partial_n\rho^{\alpha}(x)\,. \tag{45}$$

Taking the trace of (45) we get

$$\partial^2\,\rho^{\alpha} = G^{\alpha p}{}_{p}\,. \tag{46}$$

Hence, modulo the zero modes $\rho^{\alpha}_0(x)$ of the Klein–Gordon operator $\partial^2\,\rho^{\alpha}_0 = 0$ the compensator field is non–locally expressed in terms of the trace of G^{α}_{mn}

$$\rho^{\alpha} = \frac{1}{\partial^2}G^{\alpha p}{}_{p}\,. \tag{47}$$

Substituting (47) into (45) we get the Francia–Sagnotti equation for the fermionic field of spin 5/2 in the following form

$$G^{\alpha}_{mn} := \partial\!\!\!/\,\psi^{\alpha}_{mn} - 2\partial_{(m}(\gamma^{q}\psi)^{\alpha}_{n)q} = \frac{1}{\partial^2}\partial_m\partial_n G^{\alpha p}{}_{p}\,. \tag{48}$$

In the same way one can relate the compensator equations for an arbitrary half integer spin field to the corresponding non–local field equation. As in the bosonic case, one can find that for $s \geq \frac{7}{2}$ the triple–gamma trace of the fermionic gauge field potential is expressed in terms of the γ–trace and by the divergence of the compensator field and thus vanish in the 'Fronsdal gauge' $\rho^{\alpha} = 0$.

We have thus reviewed various formulations of free higher spin field dynamics.

THE INTERACTION PROBLEM

As we have seen, the free theory of higher spin fields, both massless and massive, exists and can be formulated in a conventional field–theoretical fashion using the action principle. An important problem, which still has not been completely solved, is to introduce interactions of the higher spin fields. Probably, String Field Theory should give the answer to this problem if someone manages to extract the corresponding information from the String Field Theory action. This itself is a highly non–trivial problem which has not been realized yet.

So far the study of the problem of higher spin interactions has been undertaken mainly in the framework of the standard field–theoretical approach, and I would now like to review main obstacles which one encounters in the way of constructing an interacting *massless* higher spin field theory.

One may consider self–interactions of fields of the same spin, or interactions among fields of different spin. In the first case, for example, the construction of consistent self–interactions of massless vector fields results in either the non–abelian Yang–Mills theory [21] or in the non–linear Dirac–Born–Infeld generalization of Maxwell theory. A consistent way of introducing the self–interaction of the spin 2 field leads to the Einstein theory of gravity [19]. Consistency basically means that the introduction of interactions should not break the gauge symmetry, but may only modify it in a suitably way.

An example of interactions among fields of different spin is the universal gravitational interaction of the matter and gauge fields. So, the construction of the theory of interactions of higher spin fields with gravity is an important part of the general interaction problem and it actually exhibits all aspects of the general problem.

Simple supergravity

To see what kind of problems with the higher spin interactions arise, let us first consider the example of coupling to four–dimensional gravity the field of spin 3/2 [10, 11] historically called the Rarita–Schwinger field. This is an instructive example which shows how the consistency of gravitational coupling leads to supergravity [51, 52], the theory invariant under local supersymmetry in which the spin 3/2 field becomes the superpartner of the graviton, called gravitino.

The general coordinate invariance of the complete non–linear gravitational interactions requires that in the free field equations partial derivatives get replaced with covariant derivatives and the vector indices are contracted with the gravitational metric $g_{mn}(x)$. So the free Rarita–Schwinger equation

$$\gamma^{mnp}\,\partial_n\psi_p = 0$$

should be generalized to include the interaction with gravity as follows

$$G^m = \gamma^{mnp}\,D_n\psi_p = 0\,, \tag{49}$$

where $\gamma_n = e^a_n(x)\gamma_a$ are the gamma–matrices contracted with the vielbein $e^a_n(x)$ which is related to the metric in the standard way $g_{mn} = e^a_m e^b_n \eta_{ab}$, and $D_m = \partial_m + \Gamma_{mn}{}^p + \omega_{m\alpha}{}^\beta$ is the covariant derivative which contains the Christoffel symbol Γ^p_{mn} and a spin connection $\omega_{m\alpha}{}^\beta$ acting on spinor indices.

In the presence of the Rarita–Schwinger field the right–hand–side of the Einstein equations acquires the contribution of an energy–momentum tensor of the spin 3/2 field

$$R_{mn} - \frac{1}{2}g_{mn}R = T_{mn}(\psi)\,, \quad \text{or} \quad R_{mn} - T_{mn} + \frac{1}{2}g_{mn}\,T_l{}^l = 0\,. \tag{50}$$

The explicit form of T_{mn} is not known until a consistent interacting theory is constructed.

Consider now the variation of the Rarita–Schwinger field equation under the gauge transformations

$$\delta \psi_m^\alpha = D_m \xi^\alpha, \tag{51}$$

which are the general covariant extension of the free field gauge transformations. The variation of Eq. (49) is

$$\delta G^m = \gamma^{mnp} D_n D_p \xi = \frac{1}{2}\gamma^{mnp}[D_n, D_p]\xi \sim \gamma^{mnp} R_{np,qr}\gamma^{qr}\xi \sim R^m{}_n \gamma^n \xi, \tag{52}$$

where the commutator of D_n produces the Riemann curvature which I schematically write as

$$[D_m, D_n] \sim R_{mn,pq}\gamma^{pq}, \tag{53}$$

and the last term in (52) is obtained by use of γ–matrix identities.

We thus see that the gauge variation of the Rarita–Schwinger equation is proportional to the Ricci tensor. It is zero if the Ricci tensor is zero, i.e. when the gravitational field satisfies the Einstein equations in the absence of the matter fields. This is satisfactory if we are interested in the dynamics of the spin 3/2 field in the external background of a gravitational field, such as free gravitational waves, for example. But if we would like to consider a closed graviton–spin 3/2 system, then the Ricci tensor is non–zero, since the Einstein equations take the form (50), and the variation (52) does not vanish. To improve the situation we should require that the graviton also non–trivially varies under the gauge transformations with the spinorial parameter $\xi^\alpha(x)$ as follows

$$\delta g_{mn} = \frac{i}{2}(\bar{\psi}_n \gamma_m \xi + \bar{\psi}_m \gamma_n \xi), \tag{54}$$

and we should take into account this variation of the graviton in the Rarita–Schwinger equation. This will result in the following variation of the Rarita–Schwinger equation

$$\delta G_m \sim (R_{mn} - T_{mn} + \frac{1}{2} g_{mn} T_l{}^l)\gamma^n \xi, \tag{55}$$

provided we also add appropriate second–order and fourth–order fermionic terms into the definition of the covariant derivative D_m, into the variation of ψ_m (51) and into the definition of its energy–momentum tensor. We observe that the variation of the Rarita–Schwinger equation has become proportional to the Einstein equation and hence vanishes.

Thus, by modifying the gauge transformations of the Rarita–Schwinger field, of the graviton and by appropriately modifying the Rarita–Schwinger and the Einstein equations we have achieved that under the gauge (supersymmetry) transformations with the fermionic parameter ξ^α the Rarita–Schwinger and the Einstein equations transform into each other and hence consistently describe the coupling of the spin–3/2 field to gravity.

What we have actually obtained is a simple $D = 4$ supergravity [52] which is invariant under local supersymmetry transformations. It is amazing that supergravity was not discovered much earlier than 70ths by people studied the massless spin–3/2 field using the above reasoning for the construction of a consistent gravity – spin–3/2 field interacting system. In this respect let us cite what Fierz and Pauli ([10], page 226) wrote about the massless spin 3/2 field: "Whereas the theory for the spin value 2 has an important generalization for force fields, namely the gravitational theory, we here [in the case of spin 3/2] have no such a connection with a known theory. To get a generalization of the theory with interactions one would first of all have to find a physical interpretation of the gauge group, and the conservation theorem connected with this group".

Gravitational interaction of a spin $\frac{5}{2}$ field

Let us now, by analogy with supergravity, try to couple to gravity in four dimensions a field of spin $\frac{5}{2}$ [17] which is described by the spin–tensor field $\psi^{\alpha}_{m_1m_2}$. Again, the general coordinate invariance of the gravitational interactions requires that in the free field equations partial derivatives get replaced with covariant derivatives and the vector indices are contracted with the gravitational metric $g_{mn}(x) = \eta_{mn} + \phi_{mn}(x)$, where $\phi_{mn}(x)$ is the deviation of the metric from the flat background which at the moment, for simplicity, we consider to be small and satisfy free spin 2 equations of motion. Thus the straightforward generalization of the equations of motion of the spin $\frac{5}{2}$ field which describes its "minimal" interaction with gravity is

$$G_{\alpha m_1 m_2} = i\gamma^n_{\alpha\beta}(D_n\psi^{\beta}_{m_1m_2} - D_{m_1}\psi^{\beta}_{nm_2} - D_{m_2}\psi^{\beta}_{nm_1}) = 0. \tag{56}$$

In (56) we have restored the imaginary unit i for further comparison with the massive Dirac equation $(i\partial\!\!\!/ - m)\psi = 0$.

We should now check whether these equations are invariant under the covariant modification of the higher spin gauge transformations

$$\delta\psi^{\alpha}_{m_1m_2}(x) = D_{m_1}\xi^{\alpha}_{m_2} + D_{m_2}\xi^{\alpha}_{m_1}. \tag{57}$$

The gauge variation of the equations of motion (56) is [41]

$$\delta G_{m_1m_2} = i[R_{m_1n}\gamma^n\xi_{m_2} + R_{m_2n}\gamma^n\xi_{m_1} - (R_{nm_1m_2}{}^{p} + R_{nm_2m_1}{}^{p})\gamma^n\xi_p], \tag{58}$$

where I have suppressed the spinor index. We see that in a general gravitational background the variation does not vanish because of the presence of the Ricci tensor in the first two terms and of the Riemann tensor in the last term. If the bare Riemann tensor did not appear the spin $\frac{5}{2}$ field equations would at least admit interactions with gravitational fields satisfying the free Einstein equations $R_{mn} = 0$. As we have discussed, this is the case for the gravitino field of spin $\frac{3}{2}$ whose local supersymmetry transformations produce only the Ricci tensor term in the variation of the Rarita–Schwinger equations. Introducing an appropriate supersymmetry variation of the graviton, one insures that the variation of the Rarita–Schwinger equations is proportional to the Einstein equations with the r.h.s. to be the energy–momentum tensor of the gravitino field. However, in the case of the higher spin fields with $s \geq \frac{5}{2}$ the bare Riemann tensor always appears (as part of the Weyl tensor) in the variation of the field equations, and no way has been found to cancel such terms by adding non–minimal interaction terms and/or modifying the higher spin symmetry transformations (including that of the graviton), when the zero limit of the gravitation field corresponds to the flat Minkowski space (i.e. when the cosmological constant is zero).

So the conclusion has been made that in a space–time with zero cosmological constant it is not possible to construct a consistent gauge theory of interacting higher spin fields, which is in agreement with the general theorem of the possible symmetries of the S–matrix. But, as happens with many no–go theorems, sooner or later people find a way to circumvent them. In the case of the higher spins the way out has been found in constructing the theory in the AdS space, which has a non–zero cosmological constant Λ.

In the bosonic case Fronsdal and in the fermionic case Fang and Fronsdal [40] have generalized the free higher–spin field equations and actions to the AdS background. For instance, the equation of motion of the spin $\frac{5}{2}$ field takes the following form

$$G^{\alpha}_{m_1m_2} = i\gamma^{n\alpha}{}_{\beta}(\nabla_n\psi^{\beta}_{m_1m_2} - \nabla_{m_1}\psi^{\beta}_{nm_2} - \nabla_{m_2}\psi^{\beta}_{nm_1}) - 2\Lambda^{\frac{1}{2}}\psi^{\alpha}_{m_1m_2} = 0. \tag{59}$$

One can notice that the last term in (59) resembles a mass term of the spin $\frac{5}{2}$ field, however the field has the number of physical degrees of freedom equal to that of the corresponding massless field in flat space, i.e. two states with helicities $\pm\frac{5}{2}$. This is because the equation of motion is invariant under the following gauge transformations

$$\delta\psi^{\alpha}_{m_1m_2}(x) = \nabla_{m_1}\xi^{\alpha}_{m_2} + \nabla_{m_2}\xi^{\alpha}_{m_1}\,. \tag{60}$$

In (59) and (60) $\nabla_m = D_m + \frac{i\Lambda^{\frac{1}{2}}}{2}\gamma_m$ is a so called $SO(2,3)$ covariant derivative, and D_m is the standard covariant derivative in the AdS space whose Riemann curvature has the well known form

$$R^{(AdS)}{}_{mnp}{}^{q} = -\Lambda(\delta_m{}^{q}g_{np} - \delta_n{}^{q}g_{mp})\,. \tag{61}$$

Remember also that the AdS metric is conformally flat $g^{AdS}_{mn} = (1 - \Lambda x^p x_p)^{-2}\eta_{mn}$. Note that the gamma–matrix γ_n entering (59) carries a curved vector index, it is hence non–constant and is related to the constant Dirac matrix γ_a (carrying a local Lorentz index) via the vielbein $e^{a}_{m}(x)$ of the AdS space $\gamma_n = e^{a}_{m}(x)\gamma_a$.

It is important to notice that when acting on a spinor field $\psi^{\alpha}(x)$ the commutator of ∇_m is zero

$$[\nabla_m, \nabla_n]\psi^{\alpha} = 0, \tag{62}$$

while acting on a vector field V_p the commutator is

$$[\nabla_m, \nabla_n]V_p = [D_m, D_n]V_p = -R^{(AdS)}{}_{mnp}{}^{q}V_q\,. \tag{63}$$

For the spin-tensor fields $\psi^{\alpha}_{m_1\cdots m_{s-\frac{1}{2}}}$ we thus have

$$[\nabla_n, \nabla_p]\psi^{\alpha}_{m_1\cdots m_{s-\frac{1}{2}}} = [D_m, D_n]\psi^{\alpha}_{m_1\cdots m_{s-\frac{1}{2}}} = -\sum R^{(AdS)}{}_{mnm_1}{}^{q}\psi^{\alpha}_{qm_2\cdots m_{s-\frac{1}{2}}}\,. \tag{64}$$

Note also that ∇_m does not annihilate the gamma matrix γ_n

$$\nabla_m \gamma_n = \frac{i\Lambda^{\frac{1}{2}}}{2}[\gamma_m, \gamma_n]\,. \tag{65}$$

Thus the gauge invariant field equations for higher spin fields in AdS do exist. If we now consider fluctuations of the gravitational field around the AdS background the gauge variation (58) of the equations (60) will again have contributions similar to (59) of the bare Riemann curvature of the fluctuating gravitational field

$$\delta G_{m_1 m_2} = -i(R_{nm_1m_2}{}^p + R_{nm_2m_1}{}^p)\gamma^n \xi_p + \cdots\,, \tag{66}$$

where $\cdots$ stand for harmless terms, which can be canceled by an appropriate modification of the gauge transformations. As has been first noticed by Fradkin and Vasiliev in 1987 [24], because of the non–zero *dimensionful* cosmological constant of the background, it is now possible to modify the field equation (59) such that the variation of an appropriate additional term will cancel the dangerous Riemann curvature term in (66), at least in the first order of the perturbation of the gravitational field. For the spin $\frac{5}{2}$ field the appropriate term describing its non–minimal coupling to a gravitational fluctuation is

$$\triangle G_{m_1 m_2} = \frac{i}{2\Lambda}(R_{pm_1m_2q} + R_{pm_2m_1q})\,\nabla\!\!\!/\,\psi^{pq}\,, \tag{67}$$

where $R_{pm_1m_2q}$ is the Riemann curvature corresponding to the deviation of the gravitational field from the AdS background. Such a term can be obtained from a cubic interaction term in the spin $\frac{5}{2}$ action

$$S_{int} = \frac{i}{\Lambda}\int d^D x\sqrt{g}\,\{\bar\psi^{m_1m_2}R_{pm_1m_2q}\,\nabla\!\!\!/\,\psi^{pq} + \bar\psi^{m_1m_2}(\nabla_r\gamma^r R_{pm_1m_2q})\psi^{pq}\}\,. \tag{68}$$

Note that the cosmological constant enters the interacting term (67), (68) in a non–polynomial way. Therefore, such terms become singular when the cosmological constant tends to zero, and, does not allow of the flat space limit.

Consider now the gauge variation of the interaction term (67)

$$\begin{aligned}\delta(\triangle G_{m_1m_2}) &= \frac{2i}{\Lambda}(R_{pm_1m_2q} + R_{pm_2m_1q})\,\gamma_n\nabla^n\nabla^p\xi^q \\ &= \frac{i}{\Lambda}(R_{pm_1m_2q} + R_{pm_2m_1q})\,\gamma_n[\nabla^n,\nabla^p]\xi^q + \cdots\,,\end{aligned} \tag{69}$$

where $\cdots$ stand for the terms with the anticommutator of ∇_n which are assumed to be harmless, i.e. can be canceled by an appropriate modification of the gauge transformations of fields and/or by adding more cubic interaction terms (with higher derivatives) in to the action and into the equation of motion. If we restrict ourselves to the consideration of only the first order in small gravitational interactions, then in (69) the commutator of derivatives should be restricted to the zero order contribution of the AdS curvature (61), (63). The AdS curvature (61) is proportional to the cosmological constant which cancels that in the denominator of (69). So in this approximation the form of the variation of the non–minimal interaction term (58) reduces to that of the rest of the field equation (66) with the opposite sign and thus cancels the latter.

Towards a complete non–linear higher spin field theory

It turns out that beyond the linear approximation of gravitational fluctuations the situation with gauge invariance becomes much more complicated. As the analysis carried out by different people showed [20, 21, 24, 25], a gauge invariant interacting theory of massless higher spin fields should contain

- infinite number of fields of increasing spins involved in the interaction and in symmetry transformations and
- terms with higher derivatives of fields both in the action and in gauge transformations.

No complete action has been constructed so far to describe such an interacting theory of infinite number of higher spin fields, though generic non–linear equations of motion describing higher spin interactions do exist [26, 2]. A main problem is in finding and understanding the (non–abelian) algebraic structure of the gauge transformations of the higher spin fields modified by their interactions. In other words, the question is what is the gauge symmetry algebra which governs the interacting higher spin theory? Note that in the case of Yang–Mills vector fields and gravity the knowledge of the structure of non–abelian gauge symmetries and general coordinate invariance was crucial for the construction of the complete non–linear actions for these fields.

To deal simultaneously with the whole infinite tower of higher spins and to analyze their symmetry and geometrical properties, one may try to cast them into a finite number of 'hyperfields' by extending space–time with additional directions associated with infinitely many spin degrees of freedom. In the formulation discussed above this can be done by introducing auxiliary vector coordinates y^m [39].

Consider in a $(D+D)$-dimensional space parametrized by x^m and y^n a scalar field $\Phi(x,y)$ which is analytic in y^n. Then $\Phi(x,y)$ can be presented as a series expansion in powers of y^n

$$\Phi(x,y) = \phi(x) + \phi_m(x)\,y^m + \phi_{m_1m_2}(x)\,y^{m_1}y^{m_2} + \sum_{s=3}^{\infty} \phi_{m_1\cdots m_s}(x)\,y^{m_1}\cdots y^{m_s}\,. \tag{70}$$

We see that the components of this expansion are D-dimensional symmetric tensor fields $\phi_{m_1\cdots m_s}(x)$. We would like $\Phi(x,y)$ to have symmetry properties and to satisfy field equations which would produce the gauge transformations (1), the traceless conditions (5), (11) and the equations of motion (2) of the higher spin fields $\phi_{m_1\cdots m_s}(x)$. It is not hard to check that the gauge transformation of the hyperfield $\Phi(x,y)$ should have the form

$$\delta\Phi(x,y) = y^m\,\partial_m\,\Xi(x,y)\,, \tag{71}$$

where as above $\partial_m = \frac{\partial}{\partial x^m}$, and higher components of the gauge parameter $\Xi(x,y) = \sum_{s=0}^{\infty} \xi_{m_1\cdots m_s}(x)y^{m_1}\cdots y^{m_s}$ are traceless, which is ensured by imposing the condition

$$\partial_y^2\,\Xi(x,y) = 0\,, \qquad \partial_y^2 \equiv \eta^{mn}\frac{\partial}{\partial y^m}\frac{\partial}{\partial y^n}\,. \tag{72}$$

The double tracelessness (11) is encoded in the equation

$$\partial_y^2 \partial_y^2 \Phi(x,y) = 0, \tag{73}$$

and the higher spin field equations (2) are derived from the following equations of motion of the hyperfield $\Phi(x,y)$

$$\left[\eta^{mn} - y^m \frac{\partial}{\partial y_n} + y^m y^n \partial_y^2\right] \partial_m \partial_n \Phi(x,y) = 0. \tag{74}$$

Analogously, to describe the fields with half integer spins let us introduce a spinorial hyperfield

$$\Psi^\alpha(x,y) = \psi^\alpha(x) + \psi_m^\alpha(x) y^m + \sum_{s-\frac{5}{2}}^{\infty} \psi^\alpha_{m_1 \cdots m_{s-\frac{1}{2}}}(x) y^{m_1} \cdots y^{m_{s-\frac{1}{2}}}. \tag{75}$$

The gauge transformations of $\Psi^\alpha(x,y)$ are

$$\delta \Psi^\alpha(x,y) = y^m \partial_m \Xi^\alpha(x,y), \quad \gamma^m \frac{\partial}{\partial y^m} \Xi^\alpha(x,y) = 0\,, \tag{76}$$

$\Psi^\alpha(x,y)$ satisfies the 'triple' gamma–traceless condition

$$\gamma^m \frac{\partial}{\partial y^m} \partial_y^2 \Psi(x,y) = 0, \tag{77}$$

and the equations of motion

$$\gamma^m \left[\partial_m - y^n \frac{\partial}{\partial y^m} \partial_n\right] \Psi(x,y) = 0. \tag{78}$$

The equations (76), (77) and (78) comprise those for the half–integer spin fields.

The above construction is a simple example of how one can formulate the free theory of infinite number of higher spin fields in terms of a finite number of fields propagating in extended space. In contrast to the Fronsdal formulation, this construction is on the mass shell. It is not clear how to construct an action in the extended space which would produce the equations (74) and (78), neither how to generalize these equations to include non–linear terms.

Much more sophisticated on–shell formulations which involve either vector or spinor auxiliary variables and are based on a solid group–theoretical ground have been developed in [26, 27].

For instance, to find and study the algebraic and geometrical structure of higher spin symmetries (at least in $D = 4$ and $D = 6$), an alternative description of the higher spin fields has proved to be useful. It has been mainly developed by Vasiliev with collaborators (see [26], [2] and references therein) and by Sezgin and Sundell [5]. This is a so called unfolded formulation of the equations of motion of higher spin field theory with the use of spin–tensor representations of the Lorentz group and auxiliary commuting spinor coordinates. "Unfolded" basically means that all fields (including scalars and spinors) enter into the game with their descendants, i.e. auxiliary fields which on the mass shell are higher derivatives of the physical fields. From the algebraic point of view the unfolded formulation is a particular realization and an extension of a so called free differential algebra which is also a basis of the group manifold approach [53]. The field equations are formulated as a zero curvature condition which requires also 0–forms to be involved into the description of systems with infinite number of degrees of freedom.

Unfolded field dynamics

In the unfolded formulation [2] the fields of spin $s \geq 1$ in $D = 4$ are described by a generalized vielbein and connection one–form

$$\omega(x,Y) = \sum_{n,p=0}^{\infty} dx^m \omega_m^{A_1 \cdots A_n, \dot{A}_1 \cdots \dot{A}_p}(x)\, y_{A_1} \cdots y_{A_p} \bar{y}_{\dot{A}_1} \cdots \bar{y}_{\dot{A}_p} \qquad (79)$$

and by its curvature two–form

$$R(x,Y) = d\omega(x,Y) - (\omega \wedge \star \omega)(x,Y)\,, \qquad (80)$$

where $(y_A, \bar{y}_{\dot{A}}) = Y_\alpha$ $(A, \dot{A} = 1,2)$ are auxiliary two–component Weyl spinor variables with even Grassmann parity which resemble twistors and satisfy the oscillator (or Moyal star–product) commutation relations

$$y_A \star y_B - y_B \star y_A = \varepsilon_{AB}, \quad \bar{y}_{\dot{A}} \star \bar{y}_{\dot{B}} - \bar{y}_{\dot{B}} \star \bar{y}_{\dot{A}} = \varepsilon_{\dot{A}\dot{B}}\,, \qquad (81)$$

and the star–product of the connection has been used in the definition of the curvature (80).

Another object of the unfolded formulation is the zero–form

$$C(x,Y) = \sum_{n,p=0} C^{A_1 \cdots A_n, \dot{A}_1 \cdots \dot{A}_p}(x)\, y_{A_1} \cdots y_{A_n} \bar{y}_{\dot{A}_1} \cdots \bar{y}_{\dot{A}_p}\,, \qquad (82)$$

which contains the scalar field $\phi(x) = C(x,Y)|_{Y=0}$, the spinor field $\psi^\alpha(x) = \frac{\partial}{Y_\alpha} C(x,Y)|_{Y=0}$ and (Weyl) curvature tensors of the higher spin fields. The field $C(x,Y)$ is introduced to incorporate the spin–0 and spin–$\frac{1}{2}$ matter fields, and its higher components in the Y–series expansion are either gauge field curvature tensors related to (80) or higher derivatives of the matter fields and of the gauge field curvatures.

The higher spin gauge transformations are

$$\delta\,\omega(x,Y) = d\xi(x,Y) - (\omega \star \xi)(x,Y) + (\xi \star \omega)(x,Y)\,, \qquad (83)$$

$$\delta C(x,Y) = (\xi \star C)(x,Y) - (C \star \tilde{\xi})(x,Y)\,, \qquad (84)$$

where $\tilde{\xi} = \xi(x, y, -\bar{y})$.

In the free (linearized) higher spin theory, the following relation holds

$$R_{linear}(x,Y) = \left\{\partial^A\bar{\partial}^{\dot{B}}\omega(x,Y)\wedge\partial_A\bar{\partial}^{\dot{C}}\omega(x,Y)\right\}|_{Y=0}\,\bar{\partial}_{\dot{B}}\bar{\partial}_{\dot{C}}C(x,0,\bar{y})$$
$$+\left\{\partial^A\bar{\partial}^{\dot{B}}\omega(x,Y)\wedge\partial^C\bar{\partial}_{\dot{B}}\omega(x,Y)\right\}|_{Y=0}\,\partial_A\partial_C C(x,y,0)\,, \qquad (85)$$

and $C(x,Y)$ satisfies the unfolded field equations

$$i\sigma^m_{A\dot{A}}\frac{\partial}{\partial x^m}C(x,Y)=\frac{\partial}{\partial y^A}\frac{\partial}{\partial\bar{y}^{\dot{A}}}C(x,Y)\,, \qquad (86)$$

which when written in components are equivalent to the free higher spin field equations considered in the symmetric tensor formulation. For details on the non–linear generalization of the unfolded equations (85) and (86) we refer the reader to [2] and references therein.

An advantage of the unfolded formalism is that it allows one to treat the whole infinite tower of the higher spin fields simultaneously and provides a compact form of the higher spin symmetry transformations which form an infinite dimensional associative Lie (super)algebra. All this is required, as we have discussed above, for the construction of the consistent interacting higher spin theory. The action describing the unfolded dynamics is still to be found though.

OTHER DEVELOPMENTS

Higher spin field theory from dynamics in tensorial spaces. An alternative to Kaluza–Klein.

The experience of studying various field theories teaches us that in many cases a new insight into their structure can be gained by finding and analyzing a classical dynamical object whose quantization would reproduce the field theory of interest. The well known examples are various spinning particle and superparticle models whose quantum dynamics is described by a corresponding (supersymmetric) field theory. In all the conventional cases only finite number of states of different spins can be produced by quantizing particle models. But, as we have mentioned, for a consistent interacting higher spin field theory we need an infinite number of states. These are produced by strings, but as it has been noted, the associated string field theory is rather complicated. It seems desirable at first to find and analyze a simpler model with an infinite number of quantum higher spin states.

Such a superparticle model does exist [55]. In addition to the relation to higher spins, this model reveals other interesting features, such as the invariance under supersymmetry with tensorial charges (which are usually associated with brane solutions of Superstring and M–Theory), and it has been the first example of a dynamical BPS system which preserves more than one half supersymmetry of the bulk. The study of these features was a main motivation for the original paper [55]. BPS states preserving $\frac{2n-1}{2n}$ supersymmetries (with $n=16$ for $D=10,11$) have later on been shown to be building blocks of any BPS state and conjectured to be hypothetical constituents or 'preons' of M-theory [56]. The relation of the model of [55] to the theory of massless higher spin fields in the unfolded formulation (86) was assumed in [57], where the quantum states of the superparticle was shown to form an infinite tower of the massless higher spin fields. This relation has been analyzed in detail in [59, 60, 61].

Probably, the first who suggested a physical application of tensorial spaces to the theory of higher spins was C. Fronsdal.

In his Essay of 1985 [54] Fronsdal conjectured that four–dimensional higher spin field theory can be realized as a field theory on a ten–dimensional tensorial manifold parametrized by the coordinates

$$x^{\alpha\beta} = x^{\beta\alpha} = \frac{1}{2} x^m \gamma_m^{\alpha\beta} + \frac{1}{4} y^{mn} \gamma_{mn}^{\alpha\beta}, \quad m,n = 0,1,2,3; \quad \alpha,\beta = 1,2,3,4 \tag{87}$$

where x^m are associated with four coordinates of the conventional $D = 4$ space–time and six $y^{mn} = -y^{mn}$ describe spinning degrees of freedom.

The assumption was that by analogy with, for example, $D = 10$ or $D = 11$ supergravities, which are relatively simple theories but whose dimensional reduction to four dimensions produces very complicated extended supergravities, there may exist a theory in ten–dimensional tensorial space whose alternative Kaluza–Klein reduction may lead in $D = 4$ to an infinite tower of fields with increasing spins instead of the infinite tower of Kaluza–Klein particles of increasing mass. The assertion was based on the argument that the symmetry group of the theory should be $OSp(1|8) \supset SU(2,2)$, which contains the $D = 4$ conformal group as a subgroup such that an irreducible (oscillator) representation of $OSp(1|8)$ contains each and every massless higher spin representation of $SU(2,2)$ only once. So the idea was that using a single representation of $OSp(1|8)$ in the ten-dimensional tensorial space one could describe an infinite tower of higher spin fields in $D = 4$ space–time in a simpler way. Fronsdal regarded the tensorial space as a space on which $Sp(8)$ acts like a group of generalized conformal transformations. Ten is the minimal dimension of such a space which can contain D=4 space–time as a subspace. For some reason Fronsdal gave only a general definition and did not identify this ten–dimensional space with any conventional manifolds, like the ones mentioned above.

In his Essay Fronsdal also stressed the importance of $OSp(1|2n)$ supergroups for the description of theories with superconformal symmetry. In the same period and later on different people studied $OSp(1|2n)$ supergroups in various physical contexts. For instance, $OSp(1|32)$ and $OSp(1|64)$ have been assumed to be underlying superconformal symmetries of string- and M-theory.

The tensorial superparticle model of Bandos and Lukierski [55] turned out to be the first dynamical realization of the Fronsdal proposal.

The tensorial particle action has the following form

$$S[X,\lambda] = \int E^{\alpha\beta}(X(\tau))\, \lambda_\alpha(\tau)\, \lambda_\beta(\tau)\,, \tag{88}$$

where $\lambda_\alpha(\tau)$ is an auxiliary commuting real spinor, a *twistor–like* variable, and $E^{\alpha\beta}(x(\tau))$ is the pull back on the particle worldline of the tensorial space vielbein. In flat tensorial space

$$E^{\alpha\beta}(X(\tau)) = d\tau\,\partial_\tau X^{\alpha\beta}(\tau) = dX^{\alpha\beta}(\tau). \tag{89}$$

The dynamics of particles on the supergroup manifolds $OSp(N|n)$ (which are the tensorial extensions of AdS superspaces) was considered for $N = 1$ in [58, 61] and for a generic N in [59, 60]. The twistor–like superparticle in $n = 32$ tensorial superspace was considered in [63] as a point–like model for BPS preons [56], the hypothetical $\frac{31}{32}$–supersymmetric constituents of M–theory.

The action (88) is manifestly invariant under global $GL(n)$ transformations. Without going into details which the reader may find in [55, 59, 61], let us note that the action (88) is invariant under global $Sp(2n)$ transformations, acting non–linearly on $X^{\alpha\beta}$ and on λ_α, *i.e.* it possesses the symmetry considered by Fronsdal to be an underlying symmetry of higher spin field theory in the case $n = 4, D = 4$ [54].

Applying the Hamiltonian analysis to the particle model described by (88) and (89), one finds that the momentum conjugate to $X^{\alpha\beta}$ is related to the twistor–like variable λ_α via the constraint

$$P_{\alpha\beta} = \lambda_\alpha\lambda_\beta\,. \tag{90}$$

This expression is the direct analog and generalization of the Cartan–Penrose (twistor) relation for the particle momentum $P_m = \lambda\gamma_m\lambda$. In virtue of the Fierz identity $\gamma_{m(\alpha\beta}\gamma^m_{\gamma)\delta} = 0$ held in $D = 3,4,6$ and 10 space–time, the twistor particle momentum is light–like in these dimensions. Therefore, in the tensorial spaces corresponding to these dimensions of space–time the first–quantized particles are massless [55, 57].

The quantum counterpart of (90) is the equation [57]

$$D_{\alpha\beta}\Phi(X,\lambda) = \left(\frac{\partial}{\partial X^{\alpha\beta}} - i\lambda_\alpha\lambda_\beta\right)\Phi(X,\lambda) = 0, \tag{91}$$

where the wave function $\Phi(X,\lambda)$ depends on $X^{\alpha\beta}$ and λ_α. The general solution of (91) is the plane wave

$$\Phi(X,\lambda) = e^{iX^{\alpha\beta}\lambda_\alpha\lambda_\beta}\varphi(\lambda), \tag{92}$$

where $\varphi(\lambda)$ is a generic function of λ_α.

One can now Fourier transform the function (92) to another representation

$$C(X,Y) = \int d^4\lambda\, e^{-iY^\alpha\lambda_\alpha}\Phi(X,\lambda) = \int d^4\lambda\, e^{-iY^\alpha\lambda_\alpha + iX^{\alpha\beta}\lambda_\alpha\lambda_\beta}\varphi(\lambda)\,. \tag{93}$$

The wave function $C(X,Y)$ satisfies the Fourier transformed equation

$$\left(\frac{\partial}{\partial X^{\alpha\beta}} + i\frac{\partial^2}{\partial Y^\alpha \partial Y^\beta}\right)C(x,Y) = 0, \tag{94}$$

which is similar to the unfolded equation (86) and which actually reduces to the latter [59, 61].

Quantum states of the tensorial superparticle satisfying Eq. (91) was shown to form an infinite series of massless higher spin states in $D = 4,6$ and 10 space–time [57]. In

[58] quantum superparticle dynamics on $OSp(1|4)$ was assumed to describe higher spin field theory in $N=1$ super AdS_4.

In [57] it was shown explicitly how the alternative Kaluza–Klein compactification produces higher spin fields. It turns out that in the tensorial superparticle model, in contrast to the conventional Kaluza–Klein theory, the compactification occurs in the momentum space and not in the coordinate space. The coordinates conjugate to the compactified momenta take discrete (integer and half integer values) and describe spin degrees of freedom of the quantized states of the superparticle in conventional space–time.

In [59] M. Vasiliev has extensively developed this subject by having shown that the first–quantized field equations (94)in tensorial superspace of a bosonic dimension $\frac{n(n+1)}{2}$ and of a fermionic dimension nN are $OSp(N|2n)$ invariant, and for $n=4$ correspond to the unfolded higher spin field equations in $D=4$. It has also been shown [60] that the theory possesses properties of causality and locality.

As was realized in [59, 60], the field theory of quantum states of the tensorial particle is basically a classical theory of two fields in the tensorial space, a scalar field $b(X^{\alpha\beta})$ and a spinor field $f^{\alpha}(X^{\beta\gamma})$. These fields form a fundamental linear representation of the group $OSp(1|2n)$ and satisfy the following tensorial equations

$$(\partial_{\alpha\beta}\partial_{\gamma\delta}-\partial_{\alpha\gamma}\partial_{\beta\delta})b(X)=0, \quad \partial_{\alpha\beta}f_{\gamma}(X)-\partial_{\alpha\gamma}f_{\beta}(X)=0. \tag{95}$$

In the case of $n=4$ (87) the fields $b(X)$ and $f_{\alpha}(X)$ subject to eqs. (95) describe the infinite tower of the massless (conformally invariant) fields of all possible integer and half–integer spins in the physical four–dimensional subspace of the ten–dimensional tensorial space [54, 59]. In the cases of $n=8$ and $n=16$ which correspond to $D=6$ and $D=10$ space–time, respectively, the equations (95) describe conformally invariant higher spin fields with self–dual field strengths [50].

If in the case of $n=4$ and $D=4$ we split $X^{\alpha\beta}$ onto x^m and y^{mn} as in Eq. (87), the system of equations (95) takes the form

$$\begin{gathered}\partial_p\,\partial^p\,b(x^l,y^{mn})=0,\\ \partial_p\,\partial_q\,b(x^l,y^{mn})-4\partial_{pr}\,\partial^r{}_q\,b(x^l,y^{mn})=0, \quad \partial_q{}^p\,\partial_p\,b(x^l,y^{mn})=0,\\ \varepsilon^{pqrt}\,\partial_q\,\partial_{rt}\,b(x^l,y^{mn})=0, \quad \varepsilon^{pqrt}\,\partial_{pq}\,\partial_{rt}\,b(x^l,y^{mn})=0,\end{gathered} \tag{96}$$

$$\gamma^p\,\partial_p\,f(x^l,y^{mn})=0, \quad [\partial_p-2\gamma^r\,\partial_{rp}]\,f(x^l,y^{mn})=0, \tag{97}$$

where ∂_p and ∂_{rp} are the derivatives along x^p and y^{rp}, respectively.

Then let us expand $b(x,y)$ and $f_{\alpha}(x,y)$ in series of y^{mn}

$$\begin{aligned} b(x^l,y^{mn}) &= \phi(x)+y^{m_1n_1}F_{m_1n_1}(x)+y^{m_1n_1}\,y^{m_2n_2}\,[R_{m_1n_1,m_2n_2}(x)-\tfrac{1}{2}\eta_{m_1m_2}\partial_{n_1n_2}\phi(x)]\\ &\quad+\textstyle\sum_{s=3}^{\infty}y^{m_1n_1}\cdots y^{m_sn_s}\,[R_{m_1n_1,\cdots,m_sn_s}(x)+\cdots], \\ f^{\alpha}(x^l,y^{mn}) &= \psi^{\alpha}(x)+y^{m_1n_1}[\mathscr{R}^{\alpha}_{m_1n_1}(x)-\tfrac{1}{2}\partial_{m_1}(\gamma_{n_1}\psi)^{\alpha}]\\ &\quad+\textstyle\sum_{s=\frac{5}{2}}^{\infty}y^{m_1n_1}\cdots y^{m_{s-\frac{1}{2}}n_{s-\frac{1}{2}}}\,[\mathscr{R}^{\alpha}_{m_1n_1,\cdots,m_{s-\frac{1}{2}}n_{s-\frac{1}{2}}}(x)+\cdots].\end{aligned} \tag{98}$$

In (98) $\phi(x)$ and $\psi^{\alpha}(x)$ are scalar and spin 1/2 field, $F_{m_1n_1}(x)$ is the Maxwell field strength, $R_{m_1n_1,m_2n_2}(x)$ is the curvature tensor of the linearized gravity, $\mathscr{R}^{\alpha}_{m_1n_1}(x)$ is the Rarita–Schwinger field strength and other terms in the series stand for generalized Riemann curvatures of spin-s fields (which also contain contributions of derivatives of lower spin fields denoted by dots, as in the case of the Rarita–Schwinger field and gravity). The scalar and the spinor field satisfy, respectively, the Klein–Gordon and the Dirac equation, and the higher spin field curvatures satisfy the Bianchi identities (21), (22) and the linearized higher spin field equations (26) and (27) in $D=4$ space–time. Similar equations also follow from the unfolded equations (86). In the model under consideration they are consequences of the field equations (95), or equivalently of (96) and (97) in the flat tensorial space. The generalization of the equations (95) to a field theory on the tensorial manifold $OSp(1|n)$, which for $n=4$ corresponds to the theory of higher spin fields in AdS_4, has been derived in [61].

An interesting and important problem is to find a simple and appropriate non–linear generalization of equations (refbf which would correspond to an interacting higher spin field theory. An attempt to construct such a generalization in the framework of tensorial superspace supergravity was undertaken in [62].

Massless higher spin field theory as a tensionless limit of superstring theory

In these lectures we have considered the formulations of massless higher spin field theory which are not directly related to String Theory. A natural question arises which formulation one can derive from String Theory at the tensionless limit $T\sim\frac{1}{\alpha'}\to 0$. This has been a subject of a number of papers [64]–[67] (and references therein) which we briefly sketch below.

Consider, for instance a free open bosonic string in flat space–time, whose worldsheet is parametrized by a 'spatial' coordinate $\sigma\in[0,\pi]$ and a 'time' coordinate τ. String dynamics is described by the coordinates

$$X^m(\tau,\sigma)=x^m+2\alpha' p^m\tau+i\sqrt{2\alpha'}\sum_{n\neq 0}^{\infty}\frac{1}{n}a^m_n e^{-in\tau}cos(n\sigma) \tag{99}$$

and momenta

$$P^m(\tau,\sigma)=p^m+\frac{1}{\sqrt{2\alpha'}}\sum_{n\neq 0}^{\infty}a^m_n e^{-in\tau}cos(n\sigma)\,, \tag{100}$$

where x^m and p^m are the center of mass variables and a^m_n are the string oscillator modes satisfying (upon quantization) the commutation relations $[p^m,x^p]=-i\eta^{mp}$, $[a^m_n,a^p_l]=n\delta_{n+l}\eta^{mp}$.

String dynamics is subject to the Virasoro constraints

$$L_k=\frac{1}{2}\sum_{n=-\infty}^{+\infty}a^m_{k-n}a_{mn}=\sqrt{2\alpha'}\,p^m a_{mk}+\frac{1}{2}\sum_{n\neq k,0}a^m_{k-n}a_{mn}\,,\quad k\neq 0, \tag{101}$$

$$L_0 = 2\alpha' p^m p_m + \sum_{n>0} a^m_{-n} a_{mn} \,. \tag{102}$$

The latter produces the mass shell condition for the string states

$$M^2 = -p^m p_m = \frac{1}{2\alpha'} \sum_{n>0} a^m_{-n} a_{mn} \,. \tag{103}$$

We observe that in the tensionless limit $\alpha' \to \infty$ all string states become massless, while the properly rescaled Virasoro constraints become at most linear in the oscillator modes

$$l_0 = \frac{1}{2\alpha'} L_0|_{\alpha' \to \infty} = p^m p_m \,, \qquad l_k = \frac{1}{\sqrt{2\alpha'}} L_k|_{\alpha' \to \infty} = p_m a^m_k \tag{104}$$

and satisfy a simple algebra without any central charge

$$[l_0, l_k] = 0 \,, \qquad [l_j, l_k] = \delta_{j+k}\, l_0 \,. \tag{105}$$

Thus in the tensionless limit the quantum consistency of string theory does not require any critical dimension for the string to live in. Note that at $\alpha' \to \infty$ the string coordinate (99) blows up and is not well defined, while the oscillator modes remain appropriate variables for carrying out the quantization of the theory.

The corresponding nilpotent BRST charge takes the form

$$Q = \sum_{n=-\infty}^{+\infty} (c_{-n} l_n - \frac{n}{2} b_0 c_{-n} c_n) \,, \tag{106}$$

where c_n and b_n are the ghosts and anti–ghosts associated with the constraint algebra (105).

The BRST charge can be used to construct a free action for the string field states $|\Phi>$, obtained by acting on the Fock vacuum by the creating operators,

$$S = \frac{1}{2} \int <\Phi|Q|\Phi> \,. \tag{107}$$

The action (107) can be used for the derivation of a corresponding action and equations of motions of the higher spin fields encoded in $|\Phi>$. As has been shown in [67] such an action and equations of motion are more involved than eqs. (15), (16), (2) and (3) since they contain intertwined (triplet) fields of different spins. The equations (15), (16), (2) and (3) are obtained from (107) upon gauge fixing part of the available local symmetry and by eliminating auxiliary fields.

Further details the interested reader can find in [66, 67] and references therein. We should note that the tensionless limit of the string considered here differs from the so called null string models [68]. In these models, in contrast to the way of getting the tensionless string discussed above the limit is taken in such a way that the string coordinate $X^m(\sigma, \tau)$ remains a well defined variable, while the oscillator modes disappear. As a result the quantum states of the null strings correspond to a continuous set of massless particles without (higher) spin.

CONCLUSION

In these lectures we have described main features and problems of higher spin field theory and have flashed some ways along which it has been developed over last years.

ACKNOWLEDGMENTS

I am grateful to Igor Bandos, Jose de Azcárraga, Alexander Filippov, Jerzy Lukierski, Paolo Pasti, Mario Tonin and attenders of the courses and seminars for the encouragement to write these lecture notes. I would also like to thank Misha Vasiliev for useful comments. This work was partially supported by the Grant N 383 of the Ukrainian State Fund for Fundamental Research, by the INTAS Research Project N 2000-254, by the European Community's Human Potential Programme under contract HPRN-CT-2000-00131 Quantum Spacetime, by the EU MRTN-CT-2004-005104 grant 'Forces Universe', and by the MIUR contract no. 2003023852.

REFERENCES

1. M. A. Vasiliev, Int. J. Mod. Phys. **D5**, 763 (1996), hep-th/9910096.
   M. A. Vasiliev, *"Higher spin symmetries, star-product and relativistic equations in AdS Space"*, hep-th/0002183.
2. M. A. Vasiliev, *"Progress in higher spin gauge theories"*, hep-th/0104246 .
3. B. Sundborg, Nucl. Phys **102**, 113 (*Proc. Suppl.*) (2001), hep-th/0103247.
4. L. Dolan, C. Nappi and E. Witten, JHEP **0110**, 016 (2001), hep-th/0109096.
5. E. Sezgin and P. Sundell, JHEP **0207**, 055 (2002), hep-th/0205132; Nucl. Phys. **B644**, 303 (2002), hep-th/0205131.
6. S. Deser and A. Waldron, Nucl.Phys. **B662**, 379 (2003), hep-th/0301068.
7. M. Bianchi, *"Higher spins and stringy AdS(5) x S(5)"*, hep-th/0409304.
8. P. A. M. Dirac, Proc. Roy. Soc. Lond. **155A**, 447 (1936).
9. E. P. Wigner, Annals Math. **40**, 149 (1939); Nucl. Phys. Proc. Suppl. **6**, 9 (1989).
10. M. Fierz and W. Pauli, Proc. Roy. Soc. Lond. **A173**, 211 (1939).
11. W. Rarita and J. S. Schwinger, Phys. Rev. **60**, 61 (1941).
12. V. Bargmann, E. P. Wigner, Proc. Nat. Acad. Sci. **34**, 211 (1948).
13. C. Fronsdal, Nuovo Cimento Suppl. **9**, 416 (1958).
14. S. Weinberg, Phys. Rev. **133**, B1319 (1964); Phys. Rev. **134**, B882 (1964).
15. S.-J. Chang, Phys. Rev. **161**, 1308 (1967).
16. L. P. S. Singh and C. R. Hagen, Phys. Rev. **D9**, 898 (1974); Phys. Rev. **D9**, 910 (1974).
17. C. Aragone and S. Deser, Nuovo Cim. **A3**, 709 (1971); Nuovo Cim. **B57**, 33 (1980);
    C. Aragone and S. Deser, Phys. Lett. **B86**, 161 (1979).
18. F.A. Berends, J.W. van Holten, P. van Nieuwenhuizen and B. de Wit, J. Phys. **A13**, 1643 (1980).
19. S. Deser, Gen. Rel. Grav. **1**, 9 (1970);
    D.G. Boulware and S. Deser, Annals Phys. **89**, 193 (1975).
20. A.K.H. Bengtsson, I. Bengtsson and L. Brink, Nucl. Phys. **B227**, 31 (1980); Nucl. Phys. **B227**, 41 (1980).
21. F.A. Berends, G.J.H. Burgers and H. Van Dam, Z. Phys. **C24**, 247 (1984); Nucl. Phys. **B260**, 295 (1985).
22. C. Aragone, S. Deser and Z. Yang, Annals Phys. **179**, 76 (1987).
23. S. Coleman and J. Mandula, Phys. Rev. **159**, 1251 (1967);
    R. Haag, J. Lopuszanski and M. Sohnius, Nucl. Phys. **B88**, 61 (1975).

24. E.S. Fradkin and M.A. Vasiliev, Phys. Lett. **B189**, 89 (1987); Nucl. Phys. **B291**, 141 (1987).
25. T. Damour and S. Deser, Class. Quant. Grav. **4**, L95 (1987).
26. M.A. Vasiliev, Phys. Lett. **209B**, 491 (1988);
M. A. Vasiliev, Ann. Phys. (NY) **190**, 59 (1989).
27. M.A. Vasiliev, *"Higher spin gauge theories in various dimensions"*, hep-th/0401177 and references therein.
28. T. Curtright, Phys. Lett. **B165**, 304 (1985);
C.S. Aulakh, I.G. Koh and S. Ouvry, Phys. Lett. **B173**, 284 (1986);
W. Siegel and B. Zwiebach, Nucl. Phys. **B282**, 125 (1987);
J.M.F. Labastida, Phys. Rev. Lett. **58**, 531 (1987); Nucl. Phys. **B322**, 185 (1989).
29. A. Pashnev and M.M. Tsulaia, Mod. Phys. Lett. **A12**, 861 (1997), hep-th/9703010; Mod. Phys. Lett. **A13**, 1853 (1998), hep-th/9803207;
C. Burdik, A. Pashnev and M. Tsulaia, Mod. Phys. Lett. **A16**, 731 (2001), hep-th/0101201; Nucl. Phys. Proc. Suppl **102**, 285 (2001), hep-th/0103143.
30. R.R. Metsaev, Phys. Lett. **B354**, 78 (1995); Phys. Lett. **B419**, 49 (1998), hep-th/9802097.
31. L.Brink, R.R.Metsaev and M.A.Vasiliev, Nucl. Phys. **B586**, 183 (2000), hep-th/0005136.
32. X. Bekaert and N. Boulanger, *"Tensor gauge fields in arbitrary representations of GL(D,R): duality and Poincare lemma"*, hep-th/0208058; Phys. Lett. **B561**, 183 (2003), hep-th/0301243.
33. P. de Medeiros and C. Hull, JHEP **0305**, 019 (2003), hep-th/0303036.
P. de Medeiros, *"Massive gauge-invariant field theories on spaces of constant curvature"*, hep-th/0311254.
34. Y.M. Zinoviev, *"On massive mixed symmetry tensor fields in Minkowski space and (A)dS"*, hep-th/0211233; *"First Order Formalism for Mixed Symmetry Tensor Fields"*, hep-th/0304067.
35. S. Deser and A. Waldron, Phys. Rev. Lett. **87**, 031601 (2001), hep-th/0102166;
Nucl. Phys. **B607**, 577 (2001), hep-th/0103198; Phys. Lett. **B513**, 137 (2001), hep-th/0105181; Nucl. Phys. **B631**, 369 (2002), hep-th/0112182.
36. H.A. Buchdahl, J. Phys. **A15**, 1057 (1982);
V. Wunsch, Gen. Rel. Grav. **17**, 15 (1985);
R. Illge, Commun. Math. Phys. **158**, 433 (1993).
37. K.B. Alkalaev, O.V. Shaynkman and M.A. Vasiliev, *"On the Frame-Like Formulation of Mixed-Symmetry Massless Fields in* $(A)dS_d$*"*, hep-th/0311164.
38. K.B. Alkalaev, *"Two-column fields in* AdS_d*"*, hep-th/0311212.
39. C. Fronsdal, Phys. Rev. **D18**, 3624 (1978);
J. Fang and C. Fronsdal, Phys. Rev. **D18**, 3630 (1978).
40. C. Fronsdal, Phys. Rev. **D20**, 848 (1979);
J. Fang and C. Fronsdal, Phys. Rev. **D22**, 1361 (1980).
41. D. de Wit and D.Z. Freedman, Phys. Rev. **D21**, 358 (1980).
42. D. Francia and A. Sagnotti, Phys. Lett. **B543**, 303 (2002); Class. Quant. Grav. **20**, S473–S486 (2003).
43. N. Bouatta, G. Compere and A. Sagnotti, *"An introduction to free higher-spin fields"*, hep-th/0409068.
44. P.J. Olver, *Differential hyperforms I,* Univ. of Minnesota report 82-101.
45. M. Dubois-Violette and M. Henneaux, Lett. Math. Phys. **49**, 245 (1999), math.qa/9907135;
Commun. Math. Phys. **226**, 393 (2002), math-qa/0110088.
46. X. Bekaert and N. Boulanger, Commun. Math. Phys. **245**, 27 (2004), hep-th/0208058.
47. T. Damour and S. Deser, Annales Poincare Phys. Theor. **47**, 277 (1987).
48. X. Bekaert and N. Boulanger, Phys. Lett. **B561**, 183 (2003), hep-th/0301243.
49. G.P. Collins and A. Doughty, J. Math. Phys. **28**, 448 (1987).
50. I. Bandos, X. Bekaert, J. de Azcárraga, D. Sorokin and M. Tsulaia, *"Dynamics of Higher Spin Fields and Tensorial Space"*, hep-th/0501113.
51. D.V. Volkov and V.A. Soroka, JETP Lett. **18**, 312 (1973); [Pisma Zh. Eksp. Teor. Fiz. **18**, 529 (1973)].
52. D.Z. Freedman, P. van Nieuwenhuizen and S. Ferrara, Phys. Rev. **D13**, 3214 (1976);
S. Deser and B. Zumino, Phys. Lett. **B62**, 335 (1976).
53. L. Castellani, R. D'Auria and P. Fre, *"Supergravity And Superstrings: A Geometric Perspective"*, Singapore: World Scientific, 1991.

54. C. Fronsdal, *"Massless Particles, Ortosymplectic Symmetry and Another Type of Kaluza–Klein Theory"*, Preprint UCLA/85/TEP/10, in *"Essays on Supersymmetry"*, Reidel, 1986 (Mathematical Physics Studies, v. 8).
55. I. Bandos and J. Lukierski, Mod. Phys. Lett. **A14**,1257 (1999), hep-th/9811022.
56. I. A. Bandos, J.A. de Azcarraga, J.M. Izquierdo and J. Lukierski, Phys. Rev. Lett. **86**, 4451 (2001), hep-th/0101113.
57. I. Bandos, J. Lukierski and D. Sorokin, Phys. Rev. **D61**, 045002 (2000);
"The OSp(1|4) superparticle and exotic BPS states", hep-th/9912264.
58. I. Bandos, J. Lukierski, C. Preitschopf and D. Sorokin, Phys. Rev. **D61**, 065009 (2000).
59. M.A. Vasiliev, Phys. Rev. **D66**, 066006 (2002).
60. M.A. Vasiliev, *"Relativity, causality, locality, quantization and duality in the Sp(2M) invariant generalized space-time"*, hep-th/0111119.
61. M. Plyushchay, D. Sorokin and M. Tsulaia, JHEP **0304**, 013 (2003);
"GL flatness of OSp(1|2n) and higher spin field theory from dynamics in tensorial spaces", hep-th/0310297.
62. I. Bandos, P. Pasti, D. Sorokin and M. Tonin, **0411**, 023 (2004), hep-th/0407180.
63. I.A. Bandos, J.A. de Azcárraga, M. Picón and O. Varela, Phys. Rev. **D69**, 085007 (2004), hep-th/0307106.
64. M. Henneaux and C. Teitelboim, *"First and Second Quantized Point Particles of Any Spin"*, in *"Santiago 1987, Proceedings, Quantum Mechanics of Fundamental Systems 2"*, pp. 113–152. C. Teitelboim and J. Zanelli Eds., Plenum Press.
65. U. Lindstrom and M. Zabzine, Phys. Lett. **B584**, 178 (2004).
66. G. Bonelli, Nucl. Phys. **B669**, 159 (2003);
G. Bonelli, JHEP **0311**, 028 (2003).
67. A. Sagnotti and M. Tsulaia, Nucl. Phys. **B682**, 83 (2004).
68. A. Schild, Phys. Rev. **D16**, 1722 (1977);
A. Karlhede and U. Lindstrom, Class. Quant. Grav. **3**, L73 (1986);
U. Lindstrom, B. Sundborg and G. Theodoridis, Phys. Lett. **B253**, 319 (1991); Phys. Lett. **B258**, 331 (1991);
A. A. Zheltukhin, Sov. J. Nucl. Phys. **48**, 375 (1988), [Yad. Fiz. **48**, 587 (1988)];
I.A. Bandos and A. A. Zheltukhin, Sov. J. Nucl. Phys. **50**, 556 (1989), [Yad. Fiz. **50**, 893 (1989)]; Phys. Lett. **B261**, 245 (1991); Fortsch. Phys. **41**, 619 (1993);
J. Barcelos-Neto and M. Ruiz-Altaba, Phys. Lett. **B228**, 193 (1989);
J. Gamboa, C. Ramirez and M. Ruiz-Altaba, Nucl. Phys. **B338**, 143 (1990).

Quantum BRST Charge and $OSp(1|8)$ Superalgebra of Twistor–Like p-branes with Exotic Supersymmetry and Weyl Symmetry

D.V. Uvarov* and A.A. Zheltukhin†

*Kharkov Institute of Physics and Technology, 61108 Kharkov, Ukraine
†Kharkov Institute of Physics and Technology, 61108 Kharkov, Ukraine
Institute of Theoretical Physics, University of Stockholm, Albanova, SE-10691 Stockholm, Sweden

INTRODUCTION

The physical interpretation of the central charges in supersymmetry algebra as topological charges carried by branes [1] advanced understanding of the phenomena of partial spontaneous breaking of supersymmetry [2]. Because branes are constituents of M-theory, spontaneously breaking supersymmetry, their global and local symmetries correlate with the symmetries of M-theory [3], [4]. Studying these symmetries resulted in the model independent classical analysis of BPS states preserving $\frac{1}{4}$, $\frac{1}{2}$ or $\frac{3}{4}$ fractions of the partially spontaneously broken $D=4$ $N=1$ supersymmetry [5]. A special interest to construction of a physical model with domain wall configurations preserving $\frac{3}{4}$ fraction of the $D=4$ $N=1$ supersymmetry against spontaneous breaking was subscribed there. That configurations were earlier studied in superparticle dynamics [6] and algebraically realized as the brane intersections in [7]. Then the tensionless string/brane model preserving $\frac{3}{4}$ fraction of the $D=4$ $N=1$ supersymmetry and generating static solutions for these tensionless objects was proposed in [8]. These results have sharpen the general question: whether quantum exotic BPS states saturated by the p-brane states protect the same high $\frac{\mathsf{M}-1}{\mathsf{M}}$ fraction of $N=1$ global supersymmetry against spontaneous breaking as in the classical case?

We have started studying the question in [9] on the example of the p-branes preserving $\frac{3}{4}$ fraction of the partially spontaneously broken $D=4$ $N=1$ supersymmetry and found some obstacles for the quantization in the $\hat{\mathcal{Q}}\hat{\mathcal{P}}$-ordering previously studied in [10] (see also [11]). Here we analyze the quantization problem applying the BFV approach [12] and construct quantum Hermitian BRST operator and the generators of gauge Weyl, Virasoro and global $OSp(1|8)$ symmetries extended by the ghost contributions. We prove the nilpotency of the quantum Hermitian BRST charge, its (anti)commutativity with the quantum Hermitian generators of the $OSp(1|8)$ superalgebra and the closure of this quantized superalgebra. At the same time we show that the quantum $\hat{\mathcal{Q}}\hat{\mathcal{P}}$-ordered BRST operator and the $OSp(1|8)$ generators are nonHermitian and differ from the Hermitian ones by the presence of the divergent terms. We discuss a possibility to overcome this

CP767, *Fundamental Interactions and Twistor-Like Methods*, edited by J. Lukierski and D. Sorokin

obstacle by the choice of the special regularization prescription for the p-brane world-volume delta function $\delta^p(\vec{\sigma}-\vec{\sigma}')|_{\vec{\sigma}=\vec{\sigma}'}$ and its derivative $\partial_M\delta^p(0)$. A possibility of the exact cancellation of the divergent terms in the $\hat{\mathfrak{Q}}\hat{\mathfrak{P}}$-ordered quantum operators for other dimensions $D=2,3,4(mod 8)$ is also discussed.

CONVERSION OF TENSIONLESS SUPER *P*-BRANE CONSTRAINTS

New models of tensionless string and p-branes evolving in the symplectic superspace $\mathcal{M}^{susy}_{\mathsf{M}}$ and preserving all but one fractions of $N=1$ supersymmetry were recently studied in [8], [13]. For $\mathsf{M}=2^{[\frac{D}{2}]}$ ($D=2,3,4$ *mod* 8) the space $\mathcal{M}^{susy}_{\mathsf{M}}$ extends the standard D-dimensional super space-time (x_{ab},θ_a), (where $a,b=1,2,...,2^{[\frac{D}{2}]}$) by the tensor central charge (TCC) coordinates z_{ab}. The coordinates $x_{ab}=x^m(\gamma_m C^{-1})_{ab}$ and $z_{ab}=iz^{mn}(\gamma_{mn}C^{-1})_{ab}+z^{mnl}(\gamma_{mnl}C^{-1})_{ab}+...$ constitute components of the symmetric spin-tensor Y_{ab}. In terms of Y_{ab} and the Majorana spinor θ_a the action [8], invariant under the spontaneously broken $N=1$ supersymmetry and world-volume reparametrizations, is given by

$$S_p=\frac{1}{2}\int d\tau d^p\sigma\ \rho^\mu U^a W_{\mu ab}U^b\,, \tag{1}$$

where $W_{ab}=W_{\mu ab}d\xi^\mu$ is the supersymmetric Cartan differential one-form

$$W_{\mu ab}=\partial_\mu Y_{ab}-2i(\partial_\mu\theta_a\theta_b+\partial_\mu\theta_b\theta_a)\,, \tag{2}$$

and $\partial_\mu\equiv\frac{\partial}{\partial\xi^\mu}$ with $\xi^\mu=(\tau,\sigma^M)$, $(M=1,2,...,p)$ parametrizing the p-brane world volume. The local auxiliary Majorana spinor $U^a(\tau,\sigma^M)$ parametrizes the generalized momentum $P^{ab}=\frac{1}{2}\rho^\tau U^aU^b$ of tensionless p-brane and $\rho^\mu(\tau,\sigma^M)$ is the world-volume vector density providing the reparametrization invariance of S_p similarly to the null branes [14].

The action (1) has $(\mathsf{M}-1)$ $\kappa-$symmetries

$$\delta_\kappa\theta_a=\kappa_a,\quad \delta_\kappa Y_{ab}=-2i(\theta_a\kappa_b+\theta_b\kappa_a),\quad \delta_\kappa U^a=0\,, \tag{3}$$

which protect $\frac{\mathsf{M}-1}{\mathsf{M}}$ fraction of the $N=1$ global supersymmetry to be spontaneously broken, because of the one real condition $U^a\kappa_a=0$ for the transformation parameters $\kappa_a(\tau,\vec{\sigma})$[1].

For the four-dimensional space-time the action (1) takes the form

$$S_p=\frac{1}{2}\int d\tau d^p\sigma\ \rho^\mu\left(2u^\alpha\omega_{\mu\alpha\dot\alpha}\bar{u}^{\dot\alpha}+u^\alpha\omega_{\mu\alpha\beta}u^\beta+\bar{u}^{\dot\alpha}\bar\omega_{\mu\dot\alpha\dot\beta}\bar{u}^{\dot\beta}\right)\,. \tag{4}$$

---

[1] To remove a possible misunderstanding in this terminology let us remind that from the world-volume perspective the last fraction of the $N=1$ supersymmetry is also the symmetry of the action (1), but it is spontaneously broken, because of the presence of Goldstone fermion $\tilde{\eta}=-2i(U^a\theta_a)$ encoding the single physical fermionic degree of freedom associated with θ_a (see [13] for details).

This action is invariant under the $OSp(1|8)$ symmetry, which is global supersymmetry of the massless fields of all spins in $D=4$ space-time extended by TCC coordinates [15], [16].

The Hamiltonian structure of the action (4), described in [17], is characterized by 3 fermionic and $2p+7$ bosonic first-class constraints that generate its local symmetries, as well as, 1 fermionic and 8 bosonic second-class constraints taken into account by the construction of the Dirac bracket. We found that the D.B. algebra of the first-class constraints has the rank equal two and it gives rise to the higher powers of the ghosts in the BRST generator.

To simplify transition to the quantum theory the conversion method [18]-[23], transforming all the primary and secondary constraints to the first class, has been applied in [9]. To this end the additional canonically conjugate pairs (P_q^α, q^α), $(\bar{P}_q^{\dot\alpha}, \bar{q}^{\dot\alpha})$, $(P_\tau^{(\varphi)}, \varphi^\tau)$ and the self-conjugate Grassmannian variable f have been introduced. As a result, all the constraints have been converted to the effective first-class constraints in the extended phase space.

The converted constraints for the auxiliary fields are the following

$$\widetilde{P}_u^\alpha = P_u^\alpha + P_q^\alpha \approx 0, \quad \widetilde{\bar{P}}_u^{\dot\alpha} = \bar{P}_u^{\dot\alpha} + \bar{P}_q^{\dot\alpha} \approx 0, \tag{5}$$

$$\widetilde{P}_\tau^{(\rho)} = P_\tau^{(\rho)} + P_\tau^{(\varphi)} \approx 0, \quad P_M^{(\rho)} \approx 0. \tag{6}$$

The converted bosonic constraints $\widetilde{\Phi} \equiv (\widetilde{\Phi}^{\dot\alpha\alpha}, \widetilde{\Phi}^{\alpha\beta}, \widetilde{\bar{\Phi}}^{\dot\alpha\dot\beta})$ originating from the Φ-constraints [9] are given by

$$\begin{aligned}
\widetilde{\Phi}^{\dot\alpha\alpha} &= p^{\dot\alpha\alpha} - \tilde{\rho}^\tau \tilde{u}^\alpha \tilde{\bar{u}}^{\dot\alpha} \approx 0, \\
\widetilde{\Phi}^{\alpha\beta} &= \pi^{\alpha\beta} + \frac{1}{2}\tilde{\rho}^\tau \tilde{u}^\alpha \tilde{u}^\beta \approx 0, \\
\widetilde{\bar{\Phi}}^{\dot\alpha\dot\beta} &= \bar{\pi}^{\dot\alpha\dot\beta} + \frac{1}{2}\tilde{\rho}^\tau \tilde{\bar{u}}^{\dot\alpha} \tilde{\bar{u}}^{\dot\beta} \approx 0
\end{aligned} \tag{7}$$

and have zero Poisson brackets (P.B.) with the constraints (5), (6) and among themselves. The converted fermionic constraints $\widetilde{\Psi} = (\widetilde{\Psi}^\alpha, \widetilde{\bar{\Psi}}^{\dot\alpha})$ originating from the primary Ψ-constraints and generating four $\kappa-$symmetries take the form

$$\begin{aligned}
\widetilde{\Psi}^\alpha &= \pi^\alpha - 2i\bar{\theta}_{\dot\alpha} P^{\dot\alpha\alpha} - 4i\pi^{\alpha\beta}\theta_\beta + 2(\tilde{\rho}^\tau)^{1/2}\tilde{u}^\alpha f \approx 0, \\
\widetilde{\bar{\Psi}}^{\dot\alpha} &= -(\widetilde{\Psi}^\alpha)^* = \bar{\pi}^{\dot\alpha} - 2iP^{\dot\alpha\alpha}\theta_\alpha - 4i\bar{\pi}^{\dot\alpha\dot\beta}\bar{\theta}_{\dot\beta} - 2(\tilde{\rho}^\tau)^{1/2}\tilde{\bar{u}}^{\dot\alpha} f \approx 0,
\end{aligned} \tag{8}$$

where $f^* = f$ is an auxiliary Grassmannian variable characterized by the P.B.

$$\{f(\vec\sigma), f(\vec\sigma')\}_{P.B.} = -i\delta^p(\vec\sigma - \vec\sigma'). \tag{9}$$

The addition of the field $f(\tau,\vec\sigma)$ restores the forth κ-symmetry and transforms all $\widetilde{\Psi}$-constraints to the first class. The Weyl symmetry constraint $\widetilde{\Delta}_W$ in the extended phase space is

$$\widetilde{\Delta}_W = (\tilde{P}_u^\alpha \tilde{u}_\alpha + \tilde{\bar{P}}_u^{\dot\alpha} \tilde{\bar{u}}_{\dot\alpha}) - 2\tilde{\rho}^\tau \tilde{P}_\tau^{(\rho)} - 2\rho^M P_M^{(\rho)} \approx 0, \tag{10}$$

where the variables $(\tilde{u}^{\alpha} = u^{\alpha} - q^{\alpha}, \tilde{\tilde{P}}_u^{\alpha} = \frac{1}{2}(P_u^{\alpha} - P_q^{\alpha}))$ and $(\tilde{\rho}^{\tau} = \rho^{\tau} - \varphi^{\tau}, \tilde{\tilde{P}}_{\tau}^{(\rho)} = \frac{1}{2}(P_{\tau}^{(\rho)} - P_{\tau}^{(\varphi)}))$ form canonically conjugate pairs [9]. Finally, the converted constraints $\widetilde{L}_M$ of the world-volume $\vec{\sigma}-$reparametrizations are

$$\begin{aligned}\widetilde{L}_M &= \partial_M x_{\alpha\dot{\alpha}} P^{\dot{\alpha}\alpha} + \partial_M z_{\alpha\beta}\pi^{\alpha\beta} + \partial_M \bar{z}_{\dot{\alpha}\dot{\beta}}\bar{\pi}^{\dot{\alpha}\dot{\beta}} + \partial_M\theta_{\alpha}\pi^{\alpha} + \partial_M\bar{\theta}_{\dot{\alpha}}\bar{\pi}^{\dot{\alpha}} \\ &\quad + \partial_M\tilde{u}_{\alpha}\tilde{\tilde{P}}_u^{\alpha} + \partial_M\tilde{\bar{u}}_{\dot{\alpha}}\tilde{\tilde{\bar{P}}}_u^{\dot{\alpha}} - \tilde{\rho}^{\tau}\partial_M\tilde{\tilde{P}}_{\tau}^{(\rho)} - \rho^N\partial_M P_N^{(\rho)} - \frac{i}{2}f\partial_M f \approx 0. \qquad (11)\end{aligned}$$

The P.B. superalgebra of the converted first-class constraints (5)–(8), (10), (11) is described by the following non zero relations

$$\{\widetilde{\Psi}^{\alpha}(\vec{\sigma}), \widetilde{\Psi}^{\beta}(\vec{\sigma}')\}_{P.B.} = -8i\widetilde{\Phi}^{\alpha\beta}\delta^p(\vec{\sigma}-\vec{\sigma}'), \qquad (12)$$

$$\{\widetilde{\Psi}^{\alpha}(\vec{\sigma}), \widetilde{\bar{\Psi}}^{\dot{\beta}}(\vec{\sigma}')\}_{P.B.} = -4i\widetilde{\Phi}^{\dot{\beta}\alpha}\delta^p(\vec{\sigma}-\vec{\sigma}'), \qquad (13)$$

$$\{\widetilde{\Delta}_W(\vec{\sigma}), P_M^{(\rho)}(\vec{\sigma}')\}_{P.B.} = 2P_M^{(\rho)}\delta^p(\vec{\sigma}-\vec{\sigma}'), \qquad (14)$$

$$\{\widetilde{L}_M(\vec{\sigma}), P_N^{(\rho)}(\vec{\sigma}')\}_{P.B.} = \partial_M P_N^{(\rho)}\delta^p(\vec{\sigma}-\vec{\sigma}'), \qquad (15)$$

$$\{\widetilde{L}_M(\vec{\sigma}), \widetilde{L}_N(\vec{\sigma}')\}_{P.B.} = (\widetilde{L}_M(\vec{\sigma}')\partial_{N'} - \widetilde{L}_N(\vec{\sigma})\partial_M)\delta^p(\vec{\sigma}-\vec{\sigma}'), \qquad (16)$$

$$\{\widetilde{L}_M(\vec{\sigma}), \chi(\vec{\sigma}')\}_{P.B.} = -\chi(\vec{\sigma})\partial_M\delta^p(\vec{\sigma}-\vec{\sigma}'), \qquad (17)$$

where χ are $\widetilde{\Phi}$, $\widetilde{\Psi}$ and $\widetilde{\Delta}_W$ constraints. The complex conjugate relations have to be added to (12)–(17). The remaining P.B.'s of the constraints are equal to zero in the strong sense. Having the algebra (12)–(17) one can construct BRST charge of the tensionless super p-brane.

BRST CHARGE AND $OSP(1|8)$ SYMMETRY GENERATORS

The algebra(12)–(17) has the rank equal unity and may be presented in the generalized canonical form

$$\{\Upsilon^{\mathcal{A}}(\vec{\sigma}), \Upsilon^{\mathcal{B}}(\vec{\sigma}')\}_{P.B.} = \int d^p\sigma'' f^{\mathcal{AB}}{}_{\mathcal{C}}(\vec{\sigma},\vec{\sigma}'|\vec{\sigma}'')\Upsilon^{\mathcal{C}}(\vec{\sigma}''), \qquad (18)$$

where $f^{\mathcal{AB}}{}_{\mathcal{C}}$ are structure functions. Let us note that the algebra (18) generalizes the original algebra [12] by the taking into account $\partial_M\delta^p(\vec{\sigma}-\vec{\sigma}')$ in the structure functions following from the P.B.'s including the Virasoro constraints $\widetilde{L}_M(\vec{\sigma})$ such as

$$\begin{aligned} f^{\widetilde{L}_M\widetilde{\Phi}^{\alpha\beta}}{}_{\widetilde{\Phi}^{\dot{\gamma}\delta}} &= -\delta^{\dot{\alpha}}_{\dot{\gamma}}\delta^{\beta}_{\delta}\partial_M\delta^p(\vec{\sigma}-\vec{\sigma}')\delta^p(\vec{\sigma}-\vec{\sigma}''), \\ f^{\widetilde{L}_M\widetilde{L}_N}{}_{\widetilde{L}_Q} &= -\delta^Q_N\partial_M\delta^p(\vec{\sigma}-\vec{\sigma}')\delta^p(\vec{\sigma}-\vec{\sigma}'') + \delta^Q_M\partial_{N'}\delta^p(\vec{\sigma}-\vec{\sigma}')\delta^p(\vec{\sigma}'-\vec{\sigma}'') \end{aligned} \qquad (19)$$

and other ones.

The canonically conjugate ghost pairs of the minimal sector corresponding to the first-class constraints may be introduced forming the following triads

$$
\begin{array}{lll}
(\widetilde{\Phi}^{\alpha\beta}, C_{\alpha\beta}, \check{P}^{\alpha\beta}); & (\widetilde{\bar{\Phi}}^{\dot\alpha\dot\beta}, \bar{C}_{\dot\alpha\dot\beta}, \check{\bar{P}}^{\dot\alpha\dot\beta}); & (\widetilde{\Phi}^{\dot\alpha\beta}, C_{\beta\dot\alpha}, \check{P}^{\dot\alpha\beta}); \\
(\widetilde{\Psi}^{\alpha}, C_{\alpha}, \check{P}^{\alpha}); & (\widetilde{\bar{\Psi}}^{\dot\alpha}, \bar{C}_{\dot\alpha}, \check{\bar{P}}^{\dot\alpha}); & \\
(\tilde{P}^{\alpha}_{u}, C_{u\alpha}, \check{P}^{\alpha}_{u}); & (\tilde{\bar{P}}^{\dot\alpha}_{u}, \bar{C}_{u\dot\alpha}, \check{\bar{P}}^{\dot\alpha}_{u}); & \\
(\tilde{P}^{(\rho)}_{\tau}, C^{(\rho)\tau}, \check{P}^{(\rho)}_{\tau}); & (P^{(\rho)}_{M}, C^{(\rho)M}, \check{P}^{(\rho)}_{M}); & \\
(\widetilde{L}_{M}, C^{M}, \check{P}_{M}); & (\widetilde{\Delta}_{W}, C^{(W)}, \check{P}^{(W)}) . &
\end{array}
\tag{20}
$$

Utilizing nonzero structure functions of the superalgebra (12)–(17) one can present the corresponding BRST generator Ω of the minimal sector [12]

$$
\Omega = \int d^{p}\sigma (C_{\mathcal{A}} \Upsilon^{\mathcal{A}} + \tfrac{1}{2}(-)^{b} C_{\mathcal{B}} C_{\mathcal{A}} \check{f}^{\mathcal{A}\mathcal{B}}{}_{\mathcal{C}} \check{P}^{\mathcal{C}})(\vec{\sigma}), \tag{21}
$$

by the following integral along the hypersurface of the closed super p-brane

$$
\begin{aligned}
\Omega \; = \; & \int d^{p}\sigma \Big[(C_{\alpha\beta} \widetilde{\Phi}^{\alpha\beta} + \bar{C}_{\dot\alpha\dot\beta} \widetilde{\bar{\Phi}}^{\dot\alpha\dot\beta} + C_{\alpha} \widetilde{\Psi}^{\alpha} - \bar{C}_{\dot\alpha} \widetilde{\bar{\Psi}}^{\dot\alpha} + C_{u\alpha} \widetilde{P}^{\alpha}_{u} + \bar{C}_{u\dot\alpha} \widetilde{\bar{P}}^{\dot\alpha}_{u}) \\
& + C_{\alpha\dot\beta} \widetilde{\Phi}^{\dot\beta\alpha} + C^{(\rho)\tau} \widetilde{P}^{(\rho)}_{\tau} + C^{(\rho)M} P^{(\rho)}_{M} + C^{(W)} \widetilde{\Delta}^{ext}_{W} + C^{M} \widetilde{L}^{ext}_{M} \\
& + 4i (C_{\alpha} C_{\beta} \check{P}^{\alpha\beta} + \bar{C}_{\dot\alpha} \bar{C}_{\dot\beta} \check{\bar{P}}^{\dot\alpha\dot\beta}) - 4i C_{\alpha} \bar{C}_{\dot\beta} \check{P}^{\dot\beta\alpha} \Big] .
\end{aligned}
\tag{22}
$$

$\widetilde{\Delta}^{ext}_{W}$ in Eq. (22) is the generator of the gauge world-volume Weyl symmetry

$$
\widetilde{\Delta}^{ext}_{W} = \widetilde{\Delta}_{W} - 2 C^{(\rho)M} \check{P}^{(\rho)}_{M}, \tag{23}
$$

and $\widetilde{L}^{ext}_{M}$ is the generalized Virasoro generator

$$
\begin{aligned}
\widetilde{L}^{ext}_{M} \; = \; & \widetilde{L}_{M} + \partial_{M} C_{\alpha\beta} \check{P}^{\alpha\beta} + \partial_{M} \bar{C}_{\dot\alpha\dot\beta} \check{\bar{P}}^{\dot\alpha\dot\beta} + \partial_{M} C_{\alpha\dot\beta} \check{P}^{\dot\beta\alpha} \\
& + \partial_{M} C_{\alpha} \check{P}^{\alpha} + \partial_{M} \bar{C}_{\dot\alpha} \check{\bar{P}}^{\dot\alpha} + \partial_{M} C^{(W)} \check{P}^{(W)} - C^{(\rho)N} \partial_{M} \check{P}^{(\rho)}_{N} + \partial_{M} C^{N} \check{P}_{N} ,
\end{aligned}
\tag{24}
$$

extended by the ghost contributions.

Using the P.B.'s of the superalgebra (12)–(17) one can show that the P.B. of the BRST generator density $\Omega(\tau, \vec{\sigma})$, defined by the integrand (22), with itself is equal to the total derivative

$$
\begin{aligned}
\{\Omega(\vec{\sigma}), \Omega(\vec{\sigma}')\}_{P.B.} \; = \; & \partial_{M} (C^{M} (C_{\alpha\beta} \widetilde{\Phi}^{\alpha\beta} + \bar{C}_{\dot\alpha\dot\beta} \widetilde{\bar{\Phi}}^{\dot\alpha\dot\beta} + C_{\alpha\dot\beta} \widetilde{\Phi}^{\dot\beta\alpha} \\
& + C_{\alpha} \widetilde{\Psi}^{\alpha} - \bar{C}_{\dot\alpha} \widetilde{\bar{\Psi}}^{\dot\alpha} + C^{(\rho)N} P^{(\rho)}_{N} + C^{(W)} \widetilde{\Delta}^{ext}_{W} + C^{N} \widetilde{L}^{ext}_{N} \\
& + 4i (C_{\alpha} C_{\beta} \check{P}^{\alpha\beta} + \bar{C}_{\dot\alpha} \bar{C}_{\dot\beta} \check{\bar{P}}^{\dot\alpha\dot\beta} - C_{\alpha} \bar{C}_{\dot\beta} \check{P}^{\dot\beta\alpha})) \delta^{p}(\vec{\sigma} - \vec{\sigma}') ,
\end{aligned}
\tag{25}
$$

because of the presence of $\partial_M \delta^p(\vec{\sigma}-\vec{\sigma}')$ in the structure functions of the superalgebra (12)–(17). But, the contribution of the total derivative in the r.h.s. of (25) vanishes after integration in $\vec{\sigma}$ and $\vec{\sigma}'$ due to the periodical boundary conditions for the closed p-brane. It results in the P.B.-anticommutativity of the BRST charge $\Omega \equiv \int d^p\sigma \Omega(\tau,\vec{\sigma})$ (21) with itself

$$\{\Omega,\Omega\}_{P.B.} = 0. \qquad (26)$$

The introduction of the ghost variables leads to the extension of the $OSp(1|8)$ symmetry generators providing the P.B.-(anti)commutativity of the $OSp(1|8)$ generators with Ω (21). The ghost extended "square roots" $\widetilde{S}_{\gamma}(\tau,\vec{\sigma})$ and $\bar{\widetilde{S}}_{\dot\gamma}(\tau,\vec{\sigma})$ of the ghost extended conformal boost densities $\widetilde{K}_{\gamma\dot\gamma}(\tau,\vec{\sigma})$ and $\widetilde{K}_{\gamma\lambda}(\tau,\vec{\sigma})$ are given by

$$\begin{aligned}
\widetilde{S}_{\gamma}(\tau,\vec{\sigma}) \;=\;& (z_{\gamma\delta} - 2i\theta_{\gamma}\theta_{\delta})Q^{\delta} + (x_{\gamma\dot\delta} - 2i\theta_{\gamma}\bar\theta_{\dot\delta})\bar{Q}^{\dot\delta} \\
& +4i(\tilde{u}^{\delta}\theta_{\delta} - \bar{\tilde{u}}^{\dot\delta}\bar\theta_{\dot\delta})\bar{\tilde{P}}_{u\gamma} + \frac{2}{(\tilde{\rho}^{\tau})^{1/2}}\bar{\tilde{P}}_{u\gamma}f \\
& +C_{\gamma}{}^{\beta}\breve{P}_{\beta} + C_{\gamma\dot\beta}\bar{\breve{P}}^{\dot\beta} + 4iC_{\gamma}(\theta_{\beta}\breve{P}^{\beta} - \bar\theta_{\dot\beta}\bar{\breve{P}}^{\dot\beta}) \\
& -8i\theta_{\delta}C_{\gamma}{}^{\beta}\breve{P}_{\beta}{}^{\delta} + 4i\bar\theta^{\dot\delta}C_{\gamma}{}^{\beta}\breve{P}_{\beta\dot\delta} + 4i\theta_{\delta}C_{\gamma\dot\beta}\breve{P}^{\dot\beta\delta} - 8i\bar\theta^{\dot\delta}C_{\gamma\dot\beta}\bar{\breve{P}}^{\dot\beta}{}_{\dot\delta},
\end{aligned} \qquad (27)$$

$$\begin{aligned}
\bar{\widetilde{S}}_{\dot\gamma}(\tau,\vec{\sigma}) \;=\;& (\bar{z}_{\dot\gamma\dot\delta} - 2i\bar\theta_{\dot\gamma}\bar\theta_{\dot\delta})\bar{Q}^{\dot\delta} + (x_{\delta\dot\gamma} + 2i\theta_{\delta}\bar\theta_{\dot\gamma})Q^{\delta} \\
& -4i(\tilde{u}^{\delta}\theta_{\delta} - \bar{\tilde{u}}^{\dot\delta}\bar\theta_{\dot\delta})\bar{\bar{\tilde{P}}}_{u\dot\gamma} - \frac{2}{(\tilde{\rho}^{\tau})^{1/2}}\bar{\bar{\tilde{P}}}_{u\dot\gamma}f \\
& -C_{\beta\dot\gamma}\breve{P}^{\beta} - \bar{C}^{\dot\beta}{}_{\dot\gamma}\bar{\breve{P}}_{\dot\beta} - 4i\bar{C}_{\dot\gamma}(\theta_{\beta}\breve{P}^{\beta} - \bar\theta_{\dot\beta}\bar{\breve{P}}^{\dot\beta}) \\
& -8i\theta^{\delta}C_{\beta\dot\gamma}\breve{P}_{\delta}{}^{\beta} + 4i\bar\theta_{\dot\delta}C_{\beta\dot\gamma}\breve{P}^{\dot\delta\beta} + 4i\theta^{\delta}\bar{C}^{\dot\beta}{}_{\dot\gamma}\breve{P}_{\delta\dot\beta} - 8i\bar\theta_{\dot\delta}\bar{C}^{\dot\beta}{}_{\dot\gamma}\bar{\breve{P}}^{\dot\delta}{}_{\dot\beta}.
\end{aligned} \qquad (28)$$

Using the densities $\Omega(\tau,\vec{\sigma})$ (22) and $\widetilde{S}_{\gamma}(\tau,\vec{\sigma}')$ (27) we find their P.B.

$$\{\Omega(\sigma),\widetilde{S}_{\gamma}(\vec{\sigma}')\}_{P.B.} = -(C^M\widetilde{S}_{\gamma})(\vec{\sigma})\partial_M\delta^p(\sigma-\sigma'), \qquad (29)$$

and conclude that the contribution of the total derivative in the r.h.s. of (29) vanishes after integration with respect to $\vec{\sigma}$ and $\vec{\sigma}'$. Thus, the BRST charge Ω (22) has zero P.B.'s with the conformal supercharges $\widetilde{S}_{\gamma}, \bar{\widetilde{S}}_{\dot\gamma}$

$$\{\Omega,\widetilde{S}_{\gamma}\}_{P.B.} = 0, \quad \{\Omega,\bar{\widetilde{S}}_{\dot\gamma}\}_{P.B.} = 0. \qquad (30)$$

The same P.B.-commutativity

$$\{\Omega,G\}_{P.B.} = 0, \qquad (31)$$

between Ω and other $OSp(1|8)$ symmetry charges $G \equiv \int d^p\sigma G(\tau,\vec{\sigma})$ extended by the ghost contributions will also be preserved, because of the general relation (17) for the generator densities: $\{\widetilde{L}_M(\vec{\sigma}),G(\vec{\sigma}')\}_{P.B.} = -G(\vec{\sigma})\partial_M\delta^p(\vec{\sigma}-\vec{\sigma}')$.

The above mentioned expressions for the generator densities of the generalized conformal transformations extended by the ghost contributions take the form[2]

$$\begin{aligned}
\widetilde{K}_{\gamma\lambda}(\tau,\vec{\sigma}) &= 2z_{\gamma\beta}z_{\lambda\delta}\pi^{\beta\delta}+2x_{\gamma\dot{\beta}}x_{\lambda\dot{\delta}}\bar{\pi}^{\dot{\beta}\dot{\delta}}+z_{\gamma\beta}x_{\lambda\dot{\delta}}P^{\dot{\delta}\beta}+x_{\gamma\dot{\beta}}z_{\lambda\delta}P^{\dot{\beta}\delta} \\
&+\theta_{\lambda}(z_{\gamma\delta}\pi^{\delta}+x_{\gamma\dot{\delta}}\bar{\pi}^{\dot{\delta}})+\theta_{\gamma}(z_{\lambda\delta}\pi^{\delta}+x_{\lambda\dot{\delta}}\bar{\pi}^{\dot{\delta}}) \\
&+(\tilde{u}^{\delta}z_{\delta\lambda}-\bar{\tilde{u}}^{\dot{\delta}}x_{\lambda\dot{\delta}})\tilde{\bar{P}}_{u\gamma}+(\tilde{u}^{\delta}z_{\delta\gamma}-\bar{\tilde{u}}^{\dot{\delta}}x_{\gamma\dot{\delta}})\tilde{\bar{P}}_{u\lambda} \\
&-2i(\tilde{u}^{\delta}\theta_{\delta}-\bar{\tilde{u}}^{\dot{\delta}}\bar{\theta}_{\dot{\delta}})(\theta_{\lambda}\tilde{\bar{P}}_{u\gamma}+\theta_{\gamma}\tilde{\bar{P}}_{u\lambda}) \\
&+\tfrac{2}{(\tilde{p}^{\tau})^{1/2}}(\theta_{\lambda}\tilde{\bar{P}}_{u\gamma}+\theta_{\gamma}\tilde{\bar{P}}_{u\lambda})f-\tfrac{1}{\tilde{p}^{\tau}}\tilde{\bar{P}}_{u\gamma}\tilde{\bar{P}}_{u\lambda} \\
&+\theta_{\lambda}C_{\gamma}{}^{\beta}\check{P}_{\beta}+\theta_{\gamma}C_{\lambda}{}^{\beta}\check{P}_{\beta}+\theta_{\lambda}C_{\gamma\dot{\beta}}\bar{\check{P}}^{\dot{\beta}}+\theta_{\gamma}C_{\lambda\dot{\beta}}\bar{\check{P}}^{\dot{\beta}} \\
&-(z_{\lambda}{}^{\beta}+2i\theta_{\lambda}\theta^{\beta})C_{\gamma}\check{P}_{\beta}-(z_{\gamma}{}^{\beta}+2i\theta_{\gamma}\theta^{\beta})C_{\lambda}\check{P}_{\beta} \\
&-(x_{\lambda\dot{\beta}}+2i\theta_{\lambda}\bar{\theta}_{\dot{\beta}})C_{\gamma}\bar{\check{P}}^{\dot{\beta}}-(x_{\gamma\dot{\beta}}+2i\theta_{\gamma}\bar{\theta}_{\dot{\beta}})C_{\lambda}\bar{\check{P}}^{\dot{\beta}} \\
&-2(z_{\gamma}{}^{\beta}+2i\theta_{\gamma}\theta^{\beta})C_{\lambda\delta}\check{P}_{\beta}{}^{\delta}-2(z_{\lambda}{}^{\beta}+2i\theta_{\lambda}\theta^{\beta})C_{\gamma\delta}\check{P}_{\beta}{}^{\delta} \\
&+(z_{\gamma\beta}+2i\theta_{\gamma}\theta_{\beta})C_{\lambda\dot{\delta}}\check{P}^{\dot{\delta}\beta}+(z_{\lambda\beta}+2i\theta_{\lambda}\theta_{\beta})C_{\gamma\dot{\delta}}\check{P}^{\dot{\delta}\beta} \\
&+(x_{\gamma\dot{\beta}}+2i\theta_{\gamma}\bar{\theta}_{\dot{\beta}})C_{\lambda\delta}\check{P}^{\dot{\beta}\delta}+(x_{\lambda\dot{\beta}}+2i\theta_{\lambda}\bar{\theta}_{\dot{\beta}})C_{\gamma\delta}\check{P}^{\dot{\beta}\delta} \\
&+2(x_{\gamma\dot{\beta}}+2i\theta_{\gamma}\bar{\theta}_{\dot{\beta}})C_{\lambda\dot{\delta}}\bar{\check{P}}^{\dot{\beta}\dot{\delta}}+(x_{\lambda\dot{\beta}}+2i\theta_{\lambda}\bar{\theta}_{\dot{\beta}})C_{\gamma\dot{\delta}}\bar{\check{P}}^{\dot{\beta}\dot{\delta}}, \qquad (32)
\end{aligned}$$

for $\widetilde{K}_{\gamma\lambda}(\tau,\vec{\sigma})$ and respectively for $\widetilde{K}_{\gamma\dot{\gamma}}(\tau,\vec{\sigma})$

$$\begin{aligned}
\widetilde{K}_{\gamma\dot{\gamma}}(\tau,\vec{\sigma}) &= z_{\gamma\delta}\bar{z}_{\dot{\gamma}\dot{\delta}}P^{\dot{\delta}\delta}+x_{\gamma\dot{\delta}}x_{\delta\dot{\gamma}}P^{\dot{\delta}\delta}+2(z_{\gamma\delta}x_{\lambda\dot{\gamma}}\pi^{\delta\lambda}+x_{\gamma\dot{\delta}}\bar{z}_{\dot{\gamma}\dot{\lambda}}\bar{\pi}^{\dot{\delta}\dot{\lambda}}) \\
&+\theta_{\gamma}(\bar{z}_{\dot{\gamma}\dot{\delta}}\bar{\pi}^{\dot{\delta}}+x_{\delta\dot{\gamma}}\pi^{\delta})+\bar{\theta}_{\dot{\gamma}}(z_{\gamma\delta}\pi^{\delta}+x_{\gamma\dot{\delta}}\bar{\pi}^{\dot{\delta}}) \\
&+(\tilde{u}^{\delta}x_{\delta\dot{\gamma}}-\bar{\tilde{u}}^{\dot{\delta}}\bar{z}_{\dot{\delta}\dot{\gamma}})\tilde{\bar{P}}_{u\gamma}+(x_{\gamma\dot{\delta}}\bar{\tilde{u}}^{\dot{\delta}}-z_{\gamma\delta}\tilde{u}^{\delta})\bar{\tilde{\bar{P}}}_{u\dot{\gamma}}
\end{aligned}$$

[2] The discussed formulae would look more compact in the Majorana spinor representation. However, it makes more obscure the contribution of the TCC coordinates $z_{\alpha\beta}$ which presence is crucial for the exotic supersymmetry protection and generation of the new bosonic gauge symmetries (see [13]) generalizing the well known symmetries of the Penrose twistor approach originally formulated in $D=4$ [24]. In the Majorana representation $z_{\alpha\beta}$ are encoded in the symmetric 4×4 matrix $Y_a{}^b \equiv Y_{ad}C^{db} = \begin{pmatrix} z_\alpha{}^\beta & x_{\alpha\dot{\beta}} \\ \bar{x}^{\dot{\alpha}\beta} & \bar{z}^{\dot{\alpha}}{}_{\dot{\beta}} \end{pmatrix}$ together with the space-time coordinates $x_{\alpha\dot{\alpha}}$ (see [8], [17]).

$$-2i(\tilde{u}^{\delta}\theta_{\delta}-\bar{\tilde{u}}^{\dot{\delta}}\bar{\theta}_{\dot{\delta}})(\bar{\theta}_{\dot{\gamma}}\tilde{\bar{P}}_{u\gamma}-\theta_{\gamma}\bar{\tilde{\bar{P}}}_{u\dot{\gamma}})+\frac{2}{(\bar{P}^{\tau})^{1/2}}(\bar{\theta}_{\dot{\gamma}}\tilde{\bar{P}}_{u\gamma}-\theta_{\gamma}\bar{\tilde{\bar{P}}}_{u\dot{\gamma}})f$$
$$+\frac{1}{\bar{P}^{\tau}}\tilde{\bar{P}}_{u\gamma}\bar{\tilde{\bar{P}}}_{u\dot{\gamma}}+\bar{\theta}_{\dot{\gamma}}C_{\gamma}{}^{\beta}\check{P}_{\beta}-\theta_{\gamma}\bar{C}_{\dot{\gamma}}{}^{\dot{\beta}}\bar{\check{P}}_{\dot{\beta}}+\bar{\theta}_{\dot{\gamma}}C_{\gamma\dot{\beta}}\bar{\check{P}}^{\dot{\beta}}-\theta_{\gamma}C_{\beta\dot{\gamma}}\check{P}^{\beta}$$
$$+(\bar{z}_{\dot{\gamma}}{}^{\dot{\beta}}+2i\bar{\theta}_{\dot{\gamma}}\bar{\theta}^{\dot{\beta}})C_{\gamma}\bar{\check{P}}_{\dot{\beta}}+(z_{\gamma}{}^{\beta}+2i\theta_{\gamma}\theta^{\beta})\bar{C}_{\dot{\gamma}}\check{P}_{\beta}$$
$$+(x_{\beta\dot{\gamma}}-2i\theta_{\beta}\bar{\theta}_{\dot{\gamma}})C_{\gamma}\check{P}^{\beta}+(x_{\gamma\dot{\beta}}+2i\theta_{\gamma}\bar{\theta}_{\dot{\beta}})\bar{C}_{\dot{\gamma}}\bar{\check{P}}^{\dot{\beta}}$$
$$-2(z_{\gamma}{}^{\beta}+2i\theta_{\gamma}\theta^{\beta})C_{\delta\dot{\gamma}}\check{P}_{\beta}{}^{\delta}-2(\bar{z}_{\dot{\gamma}}{}^{\dot{\beta}}+2i\bar{\theta}_{\dot{\gamma}}\bar{\theta}^{\dot{\beta}})\bar{C}_{\gamma\dot{\delta}}\bar{\check{P}}^{\dot{\delta}}{}_{\dot{\beta}}$$
$$+(z_{\gamma}{}^{\beta}+2i\theta_{\gamma}\theta^{\beta})\bar{C}^{\dot{\delta}}{}_{\dot{\gamma}}\check{P}_{\beta\dot{\delta}}+(\bar{z}^{\dot{\beta}}{}_{\dot{\gamma}}+2i\bar{\theta}_{\dot{\gamma}}\bar{\theta}^{\dot{\beta}})C_{\gamma}{}^{\delta}\bar{\check{P}}_{\delta\dot{\beta}}$$
$$+(x_{\gamma\dot{\beta}}+2i\theta_{\gamma}\bar{\theta}_{\dot{\beta}})C_{\delta\dot{\gamma}}\check{P}^{\dot{\beta}\delta}+(x_{\beta\dot{\gamma}}-2i\theta_{\beta}\bar{\theta}_{\dot{\gamma}})C_{\gamma\dot{\delta}}\check{P}^{\dot{\delta}\beta}$$
$$-2(x_{\gamma\dot{\beta}}+2i\theta_{\gamma}\bar{\theta}_{\dot{\beta}})\bar{C}^{\dot{\delta}}{}_{\dot{\gamma}}\bar{\check{P}}^{\dot{\beta}}{}_{\dot{\delta}}-2(x_{\beta\dot{\gamma}}-2i\theta_{\beta}\bar{\theta}_{\dot{\gamma}})C_{\gamma}{}^{\delta}\check{P}_{\delta}{}^{\beta}\,. \tag{33}$$

The remaining 16 generator densities of $OSp(1|8)$ supergroup extended by the ghosts are the following

$$\begin{aligned}\widetilde{L}^{\alpha}{}_{\beta}(\tau,\vec{\sigma}) &= x_{\beta\dot{\beta}}P^{\dot{\beta}\alpha}+2z_{\gamma\beta}\pi^{\alpha\gamma}+\theta_{\beta}\pi^{\alpha}+\tilde{u}^{\alpha}\bar{\tilde{P}}_{u\beta}-2C_{\beta}{}^{\gamma}\check{P}_{\gamma}{}^{\alpha}+C_{\beta\dot{\gamma}}\check{P}^{\dot{\gamma}\alpha}+C_{\beta}\check{P}^{\alpha},\\ \widetilde{L}^{\alpha}{}_{\dot{\beta}}(\tau,\vec{\sigma}) &= 2x_{\gamma\dot{\beta}}\pi^{\alpha\gamma}+\bar{z}_{\dot{\beta}\dot{\gamma}}P^{\dot{\gamma}\alpha}+\bar{\theta}_{\dot{\beta}}\pi^{\alpha}-\tilde{u}^{\alpha}\bar{\tilde{\bar{P}}}_{u\dot{\beta}}+2C_{\gamma\dot{\beta}}\check{P}^{\gamma\alpha}+\bar{C}_{\dot{\beta}\dot{\gamma}}\check{P}^{\dot{\gamma}\alpha}-\bar{C}_{\dot{\beta}}\check{P}^{\alpha}\,.\end{aligned} \tag{34}$$

The adduced expressions should be complemented by their complex conjugate.

Note that supersymmetry and generalized translation generator densities do not contain any ghost contribution. One can check that the P.B.-commutation relations of the $OSp(1|8)$ superalgebra extended by the ghost contributions coincide with the P.B.-commutation relations of the original $OSp(1|8)$ superalgebra [17].

QUANTIZATION: NILPOTENT BRST OPERATOR AND QUANTUM $OSP(1|8)$ ALGEBRA

Upon transition to quantum theory all the quantities entering the converted constraints and $OSp(1|8)$ generator densities are treated as operators that implies a choice of certain ordering for products of noncommuting operators. At the same time the canonical Poisson brackets $\{\mathcal{P}^{\mathfrak{M}}(\vec{\sigma}),\mathcal{Q}_{\mathfrak{N}}(\vec{\sigma}')\}_{P.B.}=\delta^{\mathfrak{M}}_{\mathfrak{N}}\delta^{p}(\vec{\sigma}-\vec{\sigma}')$ used in [17] transform into (anti)commutators

$$[\hat{\mathcal{P}}^{\mathfrak{M}}(\vec{\sigma}),\hat{\mathcal{Q}}_{\mathfrak{N}}(\vec{\sigma}')\}=-i\delta^{\mathfrak{M}}_{\mathfrak{N}}\delta^{p}(\vec{\sigma}-\vec{\sigma}')\,. \tag{35}$$

It is necessary to provide further nilpotency of the BRST operator, fulfilment of (anti)commutation relations of the $OSp(1|8)$ superalgebra and its generator (anti)-commutativity with the BRST operator ensuring the global quantum invariance of the model. In addition, the Hermiticity of the quantum BRST operator and $OSp(1|8)$

generators has to be supported. The Hermiticity requirement may be manifestly satisfied if we start from the above constructed classical representations for the $OSp(1|8)$ generators and BRST charge in which all coordinates are disposed from the left of momenta, i.e. in the form $\mathcal{QP}$, where $\mathcal{Q}$ and $\mathcal{P}$ are the products of the coordinates and momenta contained in Ω and the generators. Then the operator expressions for the latter are presented in the manifestly Hermitian form composed of the operator products $\frac{1}{2}(\hat{\mathcal{Q}}\hat{\mathcal{P}} + (-)^{\varepsilon(\mathcal{Q})\varepsilon(\mathcal{P})}(\hat{\mathcal{Q}}\hat{\mathcal{P}})^\dagger)$, where $\varepsilon(\mathcal{Q})$ and $\varepsilon(\mathcal{P})$ are Grassmannian gradings of these coordinate and momentum monomials.

In particular, we obtain the following Hermitian operator representations

$$
\begin{aligned}
\hat{\tilde{\mathcal{S}}}_\gamma(\tau,\vec{\sigma}) \;=\;& \tfrac{1}{2}(\hat{z}_{\gamma\delta} - 2i\hat{\theta}_\gamma\hat{\theta}_\delta)\hat{Q}^\delta + \tfrac{1}{2}\hat{Q}^\delta(\hat{z}_{\gamma\delta} - 2i\hat{\theta}_\gamma\hat{\theta}_\delta) \\
&+ \tfrac{1}{2}(\hat{x}_{\gamma\dot{\delta}} - 2i\hat{\theta}_\gamma\hat{\bar{\theta}}_{\dot{\delta}})\hat{\bar{Q}}^{\dot{\delta}} + \tfrac{1}{2}\hat{\bar{Q}}^{\dot{\delta}}(\hat{x}_{\gamma\dot{\delta}} - 2i\hat{\theta}_\gamma\hat{\bar{\theta}}_{\dot{\delta}}) \\
&+ 2i\hat{\theta}_\delta(\hat{u}^\delta\hat{\tilde{P}}_{u\gamma} + \hat{\tilde{P}}_{u\gamma}\hat{u}^\delta) - 4i\hat{\bar{u}}^{\dot{\delta}}\hat{\bar{\theta}}_{\dot{\delta}}\hat{\tilde{P}}_{u\gamma} + \tfrac{2}{(\hat{\rho}^\tau)^{1/2}}\hat{f}\hat{\tilde{P}}_{u\gamma} \\
&+ \hat{C}_\gamma{}^\beta\hat{P}_\beta + \hat{C}_{\gamma\dot{\beta}}\hat{\bar{P}}^{\dot{\beta}} + 2i\hat{\theta}_\delta(\hat{C}_\gamma\hat{P}^\delta + \hat{P}^\delta\hat{C}_\gamma) + 4i\hat{\bar{\theta}}^{\dot{\delta}}\hat{C}_\gamma\hat{\bar{P}}_{\dot{\delta}} \\
&- 4i\hat{\theta}_\delta(\hat{C}_\gamma{}^\beta\hat{P}_\beta{}^\delta - \hat{P}_\beta{}^\delta\hat{C}_\gamma{}^\beta) + 4i\hat{\bar{\theta}}^{\dot{\delta}}\hat{C}_\gamma{}^\beta\hat{P}_{\beta\dot{\delta}} \\
&+ 2i\hat{\theta}_\delta(\hat{C}_{\gamma\dot{\beta}}\hat{\bar{P}}^{\dot{\beta}\delta} - \hat{\bar{P}}^{\dot{\beta}\delta}\hat{C}_{\gamma\dot{\beta}}) - 8i\hat{\bar{\theta}}^{\dot{\delta}}\hat{C}_{\gamma\dot{\beta}}\hat{\bar{P}}^{\dot{\beta}}{}_{\dot{\delta}}\,,
\end{aligned}
\tag{36}
$$

for the classical density $\widetilde{S}_\gamma(\tau,\vec{\sigma})$ (27) and

$$
\begin{aligned}
\hat{\tilde{\mathcal{L}}}^\alpha{}_\beta(\tau,\vec{\sigma}) \;=\;& \tfrac{1}{2}(\hat{x}_{\beta\dot{\beta}}\hat{P}^{\dot{\beta}\alpha} + \hat{P}^{\dot{\beta}\alpha}\hat{x}_{\beta\dot{\beta}}) + (\hat{z}_{\beta\gamma}\hat{\pi}^{\gamma\alpha} + \hat{\pi}^{\gamma\alpha}\hat{z}_{\beta\gamma}) \\
&+ \tfrac{1}{2}(\hat{\theta}_\beta\hat{\pi}^\alpha - \hat{\pi}^\alpha\hat{\theta}_\beta) + \frac{1}{2}(\hat{u}^\alpha\hat{\tilde{P}}_{u\beta} + \hat{\tilde{P}}_{u\beta}\hat{u}^\alpha) \\
&- (\hat{C}_\beta{}^\gamma\hat{P}_\gamma{}^\alpha - \hat{P}_\gamma{}^\alpha\hat{C}_\beta{}^\gamma) + \tfrac{1}{2}(\hat{C}_{\beta\dot{\gamma}}\hat{\bar{P}}^{\dot{\gamma}\alpha} - \hat{\bar{P}}^{\dot{\gamma}\alpha}\hat{C}_{\beta\dot{\gamma}}) + \tfrac{1}{2}(\hat{C}_\beta\hat{P}^\alpha + \hat{P}^\alpha\hat{C}_\beta)\,,
\end{aligned}
\tag{37}
$$

for the generalized Lorentz density $\widetilde{L}^\alpha{}_\beta(\tau,\vec{\sigma})$ (34).

By the same way can be constructed quantum Hermitian generators $\hat{\tilde{L}}^{ext}_M$ of the $\vec{\sigma}$-reparametrizations and the Weyl symmetry generator $\hat{\tilde{\Delta}}^{ext}_W$

$$
\begin{aligned}
\hat{\tilde{\Delta}}^{ext}_W \;=\;& \tfrac{1}{2}(\hat{u}_\alpha\hat{\tilde{P}}^\alpha_u + \hat{\tilde{P}}^\alpha_u\hat{u}_\alpha + \hat{\bar{u}}_{\dot{\alpha}}\hat{\tilde{\bar{P}}}^{\dot{\alpha}}_u + \hat{\tilde{\bar{P}}}^{\dot{\alpha}}_u\hat{\bar{u}}_{\dot{\alpha}}) \\
&- \hat{\rho}^\tau\hat{\tilde{P}}^{(\rho)}_\tau - \hat{\tilde{P}}^{(\rho)}_\tau\hat{\rho}^\tau - \hat{\rho}^M\hat{P}^{(\rho)}_M - \hat{P}^{(\rho)}_M\hat{\rho}^M - \hat{C}^{(\rho)M}\hat{\mathcal{P}}^{(\rho)}_M + \hat{\mathcal{P}}^{(\rho)}_M\hat{C}^{(\rho)M}\,,
\end{aligned}
\tag{38}
$$

entering the BRST operator.

Because other converted first-class constraints are Hermitian by construction, the quantum Hermitian BRST generator

$$\hat{\mathbf{\Omega}} = \frac{1}{2}\int d^p\sigma(\hat{\Omega}(\tau,\vec{\sigma}) + \hat{\Omega}^\dagger(\tau,\vec{\sigma})), \tag{39}$$

will coincide with its classical expression Ω (22) after the substitution of (38) and the Hermitian representation for $\hat{\tilde{L}}_M(\vec{\sigma})$ originated from (24) in Eq. (22).

Now we are ready to prove that this realization of $\hat{\mathbf{\Omega}}$ preserves its nilpotency and (anti)commutativity with the Hermitian operators (36), (37) and other ones generating a quantum realization of the classical $OSp(1|8)$ superalgebra. The proof is obvious and is based on the observation that $\hat{\mathbf{\Omega}}$ and other considered Hermitian operators are linear in the momentum operators $\hat{\mathfrak{P}}^{\mathfrak{M}}(\tau,\vec{\sigma})$ of the original coordinates and ghost fields. The remarkable property of the ordered polynomial operators composed of $\hat{\mathfrak{Q}}_{\mathfrak{M}}(\tau,\vec{\sigma})$ and $\hat{\mathfrak{P}}^{\mathfrak{M}}(\tau,\vec{\sigma})$, which form the Weyl-Heisenberg algebra (35), and are linear in $\hat{\mathfrak{P}}^{\mathfrak{M}}$ is the preservation of the chosen ordering in course of calculations of their (anti)commutators. As a result, the transition from the P.B.'s to (anti)commutators will preserve all classical results obtained in the P.B. realization of the extended algebra of the $OSp(1|8)$ generators and classical BRST charge of the super p-brane. So, the quantum Hermitian BRST operator (39) occurs to be nilpotent

$$\{\hat{\mathbf{\Omega}}, \hat{\mathbf{\Omega}}\} = 0. \tag{40}$$

However, the Hermiticity of $\hat{\mathbf{\Omega}}$ and the $OSp(1|8)$ generating operators by itself is only a necessary condition for the quantum realization of the physical operators, because the relevant vacuum and physical states have also to be constructed. So, the problem of existence of the selfconsistent quantum realization of the exotic BPS states by the states of quantum tensionless super p-brane is reduced to the proof of existence of the relevant vacuum and the corresponding physical space of quantum states. At the present time we investigate this problem. However, we should like to discuss here a possible way to solve this problem based on the consideration of the $\hat{\mathfrak{Q}}\hat{\mathfrak{P}}$-ordering studied in [10].

$\hat{\mathfrak{Q}}\hat{\mathfrak{P}}$-ORDERING AND REGULARIZATION

It was motivated in [10] that the coordinate and momenta operators have to be used for the tensionless string quantization instead of the creation and annihilation operators relevant for the tensile string quantization. This motivation is physically justified by the absence of oscillator excitations for the tensionless string which makes its dynamics resembling that of collection of free particles and results in the choice of the physical vacuum as a state annihilated by the string momentum operator. In that case the coordinate $\hat{\mathfrak{Q}}$ and momentum $\hat{\mathfrak{P}}$ monomials forming the above discussed Hermitian operators have to be ordered by the shifts of all the $\hat{\mathfrak{Q}}$ monomials to the left of $\hat{\mathfrak{P}}$. To achieve that $\hat{\mathfrak{Q}}\hat{\mathfrak{P}}$-ordering we have, in particular, to permutate some noncommuting coordinate and momentum operators in the $\hat{\mathfrak{Q}}\hat{\mathfrak{P}}$-disordered Hermitian expressions of the generator densities (36), (37) and others, constraints $\hat{\tilde{\Delta}}^{ext}_W$ (38), $\hat{\tilde{L}}^{ext}_M$ and the Hermitian BRST operator.

In view of that permutations divergent terms will appear in some monomials composed from canonically conjugate operators at coinciding points of p-brane. A typical form of such divergent terms is illustrated by the relation

$$\frac{1}{2}(\hat{z}_{\lambda\delta}(\vec{\sigma})\hat{\pi}^{\delta\varepsilon}(\vec{\sigma})+\hat{\pi}^{\delta\varepsilon}(\vec{\sigma})\hat{z}_{\lambda\delta}(\vec{\sigma}))=\hat{z}_{\lambda\delta}(\vec{\sigma})\hat{\pi}^{\delta\varepsilon}(\vec{\sigma})-\frac{3i}{4}\delta_{\lambda}^{\varepsilon}\delta^{p}(0), \tag{41}$$

encoding the permutation effect of the TCC coordinates with their momenta. The r.h.s. of ambiguously defined relation (41) includes the divergent term $\delta^{p}(0)=\delta^{p}(\vec{\sigma}-\vec{\sigma}')|_{\vec{\sigma}=\vec{\sigma}'}$ and the problem appears how to deal with the $\hat{\mathfrak{Q}}\hat{\mathfrak{P}}$-ordered representation of the symmetric operator in the l.h.s. of (41). Such type a problem is typical in quantum field theory due to its inherit divergencies and the regularization procedure should be applied. Thus, in general, the ordering problem has to be analyzed together with the divergency problem. It might have happened that a regularization prescribes to use the image of the delta function at the zero point (here $\delta^{p}(0)$) to be equal to zero. Then the choice of that regularization would solve the ordering problem. Taking into account such a possibility requires study of the structure of the divergent terms appearing in the total quantum algebra of our model. To this end we firstly analyze the $\hat{\mathfrak{Q}}\hat{\mathfrak{P}}$-ordered realization of the $OSp(1|8)$ algebra.

We find that the application of the described $\hat{\mathfrak{Q}}\hat{\mathfrak{P}}$-ordering procedure to the Hermitian $OSp(1|8)$ generator densities (36), (37) and others yields the relations

$$\begin{aligned}
\hat{\hat{\mathcal{S}}}_{\gamma}(\tau,\vec{\sigma}) &= \hat{\hat{S}}_{\gamma}-2\hat{\theta}_{\gamma}\delta^{p}(0), \\
\hat{\hat{\mathcal{S}}}_{\dot{\gamma}}(\tau,\vec{\sigma}) &= \hat{\hat{S}}_{\dot{\gamma}}-2\hat{\bar{\theta}}_{\dot{\gamma}}\delta^{p}(0), \\
\hat{\hat{\mathcal{K}}}_{\gamma\lambda}(\tau,\vec{\sigma}) &= \hat{\hat{K}}_{\gamma\lambda}+i\hat{z}_{\gamma\lambda}\delta^{p}(0), \\
\hat{\hat{\mathcal{K}}}_{\dot{\gamma}\dot{\lambda}}(\tau,\vec{\sigma}) &= \hat{\hat{K}}_{\dot{\gamma}\dot{\lambda}}+i\hat{\bar{z}}_{\dot{\gamma}\dot{\lambda}}\delta^{p}(0), \\
\hat{\hat{\mathcal{K}}}_{\gamma\dot{\gamma}}(\tau,\vec{\sigma}) &= \hat{\hat{K}}_{\gamma\dot{\gamma}}+i\hat{x}_{\gamma\dot{\gamma}}\delta^{p}(0), \\
\hat{\hat{\mathcal{L}}}_{\alpha\beta}(\tau,\vec{\sigma}) &= \hat{\hat{L}}_{\alpha\beta}+\frac{i}{2}\varepsilon_{\alpha\beta}\delta^{p}(0), \\
\hat{\hat{\mathcal{L}}}_{\dot{\alpha}\dot{\beta}}(\tau,\vec{\sigma}) &= \hat{\hat{L}}_{\dot{\alpha}\dot{\beta}}+\frac{i}{2}\bar{\varepsilon}_{\dot{\alpha}\dot{\beta}}\delta^{p}(0),
\end{aligned} \tag{42}$$

connecting the Hermitian and $\hat{\mathfrak{Q}}\hat{\mathfrak{P}}$-ordered representations for the generator densities. The operators $\hat{\hat{S}}_{\gamma}$, $\hat{\hat{S}}_{\dot{\gamma}}$, $\hat{\hat{K}}_{\gamma\lambda}$, $\hat{\hat{K}}_{\dot{\gamma}\dot{\lambda}}$, $\hat{\hat{K}}_{\gamma\dot{\gamma}}$, $\hat{\hat{L}}_{\alpha\beta}$, $\hat{\hat{L}}_{\dot{\alpha}\dot{\beta}}$, collectively denoted by $\hat{\hat{G}}_{qp}$, in the r.h.s. of (42) coincide with the classical $\hat{\mathfrak{Q}}\hat{\mathfrak{P}}$-ordered representations (27), (28), (32)–(34), where the corresponding operators are substituted for the classical coordinates and momenta. These $\hat{\mathfrak{Q}}\hat{\mathfrak{P}}$-ordered operators $\hat{\hat{G}}_{qp}$ form another representation of the quantum algebra $OSp(1|8)$ similarly to the Hermitian operators, because they originate from the classical generators and preserve the $\hat{\mathfrak{Q}}\hat{\mathfrak{P}}$-ordering in the course of the calculation of their (anti)commutators. But, the $\hat{\mathfrak{Q}}\hat{\mathfrak{P}}$-ordered generators $\hat{\hat{G}}_{qp}$ are nonHermitian and their (anti)commutation with the divergent nonHermitian terms in the r.h.s. of the representation (42) contributes to the closure of the $OSp(1|8)$ algebra presented by the Hermitian generators in the l.h.s. of (42). So, we obtain two quantum realizations of the $OSp(1|8)$ algebra and one of them $\hat{\hat{G}}_{qp}$ is nonHermitian, because of its $\hat{\mathfrak{Q}}\hat{\mathfrak{P}}$-ordering. But, namely, action of the nonHermitian generators $\hat{\hat{G}}_{qp}$ on the corresponding vacuum state is well defined in accordance with [10]. Then the regularization assumption, which prescribes to accept the regularized image of $\delta^p(0)$ as equal zero, removes the nonHermiticity of the $\hat{\mathfrak{Q}}\hat{\mathfrak{P}}$-ordered generators $\hat{\hat{G}}_{qp}$. In the result we obtain the desired vacuum state for the discussed Hermitian realization of the physical operators. So, the choice of such a regularization would allow to overcome the problem of construction of the quantum space of physical states. To realize such a scenario one needs to analyze the $\hat{\mathfrak{Q}}\hat{\mathfrak{P}}$-ordered realization of Hermitian BRST generator (39).

The correspondent $\hat{\mathfrak{Q}}\hat{\mathfrak{P}}$-ordered BRST operator $\hat{\Omega}$ turns out to be nonHermitian, as follows from the relation connecting $\hat{\Omega}$ with the Hermitian BRST operator $\hat{\boldsymbol{\Omega}}$ (39)

$$\hat{\boldsymbol{\Omega}} = \hat{\Omega} - i\int d^p\sigma[\hat{C}^{(W)} + (\tfrac{p}{2} - \tfrac{7}{4})\hat{C}^M\partial_M + \tfrac{1}{2}\partial_M\hat{C}^M]\delta^p(0)\,, \tag{43}$$

and is supplemented by three antiHermitian divergent additions in the r.h.s. compensating the nonHermiticity of $\hat{\Omega}$. The first of them, proportional to $\delta^p(0)$, follows from the $\hat{\mathfrak{Q}}\hat{\mathfrak{P}}$-ordering of the Weyl symmetry generator $\hat{\hat{\Delta}}^{ext}_W$ (38) and is contributed only by the auxiliary variables $\hat{\hat{\rho}}^\tau, \hat{\hat{u}}^\alpha$ and $\hat{\hat{\bar{u}}}^{\dot\alpha}$ partially cancelling each other during the ordering with their momenta. The contribution of other auxiliary pair $(\hat{\rho}^M, \hat{P}^{(\rho)}_M)$ in here is cancelled by the ghost pair $(\hat{C}^{(\rho)M}, \hat{\hat{P}}^{(\rho)}_M)$ contribution. We observe that only the auxiliary twistor-like fields and the component $\hat{\hat{\rho}}^\tau$ of the world-volume density $\hat{\rho}^\mu$, introduced to provide the Weyl and reparametrization gauge symmetries of the brane action, contributed to the first singular term. The similar story concerns the second divergent term proportional to $\partial_M\delta^p(0)$ [3]. This term appears from the $\hat{\mathfrak{Q}}\hat{\mathfrak{P}}$-ordering of the extended Virasoro operators $\hat{\hat{L}}^{ext}_M$ and only the above mentioned auxiliary fields together with the ghost pairs $(\hat{C}^M, \hat{\hat{P}}_M)$, $(\hat{C}^{(W)}, \hat{\hat{P}}^{(W)})$ and the auxiliary fermionic field $\hat{f}$ contribute to here, because the contributions of the propagating phase-space variables are cancelled by the corresponding ghosts. This cancellation illustrates the boson-fermion cancellation mechanism provided by the BRST symmetry. The third singular addition, proportional to the total derivative, restores Hermiticity of the cubic term $\hat{C}^M\partial_M\hat{C}^N\hat{\hat{P}}_N$ but, it vanishes in view of the periodical boundary conditions. The latter could contribute in the case on a nontrivial topology of the ghost-field space.

Now let us note that the $\hat{\mathfrak{Q}}\hat{\mathfrak{P}}$-ordered operator $\hat{\Omega}$ in the r.h.s. of Eq. (43) is nilpotent

$$\{\hat{\Omega}, \hat{\Omega}\} = 0\,, \tag{44}$$

---

[3] In the symmetric regularization, where $\delta^p_\varepsilon(-\vec{\sigma}) = \delta^p_\varepsilon(\vec{\sigma})$, the derivative $\partial_M\delta^p_\varepsilon(\vec{\sigma})$ vanishes at $\vec{\sigma} = 0$. As a result the symmetric regularization does not capture the divergencies following from the reordering of terms with derivatives like $\hat{\pi}^{\delta\varepsilon}(\vec{\sigma})\partial_M\hat{z}_{\lambda\delta}(\vec{\sigma})$ (cf. (41)). To capture such a type singularity one can use more general regularization of delta function considered by Hörmander [25]. As a result, we find the r.h.s. in the regularized (anti)commutators $[\hat{\mathfrak{P}}^{\mathfrak{M}}(\vec{\sigma}), \partial_M\hat{\mathfrak{Q}}_{\mathfrak{N}}(\vec{\sigma})\} = i\delta^{\mathfrak{M}}_{\mathfrak{N}}\partial_M\delta^p(\vec{\sigma},\vec{\sigma})$, where $\partial_M\delta^p(\vec{\sigma},\vec{\sigma})$ is a regularized analogue of $\partial_M\delta^p(0)$, to be non zero.

because of the nilpotency of the classical BRST charge Ω and its linearity in momenta. Thus, the singular representation (43) contains two nilpotent operators $\hat{\boldsymbol{\Omega}}$ and $\hat{\Omega}$ in view of (40) and (44). Then, the calculation of the anticommutators of the l.h.s. of Eq. (43) with itself and the r.h.s. with itself results in the condition

$$\{\hat{\boldsymbol{\Omega}},\hat{\boldsymbol{\Omega}}\} = 0 = -\int d^p\sigma[2\hat{C}^M\partial_M\hat{C}^{(W)} + (p-\tfrac{7}{2})\hat{C}^M\partial_M\hat{C}^N\partial_N]\delta^p(0)\,. \tag{45}$$

To clarify the expression (45) consider, for simplicity, the contribution of the canonical pair $(\hat{\rho}^\tau,\hat{P}^{(\rho)}_\tau)$ to the anticommutator (45) of the BRST operator (43) using its different representations by the left and right hand sides of (43). For our purpose it is sufficient to consider the contribution to the square of the BRST operator proportional to the product of ghost operators $\hat{C}^{(W)}$ and $\hat{C}^M$. Then for the BRST operator $\hat{\boldsymbol{\Omega}}$ (39) in the l.h.s. of (43) we obtain

$$\begin{aligned}\tfrac{1}{4}\{\hat{C}^{(W)}(\hat{\rho}^\tau\hat{P}^{(\rho)}_\tau + \hat{P}^{(\rho)}_\tau\hat{\rho}^\tau)(\vec{\sigma}), -\hat{C}^M(\hat{\rho}^\tau\partial_{M'}\hat{P}^{(\rho)}_\tau + \partial_{M'}\hat{P}^{(\rho)}_\tau\hat{\rho}^\tau)(\vec{\sigma}')\\ + \hat{C}^M(\partial_{M'}\hat{C}^{(W)}\hat{\mathcal{P}}^{(W)} - \hat{\mathcal{P}}^{(W)}\partial_{M'}\hat{C}^{(W)})(\vec{\sigma}')\}\,.\end{aligned} \tag{46}$$

The desired contributions come from the two anticommutators in (46)

$$\begin{aligned}&\tfrac{1}{4}\{\hat{C}^{(W)}(\hat{\rho}^\tau\hat{P}^{(\rho)}_\tau + \hat{P}^{(\rho)}_\tau\hat{\rho}^\tau)(\vec{\sigma}), -\hat{C}^M(\hat{\rho}^\tau\partial_{M'}\hat{P}^{(\rho)}_\tau + \partial_{M'}\hat{P}^{(\rho)}_\tau\hat{\rho}^\tau)(\vec{\sigma}')\}\\ &\quad = -\tfrac{i}{2}\hat{C}^{(W)}(\vec{\sigma})\hat{C}^M(\vec{\sigma}')(\hat{\rho}^\tau(\vec{\sigma}')\hat{P}^{(\rho)}_\tau(\vec{\sigma}) + \hat{P}^{(\rho)}_\tau(\vec{\sigma})\hat{\rho}^\tau(\vec{\sigma}'))\partial_{M'}\delta^p(\vec{\sigma}-\vec{\sigma}')\\ &\quad\quad + \tfrac{i}{2}\hat{C}^{(W)}(\vec{\sigma})\hat{C}^M(\vec{\sigma}')(\hat{\rho}^\tau(\vec{\sigma})\partial_{M'}\hat{P}^{(\rho)}_\tau(\vec{\sigma}') + \partial_{M'}\hat{P}^{(\rho)}_\tau(\vec{\sigma}')\hat{\rho}^\tau(\vec{\sigma}))\delta^p(\vec{\sigma}-\vec{\sigma}')\,.\end{aligned} \tag{47}$$

Transforming the arguments of the multipliers in front of the partial derivative $\partial_{M'}\delta^p(\vec{\sigma}-\vec{\sigma}')$ to be coincident and equal $\vec{\sigma}'$, we find the equivalent representation for the r.h.s. of Eq. (47) as follows

$$\begin{aligned}-\tfrac{i}{2}\hat{C}^{(W)}\hat{C}^M(\hat{\rho}^\tau\hat{P}^{(\rho)}_\tau + \hat{P}^{(\rho)}_\tau\hat{\rho}^\tau)(\vec{\sigma}')\partial_{M'}\delta^p(\vec{\sigma}-\vec{\sigma}')\\ -\tfrac{i}{2}\partial_M\hat{C}^{(W)}\hat{C}^M(\hat{\rho}^\tau\hat{P}^{(\rho)}_\tau + \hat{P}^{(\rho)}_\tau\hat{\rho}^\tau)\delta^p(\vec{\sigma}-\vec{\sigma}')\,.\end{aligned} \tag{48}$$

Correspondingly, for the second term we obtain

$$\begin{aligned}\tfrac{1}{4}\{\hat{C}^{(W)}(\hat{\rho}^\tau\hat{P}^{(\rho)}_\tau + \hat{P}^{(\rho)}_\tau\hat{\rho}^\tau)(\vec{\sigma}),\hat{C}^M(\partial_{M'}\hat{C}^{(W)}\hat{\mathcal{P}}^{(W)} - \hat{\mathcal{P}}^{(W)}\partial_{M'}\hat{C}^{(W)})(\vec{\sigma}')\}\\ = \tfrac{i}{2}\partial_M\hat{C}^{(W)}\hat{C}^M(\hat{\rho}^\tau\hat{P}^{(\rho)}_\tau + \hat{P}^{(\rho)}_\tau\hat{\rho}^\tau)\delta^p(\vec{\sigma}-\vec{\sigma}')\,.\end{aligned} \tag{49}$$

The sum of the contributions (48) and (49) is the total divergence

$$\frac{i}{2}\partial_M\left(\hat{C}^{(W)}\hat{C}^M(\hat{\rho}^\tau\hat{P}^{(\rho)}_\tau + \hat{P}^{(\rho)}_\tau\hat{\rho}^\tau)\right)\delta^p(\vec{\sigma}-\vec{\sigma}')\,, \tag{50}$$

after the transfer of the derivative $\partial_{M'}$ from the δ-function to the phase-space operators that is a correct operation for distributions.

Thus for the Hermitian representation, the contribution of the canonical pair $(\hat{\rho}^\tau,\hat{P}^{(\rho)}_\tau)$ entering the Weyl generator to the anticommutator $\{\hat{C}^{(W)}\hat{\Delta}^{ext}_W(\vec{\sigma}),\hat{C}^M\hat{L}^{ext}_M(\vec{\sigma}')\}$ vanishes in agreement with the nilpotency of the BRST operator. The same result can be found for the contributions of other canonical pairs.

Now we are interested in the same contribution to the square of the BRST operator proportional to the product of $\hat{C}^{(W)}$ and $\hat{C}^M$ ghosts, but in the QP-ordered and singular realization given by

$$\left\{\left(\hat{C}^{(W)}\hat{\rho}^{\tau}\hat{P}^{(\rho)}_{\tau}-i\hat{C}^{(W)}\delta^p(0)\right)(\vec{\sigma}),-\left(\hat{C}^M\hat{\rho}^{\tau}\partial_{M'}\hat{P}^{(\rho)}_{\tau}-\hat{C}^M\partial_{M'}\hat{C}^{(W)}\hat{\mathcal{P}}^{(W)}\right)(\vec{\sigma}')\right\}, \tag{51}$$

where we omitted the singular terms proportional to $\hat{C}^M\partial_M\delta^p(0)$ in the r.h.s. of the anticommutator (51) which do not contribute into the anticommutator. Then for the r.h.s. of Eq. (43) we obtain the three summands

$$\begin{aligned}
\{(\hat{C}^{(W)}\hat{\rho}^{\tau}\hat{P}^{(\rho)}_{\tau})(\vec{\sigma}),-(\hat{C}^M\hat{\rho}^{\tau}\partial_{M'}\hat{P}^{(\rho)}_{\tau})(\vec{\sigma}')\} &= -i(\hat{C}^{(W)}\hat{C}^M\hat{\rho}^{\tau}\hat{P}^{(\rho)}_{\tau})(\vec{\sigma}')\partial_{M'}\delta^p(\vec{\sigma}-\vec{\sigma}') \\
&\quad -i\partial_M\hat{C}^{(W)}\hat{C}^M\hat{\rho}^{\tau}\hat{P}^{(\rho)}_{\tau}\delta^p(\vec{\sigma}-\vec{\sigma}'), \\
\{(\hat{C}^{(W)}\hat{\rho}^{\tau}\hat{P}^{(\rho)}_{\tau})(\vec{\sigma}),(\hat{C}^M\partial_{M'}\hat{C}^{(W)}\hat{\mathcal{P}}^{(W)})(\vec{\sigma}')\} &= i\partial_M\hat{C}^{(W)}\hat{C}^M\hat{\rho}^{\tau}\hat{P}^{(\rho)}_{\tau}\delta^p(\vec{\sigma}-\vec{\sigma}').
\end{aligned} \tag{52}$$

The sum of the above two summands is again the total divergence

$$i\partial_M\left(\hat{C}^{(W)}\hat{C}^M\hat{\rho}^{\tau}\hat{P}^{(\rho)}_{\tau}\right)\delta^p(\vec{\sigma}-\vec{\sigma}'), \tag{53}$$

which can be omitted for the discussed model of the closed p-brane. So, the value of the anticommutator is determined by the singular term $i\hat{C}^{(W)}\delta^p(0)$ contribution equal

$$\{-i\hat{C}^{(W)}(\vec{\sigma})\delta^p(0),(\hat{C}^M\partial_{M'}\hat{C}^{(W)}\hat{\mathcal{P}}^{(W)})(\vec{\sigma}')\}=-\delta^p(0)\hat{C}^M\partial_M\hat{C}^{(W)}\delta^p(\vec{\sigma}-\vec{\sigma}'). \tag{54}$$

One can see that the term does not transform to the total divergence.

Analogous calculations of the contributions of other canonical pairs entering the Weyl generator will also give the same divergent answer modulo numerical coefficients. Summarizing all that contributions we arrive at the first summand in the r.h.s. of (45).

Similarly, calculation of the (anti)commutators of the $OSp(1|8)$ operators (42) with $\hat{\mathbf{\Omega}}$ (43) gives the relations

$$\begin{aligned}
\{\hat{\mathbf{\Omega}},\hat{\hat{\mathbb{S}}}_{\gamma}\} &= 0 = \ 2i\int d^p\sigma(\hat{C}_{\gamma}+\hat{C}^M\partial_M\hat{\theta}_{\gamma})\delta^p(0), \\
\{\hat{\mathbf{\Omega}},\hat{\hat{\mathbb{S}}}_{\dot{\gamma}}\} &= 0 = -2i\int d^p\sigma(\hat{\bar{C}}_{\dot{\gamma}}-\hat{C}^M\partial_M\hat{\bar{\theta}}_{\dot{\gamma}})\delta^p(0), \\
[\hat{\mathbf{\Omega}},\hat{\hat{\mathcal{K}}}_{\gamma\lambda}] &= 0 = -2i\int d^p\sigma[(\hat{C}_{\gamma}\hat{\theta}_{\lambda}+\hat{\theta}_{\gamma}\hat{C}_{\lambda})+\tfrac{i}{2}\hat{C}^M\partial_M\hat{z}_{\gamma\lambda}]\delta^p(0), \\
[\hat{\mathbf{\Omega}},\hat{\hat{\mathcal{K}}}_{\gamma\dot{\gamma}}] &= 0 = -2i\int d^p\sigma[(\hat{C}_{\gamma}\hat{\bar{\theta}}_{\dot{\gamma}}-\hat{\theta}_{\gamma}\hat{\bar{C}}_{\dot{\gamma}})+\tfrac{i}{2}\hat{C}^M\partial_M\hat{x}_{\gamma\dot{\gamma}}]\delta^p(0).
\end{aligned} \tag{55}$$

The conditions (45) and (55) contain the divergent terms proportional to $\delta^p(0)$ and $\partial_M\delta^p(0)$, so to attach meaning to them one should assign values to the singular limits of $\delta^p(\vec{\sigma}-\vec{\sigma}')|_{\vec{\sigma}=\vec{\sigma}'}$ and its derivatives, i.e. to make their regularization. In accordance with the regularization the (anti)commutation relations (35) have to be changed by the regularized relations

$$[\hat{\mathfrak{P}}^{\mathfrak{M}}(\vec{\sigma}), \hat{\mathfrak{Q}}_{\mathfrak{N}}(\vec{\sigma}')\} = -i\delta^{\mathfrak{M}}_{\mathfrak{N}}\delta^{p}(\vec{\sigma},\vec{\sigma}'), \tag{56}$$

where the distribution $\delta^p(\vec{\sigma},\vec{\sigma}')$ is a regularized image of the Dirac delta function $\delta^p(\vec{\sigma}-\vec{\sigma}')$. The substitution of the distributions $\delta^p(\vec{\sigma},\vec{\sigma}')$ and $\partial_M\delta^p(\vec{\sigma},\vec{\sigma}')$ in Eqs. (45) and (55) results in their solutions

$$\delta^p(\vec{\sigma},\vec{\sigma}) = 0, \quad \partial_M\delta^p(\vec{\sigma},\vec{\sigma}) = 0. \tag{57}$$

The regularization that requires $\delta^p(\vec{\sigma},\vec{\sigma}) = 0$ is well known in the physical literature (see, for instance, [26]). This regularization prescription leads to the closure of the $\hat{\mathfrak{Q}}\hat{\mathfrak{P}}$-ordered and regularized superalgebra of local and global symmetries of the considered brane model. The regularization (57) removes the problem of ordering in the operators obtained from the classical expressions and differing from each other by the terms proportional to $\delta^p(0)$ and/or $\partial_M\delta^p(0)$. It shows the possibility of quantization of the super p-brane model in this regularization.

CONCLUSION

The general problem of quantum brane realization of the BPS states preserving $\frac{\mathsf{M}-1}{\mathsf{M}}$ fraction of partially spontaneously broken global $N=1$ supersymmetry at the classical level was analyzed on example of the twistor-like p-brane in four-dimensional space-time. Twistor-like brane models are characterized by a fine tuning of a large set of classical local and global symmetries caused by the absence of tension. We constructed classical BRST charge and generators of these global and gauge symmetries and proved the closure of their unified P.B. superalgebra. The P.B. realization of the nilpotency condition for the BRST charge and its (anti)commutativity with the unified symmetry generators were proved. After that we considered quantization of the model and proved the preservation of the above classical results using Hermitian operator realization of the symmetry generators and BRST operator. This Hermitian realization gives the relevant physical foundation for the solution of the quantization problem. The remaining problem here is the construction of the vacuum and the Hilbert space of quantum states of tensionless super p-branes.

Otherwise, such a type problem was earlier considered in [10] using the general construction of the finite inner product [11] and the $\hat{\mathfrak{Q}}\hat{\mathfrak{P}}$-ordering in the quantum BRST charge and generators of the symmetry algebra of bosonic tensionless string. There was shown the existence of the full physical vacuum that is annihilated by the BRST and string momentum operators. The latter condition has picked up the $\hat{\mathfrak{Q}}\hat{\mathfrak{P}}$-ordered representation as a physical one for tensionless objects. Motivated by these results we studied the $\hat{\mathfrak{Q}}\hat{\mathfrak{P}}$-ordered operator realization of the brane symmetry generators and the BRST operator. It was found the closure of the quantum generalized superconformal algebra $OSp(1|8)$ in this realization and the (anti)commutativity of the quantum $OSp(1|8)$ generators with nilpotent BRST operator, because of the linearity of the quantum operators in the momentum operators $\hat{\mathfrak{P}}^{\mathfrak{M}}$. But, the $\hat{\mathfrak{Q}}\hat{\mathfrak{P}}$-ordered representation for the $OSp(1|8)$ generators $\hat{\hat{G}}_{qp}$ and the BRST operator $\hat{\Omega}$ turn out to be nonHermitian. We remind that the latter problem might be solved by the definite choice of the regularization assumption for the delta function $\delta^p(0)$ and its derivatives $\partial_M \delta^p(0)$. It shows a possibility to quantize the classical super p-brane preserving $\frac{3}{4}$ fraction of partially spontaneously broken $D = 4$ $N = 1$ supersymmetry in the special regularization.

Our consideration admits straightforward generalization to the case of the space-time dimensions D belonging to the set $2, 3, 4(mod8)$ consistent with the Majorana spinor existence. Taking into account that the number coefficients in front of the divergent terms containing $\delta^p(0)$ and $\partial_M \delta^p(0)$ depend on D one can hope that they might be equal to zero for some D from the above mentioned set. It would prove the existence of the quantized model outside the regularization prescription. The above mentioned problems are under our investigation.

ACKNOWLEDGMENTS

The authors are grateful to Jerzy Lukierski for inviting this contribution and for the warm hospitality at the University of Wrocław. A.Z. thanks Fysikum at the Stockholm University and the Mittag-Leffler Institute for kind hospitality and I. Bengtsson, V. Gershun, S. Hwang, M. Kontsevich, O. Laudal, U. Lindström, M. Movshev, D. Piontkovskii, A. Rashkovskii and B. Sundborg for useful discussions. The work was partially supported by the grant of the Royal Swedish Academy of Sciences and by the SFFR of Ukraine under Project 02.07/276.

REFERENCES

1. J.A. de Azcarraga, J.P. Gauntlett, J.M. Izquierdo and P.K. Townsend, Phys. Rev. Lett. **63**, 2443 (1989).
2. J. Hughes and J. Polchinski, Nucl. Phys. **B278**, 147 (1986);
   J. Hughes, J. Liu and J. Polchinski, Phys. Lett. **B180**, 370 (1986).
3. M.J. Duff and J.M. Liu, Nucl. Phys. **B674**, 217 (2003), hep-th/0303140.
4. C.M. Hull, *Holonomy and Symmetry in M-theory*, hep-th/0305039.
5. J.P. Gauntlett, G.W. Gibbons, C.M. Hull and P.K. Townsend, Comm. Math. Phys. **216**, 431 (2001), hep-th/0001024.
6. I. Bandos and J. Lukierski, Mod. Phys. Lett. **A14**, 1257 (1999), hep-th/9811022.
7. J.P. Gauntlett and C.M. Hull, JHEP **0001**, 004 (2000), hep-th/9909098.
8. A.A. Zheltukhin and D.V. Uvarov, JHEP **0208**, 008 (2002), hep-th/0206214;
   Phys. Lett. **B545**, 183 (2002).
9. D.V. Uvarov and A.A. Zheltukhin, *Whether conformal supersymmetry is broken by quantum p-branes with exotic supersymmetry?*, hep-th/0401059.
10. H. Gustafsson, U. Lindström, P. Saltsidis, B. Sundborg and R. von Unge, Nucl. Phys. **B440**, 495 (1995), hep-th/9410143.
11. S. Hwang and R. Marnelius, Nucl. Phys. **B315**, 638 (1989).
12. E.S. Fradkin and G.A. Vilkovisky, Phys. Lett. **B55**, 224 (1975);
    I.A. Batalin and G.A. Vilkovisky, Phys. Lett. **B69**, 309 (1977).
13. A.A. Zheltukhin and D.V. Uvarov, Phys. Lett. **B565**, 22 (2003), hep-th/0304151;
    I. Bengtsson and A.A. Zheltukhin, *Wess-Zumino actiona and Dirichlet boundary conitions for super*

p-brane with exotic fractions of supersymmetry, ibid. **B570**, 222 (2003).
14. A.A. Zheltukhin, Sov. J. Nucl. Phys. **48**, 326 (1988);
I.A. Bandos and A.A. Zheltukhin, Fortschr. Phys. **61**, 619 (1993).
15. C. Fronsdal, in *"Essays on supersymmetry"*, Dordrecht, Reidel, 1986, Math. Phys. Studies, V.8, 164.
16. M.A. Vasiliev, Phys. Rev. **D66**, 066006 (2002), hep-th/0106149.
17. D.V. Uvarov and A.A. Zheltukhin, JHEP **0403**, 063 (2004), hep-th/0310284.
18. L.D. Faddeev and S.L. Shatashvili, Phys. Lett. **B167**, 225 (1986).
19. I.A. Batalin, E.S. Fradkin and T.E. Fradkina, Nucl. Phys. **B314**, 158 (1989).
20. Y. Eisenberg and S. Solomon, Nucl. Phys. **B309**, 709 (1988).
21. M.S. Plyushchay, Mod. Phys. Lett. **A4**, 1827 (1989).
22. I. Bandos, A. Maznytsia, I. Rudychev and D. Sorokin, J. Mod. Phys. **A12**, 3259 (1997), hep-th/9609107.
23. I.A. Bandos, J. Lukierski and D.P. Sorokin, Phys. Rev. **D61**, 045002 (2000), hep-th/9904109.
24. R. Penrose and W. Rindler, in *Spinors and space-time*, V.2; *Spinor and twistor methods in space-time geometry*, Cambridge University Press, 1986.
25. L.A. Hörmander, in *The analysis of linear partial differential operators*, V.1; *Distributions theory and Fourier analysis*, Springer, 1990.
26. E.S. Fradkin and G.A. Vilkovisky, in *Quantization of relativistic systems with constraints, equivalence of canonical and covariant formalisms in quantum theory of gravitational field*, Preprint CERN TH2332, 1977.
D.M. Gitman and I.V. Tyutin, in *Canonical quantization of constrained fields*, Moscow, "Nauka", 1986 (Russian Edition).

RELATED TOPICS

Brane Solutions of Gravity–Dilaton–Axion Systems

E. Bergshoeff*, A. Collinucci*, U. Gran†, D. Roest† and S. Vandoren**

*Centre for Theoretical Physics, University of Groningen, Nijenborgh 4, 9747 AG Groningen, The Netherlands
†Department of Mathematics, King's College London, Strand, London WC2R 2LS, United Kingdom
**Institute for Theoretical Physics and Spinoza Institute, Utrecht University, Leuvenlaan 4, 3508 TD Utrecht, The Netherlands

Abstract. We consider general properties of brane solutions of gravity-dilaton-axion systems. We focus on the case of 7-branes and instantons. In both cases we show that besides the standard solutions there are new deformed solutions whose charges take value in any of the three conjugacy classes of $SL(2,\mathbb{R})$. In the case of 7-branes we find that for each conjugacy class the 7-brane solutions are 1/2 BPS. Next, we discuss the relation of the 7-branes with the DW/QFT correspondence. In particular, we show that the two (inequivalent) 7-brane solutions in the SO(2) conjugacy class have a nice interpretation as a distribution of (the so-called near horizon limit of) branes. This suggests a way to define the near-horizon limit of a 7-brane.

In the case of instantons only the solutions corresponding to the $\mathbb{R}$ conjugacy class are 1/2 BPS. The solutions corresponding to the other two conjugacy classess correspond to non-extremal deformations. We first discuss an alternative description of these solutions as the geodesic motion of a particle in a two-dimensional AdS_2 space. Next, we discuss the instanton-soliton correspondence. In particular, we show that for two of the conjugacy classes the instanton action in D dimensions is given by the mass of the corresponding soliton which is a (non-extremal) black hole solution in D+1 dimension. We speculate on the role of the non-extremal instantons in calculating higher-derivative corrections to the string effective action and, after a generalization from a flat to a curved AdS_5 background, on their role in the AdS/CFT corresopondence.

INTRODUCTION

Gravity coupled to the two scalars (dilaton and axion) that parametrise an $SL(2,\mathbb{R})/SO(2)$ coset space is an important subsector of the low-energy limit of type IIB superstring theory. Among the different solutions of this system are seven-brane solutions that carry magnetic charges with respect to the three generators of $SL(2,\mathbb{R})$. These magnetic charges combine into a traceless 2 x 2 charge matrix Q which transforms in the adjoint representation of $SL(2,\mathbb{R})$. The combination $\det(Q)$, being invariant under these transformations, labels the three different conjugacy classes of $SL(2,\mathbb{R})$. Each pair of solutions in the same conjugacy class is related via $SL(2,\mathbb{R})$. On the other hand, two solutions that belong to two different conjugacy classes can not be related via $SL(2,\mathbb{R})$. The "circular" 1/2 BPS D7^c–brane of [1] is represented by the $\det Q = 0$ conjugacy class.

It is well-known that the electric-magnetic dual of the D7–brane is the D-instanton [2]. The D-instanton is a half-supersymmetric solution of the Euclidean gravity-dilaton-

CP767, *Fundamental Interactions and Twistor-Like Methods,* edited by J. Lukierski and D. Sorokin

axion system, and carries electric charge with respect to the Euclidean $SL(2,\mathbb{R})$ symmetry. In complete analogy to the case of seven-branes, the three Euclidean $SL(2,\mathbb{R})$ charges combine into a 2 x 2 charge matrix Q that transforms in the adjoint representation of the Euclidean $SL(2,\mathbb{R})$. The D-instanton is represented by the same conjugacy class that represents the circular $D7^c$–brane, i.e. the one with $\det(Q) = 0$.

It is natural to ask whether there exist 7-branes and instantons with $SL(2,\mathbb{R})$ charges corresponding to the other two conjugacy classes of $SL(2,\mathbb{R})$. It is the aim of this work to construct and investigate such solutions. This talk is a summary of [3, 4, 5, 6, 7].

GRAVITY-DILATON-AXION SYSTEMS

In this section we briefly review some basic properties of the gravity-dilaton axion system. The basic fields are the metric $g_{\mu\nu}$, the dilaton ϕ and the axion χ. An axion, as opposed to a dilaton, only occurs via its spacetime derivative in the Lagrangian. In D-dimensional Minkowski spacetime this Lagrangian is given by

$$\mathcal{L}_{\rm M} = \tfrac{1}{2}\sqrt{|g|}\,[R - \tfrac{1}{2}(\partial\phi)^2 - \tfrac{1}{2}e^{b\phi}(\partial\chi)^2]\,, \tag{1}$$

where $b \neq 0$ is an arbitrary dilaton coupling parameter. The Lagrangian (1) is invariant under a nonlinear $SL(2,\mathbb{R})$ symmetry. For $b = 0$ this symmetry reduces to

$$SL(2,\mathbb{R}) \to ISO(2)\,. \tag{2}$$

From now on we will assume that $b \neq 0$. The case $b = 0$ should be treated separately and has been discussed in [8].

The two scalars ϕ and χ parametrize the coset

$$\frac{SL(2,\mathbb{R})}{SO(2)}\,. \tag{3}$$

This can be made manifest by combining ϕ and χ into the following $SL(2,\mathbb{R})$ matrix $\mathcal{M}$:

$$\mathcal{M} = e^{b\phi/2}\begin{pmatrix} \tfrac{1}{4}b^2\chi^2 + e^{-b\phi} & \tfrac{1}{2}b\chi \\ \tfrac{1}{2}b\chi & 1 \end{pmatrix}. \tag{4}$$

In terms of $\mathcal{M}$ the $SL(2,\mathbb{R})$ symmetry is given by

$$\mathcal{M} \to \Omega\mathcal{M}\Omega^T \quad \text{with} \quad \Omega \in SL(2,\mathbb{R})\,. \tag{5}$$

Note that we defined the gravity-dilaton-axion system in Minkowski spacetime. One can show that similar formulae hold in Euclidean space. In the Euclidean case the axion and dilaton parametrise the coset

$$\frac{SL(2,\mathbb{R})}{SO(1,1)}\,. \tag{6}$$

TABLE 1. The three conjugacy classes of $SL(2,\mathbb{R})$

$\det Q < 0$	$\det Q = 0$	$\det Q > 0$
SO(1,1)	$\mathbb{R}$	SO(2)

For more details, see [7].

In the following discussion we will keep the dimension D and the dilaton coupling parameter $b \neq 0$ arbitrary. The standard example is $D = 10$ and $b = 2$ corresponding to IIB supergravity. Other cases may correspond to (truncations of) compactifications of N=2 supergravity. For instance, the Euclidean Lagrangian for the universal hypermultiplet that arises from a Calabi-Yau compactification of type II strings [9, 10] can be truncated to a D=4 Euclidean gravity-dilaton-axion system with $b = 1$ or $b = 2$.

Given the $SL(2,\mathbb{R})$ symmetry one can define corresponding Noether currents

$$J_\mu \sim (\partial_\mu \mathcal{M})\mathcal{M}^{-1} \,. \tag{7}$$

For the 7-brane and instanton solutions that we study here one can define corresponding Noether charges[1]

$$\begin{aligned} 7-branes : Q &\sim \int_{S^1} J \,, \\ instantons : Q &\sim \int_{S^{D-1}} J \,. \end{aligned} \tag{8}$$

These charges transform under $SL(2,\mathbb{R})$ as follows:

$$Q \to \Omega Q \Omega^{-1} \,. \tag{9}$$

We now come to an important point that will be crucial for the remaining part of this work. From the above transformation rule we see that the determinant $\det Q$ is *invariant* under the $SL(2,\mathbb{R})$ transformations. This means that we have a family of distinct conjugacy classess which are labelled by the value of $\det Q$. It is natural to distinguish between the three cases indicated in table 1. In this table we have also indicated the one-dimensional subgroup of $SL(2,\mathbb{R})$ associated to each of the three types of conjugacy classes. This association means that each element g of the given conjugacy class can be written as an element h of the corresponding subgroup conjugated with an arbitrary $SL(2,\mathbb{R})$ group element Ω, i.e.

$$g = \Omega h \Omega^{-1} \,. \tag{10}$$

[1] For nonlinear symmetries, like $SL(2,\mathbb{R})$, it is *a priori* not guaranteed that the integrals of the currents are finite. In our case, where the solutions depend on only one coordinate, the integrals are finite. We thank E. Ivanov for a discussion on this point.

Any 7-brane or instanton solution of the gravity-dilaton-instanton system will carry $SL(2,\mathbb{R})$ charges that fall into one of the three types of conjugacy classes of table 1. Let us first consider the standard D7-brane solution of IIB string theory. In general we wish to consider branes with *two* transverse directions, i.e. $(D-3)$-branes in D dimensions. For simplicity we will often call these branes just 7-branes instead of $(D-3)$-branes. The position of the D7-brane is often indicated by the diagram

$$\text{D7} \;:\; \times | \times\times\times\times\times\times\times \; -- \tag{11}$$

where $\times(-)$ indicates a worldvolume (transverse) direction. The first $\times$ before the $|$ indicates the worldvolume time direction. A $(D-3)$-brane naturally couples to a C_{D-2}-form which is just the dual of the axion:

$$\partial\chi \sim (\partial C_{\mathrm{D}-2})^{\star}\,. \tag{12}$$

The special thing about a brane with *two* transverse directions is that the harmonic function over this space is given by a logarithm

$$H(r) \sim \ln r\,, \tag{13}$$

where r indicates the radial transverse direction. This solution is half-supersymmetric. It turns out that it is not possible to reduce this brane solution over one of the two transverse directions to a half-supersymmetric domain wall solution in one dimension lower. Neither can one define the near-horizon limit of the standard D7-brane solution. Nevertheless, as we will see, the analysis below will suggest a way to define the near-horizon limit which brings the 7-branes on the same footing with the other branes in the so-called DW/QFT correspondence [11, 12].

In order to dualize the 7-brane to, for instance, the D8 brane solution of type IIA string theory it is necessary to introduce the so-called *circular* D7^{c}-brane [1] which has an extra isometry in one of the two transverse directions. The harmonic in the remaining single direction r is given by

$$H(r) \sim r\,. \tag{14}$$

Indeed, one can show that this D7^{c}-brane solution is T-dual to the D8-brane of type IIA string theory. It turns out that both the D7-brane and the circular D7^{c}-brane have charges that are in the $\det Q = 0$ conjugacy class of $SL(2,\mathbb{R})$.

We next consider the Euclidean D-instanton solution of [2]. The D-instanton can be viewed as the extreme case of a D(-1) brane in the family of T-dual Dp-brane solutions with only transverse and no worldvolume directions. The corresponding diagram is given by

$$\text{D}{-1} \;:\; ---------- \tag{15}$$

The solution can be given in terms of a harmonic over the D-dimensional Euclidean transverse space:

$$H(r) \sim 1/r^{D-2}. \tag{16}$$

One can define Euclidean $SL(2,\mathbb{R})$ charges for the D-instanton. Like the D7-brane and the D7^{c}-brane these charges take value in the $\det Q = 0$ conjugacy class.

Clearly, by performing arbitrary $SL(2,\mathbb{R})$ rotations on a $\det Q = 0$ solution one only obtains solutions that fall into the same conjugacy class. The following obvious question arises that will be central theme of this investigation:

$$\boxed{\text{What about 7-branes and instantons with } \det Q > 0 \text{ and } \det Q < 0\ ?}$$

We will show that both for the 7-branes and the D-instantons deformations can be found which carry charges that belong to the $\det Q > 0$ or $\det Q < 0$ conjugacy classes. Apart from this there are also major differences between the 7-brane and instanton case. We first discuss the 7-brane solutions.

7-BRANES AND THE DW/QFT CORRESPONDENCE

It is indeed possible to find generalizations of the D7^{c}-brane solution. This extended class of solutions is characterized by two arbitrary holomorphic functions $f(z)$ and $g(z)$. Here $z = x + iy$ is the complex coordinate parametrizing the two-dimensional transverse space. We have found the following class of 1/2 BPS 7-brane solutions [3, 13]:

$$\boxed{\begin{aligned} ds^2 &= ds^2_{\mathrm{D}-2} + \Im\mathrm{m} f(z) e^{-\Re\mathrm{e}\, g(z)} dz d\bar{z}\,, \\ \tau &\equiv \chi + i e^{-\phi} = f(z)\,. \end{aligned}} \tag{17}$$

All solutions are half-supersymmetric with the Killing spinor given by

$$\varepsilon = e^{i\Im\mathrm{m}\, g}\varepsilon_0 \quad with \quad \Gamma_z \varepsilon = 0\,. \tag{18}$$

These solutions have been obtained by first considering the 1/2 BPS domain wall solutions of the maximally gauged supergravities in D=9 dimensions. These gauged supergravities can be obtained from IIB supergravity by a so-called *twisted* reduction over a circle with radius R. Assuming that x parametrizes the circle this means that the fields at x and $x + 2\pi R$ are related to each other via an $SL(2,\mathbb{R})$ matrix Ω that takes value in the one-dimensional subgroup corresponding to one of the three conjugacy classes, i.e.

$$\Phi(x) = \Omega\Phi(x + 2\pi R) \quad with \quad \Omega \in SO(1,1), \mathbb{R} \text{ or } SO(2) \tag{19}$$

for any field Φ. These twisted reductions of IIB supergravity lead to maximally-supersymmetric $SO(1,1), \mathbb{R}$ or $SO(2)$ gauged supergravities in D=9 dimensions. Note that this construction method implies that we only find 1/2 BPS 7-brane solutions in

D=10 dimensions that can be reduced to 1/2 BPS domain wall solutions in D=9 dimensions. For instance, this method will not lead to the D7-brane solution since that solution can not be reduced to a 1/2 supersymmetric domain wall solution.

For branes with two transverse directions one usually calculates the monodromy matrix Λ instead of the charge matrix Q. We assume that the two-dimensional space has the topology of a cilinder and that x, y are cylindrical coordinates with x the circle direction parametrizing a circle of radius R. The two matrices Λ and Q are related via [13]

$$\Lambda = \underbrace{\begin{pmatrix} a & b \\ c & d \end{pmatrix}}_{monodromy} = e^{2\pi R \overset{charge}{\overset{\Downarrow}{Q}}} . \tag{20}$$

For a given set of holomorphic functions $f(z), g(z)$ it is straightforward to calculate the monodromy matrix and hence the charge matrix. For example, the circular D7$^{\rm c}$-brane solution is specified by

$$f(z) = mz, \qquad\qquad g(z) = 0. \tag{21}$$

This leads to the monodromy matrix

$$\Lambda = \begin{pmatrix} 1 & 2\pi mR \\ 0 & 1 \end{pmatrix}. \tag{22}$$

Following (20) the corresponding charge matrix is given by

$$Q = \begin{pmatrix} 0 & m \\ 0 & 0 \end{pmatrix}, \tag{23}$$

which belongs to the $\det Q = 0$ conjugacy class.

It is not too difficult to find choices of holomorphic functions that lead to 1/2 BPS 7-branes with charges corresponding to the other two conjugacy classes. We found these solutions by first constructing the domain wall solutions they give rise to after a twisted reduction over the circular x direction. The result is given in table 2 [3]. Note that in the $SO(2)$ conjugacy class we find two inequivalent solutions. The second one is a locally flat spacetime which in the transverse directions has a cone-like structure, i.e. there is a non-trivial deficit angle.

Sofar we did not consider any quantization conditions. String theory requires that we must impose the following condition on the monodromy matrix Λ:

$$\Lambda \in SL(2,\mathbb{Z}). \tag{24}$$

This is in general a diophantine equation which in the case of the $\det Q = 0$ conjugacy class has the following simple general solution:

$$\Lambda = 2\pi R \begin{pmatrix} 1 & n \\ 0 & 1 \end{pmatrix}, \qquad\qquad n \in \mathbb{Z}. \tag{25}$$

TABLE 2. The choices of $f(z), g(z)$ corresponding to the three conjugacy classes of $SL(2,\mathbb{R})$. The last column indicates the one-dimensional subgroup characterizing each conjugacy class.

class	f(z)	g(z)	group
$\det Q = 0$	mz	0	$\mathbb{R}$
$\det Q < 0$	ie^{mz}	mz	SO(1,1)
$\det Q > 0$	$\tan \frac{1}{2} mz$	$\ln \cos \frac{1}{2} mz$	SO(2)
	i	imz	SO(2)

The integer n specifies how many circular 7-branes we have. For the other two conjugacy class the quantization conditions are much more difficult to work out. The general solution can be found in [14, 15].

The two solutions given in table 2 corresponding to the SO(2) conjugacy class are especially interesting in the context of the AdS/CFT correspondence since after reduction over x they lead to 1/2 supersymmetric domain wall solutions in SO(2) gauged supergravity theories. The general scheme in the DW/QFT correspondence is that the near horizon limit of a brane with n transverse directions leads, after spherical reduction, to domain wall solutions of $SO(n)$ gauged supergravities. We will show below that 7-branes naturally fit this picture in the special case that $n = 2$.

The DW/QFT correspondence is a non-conformal generalization of the AdS/CFT correspondence [16]. The standard example of the AdS/CFT correspondence is the D3-brane, see table 3. The near-horizon geometry of the D3-brane is the $AdS_5 \times S^5$ vacuum configuration. After compactification over the spherial part this leads to a SO(6) gauged supergravity in D=5 dimensions. This supergravity allows a maximally supersymmetric AdS_5 vacuum configuration. At the boundary of this AdS_5 spacetime lives the dual conformal N=4 supersymmetric Yang-Mills theory.

There are two ways to break the conformal symmetry. In each of these two cases the maximally supersymmetric AdS_5 space is replaced by a 1/2 BPS domain wall solution. Such a solution requires the coupling of a dilaton to the cosmological constant. We distinguish between two kinds of dilatons. First we have the volume dilaton ϕ which occurs as an overall factor in the potential of the gauged supergravity. This dilaton is absent in the potential, and hence does not couple to the cosmological constant, if the corresponding brane is conformal, i.e. D3, M2 and M5. The activation of this dilaton is required if we consider one of the other non-conformal branes. The maximally supersymmetric AdS space is in these cases replaced by a 1/2 BPS domain wall solution with a non-trivial profile for the volume dilaton. The field theory living at the boundary of this domain wall spacetime is not a conformal field theory but a non-conformal quantum field theory. Second, we have the remaining so-called distribution dilatons. These always couple to the cosmological constant. Each of these dilatons specifies whether the corresponding brane, be it a conformal brane or not, is distributed in a given transverse direction. It turns out that a brane with n transverse directions can be distributed in maximal $n-1$ directions. In the conformal case, the distribution of the branes means that in the dual conformal field theory some scalars have obtained a non-

TABLE 3. The standard example of the AdS/CFT correspondence: the D3-brane

brane	vacuum configuration	gauged supergravity
D3	$AdS_5 \times S^5$	D=5 SO(6)

TABLE 4. Branes and the DW/QFT correspondence. The third column indicates whether the volume dilaton occurs in the potential yes or no.

D	n	volume dilaton ϕ	supergravity	Brane
8	3	$\surd$	IIA on S^2	D6
7	5	-	11D on S^4	M5
6	5	$\surd$	IIA on S^4	D4
5	6	-	IIB on S^5	D3
4	8	-	11D on S^7	M2
3	8	$\surd$	IIA on S^7	F1A
2	9	$\surd$	IIA on S^8	D0

vanishing expectation value, i.e. we are in the (non-conformal) Coulomb branch of the gauge theory [17, 18, 19].

It is natural to extend the AdS/CFT correspondence of the conformal branes to a DW/QFT correspondence for the non-conformal branes [11, 12]. In both cases the branes can be distributed depending on whether some of the distribution dilatons are activated. This leads to the relations of table 4 which is a generalization of table 3 to the general brane case[2]

The point we want to make is that 7-branes can naturally be added to the top of tabel 4 if, instead of performing an ordinary circle reduction we perform a SO(2) *twisted* reduction. This leads to the extension given in table 5, where from now on we assume that $b = 2$ for the remaining part of this section. The two solutions of table 2 are now naturally interpreted as distributions of 7-branes. To see this it is instructive to first consider the case of D5-branes and D6-branes. D5-branes have a 4-dimensional transverse space and can be distributed into at most 3 independent transverse directions.

TABLE 5. The D7-brane and the DW/QFT correspondence

D	n	volume dilaton ϕ	supergravity	Brane
9	2	$\surd$	IIB with SO(2) twist	D7

[2] We have not indicated the branes of string theory that follow from M-branes by so-called direct or double dimensional reduction. In D=3 there is furthermore an independent possibility with the fundamental string F1B of type IIB string theory. This leads to an inequivalent maximally supersymmetric SO(8) gauged supergravity in D=3 dimensions, compare to [20].

This happens when all three distribution dilatons are activated in the domain wall solution of the D=7 SO(5) gauged supergravity. The branes are distributed as branes with positive (+) charge on the surface of a 3-dimensional ellipsoid and as branes with negative (-) charge inside this ellipsoid, see the first picture in figure 1. Note that the volume dilaton is always activated since the D5-brane is not a conformal brane.

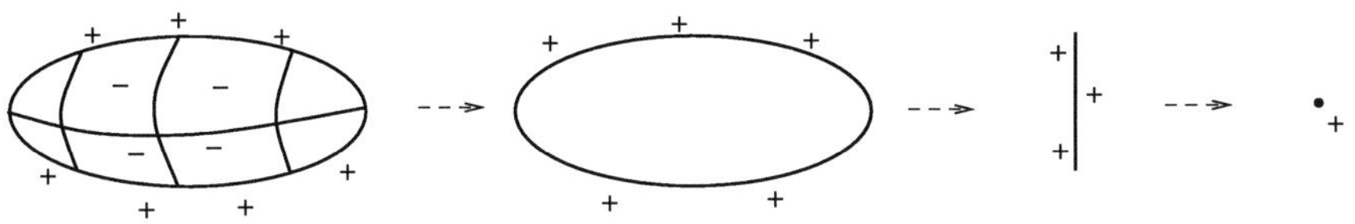

FIGURE 1. Distribution of D5-branes in 3, 2, 1 and 0 directions

Setting one of the distribution dilatons equal to zero corresponds to undoing the distribution in one of the three directions. Due to a cancellation of + branes on the boundary, except the ones at the equator, with - branes in the bulk one is left with a two-dimensional distribution with + branes only positioned at the boundary of a two-dimensional ellipsoid, see the second picture in figure 1. This undoing of the distribution in a given direction can be done two more times, see the third and fourth picture in figure 1. At the end one ends up with a set of stacked branes with no distribution at all.

We next consider the case of D6-branes, see figure 2. Since D6-branes have a 3-dimensional transverse space, we start with a maximal distribution in 2 directions, see the first picture in figure 2. The + branes are positioned at the boundary of a two-dimensional ellipsoid while the - branes are distributed inside this ellipsoid. Undoing the distribution in one of the 2 directions we end up with a two-centred solution. The uplifting of this solution to M-theory leads to the Eguchi-Hanson metric which is the near-horizon limit of the two-centred Kaluza-Klein monopole solution. Undoing the distribution in the second direction we end up with (the near-horizon limit of) a set of stacked D6-branes.

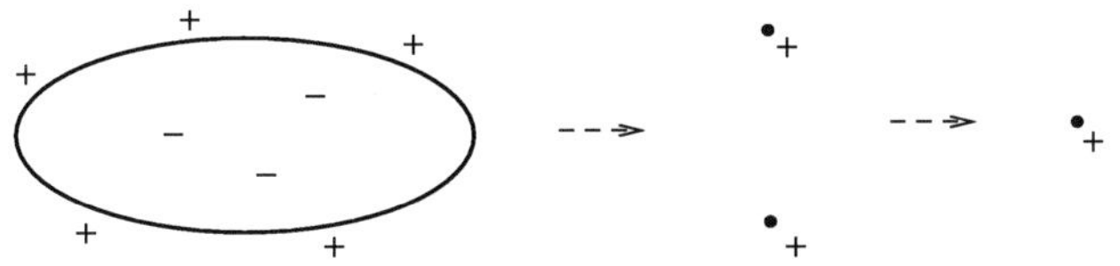

FIGURE 2. Distribution of D6-branes in 2,1 and 0 directions

Finally, we consider the case of interest, i.e. 7-branes, see figure 3. 7-branes have two transverse directions and therefore we would like to interpret the first picture of figure 3, representing the first SO(2) conjugacy class solution of table 2, as a distribution of branes in one direction with + branes positioned at the end of the line and - branes distributed along the line. Somewhat to a surprise, this is indeed possible provided we view the first picture in figure 3 as a distribution of D7-branes, which themselves belong to the $\mathbb{R}$ conjugacy class and cannot be reduced to 1/2 supersymmetric domain walls! The reason that this is possible is related to the fact that the product of two monodromy matrices in the $\mathbb{R}$ conjugacy class can yield a monodromy matrix in the SO(2) conjugacy class, i.e. one can view the SO(2) conjugacy solution as a bound state of $\mathbb{R}$ conjugacy class solutions. Undoing the distribution in the single direction leads to the second picture in figure 3 which represents the second SO(2) solution of table 2. Due to a cancellation of charges one is left with no charge at all! Indeed, this is consistent with the fact that the second solution is a locally flat spacetime where the two transverse directions have a cone-like structure with quantized deficit angle. It is very suggestive to define this configuration as the near-horizon limit of the D7-brane solution.

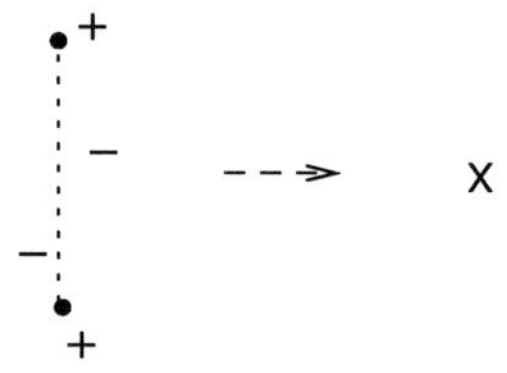

FIGURE 3. Distribution of D7-branes in 1 and 0 directions.

This concludes our discussion of the relation between 7-branes and the DW/QFT correspondence. In the remaining part of this work we will discuss the instanton solutions and their relation to the AdS/CFT corrspondence.

INSTANTONS AND THE ADS/CFT CORRESPONDENCE

To construct generalizations of the D-instanton we allow in the Ansatz for *conformally flat* metrics. Allowing a non-zero conformal factor in the Euclidean spacetime metric leads to the following class of solutions:

$$\begin{aligned} ds^2 &= \left(1 - \frac{\vec{\mathfrak{q}}^2}{r^{\frac{2}{(D-2)}}}\right)^{\frac{2}{(D-2)}} (dr^2 + r^2 d\Omega_{D-1}^2)\,, \\ e^{b\phi(r)} &= \left(\frac{q_-}{\mathfrak{q}} \sinh(H(r) + C\right)^2, \\ \chi(r) &= \frac{2}{bq_-}(\mathfrak{q}\coth(H(r)+C) - q_3)\,. \end{aligned} \tag{26}$$

Here $H(r)$ is the harmonic over the conformally Euclidean metric in (26) which is given by

$$H(r) = \frac{bc}{2} \log(f_+(r)/f_-(r)), \qquad f_\pm(r) = 1 \pm \frac{\mathsf{q}}{r^{D-2}}. \tag{27}$$

Remember that b is the dilaton coupling parameter. The numerical constant c is given by

$$c = \sqrt{2(D-1)/(D-2)}. \tag{28}$$

The solutions (26) contain 4 integration constants. The integration constant C is related to the fact that one can perform special $SL(2,\mathbb{R})$ transformations

$$\Omega = e^{\lambda Q} \tag{29}$$

that commutes with the charge matrix Q. The other 3 integration constants $\vec{\mathsf{q}}^2, q_3, q_-$ are related to the $SL(2,\mathbb{R})$ charge matrix Q in the following way:

$$Q = \begin{pmatrix} q_3 & iq_+ \\ iq_- & -q_3 \end{pmatrix}, \qquad \det Q = -\vec{\mathsf{q}}^2, \qquad q \equiv \sqrt{\vec{\mathsf{q}}^2}. \tag{30}$$

Similar solutions have been discussed in [21]. Note that $q \equiv \sqrt{\vec{\mathsf{q}}^2}$ can be imaginary for negative $\vec{\mathsf{q}}^2$. In that case a real solution can be obtained by analytic continuation in which the hyperbolic functions in (26) get replaced by goniometric ones. It is easiest to discuss the three cases $\vec{\mathsf{q}}^2 > 0$, $\vec{\mathsf{q}}^2 = 0$ and $\vec{\mathsf{q}}^2 < 0$ separately. We do this below.

• $\vec{\mathsf{q}}^2 > 0$: *Black Holes*

In this case there is a curvature singularity for $r_c \sim \vec{\mathsf{q}}^2$. There are two ways in which the singularity might be resolved. It could be that string effects will soften the singularity since $e^\phi \to \infty$ at $r \to r_c$. For certain values of the dilaton coupling parameter b the singularity is resolved by uplifting the solution to a p-brane solution in $D+p+1$ dimensions, see later in this section. Note that for $\vec{\mathsf{q}}^2 > 0$ the limit $q_- \to 0$ is well-defined. It leads to a special case in which the axion decouples.

• $\vec{\mathsf{q}}^2 = 0$: *Extremal Instantons*

This is the case of the 1/2 BPS D-instanton [2]. In this limit the solution is given by

$$\boxed{\begin{aligned} ds^2 &= dr^2 + r^2 d\Omega_{D-1}^2, \\ e^{b\phi(r)/2} &= g_s^{b/2} + \frac{bcq_-}{r^{D-2}}, \\ \chi(r) &= \frac{2}{b}(e^{-b\phi(r)/2} - \frac{q_3}{q_-}). \end{aligned}} \tag{31}$$

This limiting case can be obtained either by starting from $\vec{q}^2 > 0$ or $\vec{q}^2 < 0$ and taking the limit $\vec{q}^2 \to 0$. This limit is facilitated by first making the redefinition

$$C \to \frac{q}{q_-} C\,. \tag{32}$$

• $\vec{q}^2 < 0$: <u>*Wormholes*</u>

To obtain a real solution we redefine $q \to i\tilde{q}$ such that $\tilde{q}^2 > 0$. After an analytic continuation the solution for this case is given by

$$\boxed{\begin{aligned} ds^2 &= \Big(1+\frac{\tilde{q}^2}{r^{2(D-2)}}\Big)^{\frac{2}{(D-2)}}\,(dr^2+r^2 d\Omega^2_{D-1})\,,\\ e^{b\phi(r)} &= \left(\frac{q_-}{\tilde{q}}\,\sin(bc\arctan(\frac{\tilde{q}}{r^{D-2}})+C)\right)^2\,,\\ \chi(r) &= \frac{2}{bq_-}\,\tilde{q}\cot(bc\arctan(\frac{\tilde{q}}{r^{D-2}})+C)-q_3)\,. \end{aligned}}$$

The metric is regular for $0 < r < \infty$ and the scalars are regular for $bc < 2$. It turns out that the singularity in the scalars for $bc \geq 2$ can be understood from the fact that, after the Wick rotation from Minkowski spacetime to Euclidean space the dilaton and axion do not define a global coordinate system for the AdS_2 scalar sigma manifold. This is due to the fact that the Poincare coordinates they define do cover the *Euclidean* AdS_2 space but only half of the *Minkowskian* AdS_2 space. For more details, see [22].

We mention that, unlike the case of 7-branes only the D-instanton, belonging to the $\det Q = 0$ conjugacy class, is 1/2 BPS. The instantons with charges in the other two conjugacy classes, i.e. the ones with $\det Q > 0$ and $\det Q < 0$, are *not* supersymmetric. For this reason, and others, see below, we call these instantons *non-extremal.*

It is well-known that the D-instanton has a wormhole geometry in the string frame metric [2]. This wormhole is asymptotically flat with a neck of physical radius $\rho = \rho_{sd}$ positioned at the fixed point $r = r_{sd}$, see figure 4.

It turns out that the non-extremal $\tilde{q}^2 < 0$ instantons also have a wormhole geometry in the Einstein frame. This can be deduced from the fact that the metric given in (33) has a $\mathbb{Z}_2$ isometry corresponding to the reflection

$$r^{D-2} \to \tilde{q} r^{2-D} \tag{33}$$

which interchanges the two asymptotically flat regions. This reflection has a fixed point, corresponding to the selfdual radius

$$r_{sd}^{D-2} = \tilde{q}\,. \tag{34}$$

The thickness of the neck was computed to be [7]

$$\rho_{sd}^{D-2} = 2\tilde{q}\,. \tag{35}$$

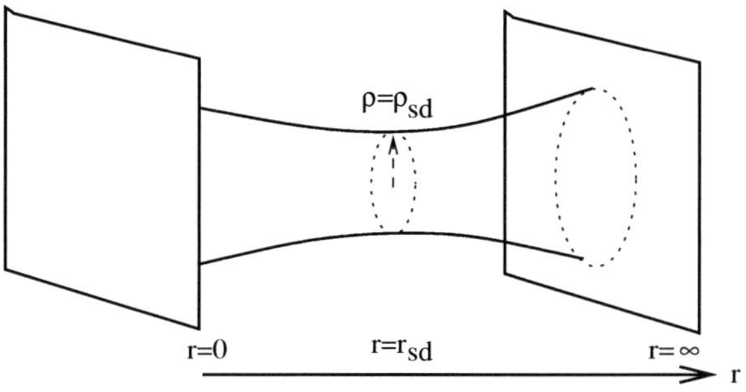

Figure 4. The geometry of a wormhole. The two asymptotically flat regions at $r = 0$ and $r = \infty$ are connected via a neck with a minimal physical radius ρ_{sd} at the self-dual radius r_{sd}.

The $\vec{q}^2 > 0$ instantons, on the other hand, have a wormhole geometry in the so-called dual frame only when the dilaton coupling parameter b is given by $bc = 2$. In summary, wormhole geometries occur in the following frames:

- $\underline{\vec{q}^2 > 0}$: dual frame (only if $bc = 2$)
- $\underline{\vec{q}^2 = 0}$: string frame
- $\underline{\vec{q}^2 < 0}$: Einstein frame

Similar wormhole geomtries have been studied in the eighties, see e.g. [23]. The new thing about the situation here is that we have been able to construct *regular* wormhole solutions for $\vec{q}^2 < 0$ and $bc < 2$. In type IIB string theory in ten dimensions this is not satisfied. Toroidal compactifications of string theory only lead to values of b for which $bc \geq 2$, so no wormholes exist for these cases. However, for the universal hypermultiplet, which descends from a Calabi-Yau compactification of type II strings, one can have the value $b = 1$ in $D = 4$, and so $bc = \sqrt{3} < 2$. The solution is then characterized by the dilaton and the RR scalar that descends from the RR three-form gauge potential in type IIA string theory in $D = 10$ dimensions. Since the extremal case $\vec{q}^2 = 0$ corresponds to a wrapped type IIA Euclidean membrane over a (supersymmetric) three-cycle, it is natural to suggest that the regular wormhole, with $\vec{q}^2 < 0$, corresponds to a wrapped non-extremal Euclidean D2–brane [24].

It turns out that there is an interesting alternative way of describing the instantons as the geodesic motion of a particle in a Minkowskian AdS_2 spacetime. The technique described below is taken from a similar particle description in the case of accelerating cosmologies [25]. In fact, the same technique can be applied to domain walls as well, see [22].

Our starting point is the following Lagrangian and Ansatz:

$$
\boxed{\begin{array}{l}\hat{\mathcal{L}} = \hat{R} + (\partial\hat{\phi})^2 + e^{b\hat{\phi}}(\partial\hat{\chi})^2\,,\\ d\hat{s}^2 = e^{c\varphi(r)}e(r)^2dr^2 + e^{c\varphi(r)/(D-1)}dS^2_{D-1}\,,\\ \hat{\phi}(r) = \phi(r)\,, \qquad\qquad \hat{\chi}(r) = \chi(r).\end{array}} \tag{36}
$$

Remember that the numerical constant c is defined in (28). The above Ansatz is a standard Kaluza-Klein Ansatz for a spherical reduction to 1 dimension. The metric contains two functions: The function $\varphi(r)$ is the would-be conformal factor of the instanton solution and the function $e(r)$ is the einbein in 1 dimension with r playing the role of Euclidean time. After reduction the Lagrangian is given by

$$
\mathcal{L} \sim e^{-1}(\dot{\phi}^2 + e^{b\phi}\dot{\chi}^2) + \underbrace{e^{-1}\dot{\varphi}^2 + e(D-1)(D-2)e^{2\varphi/c}}_{\textit{Liouville equation}}\,. \tag{37}
$$

As indicated in this equation the Kaluza-Klein scalar decouples and its equation of motion is given by the Liouville equation

$$
\ddot{\varphi} - 2(D-1)(D-2)/c\,e^{2\varphi/c} = 0\,. \tag{38}
$$

After a coordinate transformation

$$
d\tilde{r} = edr \tag{39}
$$

the general solution for the conformal factor is given by

$$
e^{2\varphi/c} = \frac{-8\mathrm{q}^2}{(D-2)\left[1 + cosh\left(\frac{2}{c}\sqrt{8(D-1)\mathrm{q}^2\,\tilde{r}}\right)\right]}\,. \tag{40}
$$

After substituting this solution back into the action we obtain

$$
\mathcal{L} \sim e^{-1}\underbrace{(\dot{\phi}^2 + e^{b\phi}\dot{\chi}^2)}_{AdS_2} + e\vec{\mathrm{q}}^2\,. \tag{41}
$$

We deduce that the dilaton coupling parameter b can be identified with the radius R of the AdS_2 space:

$$
b = 2/R\,. \tag{42}
$$

Moreover the nature of the geodesic (spacelike, timelike or null) or, equivalently, the mass of the particle (tachyonic, massless or massive) is determined by the value of $\vec{\mathrm{q}}^2$, i.e. by the conjugacy class. In summary, we obtain the following relations:

- $\underline{\vec{\mathrm{q}}^2 > 0}$: Black Holes $\leftrightarrow$ Tachyonic Particle

TABLE 6. The D-instanton and the DW/QFT correspondence

D	n	volume dilaton ϕ	supergravity	Brane
1	10	$\surd$	IIA on S^9	D-instanton

- $\vec{q}^2 = 0$: Extremal Instantons $\leftrightarrow$ Massless Particle

- $\vec{q}^2 < 0$: Wormholes $\leftrightarrow$ Massive Particle

The spherical reduction from D to 1 dimension naturally triggers the question whether the D-instanton plays a similar role in the AdS/CFT and DW/QFT correspondence like the other branes do. In other words, it is natural to extend table 4 not only with 7-branes, as indicated in table 5 but also with instantons, see table 6. Whether or not this analogy with the other branes can be made remains to be seen. For an earlier reference to this suggestion, see [26]. Below we will investigate the relation between (extremal and non-extremal) D-instantons and the AdS/CFT correspondence from a quite different point of view.

The non-extremal instanton solutions we have constructed fall naturally in the class of non-extremal Dp-brane ($0 \leq p \leq 6$) solutions which have been constructed in the literature. In fact, there are two classes of non-extremal branes available in the literature. In the first class the isometries of the worldvolume and transverse space are broken. These solutions are given by [27]

$$\boxed{ds^2 = e^{\alpha H}\left(-e^{2f}dt^2 + dx_p^2\right) + e^{\beta H}\left(e^{-2f}dr^2 + r^2 d\Omega^2\right)} \tag{43}$$

for some constants α and β. The function H is harmonic and f is the non-extremality function. In the second class the isometries remain unbroken [28]

$$\boxed{ds^2 = e^{A}\left(-dt^2 + dx_p^2\right) + e^{B}\left(dr^2 + r^2 d\Omega^2\right)} \tag{44}$$

Here $A \sim B$ is harmonic in the extremal case but $A \neq B$ and not harmonic in the non-extremal case. The above formulae are only valid for $0 \leq p \leq 6$. However, they can be analytically continuated to $p = -1$ as well. It turns out that the non-extremal instantons we constructed just fits in as the $p = -1$ member of the second class of non-extremal Dp-branes given in (44). In fact in all cases there are *two* non-extremal deformations, one corresponding to $\vec{q}^2 > 0$ and one corresponding to $\vec{q}^2 < 0$. Only when $p = -1$ has $\vec{q}^2$ an interpretation in terms of $SL(2,\mathbb{R})$ charges.

At the other side of the chain, it is not clear how to define non-extremal 7-branes. Note also that the double Wick rotation of the $\vec{q}^2 > 0$ non-extremal Dp-branes leads to Sp-branes. Recently this exercise has been repeated, via a *single* Wick rotation, to the non-extremal $\vec{q}^2 > 0$ D-instanton leading to a "S-1-brane" solution [29].

We now wish to discuss another aspect of the instanton solutions. It is well-known that there is an intricate relation between instantons and solitons in one dimension higher. The uplifting of the gravity-dilaton-axion system leads to the following Lagrangian in D+1 dimensions containing a metric, a dilaton and a vector:

$$\boxed{\mathcal{L}_{\text{D+1}} \sim \hat{R} + (\partial\hat{\phi})^2 + e^{a\hat{\phi}}\hat{F}^2} \tag{45}$$

The constant a defines the dilaton coupling in D+1 dimensions. The reduction of this Lagrangian over time leads to the gravity-dilaton-axion system (1) with $bc \geq 2$. The case $bc = 2$ corresponds to zero dilaton coupling in D+1 dimensions, i.e. $a = 0$.

It is instructive to consider the instanton-soliton correspondence for the case $bc = 2$. It turns out that for that case the uplift of the instantons are given by the (non-extremal) Reissner-Nordström black hole with mass M and charge Q. We find the following relations between the different cases:

$$\begin{array}{rll} extremal\ BH: & M^2 = Q^2 & \Leftrightarrow\ \vec{\mathsf{q}}^2 = 0\,, \\ non-extremal\ BH: & M^2 > Q^2 & \Leftrightarrow\ \vec{\mathsf{q}}^2 > 0\,, \\ singular\ BH: & M^2 < Q^2 & \Leftrightarrow\ \vec{\mathsf{q}}^2 < 0\,. \end{array} \tag{46}$$

The relation between M, Q and the $SL(2,\mathbb{R})$ charges is as follows:

$$M \sim \sqrt{\mathsf{q}_-^2 + \vec{\mathsf{q}}^2}\,, \qquad\qquad Q \sim \mathsf{q}_-\,. \tag{47}$$

Inverting these relations we find that

$$\boxed{\vec{\mathsf{q}}^2 \sim M^2 - Q^2} \tag{48}$$

This relation shows that $\vec{\mathsf{q}}^2$ acts as the non-extremality parameter.

The above uplifting can easily be extended to the uplift of the instantons to p-branes with $p > 0$. The condition fot uplifting to a p-brane, i.e. the analogue of $bc = 2$, becomes

$$(bc)^2 = \frac{4(p+1)(D-1)}{D+p-1}\,. \tag{49}$$

We are thus led to the Instanton Scan given in table 7.

The value of the action, evaluated on the instanton solution, is a key ingredient in the semiclassical approximation of the euclidean path integral. Using the relations between instantons and black holes we have obtained an elegant answer for the $\vec{\mathsf{q}}^2 \geq 0$ instanton action: the action is just the mass of the black hole! More precisely, we have found that (for more details, see [7])

TABLE 7. The Instanton Scan

bc	Dimension	Regular Solutions
<2	D	wormholes with $\vec{q}^2 < 0$
=2	D+1	RN black hole with $\vec{q}^2 \geq 0$
>2	D+1	dilatonic BH with $\vec{q}^2 = 0$
= (49)	D+p+1	non-dilatonic p-branes with $\vec{q}^2 \geq 0$
> (49)	D+p+1	dilatonic p-branes with $\vec{q}^2 = 0$

$$\boxed{S_{inst} = \frac{4}{b^2}(D-2)Vol(S^{D-1})bc\sqrt{\frac{q_-^2}{g_s^b} + \vec{q}^2}} \tag{50}$$

There are two interesting limits to consider. First, the extremal limit is given by $\vec{q}^2 = 0$ in which case the action reaches its lowest value and reduces to that of the D-instanton:

$$S_{inst} \sim \frac{1}{g_s^{b/2}}|q_-| = \frac{1}{g_s^{b/2}}Q\,. \tag{51}$$

The D-instanton of ten-dimensional IIB string theory corresponds to taking $b = 2$. Other cases in D=4 dimensions include the membrane instanton (b=1) and the NS-fivebrane instanton (b=2) [9, 10]. The other limit is the Schwarzschild limit $q_- \rightarrow 0$. In that case the action is given by

$$S_{inst} \sim |q|\,. \tag{52}$$

A natural question to ask is whether the non-extremal instantons we have been constructing give rise to extra corrections to the string effective action like the D-instanton case does. It is well-known that the D-instantons, together with contributions from tree-level graviton scattering and one-looop contributions, give rise to terms of the form [30, 31]

$$f(\tau,\bar{\tau})R^4\,. \tag{53}$$

Based on a field theory analysis one can argue that the non-extremal instantons constructed in this work give rise to terms of the form

$$f(\tau,\bar{\tau})R^8 \tag{54}$$

for some $SL(2,\mathbb{Z})$ modular forms $f(\tau,\bar{\tau})$. This remains to be investigated.

Finally, we would like to make some comments about the dual picture of the non-extremal instantons in the AdS/CFT correspondence. It is well-known that the D-instanton with a flat metric can be generalized to an instanton in a $AdS_5 \times S^5$ background. One can view this configuration as the near-horizon limit of a D3-D(-1) bound state configuration. Such $AdS_5 \times S^5$ instantons exactly correspond to the standard self-dual Yang-Mills instanton in N=4, D=4 supersymmetric Yang-Mills theory. To obtain such instanton solutions one should consider the following deformation of the gravity-dilaton-axion system [3]

$$\boxed{\mathcal{L}_{\mathrm{M}} \sim R + (\partial\phi)^2 + e^{b\phi}(\partial\chi)^2 + \Lambda} \tag{55}$$

The extremal instantons of this system, i.e. for $\vec{q}^2 = 0$, have been constructed sometime ago and are given by [32]

$$\begin{aligned} ds^2 &= d\rho^2 + l^2 sinh^2(\rho/l)d\Omega_4^2\,, & (56)\\ e^{b\phi/2} &= H(\rho)\,, & (57)\\ \chi(r) &= 2/b\big(H^{-1}(\rho) - q_3/q_-\big)\,, & (58) \end{aligned}$$

with $l = \sqrt{12/|\Lambda|}$.

We have now instantons with $\vec{q}^2 \neq 0, \Lambda = 0$ and instantons with $\vec{q}^2 = 0, \Lambda \neq 0$. The natural question to ask is: are there instantons with $\Lambda \neq 0$ *and* $\vec{q}^2 \neq 0$? Indeed they do exist and are given by [8, 23, 33]

$$\begin{aligned} ds^2 &= \big(1 + r^2/l^2 + \vec{q}^2/r^{2(D-2)}\big)^{-1} + r^2 d\Omega_4^2\,, & (59)\\ e^{b\phi(r)} &= \big(q_-/q\,sinh(H(r)+C)\big)^2\,, & (60)\\ \chi(r) &= 2/bq_-\big(q coth(H(r)+C) - q_3\big)\,. & (61) \end{aligned}$$

It would be interesting to see whether such bulk solutions correspond to non-selfdual instantons of N=4, D=4 supersymmetric Yang-Mills theory [34].

CONCLUSIONS

In this work we discussed the properties of brane solutions to the gravity-dilaton-axion system in string theory. A central theme was played by the underlying $SL(2,\mathbb{R})$ duality symmetries underlying this system. In particular we exploited the fact that the group $SL(2,\mathbb{R})$ has three conjugacy classes. Our results and conclusions can be summarized as follows.

• $7-$*branes*

Each conjugacy class is represented by a 1/2 BPS 7-brane solution. Using the monodromy relations one can view the $\det Q > 0$ and $\det Q < 0$ 7-branes as bound states

---

[3] We assume that a spherical reduction has been performed. The flux of the selfdual 5-form curvature leads to the cosmological constant Λ in (55).

of 7-brane solutions belonging to the $\det Q = 0$ conjugacy class. Our work shows that 7-branes fit nicely with the other branes in a general DW/QFT correspondence as far as the supergravity description goes. It led us to make a conjecture for a sensible definition of the near-horizon limit of a D7-brane. Whether all these nice features survive in a more rigorous string theory approach to the AdS/CFT correspondence remains to be seen. It would be interesting to see whether there is some better understanding of our results from a F-theory perspective.

• Instantons

Only the $\det Q = 0$ conjugacy class is represented by a 1/2 BPS D-instanton. The instantons in the $\det Q > 0$ and $\det Q < 0$ conjugacy classes are represented by non-supersymmetric non-extremal instantons. Whether or not these non-extremal instantons play an important role in string theory remains to be seen. Apart from the fact whether or not they give rise to corrections in the string effective action, it is also of interest to investigate what their picture is in the dual gauge theory. For the extremal D-instanton there is a well-established relation with the selfdual YM instanton. For the non-extremal instantons one is tempted to consider the idea of non-selfdual instantons. These and other issues are presently under investigation [34].

• S − branes

A third class of solutions that we did not consider in this work are the so-called S-branes. They are related, via a double Wick rotation to the non-extremal Dp-branes with $\vec{q}^2 > 0$ and $0 \le p \le 6$. Recently, it has been shown that the single Wick rotation of the $\vec{q}^2 > 0$ D-instanton leads to a S-1-brane [29]. Another issue is whether S7-branes exist for each of the three conjugacy classes. Their existence is related to the fact whether non-extremal 7-branes for $\vec{q}^2 > 0$ can be defined. We expect that each 1/2 BPS 7-brane solution indeed has such a non-extremal deformation. We hope to come back to this point in [22].

ACKNOWLEDGMENTS

We would like to thank the organizers of the XIXth Max Born Symposium for their hospitality and all the participants for providing a stimulating atmosphere. S.V. is supported by the European Commission FP6 program MRTN-CT-2004-005104 while E.B. and A.C. are supported by the European Commission FP6 program MRTN-CT-2004-005104 in which E.B. and A.C. are associated to Utrecht University. The work of U.G. is funded by the The Swedish Research Council and in addition the research of both U.G. and D.R. is funded by the PPARC grant PPA/G/O/2002/00475.

REFERENCES

1. E. Bergshoeff, M. de Roo, M.B. Green, G. Papadopoulos and P.K. Townsend, *Nucl. Phys.*, **B470**, 113–135 (1996).
2. G.W. Gibbons, M.B. Green and M.J. Perry, *Phys. Lett.*, **B370**, 37–44 (1996).
3. E. Bergshoeff, U. Gran and D. Roest, *Class. Quant. Grav.*, **19**, 4207–4226 (2002).
4. N. Alonso Alberca, E. Bergshoeff, U. Gran, R. Linares, T. Ortí n and D. Roest, *JHEP*, **06**, 038 (2003).
5. E. Bergshoeff et al., *Class. Quant. Grav.*, **20**, 3997–4014 (2003).
6. E. Bergshoeff, M. Nielsen and D. Roest, *JHEP*, **07**, 006 (2004).
7. E. Bergshoeff, A. Collinucci, U. Gran, D. Roest and S. Vandoren, *JHEP*, **10**, 031 (2004).
8. R.C. Myers, *Phys. Rev.*, **D38**, 1327 (1988).
9. U. Theis and S. Vandoren, *JHEP*, **09**, 059 (2002).
10. M. Davidse, M. de Vroome, U. Theis and S. Vandoren, *Fortsch. Phys.*, **52**, 696–701 (2004).
11. H.J. Boonstra, K. Skenderis and P.K. Townsend, *JHEP*, **01**, 003 (1999).
12. N. Itzhaki, J.M. Maldacena, J. Sonnenschein and S. Yankielowicz, *Phys. Rev.*, **D58**, 046004 (1998).
13. P. Meessen and T. Ortin, *Nucl. Phys.*, **B541**, 195–245 (1999).
14. O. DeWolfe, T. Hauer, A. Iqbal and B. Zwiebach, *Adv. Theor. Math. Phys.*, **3**, 1785–1833 (1999).
15. O. DeWolfe, T. Hauer, A. Iqbal and B. Zwiebach, *Adv. Theor. Math. Phys.*, **3**, 1835–1891 (1999).
16. J.M. Maldacena, *Adv. Theor. Math. Phys.*, **2**, 231–252 (1998).
17. P. Kraus, F. Larsen and S.P. Trivedi, *JHEP*, **03**, 003 (1999).
18. D.Z. Freedman, S.S. Gubser, K. Pilch and N.P. Warner, *JHEP*, **07**, 038 (2000).
19. M. Cvetic, S.S. Gubser, H. Lu and C.N. Pope, *Phys. Rev.*, **D62**, 086003 (2000).
20. B. de Wit, H. Nicolai and H. Samtleben, `hep-th/0403014` (2004).
21. M.B. Einhorn and L.A. Pando Zayas, *Nucl. Phys.*, **B582**, 216–230 (2000).
22. E. Bergshoeff, A. Collinucci, D. Roest and P.K. Townsend, *in preparation*.
23. S.B. Giddings and A. Strominger, *Phys. Lett.*, **B230**, 46 (1989).
24. E. Bergshoeff, A. Collinucci, U. Gran, D. Roest and S. Vandoren, `hep-th/0412183` (2004).
25. P.K. Townsend and M.N.R. Wohlfarth, *Class. Quant. Grav.*, **21**, 5375 (2004).
26. H. Ooguri and K. Skenderis, *JHEP*, **11**, 013 (1998).
27. G.T. Horowitz and A. Strominger, *Nucl. Phys.*, **B360**, 197–209 (1991).
28. H. Lu, C.N. Pope and K.W. Xu, *Mod. Phys. Lett.*, **A11**, 1785–1796 (1996).
29. J.X. Lu and S. Roy, `hep-th/0408242` (2004).
30. M.B. Green and P. Vanhove, *Phys. Lett.*, **B408**, 122–134 (1997).
31. M.B. Green and M. Gutperle, *Nucl. Phys.*, **B498**, 195–227 (1997).
32. C.-S. Chu, P.-M. Ho and Y.-Y. Wu, *Nucl. Phys.*, **B541**, 179–194 (1999).
33. M. Gutperle and W. Sabra, *Nucl. Phys.*, **B647**, 344–356 (2002).
34. E. Bergshoeff, A. Collinucci, A. Ploegh, T. Van Riet and S. Vandoren, *in preparation*.

Superbranes, $D = 11$ CJS Supergravity and Enlarged Superspace Coordinates/Fields Correspondence

J.A. de Azcárraga

Department of Theoretical Physics and IFIC (CSIC-UVEG), Facultad de Física, 46100-Burjassot (Valencia), Spain

Abstract. We discuss the rôle of enlarged superspaces in two seemingly different contexts, the structure of the p-brane actions and that of the Cremmer-Julia-Scherk eleven-dimensional supergravity. Both provide examples of a common principle: the existence of an *enlarged superspaces coordinates/fields correspondence* by which all the (worldvolume or spacetime) fields of the theory are associated to coordinates of enlarged superspaces. In the context of p-branes, enlarged superspaces may be used to construct manifestly supersymmetry-invariant Wess-Zumino terms and as a way of expressing the Born-Infeld worldvolume fields of D-branes and the worldvolume M5-brane two-form in terms of fields associated to the coordinates of these enlarged superspaces. This is tantamount to saying that the Born-Infeld fields have a superspace origin, as do the other worldvolume fields, and that they have a composite structure. In D=11 supergravity theory enlarged superspaces arise when its underlying gauge structure is investigated and, as a result, the composite nature of the A_3 field is revealed: there is a full one-parametric family of enlarged superspace groups that solve the problem of expressing A_3 in terms of spacetime fields associated to their coordinates. The corresponding enlarged supersymmetry algebras turn out to be deformations of an *expansion* of the $osp(1|32)$ algebra. The unifying mathematical structure underlying all these facts is the cohomology of the supersymmetry algebras involved.

1. INTRODUCTION

M-theory (see [1, 2] and [3] for a chronological history) is not based at present on a definite Lagrangian or on an S-matrix description; rather, its conjectured existence relies on the properties of its six perturbative and low energy limits (string models and supergravities) and by dualities [4] among them. Such dualities, including those relating apparently different models, are believed to be symmetries of M-theory. The full set of M-theory symmetries -as the full M-theory itself- is not known[1], but it should include

[1] Several groups may play a rôle, as the rank 11 Kac-Moody E_{11} group [5] as a basis for a non-linear realization approach to $D = 11$ supergravity as well as other Kac-Moody symmetries, or the $OSp(1|64)$ group [6, 7] and its subgroup $GL(32)$ [8, 9]. This last one is the automorphism group of the M-algebra $\{Q_\alpha, Q_\beta\} = P_{\alpha\beta}$; it is also a manifest symmetry of the actions [10, 11] for BPS preons [12] (see [13] for a review), the hypothetical constituents of M-theory. Clearly, in $D = 11$ supergravity one may see only a fraction of the M-theory symmetries. As it was noticed recently [14, 15] (see also [11]), a suggestive analysis of supersymmetric $D = 11$ supergravity solutions can be carried out in terms of generalized connections with holonomy group $SL(32)$ (see [16] for an early reference on generalized holonomy). The case for a $OSp(1|32) \otimes OSp(1|32)$ gauge symmetry in a Chern-Simons context was presented and

CP767, *Fundamental Interactions and Twistor-Like Methods,* edited by J. Lukierski and D. Sorokin

these dualities as well as the symmetries of the different superstring and supergravity limits. For this reason, the study of the symmetries of p-branes as well as the underlying gauge symmetry of $D = 11$ supergravity may help to understand the symmetry structure of M-theory itself.

Superalgebras going beyond the standard supersymmetry algebra were considered very early (see [20, 21] and references in [22]) and, later, in the context of brane theory. Some of these *enlarged supersymmetry algebras* generalize Green's algebra [23]. They were introduced in [24] to make *Lie* algebras out of the free differential algebras that had been introduced in [25] to recover cohomologically the classification [26] of the scalar p-branes. The authors of [24] also showed that these algebras could be used to obtain Wess-Zumino (WZ) terms for the rigid p-brane actions strictly invariant under supersymmetry. The relation between semi- or quasi-invariance (*i.e.*, invariance but for a total derivative) of lagrangians, cohomology and group extension theory, is a problem which has a fifty years long history, but we will not discuss it here (see [27] and references therein). In the case of p-branes, the additional variables of the new supersymetry groups [22] (rigid enlarged superspaces) appear in these manifestly invariant WZ terms in a trivial way, but this is not the case for all types of branes. It was shown in [22] that the enlarged superspaces, could also be used to obtain Born-Infeld (BI) fields from one-forms defined on them (for BI fields in the IIB case see [28]). In the case of the D-branes BI fields or with the worldvolume two-form of the M5 brane these fields allow for the existence of actions where all worldvolume fields entering in the theory are associated to the enlarged superspace coordinates (there are no fields 'external' to the superspace coordinates, *i.e. directly* defined on the worldvolume). This points out to the existence of an *extended superspace coordinates/fields correspondence* for branes [22], where in this case 'fields' refers to worldvolume fields. We shall devote the first two sections to review these ideas.

It turns out that, in an analogous fashion, a similar correspondence may also be established for $D = 11$ Cremmer-Julia-Scherk (CJS) [29] *supergravity*, in which case the fields are spacetime fields. This correspondence between enlarged superspace coordinates and spacetime fields for CJS supergravity is related to the problem of its 'hidden' or underlying gauge symmetry. This problem was raised already in the original CJS paper [29]. It was considered by D'Auria and Fré [21] as a search for a composite structure of the *three-form* A_3 field that enters in the $D = 11$ supergravity multiplet (see also [30, 31] for other discussions of the geometry of $D = 11$ supergravity and [32] for an overview on local supersymmetry[2]). While the graviton, the gravitino and the spin connection are one-form fields, $e^a = dx^\mu e^a_\mu(x)$, $\psi^\alpha = dx^\mu \psi^\alpha_\mu(x)$ and $\omega^{ab} = dx^\mu \omega^{ab}_\mu(x)$, and can be considered as gauge fields for the superPoincaré group [36], the fully antisymmetric $A_{\mu_1\mu_2\mu_3}(x)$ (transverse) abelian gauge field is not associated with a symmetry generator and it rather corresponds to a *three*-form A_3 on spacetime. This prevents the association of the fields of the standard $D = 11$ supergravity multiplet with the gauge fields of a Lie superalgebra, since these are associated to *one*-forms.

---

discussed in [17, 18, 19].

[2] For recent discussions in a different perspective see [33, 34, 35].

Two enlarged supersymmetry algebras with 528 bosonic and 64 fermionic generators

$$P_a \,, Q_\alpha \,;\, Z_{a_1 a_2} \,, Z_{a_1 \ldots a_5} \,;\, Q'_\alpha \,, \tag{1}$$

including the 528+32 M-algebra [2] ones plus a central fermionic generator Q'_α, were found in [21] to allow for a decomposition of A_3. The corresponding one-form fields

$$e^a \,, \psi^\alpha \,;\, B^{a_1 a_2} \,, B^{a_1 \ldots a_5} \,, \eta^\alpha \,, \tag{2}$$

were then considered as gauge fields for these larger supergroups. In this scheme, all the CJS supergravity fields can then be treated as gauge fields, with A_3 expressed in terms of them.

As we shall see, the problem studied in [21] is mathematically analogous to that of obtaining strictly invariant WZ terms for p-branes from the originally quasi-invariant (invariant up to a total derivative) ones. Both reduce to finding a Lie algebra allowing us to write a closed invariant form (on the original supergroup manifold) as the differential of an invariant one (now on the manifold of the associated enlarged supergroup). Expressed in another way, these problems correspond to finding a trivialization of certain non-trivial Chevalley-Eilenberg (CE) [37] cocycles for the cohomology of the standard supersymmetry algebra of the theory by means of enlarging it. It turns out that the underlying gauge supergroup structure of $D = 11$ CJS supergravity can be described by any representative of a *one–parametric* family of supergroups denoted $\tilde{\Sigma}(s)$ ($s \neq 0$), or by their associated superalgebras $\tilde{\mathfrak{E}}(s)$, the two D'Auria-Fré ones being two particular elements of that family (specifically, $\tilde{\mathfrak{E}}(3/2)$ and $\tilde{\mathfrak{E}}(-1)$). There have been attempts to relate these solutions to some known algebra (see [38]). We will see that the algebras $\tilde{\mathfrak{E}}(s)$ are nontrivial ($s \neq 0$) deformations of the special element $\tilde{\mathfrak{E}}(0)$. The $\tilde{\mathfrak{E}}(s), s \neq 0$, automorphism group is $SO(1,10)$. Thus, the relevant supergroup that replaces the standard superPoincaré group $\Sigma \rtimes SO(1,10)$ becomes the semidirect product $\tilde{\Sigma}(s) \rtimes SO(1,10)$, a deformation of $\tilde{\Sigma}(0) \rtimes SO(1,10)$. As for the superalgebra $\tilde{\mathfrak{E}}(0) \oplus\!\!\!\!\supset so(1,10)$ itself, it is related to $osp(1|32)$ through an *expansion*[3]. Specifically,

$$\tilde{\Sigma}(0) \rtimes SO(1,10) \approx OSp(1|32)(2,3,2) \,, \tag{3}$$

where the numbers in $OSp(1|32)(2,3,2)$ characterize the expansion (see later). For $s = 0$ the $SO(1,10)$ automorphism group is enhanced to $Sp(32)$, and one finds that $\tilde{\Sigma}(0) \rtimes Sp(32) \approx OSp(1|32)(2,3)$.

The supergroup manifolds $\tilde{\Sigma}(s)$ determine *rigid, enlarged* superspaces. The fact that all the spacetime fields in (2) may be associated to the various coordinates of the $\tilde{\Sigma}(s)$ supergroups again suggests that there is an *extended superspace coordinates/fields correspondence* principle[4].

---

[3] The *expansion method* allows us to obtain new algebras from a given one, in general of higher dimension than the original one. Under a different name, expansions were considered in [39], and the method was studied in general in [40].

[4] The idea of a 'fundamental symmetry between coordinates and fields' is explicitly stated in Berezin [41] and is implicit in earlier work of D. V. Volkov [42]; the field space democracy is also discussed

2. WESS-ZUMINO TERMS FOR SUPER-P-BRANES AND ENLARGED SUPERSYMMETRY ALGEBRAS

In this section, and in the next one, we describe the rôle of enlarged superspaces in brane theory. We will start from the scalar p-branes [26] case. The action for a p brane in rigid superspace is given by the sum of two terms,

$$I = I_0 + I_{WZ} \, , \tag{4}$$

where I_0 is the kinetic part and $I_{WZ} = \int_{\mathcal{W}} \phi^*(b)$ is the WZ term, which is given by the integral over the worldvolume $\mathcal{W}$, parametrized by $\xi^i = (\tau, \sigma^1, \ldots, \sigma^p)$ $[i = 0, \ldots, p]$ of the pull-back $\phi^*(b)$ to $\mathcal{W}$ of a $(p+1)$-form b defined on the rigid superspace $\Sigma^{(D|n)}$ (Σ for short) of the theory (the manifold of the corresponding supersymmetry group). This form b is the potential form of a $(p+2)$-form h which happens to be exact,

$$h = db \, , \tag{5}$$

and that is invariant under the transformations of the superPoincaré group $\Sigma \rtimes SO(1, D-1)$. The study of the different possible $(p+2)$-forms h in superspaces corresponding to the minimal supersymmetries in D dimensions determines the (D, p) values for which the WZ term exists. With the appropriate relative factor, I_{WZ} in (4) leads to super-p-brane κ-invariant actions. These (D, p) values determine the 'old branescan' [26].

It turns out [25] that the h's are non-trivial Chevalley-Eilenberg (CE) $(p+2)$-cocycles for the cohomology of the standard $\mathfrak{E}^{(D|n)}$ supersymmetry algebra. This means that h is a closed (obviously, $dh = 0$) and supersymmetry invariant $(p+2)$-form constructed from the Maurer-Cartan (MC) one-forms on the graded translations (supersymmetry) group Σ (namely Π^μ and Π^α, where $\mu = 0, \ldots D-1$, and the range $1, \ldots n$ of α depends on the minimal spinor considered). This CE cocycle condition depends on the known gamma matrix identities that are true only for the (D, p) values of the 'old branescan'. Furthermore, the non-triviality of these cocycles means that the potential $(p+1)$-form b in $h = db$ is not supersymmetry invariant *i.e.*, that cannot be constructed from the invariant MC forms Π^μ and Π^α on Σ. An important consequence of this fact is that the WZ lagrangian is not manifestly invariant under supersymmetry, but only quasi-invariant (hence its 'WZ' name) and that, as a result, the algebra of charge densities produces topological extensions of the original supersymmetry algebra [44].

---

in [43]. However, we are not referring here to a democracy between the fields and *their* arguments (as one might introduce by *e.g.*, considering on an equal footing the coordinates of the *total* space of a fibre bundle the cross sections of which may be used to define fields on the *base* manifold), but rather to a correspondence between the (enlarged) superspace coordinates and the fields they originate, be them worldvolume or spacetime ones. This is why is more precise to speak of a *correspondence* between (enlarged superspace) coordinates and fields rather than of 'democracy' -the term used in [22]- since its original use referred to a democracy between the fields and *its* arguments. We have conjectured [22] the existence of a correspondence between the *coordinates* of a suitable superspace and the *fields* in theory constructed on it. These appear as the pullbacks of forms, originally defined on the target enlarged superspace, to the worldvolume or spacetime manifolds. Also, the basis for such a correspondence is group theoretical: the enlarged rigid superspaces are all supergroup manifolds.

Certain enlarged rigid superspaces associated to enlarged supersymmetry groups $\tilde{\Sigma}$, with additional bosonic and fermionic variables, can be used to obtain p-brane actions that are equivalent to the standard ones but with WZ terms that are strictly invariant under supersymmetry. On the manifolds of these groups, the same $(p+2)$-forms h of the old branescan are still CE cocycles, but they are now trivial ones: $h = d\tilde{b}$, and the new potential $(p+1)$-forms $\tilde{b}$ are $\tilde{\Sigma}$-invariant. The process of obtaining these enlarged algebras or 'brane algebras' may be thus called of 'trivialization of the CE cocycles' h on $\mathfrak{E}^{(D|n)}$.

Let us look in more detail how to achieve this trivialization (the following is not, as we shall see, the only possibility). One starts with the MC equations of the standard supersymmetry algebra in D dimensions $\mathfrak{E}^{(D|n)}$ (we consider, for simplicity, the cases that allow for real spinors; wedge products are understood in this and in the next section)

$$d\Pi^{\alpha} = 0\ , \quad d\Pi^{\mu} = a_s \Pi^{\alpha}(C\Gamma^{\mu})_{\alpha\beta}\Pi^{\beta}\ , \tag{6}$$

where [5] the constant $a_s = 1/2$ for $\tilde{\mathfrak{E}}$. The $(p+2)$-form h for a p-brane may be shown to be, up to a proportionality constant which is not important in the discussion below,

$$h = \Pi^{\alpha}(C\Gamma_{\mu_1\ldots\mu_p})_{\alpha\beta}\Pi^{\beta}\Pi^{\mu_1}\ldots\Pi^{\mu_p} \tag{7}$$

which is closed for the dimensions D for which the identity

$$(C\Gamma^{\mu_1\ldots\mu_p})_{(\alpha\beta}(C\Gamma_{\mu_1})_{\gamma\delta)} = 0 \tag{8}$$

is satisfied. The bilinear in (7) suggests that in order to find an invariant potential form for h one should first extend (6) adding the form $\Pi^{\mu_1\cdots\mu_p}$ and the MC equation

$$d\Pi_{\mu_1\ldots\mu_p} = a_0 \Pi^{\alpha}(C\Gamma_{\mu_1\ldots\mu_p})_{\alpha\beta}\Pi^{\beta}\ , \tag{9}$$

so that h is the first term in the differential of

$$\Pi_{\mu_1\ldots\mu_p}\Pi^{\mu_1}\ldots\Pi^{\mu_p}\quad . \tag{10}$$

Before going to the next step, let us note that this is a sensible thing to do for two reasons. The first is that eq. (9) and the second of eqs. (6) can be put on the same footing since they are central (if one ignores the Lorentz part) extensions of the abelian odd translation algebra defined by the simple MC equations $d\Pi^{\alpha} = 0$. So if the graded supertranslations (supersymmetry) algebra is itself a central extension, it seems mathematically natural to consider other possible extensions as well. In fact, one should consider the most

[5] *A comment on conventions.* The constant a_s is real since we assume real gamma matrices (which in $D = 11$, for instance, requires mostly plus metric) and the convention used here for the complex conjugation is $(\theta_1\theta_2)^* = (\theta_1^*\theta_2^*)$, θ_1 and θ_2 being Grassmann odd. If we used the conjugation that reverses the order, as it will be the case in Secs. 4-7, the a_s in eq. (6) would be purely imaginary. Other differences in conventions between Secs. 2,3 and Secs. 4-7 are that in Secs.2,3 we write explicitly the charge conjugation matrix C; also, in Secs. 2,3 the $\wedge$ product for forms is implicit. We have kept these two sets of conventions in order to make direct contact with [22] and with [45].

general extension where the 'central' generators appear for each symmetric $(C\Gamma^{\mu_1 \ldots \mu_q})$ matrix (we shall keep, here, however, only the generators corresponding to Π^μ and one of the $\Pi^{\mu_1 \ldots \mu_p}$ for simplicity; they will be sufficient to discuss the 'scalar' brane actions). The second reason is that including these new bosonic generators is necessary to understand, from the algebraic point of view, the existence of BPS states that break some supersymmetries but not all, as known from supergravity theories.

Equation (10) does not solve yet the problem of finding an invariant $\tilde{b}$ such that $d\tilde{b} = h$ because the exterior differential also acts on the p factors Π^μ . This means that new generators and MC equations have to be added to (6) and (10). It may be shown (see [24] and [22]) that such a $\tilde{b}$ can be found if the algebra is extended in several steps, each step involving a central (if we ignore the Lorentz part) extension of the algebra resulting from the previous one. The first invariant form by which one extends is fermionic, and has the structure $\Pi_{\mu_1 \ldots \mu_{p-1} \alpha_1}$ (so that, for $p = 1$, one obtains the Green algebra [23]). The second is an extension of the algebra whose MC equations are generated by Π^α, Π^μ, $\Pi_{\mu_1 \ldots \mu_p}$ and $\Pi_{\mu_1 \ldots \mu_{p-1} \alpha_1}$, and the new invariant forms have the structure $\Pi_{\mu_1 \ldots \mu_{p-2} \alpha_1 \alpha_2}$. This process of extensions ends when the last invariant form $\Pi_{\alpha_1 \ldots \alpha_p}$ is added. At each step in the above procedure the extension made is central, but it makes non-central the former central generator of the previous step. Thus, the resulting algebra is not a central extension of the supersymmetry one but for $p = 1$ where the only step produces the Green algebra [23]. The existence of these enlarged supersymmetry or 'brane algebras' depends on the values of D and p for which the identity (8) holds; this is not surprising since these algebras allow for the existence of $\tilde{b}$ such that $d\tilde{b} = h$ and $dh = 0$ is true only when (8) is satisfied.

We shall not give here the explicit expressions for the resulting 'brane algebras' in general; these can be found in [24] and in [22]; the associated enlarged superspace groups law is also given in [22] for the most interesting ones. We shall only write explicitly two of these enlarged supersymmetry algebras, because they will be relevant in the next sections. Let us begin by the superalgebra for $p = 2$, $D = 11$ that trivializes the CE cocycle that defines the WZ term of the $D = 11$ membrane [46]. It is given by the MC equations

$$\begin{aligned}
d\Pi^\alpha &= 0\,, \quad d\Pi^\mu = \frac{1}{2}(C\Gamma^\mu)_{\alpha\beta}\Pi^\alpha\Pi^\beta\,, \\
d\Pi^{\mu\nu} &= \frac{1}{2}(C\Gamma^{\mu\nu})_{\alpha\beta}\Pi^\alpha\Pi^\beta\,, \\
d\Pi_{\mu\alpha} &= (C\Gamma_{\nu\mu})_{\alpha\beta}\Pi^\nu\Pi^\beta + (C\Gamma^\nu)_{\alpha\beta}\Pi_{\nu\mu}\Pi^\beta\,, \\
d\Pi_{\alpha\beta} &= -\frac{1}{2}(C\Gamma_{\mu\nu})_{\alpha\beta}\Pi^\mu\Pi^\nu - \frac{1}{2}(C\Gamma^\mu)_{\alpha\beta}\Pi_{\mu\nu}\Pi^\nu \\
&\quad + \frac{1}{4}(C\Gamma^\mu)_{\alpha\beta}\Pi_{\mu\delta}\Pi^\delta + (C\Gamma^\mu)_{\delta\alpha}\Pi_{\mu\beta}\Pi^\delta + (C\Gamma^\mu)_{\delta\beta}\Pi_{\mu\alpha}\Pi^\delta\,. \qquad (11)
\end{aligned}$$

They can be obtained from the general expressions of [22], particularized to the case $p = 2$, by suitably fixing the undetermined constants. The above equations allow for the existence of an invariant $\tilde{b}$ such that $d\tilde{b} = h = \Pi^\alpha(C\Gamma^{\mu\nu})_{\alpha\beta}\Pi^\beta\Pi_\mu\Pi_\nu$. The expression for $\tilde{b}$ is [24, 22]

$$\tilde{b} = \frac{2}{3}\Pi_{\mu\nu}\Pi^{\mu}\Pi^{\nu} - \frac{3}{5}\Pi_{\mu\alpha}\Pi^{\mu}\Pi^{\alpha} - \frac{2}{15}\Pi_{\alpha\beta}\Pi^{\alpha}\Pi^{\beta} \,. \tag{12}$$

The second algebra that we shall need is the one that trivializes the CE cocycle associated to the WZ term of the D=10, IIA superstring. It can be extracted from the dimensional reduction to $D = 10$ of the algebra (11) (see [22], eqs. (88) and (90)), and is given by

$$\begin{aligned} d\Pi^{\alpha} &= 0\,, \quad d\Pi^{\mu} = \frac{1}{2}(C\Gamma^{\mu})_{\alpha\beta}\Pi^{\alpha}\Pi^{\beta} \\ d\Pi^{(z)}_{\mu} &= \frac{1}{2}(C\Gamma^{\mu}\Gamma_{11})_{\alpha\beta}\Pi^{\alpha}\Pi^{\beta} \\ d\Pi^{(z)}_{\alpha} &= (C\Gamma_{\nu}\Gamma_{11})_{\alpha\beta}\Pi^{\nu}\Pi^{\beta} + (C\Gamma^{\nu})_{\alpha\beta}\Pi^{(z)}_{\nu}\Pi^{\beta}\,, \end{aligned} \tag{13}$$

where the superscript (z) refers to the fact that the forms $\Pi^{(z)}_{\mu}$ and $\Pi^{(z)}_{\alpha}$ come from the dimensional reduction of $\Pi_{\mu\nu}$ and $\Pi_{\mu\alpha}$ respectively when μ corresponds to the z coordinate in the splitting $x^0,\ldots,x^9,z$. This algebra is consistent due to the $D = 10$ gamma matrices identity

$$(C\Gamma^{\mu}\Gamma_{11})_{(\alpha\beta}(C\Gamma_{\mu})_{\gamma\delta)} = 0\,. \tag{14}$$

The corresponding $\tilde{b}$ is given by

$$\begin{aligned} \tilde{b} &= \frac{1}{2}\Pi^{\alpha}\Pi^{(z)}_{\alpha} - \Pi^{\mu}\Pi^{(z)}_{\mu}\,, \\ d\tilde{b} &= h = (C\Gamma_{\mu}\Gamma_{11})_{\alpha\beta}\Pi^{\mu}\Pi^{\alpha}\Pi^{\beta} \quad . \end{aligned} \tag{15}$$

We note finally that the coordinates of $\tilde{\Sigma}/\Sigma$ [$(\varphi^{\mu\nu}, \varphi_{\mu\alpha}, \varphi_{\alpha\beta})$ for eq. (11) and $(\varphi^{\mu}, \varphi_{\alpha})$ for eq. (13)] that, beyond the ordinary superspace $\Sigma^{(D|n)}$ ones $(x^{\mu}, \theta^{\alpha})$, complete the parametrization the enlarged superspaces $\tilde{\Sigma}$, lead to non-dynamical fields in the action. The WZ term is written in invariant form as $\phi^*(\tilde{b})$, where ϕ^* is the pullback that takes the form $\tilde{b}$ on the $\tilde{\Sigma}$ manifold to $\mathcal{W}$. Indeed, since $d\tilde{b} = h = db$, it follows that the new fields enter in the WZ part of the action $\int_{\mathcal{W}} \phi^*(\tilde{b})$ as total derivative[6]. However, they appear non trivially in the context of D-branes and in the M5-brane, as we discuss in the next section, where they also appear in the D-brane action kinetic part.

3. THE ENLARGED SUPERSPACE COORDINATES/FIELDS CORRESPONDENCE FOR SUPERBRANES

The action for D-branes [48, 49, 50] (we shall restrict ourselves to the type IIA D-branes as in [22], see [28] for the IIB case) in a rigid background for which all forms in the

[6] We shall not discuss the behaviour of the additional variables under κ-symmetry, for which we refer to [47].

R-R sector and the dilaton vanish is given, as in the case of p-branes (4), by the sum of a kinetic term I_0 and a WZ term I_{WZ} . The first one is

$$I_0 = \int d\xi^{p+1} \sqrt{-\det(g_{ij} + \mathcal{F}_{ij})} \quad . \tag{16}$$

In (16), g_{ij} is the induced metric on the worldvolume, $g_{ij}(\xi) = \Pi_i^\mu(\xi)\Pi_{\mu j}(\xi)$ ($\Pi^\mu = \Pi_i^\mu d\xi^i$), and $\mathcal{F}_{ij}$ are the worldvolume components of the form $\mathcal{F}(\xi) = dA_1(\xi) - \phi^*(B_2)$, where $\phi^*(B_2)$ is the pull-back to $\mathcal{W}$ of a two-form (B_2) defined on the $D = 10$, IIA superspace such that

$$dB_2 = -(C\Gamma_\mu\Gamma_{11})_{\alpha\beta}\Pi^\alpha\Pi^\beta\Pi^\mu \ , \tag{17}$$

and $A_1(\xi)$ is a one-form *directly* defined on the worldvolume, the BI field, that transforms in such a way that $\mathcal{F}$ is invariant under supersymmetry. The WZ term is quasi-invariant and is given by the integral of a $(p+1)$-form that depends polynomially on $\mathcal{F}$, the coefficients being forms on the $D = 10$ IIA superspace. The explicit expressions for the different D-brane WZ terms (actually for even $p = 2,4,6,8$) are not relevant for our purposes, but we note here that the search for the possible non-trivial CE cocycles that determine h also identifies the possible D-branes [22] recovering Polchinski's classification [51] (for recent work on D-branes see [52]). Similarly, the $D = 11$ M5-brane action [53] [7] contains a *two*-form $A_2(\xi)$ directly defined on the worlvolume that enters the action through the field strength $H_3(\xi) = dA_2(\xi) - \phi^*(\mathcal{A}_3)$, where $\phi^*(\mathcal{A}_3)$ is the pull back to the worldvolume of a $D = 11$ superspace three-form $\mathcal{A}_3$ such that

$$d\mathcal{A}_3 = -(C\Gamma_{\mu\nu})_{\alpha\beta}\Pi^\mu\Pi^\nu\Pi^\alpha\Pi^\beta \ ; \tag{18}$$

$A_2(\xi)$ transforms under supersymmetry in such a way that H is invariant [8]. Again we note in passing that a CE cohomological search for the possible $D = 11$ WZ terms in this case leads to the M5 brane as the only solution [22].

In contrast with p-branes, both the D-brane and M5-brane actions cannot be written in terms of forms associated to the coordinates of ordinary superspaces, due to the presence of the one- and two-forms $A(\xi)$, which are defined directly on the worldvolume. The arguments of the previous section, however, lead to the possibility of writing them solely in terms of forms defined on suitably *enlarged* superspaces as we describe now.

Let us first consider the IIA D-branes case (this includes the D2, D4, D6 and D8 cases). The two-form $\mathcal{F}$ is supersymmetry invariant and has the property that $\mathcal{F} = \phi^*((C\Gamma_\mu\Gamma_{11})_{\alpha\beta}\Pi^\alpha\Pi^\beta\Pi^\mu)$. But these conditions are also satisfied by $\tilde{b}$ in eq. (15). Moreover, as discussed in the previous section, the new superspace variables appear in $\tilde{b}$ inside a total differential. So one may consider the IIA enlarged superspace defined by the MC equations (13) and identify $\mathcal{F}$ with $\phi^*(\tilde{b})$. Since $\tilde{b}$ has a part that is a total

---

[7] The covariant equations for the D-branes and the M5-brane were found in [54] in the framework of the superembedding approach developed for the supermembrane and superstrings in [55].

[8] Clearly, eq. (18) shows that the standard superspace three-form $\mathcal{A}_3$ cannot be invariant under the transformations of standard supersymmetry. The $\mathcal{A}_3$ here is, up to a factor, the ω_3 in eq. (31) and corresponds to the A_3 form in the case of curved susperspace.

differential which contains the new superspace coordinates, one concludes that dA may be identified with this part. The result [22] is

$$A_1(\xi) = \phi^* \left(\varphi_\mu dx^\mu + \frac{1}{2}\varphi_\alpha d\theta^\alpha \right) . \tag{19}$$

In the M5-brane case, the relevant enlarged superspace is the one corresponding to the MC equations (11), and the expression for the two-form $A_2(\xi)$ that arises by identifying H with $\phi^*(\tilde{b})$ from eq. (12) turns out to be [22]

$$\begin{aligned} A_2(\xi) = \phi^* \Big(& \frac{2}{3}\varphi_{\mu\nu}dx^\mu dx^\nu - \frac{3}{5}\varphi_{\mu\alpha}dx^\mu d\theta^\alpha - \frac{2}{15}\varphi_{\alpha\beta}d\theta^\alpha d\theta^\beta \\ & + \frac{1}{30}\varphi_{\mu\nu}x^\mu (C\Gamma^\nu)_{\alpha\beta} d\theta^\alpha d\theta^\beta + \frac{11}{30}\varphi_{\mu\nu}dx^\mu (C\Gamma^\nu)_{\alpha\beta}\theta^\alpha d\theta^\beta \\ & - \frac{13}{180}\varphi_{\mu\nu}(C\Gamma^\mu)_{\alpha\beta}(C\Gamma^\nu)_{\delta\varepsilon}\theta^\alpha d\theta^\beta \theta^\delta d\theta^\varepsilon \\ & + \frac{1}{10}\varphi_{\mu\alpha}(C\Gamma^\mu)_{\delta\varepsilon}\theta^\delta d\theta^\varepsilon d\theta^\alpha + \frac{1}{20}\varphi_{\mu\alpha}(C\Gamma^\mu)_{\delta\varepsilon} d\theta^\delta d\theta^\varepsilon \theta^\alpha \Big) . \end{aligned} \tag{20}$$

So far we have argued that $\phi^*(\tilde{b})$ has the same supersymmetry properties as $\mathcal{F}(\xi)$ (for the IIA D-branes) or $H(\xi)$ (for the M5 brane). The next question to ask is whether it is legitimate to substitute the latter for the former in the actions for the D-branes and the M5-brane. Let us begin by the D-brane case. It suffices to show that the Euler-Lagrange (E-L) equations for the actions that are obtained by replacing $A_1(\xi)$ by the r.h.s. of (19), which we shall denote more explicitly by $A_1(x(\xi),\theta(\xi),\varphi(\xi)) \equiv A_1(x,\theta,\varphi)$, have the same dynamical content as those for the original one, where $I = I[x,\theta,A_1]$. Indeed, $I[x,\theta,A_1(x,\theta,\varphi)]$ has the same variation as $I[x,\theta,A_1]$, except for the fact that one has to vary also the fields inside $A_1(x,\theta,\varphi)$. From the variation of $I[x,\theta,A_1(x,\theta,\varphi)]$ with respect to the new variables, $\varphi = (\varphi_\mu, \varphi_\alpha)$, one arrives at $\Pi_i^\mu \left.\frac{\delta I[A_1]}{\delta A_{1i}(\xi)}\right|_{A_1 = A_1(x,\theta,\varphi)} = 0$, and this leads to $\left.\frac{\delta I[A_1]}{\delta A_{1i}(\xi)}\right|_{A_1 = A_1(x,\theta,\varphi)} = 0$ provided that the induced metric on the worldvolume is nondegenerate, as is always the case in tensionful brane theory. But this last equation is one of the equations for $I[x,\theta,A_1]$. If we now substitute it in the equations for $I[x,\theta,A_1(x,\theta,\varphi)]$, and use the fact that the variations through $A_1(x,\theta,\varphi)$ are proportional to $\left.\frac{\delta I[A_1]}{\delta A_i(\xi)}\right|_{A_1 = A_1(x,\theta,\varphi)}$ due to the chain rule, we recover the remaining equations of $I[x,\theta,A_1]$.

Moreover, since in $I[x,\theta,A_1(x,\theta,\varphi)]$ all the new variables appear inside A_1, the extra degrees of freedom corresponding to them have to be reduced by a gauge symmetry to those of the customary BI field $A_{1i}(\xi)$. Since $\frac{\delta I}{\delta \varphi_\alpha} = 0$ is itself a Noether identity (which follows by looking at the E-L equations as above), by the second Noether theorem there is a gauge symmetry allowing us to remove φ_α entirely. We also note that the D equations $\left.\frac{\delta I[A_1]}{\delta \varphi_\mu}\right|_{A_1 = A_1(x,\theta,\varphi)}$ produce only $(p+1)$ independent ones $\frac{\delta I[A]}{\delta A_{1i}(\xi)} = 0$ and, consequently, the remaining $[D-(p+1)]$ equations are Noether identities. Hence, of the D extra bosonic degrees of freedom variables introduced by φ_μ in $A_{1i}(x,\theta,\varphi)$, $[D-(p+1)]$ may be eliminated by a gauge transformation. To see this explicitly, let us set $\varphi_\alpha = 0$ and write $A_{1i}(x,\varphi) = \varphi_\mu \partial_i x^\mu$ in a local gauge such that $x^0 = \tau, x^1 = \sigma^1, \ldots, x^p = \sigma^p$ and let $x^K(\xi)$ be the remaining x's. Then $A_i(\varphi,x,\theta) = \varphi_i(\xi) + \varphi_K(\xi)\partial_i x^K(\xi)$, $K = (p+1), \ldots, D-1$. Thus, $A_{1i}(\varphi,x,\theta)$, and therefore the action, remains invariant under the following set of $D-(p+1)$ gauge transformations:

$$\delta\varphi_K(\xi) = \alpha_K(\xi) \quad , \quad \delta\varphi_i(\xi) = -\alpha_K(\xi)\partial_i x^K(\xi) \quad . \tag{21}$$

By taking $\alpha_K = \varphi_K$ we find that $\varphi_K = 0$ so that $A_i(\xi) = \varphi_i(\xi)$. Thus, the actual number of degrees of freedom is $(p+1)$, and no new dynamical ones are added by assuming the compositeness $A_{1i}(x,\theta,\varphi)$ of the BI fields. The case of the M5 brane can be treated similarly [22].

We show next that an analogous mechanism is at work in D=11 supergravity when it is expressed in terms of a composite three form A_3 by using the coordinates of a suitably enlarged superspace.

4. $D = 11$ SUPERGRAVITY AND COMPOSITE NATURE OF THE A_3 FIELD

We turn now to $D = 11$ supergravity, a different theory that nevertheless presents several analogies with the previous discussion on branes. Our aim here is to find a composite structure of the A_3 field of CJS supergravity. This problem is equivalent to that of trivializing a four-cocycle (ω_4 below) for the standard supersymmetry algebra cohomology on a larger superalgebra, so that $\omega_4 = d\tilde{\omega}_3$ where $\tilde{\omega}_3$ is expressed in terms of MC one-forms on the corresponding larger superspace group manifold. In this way A_3 will not be 'external' to the superspace coordinates of the theory and, at the same time, the enlarged superymmetry algebra will reveal the hidden underlying gauge symmetry of CJS supergravity.

The field content of Cremmer-Julia-Scherk $D = 11$ supergravity multiplet is the (unique) $D = 11$ supergravity one

$$(\ e^a(x)\ ,\ \psi^\alpha(x)\ ,\ A_3(x)\ ) \tag{22}$$

where $e^a(x)$ is the *elfbein*, $\psi^\alpha(x)$ (a Majorana spinor) is the gravitino field and $A_3(x)$ is an antisymmetric three-index Abelian gauge field. The first order formulation of $D = 11$ supergravity further requires an initially independent spin connection $\omega^{ab}(x)$.

As is well known, a justification for this set of fields is provided by the on-shell counting of bosonic and fermionic degrees of freedom. By considering the transverse traceless spatial ($D = 11$) components of g_{ij} and those of A_{ijk}, one finds that e^a has $\frac{(D-2)(D-1)}{2} - 1 = \frac{D(D-3)}{2} = 44$ bosonic d.o.f. and that A_3 has $\binom{D-2}{3} = 84$ bosonic ones; as for ψ^α, it has $\frac{1}{2}2^{[D/2]}(D-3) = \frac{1}{2}32(11-3) = 128$ fermionic d.o.f. Thus, as it should be the case because of the supersymmetry of the theory, the numbers of bosonic and fermionic d.o.f. match:

$$\sharp\ \text{Bosonic d.o.f.} = 44 + 84 = 128 = \sharp\ \text{Fermionic d.o.f.} \quad .$$

The supergravity one-forms e^a, ψ^α and ω^{ab} generate a free differential algebra[9] (FDA) . This is defined by the expressions for the FDA curvatures

$$\mathbf{R}^a := de^a - e^b \wedge \omega_b{}^a + i\psi^\alpha \wedge \psi^\beta \Gamma^a_{\alpha\beta} , \tag{23}$$

$$\mathbf{R}^\alpha := d\psi^\alpha - \frac{1}{4}\psi^\beta \wedge \omega^{ab}\Gamma_{ab\,\beta}{}^\alpha , \tag{24}$$

$$\mathbf{R}^{ab} := d\omega^{ab} - \omega^{ac} \wedge \omega_c{}^b , \tag{25}$$

where $T^a := De^a = de^a - e^b \wedge \omega_b{}^a$ is the torsion and $\mathbf{R}^{ab}$ coincides with the Riemann curvature, and by the Bianchi identities (which are the consistency/integrability conditions for the FDA).

For vanishing curvatures, $\mathbf{R}^a = 0$, $\mathbf{R}^\alpha = 0$, $\mathbf{R}^{ab} = 0$, eqs. (23) and (24), (25) reduce to the MC equations for the superPoincaré algebra. Removing the unessential Lorentz part one arrives to the MC equations for the graded translations (supersymmetry) algebra $\mathfrak{E}^{(11|32)}$ ($\mathfrak{E}^{(D|n)}$ in general) (see footnote 5 for conventions)

$$de^a = -i\psi^\alpha \wedge \psi^\beta \Gamma^a_{\alpha\beta} , \qquad d\psi^\alpha = 0 , \tag{26}$$

which correspond to the supersymmetry algebra commutation relations,

$$\{Q_\alpha, Q_\beta\} = \Gamma^a_{\alpha\beta} P_a , \qquad [P_a, Q_\alpha] = 0 , \quad [P_a, P_b] = 0 . \tag{27}$$

Eq. (26) is solved by

$$e^a = \Pi^a := dx^a - i d\theta^\alpha \Gamma^a_{\alpha\beta} \theta^\beta , \qquad \psi^\alpha = \Pi^\alpha := d\theta^\alpha . \tag{28}$$

When Π^a, Π^α are considered as forms on the rigid superspace $\Sigma^{(11|32)}$ ($\Sigma^{(D|n)}$ in general) parametrized by $Z^M = (x^a, \theta^\alpha)$, they define the invariant MC forms of the supertranslation algebra (27) on the *standard supersymmetry group manifold* $\Sigma^{(11|32)}$ that may be identified with rigid superspace. When e^a and ψ^α are forms on spacetime, the x^a are still spacetime coordinates while the θ^α are Grassmann functions, $\theta^\alpha(x)$, the Volkov-Akulov Goldstone fermions [59]. For one-forms defined on the standard curved superspace, $e^a = dZ^M E^a_M(Z)$, $\psi^\alpha = dZ^M E^\alpha_M(Z)$, $\omega^{ab}(Z) = dZ^M \omega^{ab}_M(Z)$ the FDA (23), (24), (25) with nonvanishing $\mathbf{R}^\alpha$ and $\mathbf{R}^{ab} = R^{ab}$ but vanishing $\mathbf{R}^a = 0$ gives a set of superspace supergravity constraints (which are kinematical or *off-shell* for $N = 1, D = 4$, and *on-shell*, *i.e.* containing equations of motion among their consequences, for higher D including $D = 11$). Nevertheless, the FDA makes also sense for forms on spacetime, where $e^a = dx^\mu e^a_\mu(x)$ and $\psi^\alpha = dx^\mu \psi^\alpha_\mu(x)$ are the gauge fields for the supertranslations group.

However, the $D = 11$ supermultiplet (22) also includes the three-form A_3, and the previous FDA generated by the one-forms e^a, ψ^α and ω^{ab} has to be completed by the

[9] In essence, a FDA (introduced in this context in [21] as a *Cartan integrable system*) is an exterior algebra of forms, with constant coefficients, that is closed under the exterior derivative d; see [56, 21, 57, 58].

definition of the four–form field strength [21]

$$\mathbf{R}_4 = dA_3 + \frac{1}{4}\psi^\alpha \wedge \psi^\beta \wedge e^a \wedge e^b \Gamma_{ab\alpha\beta} \,. \tag{29}$$

Note that, considering the FDA (23), (24), (25), (29) on the $D = 11$ superspace and setting $\mathbf{R}^a = 0$ and $\mathbf{R}_4 = F_4 := 1/4! e^{a_4} \wedge \ldots \wedge e^{a_1} F_{a_1 \ldots a_4}$ one arrives at the original on-shell $D = 11$ superspace supergravity constraints [60, 61] (see also [62, 63]).

Thus, in contrast with the $D = 4$ case, the $D = 11$ supergravity FDA for vanishing curvatures cannot be associated with the MC *one*-forms and equations of a *Lie* superalgebra due to the presence of the *three*-form A_3. On $\Sigma^{(11|32)}$, where one also sets $\mathbf{R}_4 = 0$ by consistency, dA_3 becomes the bosonic four-form

$$a_4 = -\frac{1}{4}\psi^\alpha \wedge \psi^\beta \wedge e^a \wedge e^b \Gamma_{ab\alpha\beta} \,, \tag{30}$$

which corresponds to CE Lie algebra cohomology *four-cocycle* on the standard supersymmetry algebra $\mathfrak{E}^{(11|32)}$, *i.e.*

$$\omega_4 = -\frac{1}{4}\Pi^\alpha \wedge \Pi^\beta \wedge \Pi^a \wedge \Pi^b \Gamma_{ab\alpha\beta} = d\omega_3(x,\theta) \equiv -\frac{1}{4} d\theta^\alpha \wedge d\theta^\beta \wedge \Pi^a \wedge \Pi^b \Gamma_{ab\alpha\beta} \,. \tag{31}$$

This is so because ω_4 is 1) $\Sigma^{(11|32)}$–invariant and 2) closed. The four-cocycle ω_4 is, furthermore, CE non-trivial, since ω_3 cannot be expressed in terms of the $\mathfrak{E}^{(11|32)}$ MC forms: ω_3 is not $\Sigma^{(11|32)}$-invariant. However, it will be seen that there exists [45] a one-parametric family of *extended* superalgebras $\tilde{\mathfrak{E}}(s)$, with MC forms defined on the associated extended superspace group $\tilde{\Sigma}(s)$ manifolds, on which the CE four-cocycle ω_4 becomes trivial *i.e*, $\omega_4 = d\tilde{\omega}$ and $\tilde{\omega}$ is made out of $\tilde{\Sigma}(s)$-invariant MC forms. Of course, we have already mentioned another solution for the same mathematical problem in the context of branes, eq. (12), but here we shall concentrate on $\tilde{\mathfrak{E}}(s)$ due to their closer relation with $osp(1|32)$ and with the M-theory superalgebra (itself an expansion, osp(1|32)(2,1,2), of $osp(1|32)$ [40]). Substituting in the expression of $\tilde{\omega}_3$ the gauge fields for the MC forms, the resulting expression will provide the composite structure of the A_3 form in terms of $\tilde{\mathfrak{E}}(s)$ gauge fields.

Thus, formulated in this way, the problem of writing the A_3 field in terms of one-form fields is, like the construction of WZ invariant terms for branes or the search for an (enlarged) superspace origin of the BI fields, purely geometrical: it reduces to a problem of Lie superalgebra cohomology. It is equivalent, in the spirit of the *enlarged superspace coordinates/fields correspondence*, to looking for an enlarged supergroup manifold $\tilde{\Sigma}$ on which one can find a suitable invariant three-form $\tilde{\omega}_3$ (corresponding to A_3) written in terms of products of $\tilde{\mathfrak{E}}$ MC forms on $\tilde{\Sigma}$ (which will give rise to the one-form gauge fields). The three-form $\tilde{\omega}_3$ will necessarily depend on the coordinates of the generalized (extended) superspace group manifold $\tilde{\Sigma}$; in contrast, the original $\omega_3 = \omega_3(x^a, \theta^\alpha)$ depends on the coordinates of standard superspace $\Sigma^{(11|32)}$. The MC equations of the enlarged superspace algebra $\tilde{\mathfrak{E}}$ can be 'softened' by adding the appropriate curvatures. The resulting gauge FDA for the 'soft' forms over $D = 11$ spacetime will then describe a $D = 11$ supergravity theory in which A_3 is a *composite*, not elementary, field.

To finish this section, we remark that, although the composite nature of A_3 and the superspace origin of the worldvolume fields in the D-branes and M5 brane depend on the same mathematical problem as stated above, the relevant objects involved in those two problems are different: whereas $\tilde{\omega}_3$ above, being constructed entirely in terms of the MC forms of certain enlarged supergroups is invariant under its transformations, the *e.g.* BI fields expressed through the coordinates of the enlarged superspaces are not invariant under the corresponding enlarged supersymmetry (see eqs. (19) and (20)); only the forms $\mathcal{F}$ and H are invariant.

5. TRIVIALIZATION OF THE CE FOUR-COCYCLE

Let us describe now the solution of the trivialization problem just described. We shall first write down the algebras suitable to this end, and then the expression for $\tilde{\omega}_3$, which also gives the composite structure of A_3. At the end of this section, we shall specialize the results for a particularly simple case.

5.1. A family of extended superalgebras $\tilde{\mathfrak{E}}(s)$

The three-form A_3 of the $D = 11$ supergravity FDA may be written in terms of one-forms by introducing *two* new *bosonic* tensorial one-forms, $B^{a_1a_2}$, $B^{a_1\ldots a_5}$, and *one* new *fermionic* spinorial one-form, η^α, that obey the FDA equations (23)–(25), (29) plus

$$\mathcal{B}_2^{a_1a_2} = DB^{a_1a_2} + \psi^\alpha \wedge \psi^\beta \Gamma^{a_1a_2}_{\alpha\beta}\,, \tag{32}$$

$$\mathcal{B}_2^{a_1\ldots a_5} = DB^{a_1\ldots a_5} + i\psi^\alpha \wedge \psi^\beta \Gamma^{a_1\ldots a_5}_{\alpha\beta}\,, \tag{33}$$

$$\begin{aligned}\mathcal{B}_2^{\alpha} &= D\eta^\alpha - i\delta\, e^a \wedge \psi^\beta \Gamma_{a\beta}{}^\alpha \\ &- \gamma_1 B^{ab} \wedge \psi^\beta \Gamma_{ab\beta}{}^\alpha - i\gamma_2 B^{a_1\ldots a_5} \wedge \psi^\beta \Gamma_{a_1\ldots a_5\beta}{}^\alpha\end{aligned} \tag{34}$$

where γ_1, γ_2 and δ are parameters (that are related by eq. (40) below).

For vanishing curvatures (and ignoring the spin connection) the above FDA reduces to the MC equations

$$de^a = -i\psi^\alpha \wedge \psi^\beta \Gamma^a_{\alpha\beta}\,, \quad d\psi^\alpha = 0\,, \tag{35}$$

$$dB^{a_1a_2} = -\psi^\alpha \wedge \psi^\beta \Gamma^{a_1a_2}_{\alpha\beta}\,, \quad dB^{a_1\ldots a_5} = -i\psi^\alpha \wedge \psi^\beta \Gamma^{a_1\ldots a_5}_{\alpha\beta}\,, \tag{36}$$

$$d\eta^\alpha = \psi^\beta \wedge \left(-i\delta\, e^a \Gamma_a - \gamma_1 B^{ab}\Gamma_{ab} - i\gamma_2 B^{a_1\ldots a_5}\Gamma_{a_1\ldots a_5}\right)_\beta{}^\alpha\,, \tag{37}$$

which correspond to the $D = 11$ superalgebra commutators

$$\{Q_\alpha, Q_\beta\} = P_{\alpha\beta} := \Gamma^a_{\alpha\beta}P_a + i\Gamma^{a_1a_2}_{\alpha\beta}Z_{a_1a_2} + \Gamma^{a_1\ldots a_5}_{\alpha\beta}Z_{a_1\ldots a_5}\,, \tag{38}$$

$$\begin{gathered}[P_a, Q_\alpha] = \delta\, \Gamma_{a\,\alpha}{}^\beta Q'_\beta\,, \\ [Z_{a_1a_2}, Q_\alpha] = i\gamma_1 \Gamma_{a_1a_2\,\alpha}{}^\beta Q'_\beta\,,\ [Z_{a_1\ldots a_5}, Q_\alpha] = \gamma_2 \Gamma_{a_1\ldots a_5\,\alpha}{}^\beta Q'_\beta\,.\end{gathered} \tag{39}$$

The constants δ, γ_1, γ_2 are clearly restricted by the Jacobi identities, which require

$$\delta + 10\gamma_1 - 6!\gamma_2 = 0\,. \tag{40}$$

One non-vanishing parameter (*e.g.*, γ_1) can be removed by rescaling the new fermionic generator Q'_α and it is thus inessential. As a result, eqs. (38)–(40) describe, effectively, a *one-parameter family* of Lie superalgebras, denoted $\tilde{\mathfrak{E}}(s)$. The parameter s may be introduced, *e.g.* through

$$s := \frac{\delta}{2\gamma_1} - 1 \qquad \Rightarrow \qquad \begin{cases} \delta &= 2\gamma_1(s+1)\,, \\ \gamma_2 &= 2\gamma_1(\frac{s}{6!} + \frac{1}{5!})\,. \end{cases} \tag{41}$$

In this parametrization the element corresponding to the case $\gamma_1 = 0$ may be included as the $\gamma_1 \to 0$ limit with $\gamma_1 s \to \delta/2 \neq 0$. This implies that the corresponding algebra (labelled $\tilde{\mathfrak{E}}(\infty)$) is a regular member of the family.

In terms of s, eqs. (39) read:

$$\begin{aligned} [P_a, Q_\alpha] &= 2\gamma_1(s+1)\,\Gamma_{a\,\alpha}{}^\beta Q'_\beta\,, \\ [Z_{a_1a_2}, Q_\alpha] &= i\gamma_1\Gamma_{a_1a_2\,\alpha}{}^\beta Q'_\beta\,, \\ [Z_{a_1\ldots a_5}, Q_\alpha] &= 2\gamma_1(\tfrac{s}{6!} + \tfrac{1}{5!})\Gamma_{a_1\ldots a_5\,\alpha}{}^\beta Q'_\beta\,. \end{aligned} \tag{42}$$

The family $\tilde{\mathfrak{E}}(s)$ is equivalently defined by its MC equations

$$\begin{aligned} de^a &= -i\psi^\alpha \wedge \psi^\beta \Gamma^a_{\alpha\beta}\,, \quad d\psi^\alpha = 0\,, \\ dB^{a_1a_2} &= -\psi^\alpha \wedge \psi^\beta\,\Gamma^{a_1a_2}_{\alpha\beta}\,, \\ dB^{a_1\ldots a_5} &= -i\psi^\alpha \wedge \psi^\beta\,\Gamma^{a_1\ldots a_5}_{\alpha\beta}\,, \\ d\eta^\alpha &= -2\gamma_1\,\psi^\beta \wedge \big(i(s+1)\,e^a\Gamma_{a\beta}{}^\alpha \\ &\quad + \frac{1}{2}B^{ab}\Gamma_{ab\,\beta}{}^\alpha + i\left(\frac{s}{6!} + \frac{1}{5!}\right) B^{a_1\ldots a_5}\Gamma_{a_1\ldots a_5\beta}{}^\alpha\Big)\,. \end{aligned} \tag{43}$$

The $s = 0$ is a special case. The $\tilde{\mathfrak{E}}(0)$ superalgebra is given by

$$\{Q_\alpha, Q_\beta\} = P_{\alpha\beta}\,, \quad [P_{\alpha\beta}, Q_\gamma] = 64\,\gamma_1\,C_{\gamma(\alpha}Q'_{\beta)}\,, \tag{44}$$

which are obtained from eqs. (42) with $s = 0$ by using the Fierz identity

$$\delta_{(\alpha}{}^\gamma \delta_{\beta)}{}^\delta \;=\; \frac{1}{32}\left(\Gamma^a_{\alpha\beta}\Gamma_a^{\gamma\delta} - \frac{1}{2}\Gamma^{a_1a_2}{}_{\alpha\beta}\Gamma_{a_1a_2}{}^{\gamma\delta} + \frac{1}{5!}\Gamma^{a_1\ldots a_5}{}_{\alpha\beta}\Gamma_{a_1\ldots a_5}{}^{\gamma\delta}\right).$$

Equivalently, collecting the bosonic one-forms e^a, $B^{a_1a_2}$, $B^{a_1\cdots a_5}$ in (43) for $s = 0$ in a symmetric spin-tensor one-form $\mathcal{E}^{\alpha\beta}$,

$$\mathcal{E}^{\alpha\beta} \;=\; \frac{1}{32}\left(e^a\Gamma_a^{\alpha\beta} - \frac{i}{2}B^{a_1a_2}\Gamma_{a_1a_2}{}^{\alpha\beta} + \frac{1}{5!}B^{a_1\ldots a_5}\Gamma_{a_1\ldots a_5}{}^{\alpha\beta}\right), \tag{45}$$

the MC equations of $\tilde{\mathfrak{E}}(0)$ can be written as

$$d\mathcal{E}^{\alpha\beta} = -i\psi^\alpha \wedge \psi^\beta\,, \quad d\psi^\alpha = 0\,, \quad d\eta^\alpha = -64i\gamma_1\,\psi^\beta \wedge \mathcal{E}_\beta{}^\alpha\,; \tag{46}$$

in the form given by eqs. (44) or (46) the $Sp(32)$ automorphism symmetry of $\tilde{\mathfrak{E}}(0)$ becomes manifest.

For our purposes, the relevant features of the $\tilde{\mathfrak{E}}(s)$ superalgebras are the following:

1. For $s \neq 0$, the $\tilde{\mathfrak{E}}(s)$ may be considered as deformations of $\tilde{\mathfrak{E}}(0)$.
2. The automorphism group of $\tilde{\mathfrak{E}}(0)$ is $Sp(32)$ while, for $s \neq 0$, $\tilde{\mathfrak{E}}(s)$ has the smaller $SO(1,10)$ group of automorphisms. Hence, the groups that generalize the super-Poincaré group $\Sigma^{(11|32)} \rtimes SO(1,10)$, are given by the following semidirect products
   - $\tilde{\Sigma}(s) \rtimes SO(1,10)$, $s \neq 0$, and
   - $\tilde{\Sigma}(0) \rtimes SO(1,10) \approx Osp(1|32)(2,3,2)$,
   - $\tilde{\Sigma}(0) \rtimes Sp(32) \approx Osp(1|32)(2,3)$,

where the last two right hand sides denote the appropriate *expansions* of $OSp(1|32)$ (see later).

5.2. Trivialization of ω_4

To trivialize the CE four-cocycle $\omega_4 = -\frac{1}{4}\Pi^\alpha \wedge \Pi^\beta \wedge \Pi^a \wedge \Pi^b \Gamma_{ab\alpha\beta} = d\omega_3$ (eq.(31)), over the $\tilde{\mathfrak{E}}(s)$ enlarged superalgebra one considers first the most general ansatz that expresses the three–form A_3 in terms of combinations of wedge products of the one-forms e^a, ψ^α; $B^{a_1a_2}$, $B^{a_1\ldots a_5}$, η^α, which are assumed to satisfy the MC equations (35)–(37). Using the same notation for MC forms and fields here and below, we write

$$\begin{aligned}
4A_3 \;=\; & \lambda B^{ab} \wedge e_a \wedge e_b - \alpha_1 B_{ab} \wedge B^b{}_c \wedge B^{ca} \\
- \; & \alpha_2 B_{b_1 a_1\ldots a_4} \wedge B^{b_1}{}_{b_2} \wedge B^{b_2 a_1\ldots a_4} \\
- \; & \alpha_3 \epsilon_{a_1\ldots a_5 b_1\ldots b_5 c} B^{a_1\ldots a_5} \wedge B^{b_1\ldots b_5} \wedge e^c \\
- \; & \alpha_4 \epsilon_{a_1\ldots a_6 b_1\ldots b_5} B^{a_1a_2a_3}{}_{c_1c_2} \wedge B^{a_4a_5a_6c_1c_2} \wedge B^{b_1\ldots b_5} \\
- \; & 2i\psi^\beta \wedge \eta^\alpha \wedge \big(\beta_1\, e^a \Gamma_{a\alpha\beta} \\
& \qquad -i\beta_2 B^{ab}\Gamma_{ab\,\alpha\beta} + \beta_3 B^{abcde}\Gamma_{abcde\,\alpha\beta}\big)\,.
\end{aligned} \tag{47}$$

The problem is now to find the values of the constants $\alpha_1,\ldots,\alpha_4$, $\beta_1,\ldots,\beta_3$ and λ, such that eq. (30), $dA_3 = a_4 = -\frac{1}{4}\psi^\alpha \wedge \psi^\beta \wedge e^a \wedge e^b \Gamma_{ab\alpha\beta}$, is fulfilled. This produces a set of equations for the constants $\alpha_1, \ldots, \beta_3$ and λ that includes δ, γ_1 and γ_2 as parameters:

$$\begin{array}{ll}
\lambda - 2\delta\beta_1 = 1\,, & \lambda - 2\gamma_1\beta_1 - 2\delta\beta_2 = 0\,, \\
3\alpha_1 + 8\gamma_1\beta_2 = 0\,, & \alpha_2 - 10\gamma_1\beta_3 - 10\gamma_2\beta_2 = 0\,, \\
5!\alpha_3 - \delta\beta_3 - \gamma_2\beta_1 = 0\,, & \alpha_2 - 5!\,10\gamma_2\beta_3 = 0\,, \\
\alpha_3 - 2\gamma_2\beta_3 = 0\,, & 3\alpha_4 + 10\gamma_2\beta_3 = 0\,.
\end{array} \tag{48}$$

This system has a nontrivial solution for

$$\Delta = (2\gamma_1 - \delta)^2 = 4s^2\gamma_1^2 \neq 0 \quad \Longleftrightarrow \quad s \neq 0\,, \tag{49}$$

which in terms of the parameter s reads [45]

$$\begin{gathered}\lambda = \tfrac{1}{5}\tfrac{s^2+2s+6}{s^2}\,,\\ \beta_1 = -\tfrac{1}{10\gamma_1}\tfrac{2s-3}{s^2}\,,\ \beta_2 = \tfrac{1}{20\gamma_1}\tfrac{s+3}{s^2}\,,\ \beta_3 = \tfrac{3}{10\cdot 6!\gamma_1}\tfrac{s+6}{s^2}\,,\\ \alpha_1 = -\tfrac{1}{15}\tfrac{2s+6}{s^2}\,,\ \alpha_2 = \tfrac{1}{6!}\tfrac{(s+6)^2}{s^2}\,, \alpha_3 = \tfrac{1}{5\cdot 5!}\alpha_2\,,\ \alpha_4 = -\tfrac{1}{9\cdot 5!}\alpha_2\,;\end{gathered} \tag{50}$$

note that $\alpha_{2,3,4} \propto (s+6)$ and that all denominators depend on s^2. Thus, ω_4 can be trivialized ($\omega_4 = d\tilde{\omega}_3$) over $\tilde{\mathfrak{E}}(s)$ when $s \neq 0$; the impossibility of doing it over $\tilde{\mathfrak{E}}(0)$ may be related to the fact that precisely $\tilde{\mathfrak{E}}(0)$ has an enhanced automorphism symmetry, $Sp(32)$. This implies that the A_3 field can be considered as a composite of the one-form gauge fields of any of the $\tilde{\Sigma}(s)$ with $s \neq 0$, and that A_3 is given by eq. (47) for the values (50) of $\alpha_1, \ldots, \beta_3, \lambda$. Thus, the hidden gauge symmetry of $D = 11$ supergravity can be associated with any of the $\tilde{\Sigma}(s) \rtimes SO(1,10)$ supergroups.

The two particular solutions of D'Auria-Fré for A_3 are recovered by adjusting δ, γ_1 in eq. (50) so that $\lambda = 1$, which was the starting point of [21]. These correspond to $\tilde{\mathfrak{E}}(3/2)$, given by the parameter values

$$\begin{gathered}\delta = 5\gamma_1 \neq 0 \quad , \quad \gamma_2 = \tfrac{\gamma_1}{2\cdot 4!}\,,\\ \lambda = 1\,,\ \beta_1 = 0\,,\ \beta_2 = \tfrac{1}{10\gamma_1}\,,\ \beta_3 = \tfrac{1}{6!\,\gamma_1}\,,\\ \alpha_1 = -\tfrac{4}{15}\,,\ \alpha_2 = \tfrac{25}{6!}\,,\ \alpha_3 = \tfrac{1}{6!\,4!}\,,\ \alpha_4 = -\tfrac{1}{54\,(4!)^2}\,,\end{gathered} \tag{51}$$

and to $\tilde{\mathfrak{E}}(-1)$, for which

$$\begin{gathered}\delta = 0\,,\ \gamma_1 \neq 0\,,\ \gamma_2 = \tfrac{\gamma_1}{3\cdot 4!}\,,\\ \lambda = 1\,,\ \beta_1 = \tfrac{1}{2\gamma_1}\,,\ \beta_2 \tfrac{1}{10\gamma_1}\,,\ \beta_3 = \tfrac{1}{4\cdot 5!\,\gamma_1}\,,\\ \alpha_1 = -\tfrac{4}{15}\,,\ \alpha_2 \tfrac{25}{6!}\,,\ \alpha_3 = \tfrac{1}{6!\,4!}\,,\ \alpha_4 = -\tfrac{1}{54\,(4!)^2}\,.\end{gathered} \tag{52}$$

5.3. The minimal solution $\mathfrak{E}_{min}$

A specially simple trivialization of ω_4 is achieved for the superalgebra $\tilde{\mathfrak{E}}(-6)$, characterized by

$$\tilde{\mathfrak{E}}(-6) \ : \quad \delta \neq 0\,, \quad \delta = -10\gamma_1\,,\ \gamma_2 = 0\,. \tag{53}$$

In $\tilde{\mathfrak{E}}(-6)$ the generator $Z_{a_1 \ldots a_5}$ is central (see eq. (42)) and does not play any rôle in the trivialization of the ω_4 cocycle. Furthermore, eqs. (38)–(40) allow us to use instead the

$\tilde{\mathfrak{E}}_{min}$ superalgebra whose central extension by the generator $Z_{a_1\ldots a_5}$ gives $\tilde{\mathfrak{E}}(-6)$. $\mathfrak{E}_{min}$ is the $(66+64)$-dimensional superalgebra $\tilde{\mathfrak{E}}^{(66|32+32)}$,

$$\mathfrak{E}_{min}: \qquad \{Q_\alpha, Q_\beta\} = \Gamma^a_{\alpha\beta}P_a + i\Gamma^{a_1a_2}_{\alpha\beta}Z_{a_1a_2}\,, \tag{54}$$

$$[P_a, Q_\alpha] = -10\gamma_1\,\Gamma_{a\,\alpha}{}^\beta Q'_\beta\,, \quad [Z_{a_1a_2}, Q_\alpha] = i\gamma_1\Gamma_{a_1a_2\,\alpha}{}^\beta Q'_\beta\,, \tag{55}$$

associated with the most economic $\tilde{\Sigma}_{min} \equiv \Sigma^{(66|32+32)}$ generalized supertranslation group that trivializes ω_4.

Using the values of eq. (53) in eq. (50) we get

$$\lambda = \tfrac{1}{6}\,,\ \beta_1 = \tfrac{1}{4!\gamma_1}\,,\ \beta_2 = -\tfrac{1}{2\cdot 5!\gamma_1}\,,\ \beta_3 = 0\,,$$
$$\alpha_1 = \tfrac{1}{90}\,,\ \alpha_2 = 0\,,\ \alpha_3 = 0\,,\ \alpha_4 = 0\,. \tag{56}$$

Then, all the $B^{a_1\ldots a_5}$ terms in A_3, eq. (47), are zero. This simplifies the expression for A_3 drastically,

$$\begin{aligned} A_3 &= \frac{1}{4!}B^{ab}\wedge e_a\wedge e_b - \frac{1}{3\cdot 5!}B_{ab}\wedge B^b{}_c\wedge B^{ca} \\ &- \frac{i}{4\cdot 5!\,\gamma_1}\psi^\beta\wedge\eta^\alpha\wedge\left(10e^a\Gamma_{a\alpha\beta} + iB^{ab}\Gamma_{ab\,\alpha\beta}\right)\,. \end{aligned} \tag{57}$$

Thus, $\Sigma^{(66|32+32)}$ can be regarded as a *minimal* underlying gauge supergroup of $D=11$ supergravity.

6. DEGREES OF FREEDOM IN D=11 SUPERGRAVITY WITH A COMPOSITE A_3 AND EXTRA GAUGE GAUGE SYMMETRIES

It remains to be checked that, as in the case of the BI fields of the D-branes, the composite nature of A_3 does not change the supergravity degrees of freedom. Let us first recall that, in standard CJS supergravity, $e^a_\mu(x)$ has $\frac{(D-2)(D-1)}{2} - 1 = \frac{D(D-3)}{2} = 11^2 - 55_{Lorentz} - 2\times 11_{Diff} = 44$ degrees of freedom; $\psi^\alpha_\mu(x)$ has $(9\times 32 - 32)\times\frac{1}{2} = 128$; and $A_{\mu\nu\rho}(x)$ has $\binom{9}{3} = \binom{11}{3} - \binom{10}{2} - \binom{9}{2} = 84 = 165_{\#\text{of components}} - 45_{(\#\text{gauge symm.}-\#\text{null vectors})} - 36_{\#\text{residual gauge symm.}}$.

Now, let us consider a composite A_3 in the CJS supergravity action [29] *i.e.*, by substituting

$$A_3 = A_3(B^{ab}_1, B^{a_1\ldots a_5}_1, \eta_{1\alpha}\,;\, e^a, \psi^\alpha)\,, \tag{58}$$

as given by eqs. (47) and (50), for the original three-form field in that action. Naively assuming standard linearized equations and the usual 'group theoretical' gauge symmetry transformations for the new fields,

$$\delta B^{ab}_\mu = \partial_\mu\alpha^{ab} + \ldots\,; \quad \delta B^{a_1\ldots a_5}_\mu = \partial_\mu\alpha^{a_1\ldots a_5} + \ldots\,; \quad \delta\eta_{\mu\alpha} = \partial_\mu\varepsilon'_\alpha + \ldots\,,$$

the sum of the components of B_μ^{ab} $[9 \times \binom{11}{2} = 495]$, the components of $B_\mu^{a_1 \dots a_5}$ $[9 \times \binom{11}{5} = 4158]$ and those of $\eta_{\mu\alpha}$ [128, as for ψ_μ^α] would give a huge number of 'new' degrees of freedom for the gauge invariant theory with the additional fields. Moreover, the bosonic and fermionic degrees of freedom would not match.

However, the 'new' fields B_1^{ab}, $B_1^{a_1 \dots a_5}$, $\eta_{1\alpha}$ enter in the CJS supergravity action only through the composite $A_3(\dots)$ three form field and, as a result, the theory possesses *extra gauge symmetries*. Clearly, these are the transformations of the 'new' fields that leave A_3 invariant,

$$
\begin{aligned}
\delta B_\mu^{ab} &= \partial_\mu \alpha^{ab} + \beta^{\left(\begin{smallmatrix}\Box\Box\\ \Box\end{smallmatrix}\right)}{}_\mu^{ab} + \dots \\
\delta B_\mu^{a_1 \dots a_5} &= \partial_\mu \alpha^{a_1 \dots a_5} + \beta_\mu^{a_1 \dots a_5} + \dots \\
\delta \eta_{\mu\alpha} &= \partial_\mu \varepsilon'_\alpha + \beta_{\mu\alpha} + \dots \, .
\end{aligned}
$$

They reduce to 84 the number of B_μ^{ab} degrees of freedom and to zero those of the remaining new fields since, diagrammatically (note that $\#\begin{smallmatrix}\Box\Box\\ \Box\end{smallmatrix} = D(D^2-1)/3 = 440$)

$$
B_{c\,ab} \sim \Box \otimes \begin{smallmatrix}\Box\\ \Box\end{smallmatrix} = 11 \times 55 = 605 = \begin{smallmatrix}\Box\Box\\ \Box\end{smallmatrix} \oplus \begin{smallmatrix}\Box\\ \Box\\ \Box\end{smallmatrix} = 440 + 165 \quad , \tag{59}
$$

and the equations of motion (which are the standard ones but with a composite A_3), when linearized, affect only the antisymmetric part $B_{[\mu\nu\rho]}$ of B_ν^{ab}. In this way, the antisymmetric 165-dimensional part simulates the fundamental A_3; the mixed symmetry 440-dimensional part of B_ν^{ab} as well as $B_\mu^{a_1 \dots a_5}$ and $\eta_{\mu\alpha}$ are pure gauge and do not have independent equations of motion in the CJS action with a composite A_3. Thus,

$$
\#\, d.o.f. \text{ with fundamental } A_3 = \#\, d.o.f. \text{ with composite } A_3,
$$

as stated.

7. THE SPECIAL ELEMENT $\tilde{\mathfrak{E}}(0)$ AS AN ALGEBRA EXPANSION

We have seen that the superalgebra $\tilde{\mathfrak{E}}(0)$, although it does not trivialize ω_4, may be considered as a 'parent' superalgebra for the hidden symmetries of $D = 11$ supergravity in the sense that it gives rise to the family $\tilde{\mathfrak{E}}(s)$ of superalgebras that do trivialize the standard supersymmetry algebra $\mathfrak{E}$ four-cocyle. All the corresponding $\tilde{\Sigma}(s)$ enlarged superspace groups, $s \neq 0$, may be considered as deformations of $\tilde{\Sigma}(0)$. We shall now characterize the parent algebra $\tilde{\mathfrak{E}}(0)$ in terms of Lie algebra expansions. With this aim, we first review briefly the expansion method [40, 39] for the case which is of special interest here.

7.1. The algebra expansion method

Let G be a Lie group, of local coordinates g^i, $\mathcal{G}$ its Lie algebra and $\mathcal{G}^*$ its dual coalgebra. Let $\mathcal{G}$ admit, say, the splitting $\mathcal{G} = V_0 \oplus V_1 \oplus V_2$, where V_0, V_2 (V_1), are even (odd) subspaces of dimensions $\dim V_p$, $p = 0,1,2$. Further, let V_0 be a subalgebra of $\mathcal{G}$ and $[V_1,V_1] \subset V_0 \oplus V_2$, $[V_2,V_2] \subset V_0 \oplus V_2$ (details for the general theory are given in [40]). Then, the rescaling of the group parameters $g^{i_p} \rightarrow \lambda^p g^{i_p}$, $i_p = 1,\ldots,\dim V_p$, allows us to expand the one-forms $\omega^{i_p}(\lambda, g)$, obtained from the algebra MC forms $\omega^{i_p}(g)$ that define a basis of the dual subspaces $V_p{}^*$, as a series in λ,

$$\omega^{i_p}(\lambda) = \lambda^p \omega^{i_p,p} + \lambda^{p+2}\omega^{i_p,p+2} + \lambda^{p+4}\omega^{i_p,p+4} + \ldots = \sum_{\alpha_p} \lambda^{\alpha_p}\omega^{i_p,\alpha_p} \quad (p=0,1,2)\,. \tag{60}$$

The different powers of lambda are a consequence of the above assumptions on the subspaces V_p, and follow from the fact that the canonical form $\theta(g)$ on a Lie group G is given by $\theta(g) = g^{-1}dg = \omega^i X_i$, where $g = exp\ g^i X_i$ and X_i are the generators of the algebra $\mathcal{G}$ of G. The insertion of these series expansions into the MC equations of the original algebra $\mathcal{G}$,

$$d\omega^{i_p} = -\tfrac{1}{2} c^{\,i_p}_{j_q k_s}\, \omega^{j_q} \wedge \omega^{k_s}$$
$$(p,q,s = 0,1,2\,;\ i_{p,q,s} = 1,2,\ldots,\dim V_{p,q,s})\,, \tag{61}$$

produces, identifying the terms with the same order in λ, the following set of equations

$$d\omega^{i_p,\alpha_p} = -\tfrac{1}{2} C^{\,i_p,\alpha_p}_{j_q,\beta_q\, k_s,\gamma_s}\, \omega^{j_q,\beta_q} \wedge \omega^{k_s,\gamma_s}\,,$$
$$C^{\,i_p,\alpha_p}_{j_q,\beta_q\, k_s,\gamma_s} = \begin{cases} 0, & \text{if } \beta_q + \gamma_s \neq \alpha_p \\ c^{\,i_p}_{j_q k_s}, & \text{if } \beta_q + \gamma_s = \alpha_p \end{cases}$$
$$(\alpha_p, \beta_p, \gamma_p = p, p+2, \ldots)\quad . \tag{62}$$

The question now is how to retain consistently a number of ω^{i_p,α_p} so that the equations above correspond to the MC equations of a new, by construction expanded, algebra. Cutting the expansions of the $\omega^{i_p}(\lambda)$ at certain orders $\alpha_p = N_p$, $p = 0,1,2$, one finds that eqs. (62) for $\alpha_p = p,\ldots,N_p$ will provide the MC equations of a new finite-dimensional Lie algebra provided the chosen orders satisfy the conditions

$$N_0 = N_1 + 1 = N_2 \qquad \text{or} \qquad N_0 = N_1 - 1 = N_2 \qquad \text{or} \qquad N_0 = N_1 - 1 = N_2 - 2\,. \tag{63}$$

These conditions guarantee that for the selected set of ω^{i_p,α_p}'s, eqs. (62) do not include any ω^{i_p,α_p} outside this set and that, accordingly, define new algebras [39, 40] by becoming their MC equations. These algebras, denoted $\mathcal{G}(N_0,N_1,N_2)$ in obvious notation, are called *expansions* of $\mathcal{G}$; in general, their dimension is larger than that of the original algebra $\mathcal{G}$. They also include, as a particular case and for a specific value of (N_0,N_1,N_2), the generalized Wigner-İnönü contractions [40], in which case the dimension does not change.

The dimension of the expanded $\mathcal{G}(N_0,N_1,N_2)$ algebras is given by

$$\dim \mathcal{G}(N_0,N_1,N_2) = [(N_0+2)/2]\dim V_0 + [(N_1+1)/2]\dim V_1 + [N_2/2]\dim V_2 \quad . \tag{64}$$

7.2. $\tilde{\Sigma}(0) \rtimes SO(1,10)$ as the expansion $OSp(1|32)(2,3,2)$

Let us now consider the orthosymplectic algebra $osp(1|32)$, of dimension 560, defined by the MC equations

$$\begin{aligned} d\rho^{\alpha\beta} &= -i\rho^{\alpha\gamma}\wedge\rho_\gamma{}^\beta - i\nu^\alpha\wedge\nu^\beta \,, \\ d\nu^\alpha &= -i\nu^\beta\wedge\rho_\beta{}^\alpha \,, \quad \alpha,\beta = 1,\ldots,32 \quad , \end{aligned} \tag{65}$$

where the 528 $\rho^{\alpha\beta}$ are the bosonic and the 32 ν^α the fermionic MC one-forms.

The decomposition of $\rho_{\alpha\beta}$ as

$$\rho_{\alpha\beta} = \frac{1}{32}\left(\rho^a\Gamma_a - \frac{i}{2}\rho^{ab}\Gamma_{ab} + \frac{1}{5!}\rho^{a_1\ldots a_5}\Gamma_{a_1\ldots a_5}\right)_{\alpha\beta} \,, \quad a,b = 0,1,\ldots,10 \,. \tag{66}$$

allows us to consider the splitting $osp(1|32) = V_0 \oplus V_1 \oplus V_2$, where

V_0^*	is generated by	ρ^{ab}	(55) ,
V_1^*	by	ν^α	(32) ,
V_2^*	by	ρ^a and $\rho^{a_1\ldots a_5}$	(11 + 462) .

The various forms then expand as

$$\begin{aligned} V_0^*: &\quad \rho^{ab} = \rho^{ab,0} + \lambda^2\rho^{ab,2} + \cdots; \quad V_1^*: \quad \nu^\alpha = \lambda\nu^{\alpha,1} + \lambda^3\nu^{\alpha,3} + \cdots; \\ V_2^*: &\quad \rho^a = \lambda^2\rho^{a,2} + \cdots, \qquad \rho^{a_1\ldots a_5} = \lambda^2\rho^{a_1\ldots a_5,2} + \cdots \,. \end{aligned} \tag{67}$$

Inserting the series into the MC equations and choosing $N_0 = 2$, $N_1 = 3$, $N_2 = 2$ the MC equations of the expansion $osp(1|32)(2,3,2)$ are obtained:

$$\begin{aligned} d\rho^{ab,0} &= -\frac{1}{16}\rho^{ac,0}\wedge\rho_c{}^{b,0} \\ d\rho^{a,2} &= -\frac{1}{16}\rho^{b,2}\wedge\rho_b{}^{a,0} - i\nu^{\alpha,1}\wedge\nu^{\beta,1}\Gamma^a_{\alpha\beta} \\ d\rho^{ab,2} &= -\frac{1}{16}\left(\rho^{ac,0}\wedge\rho_c{}^{b,2} + \rho^{ac,2}\wedge\rho_c{}^{b,0}\right) - \nu^{\alpha,1}\wedge\nu^{\beta,1}\Gamma^{ab}_{\alpha\beta} \\ d\rho^{a_1\ldots a_5,2} &= \frac{5}{16}\rho^{b[a_1\ldots a_4|,2}\wedge\rho_b{}^{|a_5],0} - i\nu^{\alpha,1}\wedge\nu^{\beta,1}\Gamma^{a_1\ldots a_5}_{\alpha\beta} \\ d\nu^{\alpha,1} &= -\frac{1}{64}\nu^{\beta,1}\wedge\rho^{ab,0}\Gamma_{ab\beta}{}^\alpha \\ d\nu^{\alpha,3} &= -\frac{1}{64}\nu^{\beta,3}\wedge\rho^{ab,0}\Gamma_{ab\beta}{}^\alpha \\ &\quad -\frac{1}{2}\nu^{\beta,1}\wedge\left(i\rho^{a,2}\Gamma_a + \frac{1}{2}\rho^{ab,2}\Gamma_{ab} + \frac{i}{5!}\rho^{a_1\ldots a_5,2}\Gamma^{a_1\ldots a_5}\right)_\beta{}^\alpha \,. \end{aligned} \tag{68}$$

With the identifications

$$\begin{array}{rclcccccc} \rho^{ab,0} & = & -16\omega^{ab} & , & \rho^{a,2}=e^{a} & , & \rho^{ab,2}=B^{ab}, & & \\ \rho^{a_1\cdots a_5,2} & = & B^{a_1\cdots a_5} & , & \nu^{\alpha,1}=\psi^{\alpha} & , & \nu^{\alpha,3}=\eta^{\alpha}/64\gamma_1 & , & \end{array} \quad (69)$$

and omitting the Lorentz generators ω^{ab} to simplify, these equations read

$$\begin{aligned} de^a &= -i\psi^\alpha \wedge \psi^\beta \Gamma^a_{\alpha\beta} \,, \\ dB^{a_1 a_2} &= -\psi^\alpha \wedge \psi^\beta \, \Gamma^{ab}_{\alpha\beta} \,, \\ dB^{a_1 \dots a_5} &= -i\psi^\alpha \wedge \psi^\beta \, \Gamma^{a_1 \dots a_5}_{\alpha\beta} \,, \\ d\psi^\alpha &= 0 \,, \\ d\eta^\alpha &= -2\gamma_1 \cdot \psi^\beta \wedge \left(i e^a \Gamma_a + \frac{1}{2} B^{ab} \Gamma_{ab} + \frac{i}{5!} B^{a_1 \dots a_5} \Gamma_{a_1 \dots a_5} \right)_\beta^{\ \alpha} \quad ; \end{aligned} \quad (70)$$

the inclusion of ω^{ab} produces the MC equations of the $\tilde{\mathfrak{E}}(0) \supset\!\!\!\!\times so(1,10)$ algebra. Also, one may check that

$$\begin{aligned} &\dim OSp(1|32)(2,3,2) = 2\cdot 55 + 2\cdot 32 + 1\cdot 473 = 647 = \\ &= 592 + 55 = \dim(\tilde{\Sigma}(0) \supset\!\!\!\!\times SO(1,10)) \quad . \end{aligned} \quad (71)$$

7.3. $\tilde{\Sigma}(0) \supset\!\!\!\!\times Sp(32)$ as the expansion $OSp(1|32)(2,3)$

Let us now see that the full $\tilde{\mathfrak{E}}(0) \supset\!\!\!\!+ sp(32)$ is also an expansion of $osp(1|32)$. Let us consider now the splitting $osp(1|32) = V_0 \oplus V_1$ where V_0 is the bosonic subalgebra, generated by $\rho^{\alpha\beta}$, and V_1 the fermionic part, generated by ν^α. Choosing $N_0 = 2$ and $N_1 = 3$ we obtain the expansion $osp(1|32)(2,3)$ determined by the MC equations of the one-forms $\rho^{\alpha\beta,0}$, $\rho^{\alpha\beta,2}$, $\nu^{\alpha,1}$, $\nu^{\alpha,3}$:

$$\begin{aligned} &d\rho^{\alpha\beta,0} = -i\rho^{\alpha\gamma,0} \wedge \rho_\gamma{}^{\beta,0} \\ &d\nu^{\alpha,1} = -i\nu^{\beta,1} \wedge \rho_\beta{}^{\alpha,0} \\ &d\rho^{\alpha\beta,2} = -i\left(\rho^{\alpha\gamma,0} \wedge \rho_\gamma{}^{\beta,2} + \rho^{\alpha\gamma,2} \wedge \rho_\gamma{}^{\beta,0}\right) - i\nu^{\alpha,1} \wedge \nu^{\beta,1} \\ &d\nu^{\alpha,3} = -i\nu^{\beta,3} \wedge \rho_\beta{}^{\alpha,0} - i\nu^{\beta,1} \wedge \rho_\beta{}^{\alpha,2} \,. \end{aligned} \quad (72)$$

Identifying $\rho^{\alpha\beta,0}$ in eqs. (72) with the $sp(32)$ connection $\Omega^{\alpha\beta}$, eqs, (72) coincide with those of $\tilde{\mathfrak{E}}(0) \supset\!\!\!\!+ sp(32)$ [eqs. (46)], with the identifications $\rho^{\alpha\beta,2} = \varepsilon^{\alpha\beta}$, $\nu^{\alpha,1} = \psi^\alpha$ and $\nu^{\alpha,3} = \eta^\alpha/64\gamma_1$. One can also make a dimensions check:

$$\begin{aligned} &\dim(\tilde{\mathfrak{E}}(0) \supset\!\!\!\!+ sp(32)) = 592\ (528+64) + 528 = 1120 = \\ &= 2\cdot 528 + 2\cdot 32 = \dim(osp(1|32)(2,3)) \end{aligned} \quad (73)$$

when $N_0 = 2$, $N_1 = 3$ in eq. (64).

8. CONCLUSIONS

We have given some reasons in favour of a geometrical *enlarged superspace coordinates/fields correspondence*, both for branes, in which case the correspondence is between the extended superspace coordinates and *worldvolume* fields, and for $D = 11$ CJS supergravity, where the fields are *spacetime* fields.

In the case of *branes*, the new enlarged superspace algebras appear as the result of wishing to have manifestly invariant WZ terms or an (enlarged) superspace origin for all the fields of the theory, including the otherwise 'intrinsically' worldvolume fields of the D-branes (Born-Infeld fields) and of the M5 brane. The CE cohomology arguments that lead to the WZ terms for the scalar p-branes also allow us to caracterize the D-branes as well as the WZ term of the M5-brane. Their actions (apart from the auxiliary field in the M5-brane case) do not contain fields directly defined on the worldvolume[10]; all worldvolume fields are associated to variables of certain enlarged superspaces $\tilde{\Sigma}$. Further, the number of degrees of freedom and the dynamical contents of the E-L equations remain the same once the substitution is made [22].

The fields/coordinates correspondence for $D = 11$ *CJS supergravity* has also to do with trivializing non-trivial CE cocycles. Trivializing the supersymmetry algebra $\mathfrak{E}^{(11|32)}$) CE four–cocycle ω_4 amounts to finding a composite structure for the three–form field A_3 of the standard Cremmer–Julia–Scherk supergravity in terms of one–form gauge fields of $\tilde{\Sigma}(s)$, $A_3 = A_3(e^a\,,\,\psi^\alpha\,;\,B^{a_1a_2},\,B^{a_1\ldots a_5},\eta^\alpha\,)$. The trivialization of the CE four-cocycle ω_4 may be achieved, for $s \neq 0$, on the one-parametric family of superalgebras $\tilde{\mathfrak{E}}(s)$. These are central extensions of the M-algebra (generated, ignoring the Lorentz part, by P_a, Q_α, Z_{ab}, $Z_{a_1\ldots a_5}$) by an additional fermionic central generator Q'_α. Then, $\omega_4 = d\tilde{\omega}_3(\tilde{Z})$, $\tilde{Z} \in \tilde{\Sigma}$. The Maurer-Cartan forms of $\tilde{\mathfrak{E}}(s)$ can be replaced by *soft* one-forms obeying a free differential algebra with curvatures, and thus *one may treat the standard CJS D=11 supergravity as a gauge FDA of the* $\tilde{\Sigma}(s)$ *supergroup for any* $s \neq 0$. This fact was known before for the two superalgebras [21] that here correspond to $\tilde{\mathfrak{E}}(3/2)$ and $\tilde{\mathfrak{E}}(-1)$. The novelty of the present results is that, for $s \neq 0$, any of the $\tilde{\Sigma}(s) \rtimes SO(1,10)$ supergroups may be equally treated as an underlying gauge supergroup of the $D = 11$ supergravity.

There is a special element in the $\tilde{\mathfrak{E}}(s \neq 0)$ family of trivializations, $\tilde{\mathfrak{E}}(-6)$, for which the $Z_{a_1\ldots a_5}$ generator is central. In this case, the expression for A_3 is particularly simple: it does not involve the one-form $B^{a_1\ldots a_5}$, and $\tilde{\mathfrak{E}}(-6)$ may be reduced to $\mathfrak{E}_{min}$. Thus, the smaller $\tilde{\Sigma}_{min} = \tilde{\Sigma}^{(66|32+32)}$ associated with $\mathfrak{E}_{min}$ may be considered as the minimal underlying gauge supergroup of $D = 11$ CJS supergravity. All other representatives of the family $\tilde{\mathfrak{E}}(s)$ are equivalent, although they are not isomorphic. Their significance might be related to the fact that the field $B^{a_1\ldots a_5}$ is also needed for a coupling to BPS preons [12, 11], the hypothetical basic constituents of M-theory. The presence

[10] Although in the (either rigid or non-flat) $D = 11$ covariant M5-brane action [53] the Pasti-Sorokin-Tonin (PST) scalar $a(\xi)$ field [64] is a *worldvolume* field, it is an *auxiliary* one. Further, it was shown in [65] that, when the M5-brane interacts with dynamical supergravity in a duality symmetric formulation, the rôle of the M5-brane auxiliary PST scalar $a(\xi)$ is played by the pull-back $a(x(\xi))$ to $\mathcal{W}$ of the *spacetime* supergravity PST scalar $a(x)$ and is a kind of background field.

of a full family of superalgebras $\tilde{\mathfrak{E}}(s \neq 0)$ –rather than a unique one– trivializing the standard $\mathfrak{E}^{(11|32)}$ algebra four–cocycle ω_4, suggests that the obtained underlying gauge symmetries of $D = 11$ supergravity may be incomplete (this is almost certainly the case if one considers the symmetries of M-theory).

The singularity of the $\tilde{\mathfrak{E}}(0)$ case looks a reasonable one; the $\tilde{\Sigma}(0)$ supergroup is special because it possesses an enhanced automorphism symmetry, $Sp(32)$. The full $\tilde{\Sigma}(0) \rtimes Sp(32)$, that replaces the $D = 11$ superPoincaré group, is given by the expansion $OSp(1|32)(2,3)$ of $OSp(1|32)$. All other members of the $\tilde{\Sigma}(s \neq 0)$ family have the smaller $SO(1,10)$ automorphism symmetry and are deformations of the $s = 0$ element. Thus, we may conclude that *the underlying gauge group of $D = 11$ supergravity is determined by any element $\tilde{\Sigma}(s \neq 0) \rtimes SO(1,10)$, of a one-parametric familiy of nontrivial deformations of $\tilde{\Sigma}(0) \rtimes SO(1,10) \approx OSp(1|32)(2,3,2) \subset \tilde{\Sigma}(0) \rtimes Sp(32)$* . Furthermore, we see that the number of the extended superspace coordinates are in one-to-one correspondence with the gauge fields entering the theory, and that the additional degrees of freedom may be removed by a gauge transformation. Thus, this may be considered as another example of the conjectured extended superspaces coordinates/fields correspondence principle in which the fields are spacetime fields.

Finally it is known that, unlike its lower dimensional versions, CJS supergravity forbids a cosmological term extension. The reason is cohomological and can be traced to an obstruction produced by the A_3 three-form field [66]. It is natural to ask wether this obstruction remains when A_3 is becomes a composite field.

ACKNOWLEDGMENTS

This contribution is based on research mostly done in collaboration with I. Bandos, J.M. Izquierdo, M. Picón and O. Varela, refs. [22, 40, 45], which is acknowledged with pleasure. This work has been partially supported by the research grant BFM-2002-03681 from the Spanish Ministerio de Educación y Ciencia and from EU FEDER funds, the Generalitat Valenciana (Grupos 03/124), and by the EU network MRTN–CT–2004–005104 ('Forces Universe').

REFERENCES

1. J.H. Schwarz, *The second superstring revolution*, in Proc. of the *Second Int. Sakharov conference on physics, Moscow, 1996*, I.M. Dremin and A.M. Semikhatov eds., World Sci. 1997, pp. 562-569, [arXiv:hep-th/9607067], *Lectures on superstring and M theory dualities*, Nucl. Phys. Proc. Suppl. **55B**, 1-32 (1997), [arXiv:hep-th/9607201].
2. P.K. Townsend, *p-brane democracy*, in *Particles, strings and cosmology*, J. Bagger et al. Eds., 271-285, World. Sci. (1996), [arXiv:hep-th/9507048];
M-theory from its superalgebra, 1997 Cargèse lectures, arXiv:hep-th/9712004, and refs. therein.
3. M.J. Duff, *Benchmarks on the brane*, Asim Barut Memorial Lecture, Bogazici University, Istanbul, October 2002, [arXiv:hep-th/0407175].
4. C.M. Hull and P.K. Townsend, Nucl. Phys. **B438**, 109-137 (1995), [arXiv:hep-th/9410167];
E. Witten, Nucl. Phys. **B443**, 85-126 (1995), [arXiv:hep-th/9503124].
5. P. West, Class. Quantum Grav. **18**, 4443-4460 (2001), [arXiv:hep-th/0104081]; Phys. Lett. **B**575, 333-342 (2003), [arXiv:hep-th/0307098];

F. Englert, L. Houart, A. Taormina and P. West, JHEP **0309**, 020 (2003), [arXiv:hep-th/0304206].
6. I. Bars, Phys. Rev. **D55**, 2373-2381 (1997), [arXiv:hep-th/9607112];
I. Bars, C. Deliduman and D. Minic, Phys. Lett. **B457**, 275-284 (1999), [arXiv:hep-th/9904063];
I. Bars, Phys.Rev. **D62**, 046007 (2000), [arXiv:hep-th/0003100];
Two-time physics 2001, in *Supersymmetry and fundamental interaction theories*, Proc. of the 37th Karpacz Winter School, J. Lukierski and J. Riembeliński Eds., AIP Conf. Proc. **589**, 18-30 (2001), [hep-th/0106021].
7. P. West, JHEP **0008**, 007 (2000), [arXiv:hep-th/0005270].
8. O. Baerwald and P. West, Phys. Lett. **B476**, 157-164 (2000), [arXiv:hep-th/9912226].
9. J.P. Gauntlett, G.W. Gibbons, C.M. Hull and P.K. Townsend, Commun. Math. Phys. **216**, 431-459 (2001), [arXiv:hep-th/0001024].
10. I. Bandos and J. Lukierski, Mod. Phys. Lett. **14**, 1257-1272 (1999), [arXiv:hep-th/9811022];
I.A. Bandos, Phys. Lett. **B558**, 197–204 (2003), [arXiv:hep-th/0208110].
11. I.A. Bandos, J.A. de Azcárraga, J.M. Izquierdo, M. Picón and O. Varela, Phys. Rev. **D69**, 105010-1-11 (2004), [arXiv:hep-th/0312266].
12. I.A. Bandos, J. A. de Azcárraga, J.M. Izquierdo and J. Lukierski, Phys. Rev. Lett. **86**, 4451-4454 (2001), [arXiv:hep-th/0101113].
13. I. Bandos, *title BPS preons in supergravity and higher spin theories: an overview from the hill of the twistor approach*, these proceedings
14. M.J. Duff, J.T. Liu, Nucl. Phys. **B674**, 217-230 (2003), [arXiv:hep-th/0303140].
15. C. Hull, *Holonomy and symmetry in M theory*, arXiv:hep-th/0305039,
16. M. Duff and K. Stelle, Phys. Lett. **B253**, 113-118 (1991).
17. R. Troncoso and J. Zanelli, Phys. Rev. **D58**, 101703 (1998), [arXiv:hep-th/9710180];
J. Zanelli, Braz. J. Phys. **30**, 251-267 (2000), [arXiv:hep-th/0010049].
18. P. Hořava, Phys. Rev. **D59**, 046004-1-11 (1999), [arXiv:hep-th/9712130].
19. H. Nastase, *Towards a Chern-Simons M theory of* $OSp(1|32)\times OSp(1|32)$, arXiv:hep-th/0306269 (2003).
20. W. van Holten and P. van Proeyen, J. Phys. **15**, 3763-3783 (1982).
21. R. D'Auria and P. Fré, Nucl. Phys. **B201**, 101-140 (1982) [E.: *ibid.* **B206**, 496 (1982)].
22. C. Chryssomalakos, J.A. de Azcárraga, J.M. Izquierdo and J.C. Pérez Bueno, Nucl. Phys. **B567**, 293-330 (2000), [arXiv:hep-th/9904137];
J.A. de Azcárraga and J.M. Izquierdo, *Superalgebra cohomology, the geometry of extended superspaces and superbranes*, in in *Supersymmetry and fundamental interaction theories*, Proc. of the 37th Karpacz Winter School, J. Lukierski and J. Riembeliński Eds. AIP Conf. Proc. **589**, AIP Conf. Proc. **589**, 3-17 (2001), [arXiv:hep-th/0105125]
23. M.B. Green, Phys. Lett. **B223**, 157-164 (1989).
24. E. Bergshoeff and E. Sezgin, Phys. Lett. **B354**, 256-263 (1995), [arXiv:hep-th/9504140];
see also E. Sezgin, Phys. Lett. B **392**, 323-331 (1997), [arXiv:hep-th/9609086];
Super p-form charges and a reformulation of the suepermembrane action in eleven dimensions, contribution to the 1995 Leuven workshop, hep-th/9512082.
25. J.A. de Azcárraga, P.K. Townsend, Phys. Rev. Lett. **62**, 2579-2582 (1989).
26. A. Achúcarro, J.M. Evans, P.K. Townsend and D.L. Wiltshire, Phys. Lett. **B198**, 441 (1987).
27. J.A. de Azcárraga and J.M. Izquierdo, *Lie groups, Lie algebras, cohomology and some applications in physics*, Camb. Univ. Press (1995).
28. M. Sakaguchi, Phys. Rev. **D59**, 046007 (1999), [hep-th/9809113];
JHEP **0004**, 019 (2000), [arXiv:hep-th/9909143].
29. E. Cremmer, B. Julia and J. Scherk, Phys. Lett. **B76**, 409–412 (1978).
30. I. Bars and S.W. MacDowell, Phys. Lett. **129B**, 182-184 (1983).
31. R.E. Kallosh, Phys. Lett. **143B**, 373-378 (1984).
32. M.J. Duff, *Erice lectures on 'The status of local supersymmetry'*, arXiv:hep-th/0403160.
33. V. Mathai and H. Sati, JHEP **0403**, 016 (2004), [hep-th/0312033].
34. E. Diaconescu, D.S. Freed and G. Moore, *The M-theory 3-form and E(8) gauge theory*, hep-th/0312069.
35. K. Lechner and P. Marchetti, Nucl. Phys. **B672**, 264-302 (2003), [arXiv:hep-th/0302108].
36. S.W. MacDowell and F. Mansouri, Phys. Rev. Lett. **38**, 739-742 (1977);
F. Mansouri, Phys. Rev. **D16**, 2456-2467 (1977).

37. C. Chevalley and S. Eilenberg, Trans. Am. Math. Soc. **63**, 85-124 (1948).
38. L. Castellani, P. Fré, F. Giani, K. Pilch and P. van Nieuwenhuizen, Ann. Phys. **146**, 35-77 (1983).
39. M. Hatsuda and M. Sakaguchi, Prog. Theor. Phys. **109**, 853-867 (2003), [arXiv:hep-th/0106114].
40. J. A. de Azcárraga, J. M. Izquierdo, M. Picón and O. Varela, Nucl. Phys. **B662**, 185-219 (2003), [arXiv:hep-th/0212347];
Class. Quantum Grav. **21**, S1375-S1384 (2004), [arXiv:hep-th/0401033].
41. F.A. Berezin, Sov. J. Nucl. Phys. **29**, 857-866 (1979), Sec. 5.
42. D.V. Volkov, Sov. J. Particles Nucl. **4**, 1-17 (1973); see also ref. [59].
43. A. S. Schwarz, Nucl. Phys. **B171**, 154-166 (1980);
A.V. Gayduk, V.N. Romanov and A.S. Schwarz, Commun. Math. Phys. **79**, 507-528 (1981).
44. J.A. de Azcárraga, J.P. Gauntlett, J.M. Izquierdo, P.K. Townsend, Phys. Rev. Lett. **63**, 2443-2447 (1989).
45. I.A. Bandos, J.A. de Azcárraga, J.M. Izquierdo, M. Picón and O. Varela, Phys. Lett. **B596**, 145-155 (2004), [arXiv:hep-th/0406020];
I.A. Bandos, J.A. de Azcárraga, M. Picón and O. Varela, *On the formulation of D = 11 supergravity and the composite nature of its three-form field*, arXiv:hep-th/0409100, to appear in Ann. Phys.
46. E. Bergshoeff, E. Sezgin and P.K. Townsend, Phys. Lett. **B189**, 75-78 (1987); Ann. Phys. (NY) **185**, 330-365 (1988).
47. J.A. de Azcárraga, J.M. Izquierdo and C. Miquel-Espanya, Nucl. Phys. **B706**, 181-203 (2005), [arXiv:hep-th/0407238].
48. M. Cederwall, A. von Gussich, B.E.W. Nilsson and A. Westerberg, Nucl. Phys. **B490**, 163-178 (1997), [hep-th/9610148];
M. Cederwall, A. von Gussich, B.E.W. Nilsson, P. Sundell and A. Westerberg, Nucl. Phys. **B490**, 179-201 (1997), [hep-th/9611159];
M. Cederwall, A. von Gussich, A. Miković, B.E.W. Nilsson and A. Westerberg, Phys. Lett. **B390**, 148-152 (1997), [hep-th/9606173].
49. M. Aganagic, C. Popescu and J.A. Schwarz, Phys. Lett. **B393**, 311-315 (1997), [hep-th/9610249]; Nucl. Phys. **B495**, 99-126 (1997), [hep-th/9612080].
50. E. Bergshoeff and P.K. Townsend, Phys. Lett. **B490**, 145-162 (1997), [hep-th/9611173]; Nucl.Phys. **B531**, 226-238 (1998), [hep-th/9804011].
51. J. Polchinski, Phys. Rev. Lett. **75**, 4724-4727 (1995), [arXiv:hep-th/9510017].
52. L. Anguelova and P.A. Grassi, JHEP **0311**, 010 (2003), [arXiv:hep-th/0307260];
M. Hatsuda and K. Kamimura, Nucl. Phys. **B703**, 277-292 (2004), [arXiv:hep-th/0405202].
53. I.A. Bandos, K. Lechner, A. Nurmagambetov, P. Pasti, D.P. Sorokin and M. Tonin, Phys. Rev. Lett. **78**, 4332-4334 (1997), [arXiv:hep-th/9701149].
M. Aganagic, J. Park, C. Popescu and J. H. Schwarz, Nucl. Phys. **B496**, 191224 (1997), [arXiv:hep-th/9701166].
54. P.S. Howe and E. Sezgin, Phys. Lett. **B390**, 133-142 (1997), [arXiv:hep-th/9607227]; Phys. Lett. **B394**, 62-66 (1997), [arXiv:hep-th/9611008].
55. I.A. Bandos, D.P. Sorokin, M. Tonin, P. Pasti and D.V. Volkov, Nucl. Phys. **B446**, 79-118 (1995), [arXiv:hep-th/9501113].
56. D. Sullivan, Étud. Sci. Pub. Math. **47**, 269-331 (1977).
57. P. van Nieuwenhuizen, *Free graded differential superalgebras*, in *Group theoretical methods in physics*, M. Serdaroğlu and E. İnönü Eds., Lect. Notes in Phys. **180**, 228-247 (1983).
58. L. Castellani, R. D'Auria and P. Fré, *Supergravity and superstrings: a geometric perspective*, vol. 2, World Sci. (1991).
59. D.V. Volkov and V.P. Akulov, JETP Lett. **16**, 438-440 (1972); Phys. Lett. **B46**, 109-110 (1973).
60. E. Cremmer and S. Ferrara, Phys. Lett. **B91**, 61-77 (1980).
61. L. Brink and P.S. Howe, Phys. Lett. **B91**, 384-393 (1980).
62. A. Candiello and K. Lechner, Nucl. Phys. **B412**, 479-501 (1994), [arXiv:hep-th/9309143].
63. P.S. Howe, Phys. Lett. **B415**,149-155 (1997), [arXiv:hep-th/9707184].
64. P. Pasti, D.P. Sorokin and M. Tonin, Phys. Rev. **D52**, 4277 (1995), [arXiv:hep-th/9506109].
65. I.A. Bandos, N. Berkovits and D.P. Sorokin, Nucl. Phys. **B522**, 214-233 (1998), [arXiv:hep-th/9711055].
66. K. Bautier, S. Deser, M. Henneaux and D. Seminara, Phys. Lett. **B406**, 49-53 (1997), [arXiv:hep-th/9704131].

Unitary Realizations of U-duality Groups as Conformal and Quasiconformal Groups and Extremal Black Holes of Supergravity Theories

M. Gunaydin[1]

Physics Department , Penn State University , University Park, PA 16802, e-mail:murat@phys.psu.edu

Abstract. We review the current status of the construction of unitary representations of U-duality groups of supergravity theories in five, four and three dimensions. We focus mainly on the maximal $N = 8$ supergravity theories and on the $N = 2$ Maxwell-Einstein supergravity (MESGT) theories defined by Jordan algebras of degree three in five dimensions and their descendants in four and three dimensions. Entropies of the extremal black hole solutions of these theories in five and four dimensions are given by certain invariants of their U-duality groups. The five dimensional U-duality groups admit extensions to spectrum generating generalized conformal groups which are isomorphic to the U-duality groups of corresponding four dimensional theories. Similarly, the U-duality groups of four dimensional theories admit extensions to spectrum generating quasiconformal groups that are isomorphic to the corresponding U-duality groups in three dimensions. For example, the group $E_{8(8)}$ can be realized as a quasiconformal group in the 57 dimensional charge-entropy space of BPS black hole solutions of maximal $N = 8$ supergravity in four dimensions and leaves invariant "lightlike separations" with respect to a quartic norm. Similarly $E_{7(7)}$ acts as a generalized conformal group in the 27 dimensional charge space of BPS black hole solutions in five dimensional $N = 8$ supergravity and leaves invariant "lightlike separations" with respect to a cubic norm. For the exceptional $N = 2$ Maxwell-Einstein supergravity theory the corresponding quasiconformal and conformal groups are $E_{8(-24)}$ and $E_{7(-25)}$, respectively. We outline the oscillator construction of the unitary representations of generalized conformal groups that admit positive energy representations, which include the U-duality groups of $N = 2$ MESGT's in four dimensions . We conclude with a review of the minimal unitary realizations of U-duality groups that are obtained by quantizations of their quasiconformal actions and discuss in detail the minimal unitary realization of $E_{8(8)}$.

U-DUALITY GROUPS IN SUPERGRAVITY THEORIES

Noncompact exceptional groups as symmetries of maximally extended supergravity theories in various dimensions

Eleven dimensional supergravity [1] is the effective low energy theory of strongly coupled phase of M-theory [2] and its toroidal compactifications yield the maximally extended supergravity theories in d spacetime dimensions with global non-compact

[1] Work supported in part by the National Science Foundation under grant number PHY-0245337.

CP767, *Fundamental Interactions and Twistor-Like Methods,* edited by J. Lukierski and D. Sorokin

symmetry groups of type $E_{(11-d)(11-d)}$ [3]. The discrete subgroups $E_{(11-d)(11-d)}(\mathbb{Z})$ of these groups are expected to be the symmetries of the non-perturbative spectra of toroidally compactified M-theory [4]. We shall use the term U-duality group to refer to these discrete subgroups as well as to the global noncompact symmetry groups of supergravity theories.

In five dimensions $E_{6(6)}$ is a symmetry of the Lagrangian of the maximal ($N=8$) supergravity, under which the 27 vector fields of the theory transform irreducibly while the 42 scalar fields transform nonlinearly and parameterize the coset space

$$\mathscr{M}_5 = E_{6(6)}/USp(8)\,. \tag{1}$$

On the other hand the $E_{7(7)}$ symmetry of the maximally extended supergravity in $d=4$ is an on-shell symmetry group. The 28 vector field strengths of this theory together with their "magnetic" duals ($\tilde{28}$) transform irreducibly in the 56 of $E_{7(7)}$ and 70 scalar fields parameterize the coset space

$$\mathscr{M}_4 = E_{7(7)}/SU(8)\,. \tag{2}$$

In three dimensions all the propagating bosonic degrees of the maximal $N=16$ supergravity can be dualized to scalar fields which parameterize the coset space

$$\mathscr{M}_3 = E_{8(8)}/SO(16)\,. \tag{3}$$

U-duality groups in matter coupled supergravity theories

Non-compact global U-duality groups arise in matter coupled supergravity theories as well. In this talk I will focus mainly on U-duality groups in $N=2$ Maxwell-Einstein supergravity theories (MESGT) in $d=5$ and the corresponding theories in four and three dimensions. The MESGT's describe the coupling of an arbitrary number n of (Abelian) vector fields to $N=2$ supergravity and five dimensional theories were constructed in [5]. The bosonic part of the Lagrangian can be written as [5]

$$\begin{aligned} e^{-1}\mathscr{L}_{\text{bosonic}} &= -\frac{1}{2}R - \frac{1}{4}\overset{\circ}{a}_{IJ}F^I_{\mu\nu}F^{J\mu\nu} - \frac{1}{2}g_{xy}(\partial_\mu\varphi^x)(\partial^\mu\varphi^y) \\ &+ \frac{e^{-1}}{6\sqrt{6}}C_{IJK}\varepsilon^{\mu\nu\rho\sigma\lambda}F^I_{\mu\nu}F^J_{\rho\sigma}A^K_\lambda\,, \end{aligned} \tag{4}$$

where e and R denote the fünfbein determinant and the scalar curvature in $d=5$, respectively. $F^I_{\mu\nu}$ are the field strengths of the Abelian vector fields A^I_μ, $(I=0,1,2\cdots,n)$ with A^0_μ denoting the "bare" graviphoton. The metric, g_{xy}, of the scalar manifold $\mathscr{M}$ and the "metric" $\overset{\circ}{a}_{IJ}$ of the kinetic energy term of the vector fields both depend on the scalar fields φ^x $(x,y,..=1,2,..,n)$. The invariance under Abelian gauge transformations of the vector fields requires the completely symmetric tensor C_{IJK} to be constant. Remarkably, one finds that the entire $N=2$, $d=5$ MESGT is uniquely determined by the constant tensor C_{IJK} [5]. In particular , the metrics of the kinetic energy terms of the vector and scalar fields are determined by C_{IJK}. More specifically, consider the cubic polynomial, $\mathcal{V}(h)$, in $(n+1)$ real variables h^I $(I=0,1,\ldots,n)$ defined by the C_{IJK}

$$\mathcal{V}(h) := C_{IJK}h^I h^J h^K \; . \tag{5}$$

Using this polynomial as a real " Kähler potential" for a metric, a_{IJ}, in an $n+1$ dimensional ambient space with the coordinates h^I:

$$a_{IJ}(h) := -\frac{1}{3}\frac{\partial}{\partial h^I}\frac{\partial}{\partial h^J}\ln\mathcal{V}(h) \; . \tag{6}$$

one finds that the n-dimensional target space, $\mathcal{M}$, of the scalar fields φ^x can be identified with the hypersurface [5]

$$\mathcal{V}(h) = C_{IJK}h^I h^J h^K = 1 \tag{7}$$

in this space. The metric g_{xy} of the scalar manifold is simply the pull-back of (6) to $\mathcal{M}$ and the "metric" $\overset{\circ}{a}_{IJ}(\varphi)$ of the kinetic energy term of the vector fields appearing in (4) is given by the componentwise restriction of a_{IJ} to $\mathcal{M}$:

$$\overset{\circ}{a}_{IJ}(\varphi) = a_{IJ}|_{\mathcal{V}=1} \; . \tag{8}$$

The physical requirement of positivity of kinetic energy requires that g_{xy} and $\overset{\circ}{a}_{IJ}$ be positive definite metrics. This requirement induces constraints on the possible C_{IJK}, and in [5] it was shown that any C_{IJK} that satisfy these constraints can be brought to the following form

$$C_{000} = 1, \quad C_{0ij} = -\frac{1}{2}\delta_{ij}, \quad C_{00i} = 0, \tag{9}$$

with the remaining coefficients C_{ijk} $(i,j,k = 1,2,\ldots,n)$ being completely arbitrary. This basis is referred to as the canonical basis for C_{IJK}.

Denoting the symmetry group of the tensor C_{IJK} as G one finds that the full symmetry group of $N=2$ MESGT in $d=5$ is of the form

$$G \times SU(2)_R \, , \tag{10}$$

where $SU(2)_R$ denotes the local R-symmetry group of the $N=2$ supersymmetry algebra. A MESGT is said to be *unified* if all the vector fields, including the graviphoton, transform in an irreducible representation of a *simple* symmetry group G of the lagrangian. Of all the $N=2$ MESGT's whose scalar manifolds are symmetric spaces only four are unified [5]. More recently it was shown that if one relaxes the condition that the scalar manifolds be homogeneous spaces then one finds three novel infinite families (plus an additional sporadic one) of unified MESGT's in $d=5$ [6]. If one defines a cubic form

$$\mathcal{N}(h) := C_{IJK}h^I h^J h^K \tag{11}$$

using the constant tensor C_{IJK} , one finds that the cubic forms associated with the four unified MESGT's can be identified with the norm forms of simple (Euclidean) Jordan algebras of degree three [5]. There exist only four simple (Euclidean) Jordan algebras of degree three and they can be realized in terms of 3×3 hermitian matrices over the four division algebras with the product being one-half the anticommutator. They are denoted as $J_3^{\mathbb{A}}$, where $\mathbb{A}$ stands for the underlying division algebra , which can be real numbers $\mathbb{R}$, complex numbers $\mathbb{C}$, quaternions $\mathbb{H}$ and octonions $\mathbb{O}$. The supergravity theories defined by them were referred to as magical supergravity theories [5] since their symmetry groups in five , four and three dimensions correspond precisely to the symmetry groups of the famous Magic Square . The octonionic Jordan algebra $J_3^{\mathbb{O}}$ is the unique exceptional Jordan algebra and consequently the $N=2$ MESGT defined by it is called the exceptional supergravity theory [5]. In the table below we list the scalar manifolds G/H of the magical supergravity theories in five , four and three dimensions, where G is the global noncompact symmetry and H is its maximal compact subgroup.

J	G/H in d=5	G/H in d=4	G/H in d=3
$J_3^{\mathbb{R}}$	$SL(3,\mathbb{R})/SO(3)$	$Sp(6,\mathbb{R})/U(3)$	$F_{4(4)}/USp(6)\times SU(2)$
$J_3^{\mathbb{C}}$	$SL(3,\mathbb{C})/U(3)$	$SU(3,3)/SU(3)^2\times U(1)$	$E_{6(2)}/SU(6)\times SU(2)$
$J_3^{\mathbb{H}}$	$SU^*(6)/USp(6)$	$SO^*(12)/U(6)$	$E_{7(-5)}/SO(12)\times SU(2)$
$J_3^{\mathbb{O}}$	$E_{6(-26)}/F_4$	$E_{7(-25)}/E_6\times U(1)$	$E_{8(-24)}/E_7\times SU(2)$

Note that the exceptional $N=2$ supergravity has $E_{6(-26)}$,$E_{7(-25)}$ and $E_{8(-24)}$ as its global symmetry groups in five , four and three dimensions, respectively, whereas the maximally extended supergravity theory has the maximally split real forms $E_{6(6)}$,$E_{7(7)}$ and $E_{8(8)}$ as its symmetry groups in the corresponding dimensions.

In addition to four simple Euclidean Jordan algebras $J_3^{\mathbb{A}}$ there exist an infinite family of nonsimple Jordan algebras of degree three, which are direct sums $J=\mathbb{R}\oplus\Gamma(Q)$ of a one dimensional Jordan algebra $\mathbb{R}$ with a Jordan algebra $\Gamma(Q)$ associated with a quadratic form Q [2]. This family of $N=2$ MESGT's is called the generic Jordan family and their scalar manifolds in five dimensions are

$$\mathcal{M}_5(\mathbb{R}\oplus\Gamma(Q)) = \frac{SO(1,1)\times SO(n-1,1)}{SO(n)} \tag{12}$$

and $SO(1,1)\times SO(n-1,1)$ are the global symmetry groups of their Lagrangians. The corresponding $N=2$ MESGT's in $d=4$ obtained by dimensional reduction have the scalar manifolds:

$$\mathcal{M}_4(\mathbb{R}\oplus\Gamma(Q)) = \frac{SO(2,1)\times SO(n,2)}{SO(2)\times SO(n)\times SO(2)}. \tag{13}$$

By further reduction to three dimensions the scalar manifolds become

$$\mathcal{M}_3((\mathbb{R}\oplus\Gamma(Q)) = \frac{SO(n+2,4)}{SO(n+2)\times SO(4)}. \tag{14}$$

[2] The positivity of the kinetic terms of scalars and vectors requires the metric of the quadratic form to me Minkowskian .

The term U-duality was introduced by Hull and Townsend since the discrete symmetry group $E_{7(7)}(\mathbb{Z})$ of M/superstring theory toroidally compactified to $d=4$ *unifies* the T-duality group $SO(6,6)(\mathbb{Z})$ with the S-duality group $SL(2,\mathbb{Z})$ in a simple group since

$$SO(6,6)\times SL(2,\mathbb{R})\subset E_{7(7)}\,. \tag{15}$$

The analogous decomposition of the symmetry group $E_{7(-25)}$ of the exceptional supergravity in $d=4$ is

$$SO(10,2)\times SL(2,\mathbb{R})\subset E_{7(-25)}\,, \tag{16}$$

with similar decompositions for the other magical supergravity theories. For the generic Jordan family of $N=2$ MESGT's there is no simple U-duality group that unifies the corresponding S ($SL(2,\mathbb{R})$) and T-duality $(SO(n+2,2))$ groups.

U-DUALITY GROUPS AND ENTROPY OF BPS BLACK HOLES IN SUPERGRAVITY THEORIES

The entropy of BPS black hole solutions in maximally extended supergravity as well as in matter coupled supergravity theories are invariant under the corresponding U-duality groups. For example in $d=5$, $N=8$ supergravity the entropy S of a BPS black hole solution can be written in the form [7]

$$S=\alpha\sqrt{I_3}=\alpha\sqrt{d_{IJK}q^Iq^Jq^K}\,, \tag{17}$$

where α is some fixed constant and I_3 is the cubic invariant of $E_{6(6)}$ with the $q^I, I=0,2,...,26$ denoting the charges coupling to 27 vector fields of the theory. The BPS black hole solutions with $I_3\neq 0$ preserve 1/8 supersymmetry [8] and the solutions with $I_3=0$, but with $\frac{1}{3}\partial_I I_3=d_{IJK}q^Jq^K\neq 0$ preserve 1/4 supersymmetry. The BPS black hole solutions with both $I_3=0$ and $d_{IJK}q^Jq^K=0$ preserve 1/2 supersymmetry [8]. The orbits of the BPS black hole solutions of $N=8$ supergravity in $d=5$ under the action of its U-duality group $E_{6(6)}$ were classified in [9].

The entropy of the BPS black hole solutions of five dimensional $N=2$ MESGT's is given by the cubic form defined by the constant tensor [7]

$$S=\alpha\sqrt{\mathcal{N}}=\alpha\sqrt{C_{IJK}q^Iq^Jq^K}\,. \tag{18}$$

For those $N=2$ MESGT theories defined Jordan algebras of degree three this cubic form is the norm form and the global symmetry group G is its invariance group as explained in the previous section. The orbits of the BPS black hole solutions of $5d, N=2$ MESGT's defined by Jordan algebras under the actions of their U-duality groups were classified in [9]. This was achieved by associating with a given BPS solution with charges q^I, $(I=0,1,...n)$ an element

$$J=\sum_{I=0}^{n}e_Iq^I \tag{19}$$

of the Jordan algebra of degree three, where e_I form a basis of the Jordan algebra. This establishes a correspondence between the Jordan algebra and the charge space of the BPS black hole solutions.

Similarly , the classification of the orbits of BPS black hole solutions of the $N = 8$ supergravity in $d = 5$ as given in [9] associates with a given BPS black hole solution with charges q^I an element $J = \sum_{I=0}^{26} e_I q^I$ of the split exceptional Jordan algebra with basis elements e_I [3]. The cubic invariant $I_3(q^I)$ is then simply given by the norm form $\mathcal{N}$ of the split exceptional Jordan algebra. Invariance group $E_{6(6)}$ of this norm form (known as the reduced structure group in mathematics literature) is the U-duality group of the maximal $N = 8$ supergravity theory in $d = 5$.

In four dimensional magical $N = 2$ MESGT's obtained by dimensional reduction from five dimensions as well as the maximal $4d$, $N = 8$ supergravity the entropies of BPS black hole solutions are given by the quartic invariants of their U-duality groups [10]

$$S = \beta \sqrt{I_4} = \beta \sqrt{d_{IJKL} q^I q^J q^K q^L}, \tag{20}$$

where d_{IJKL} are the completely symmetric tensors defined by the Freudenthal-Kantor triple systems associated with the corresponding simple Jordan algebras of degree three [9] and q^I now denote both electric and magnetic charges. For the generic Jordan family of $N = 2$ MESGT's in $d = 4$ the quartic invariants are defined by the completely symmetric tensors of the Freudenthal-Kantor triple systems associated with the groups $SO(n+4,4)$. The orbits of the BPS black hole solutions of these theories under the action of the corresponding U-duality groups were given in [9]. For the maximal $N = 8$ supergravity q^I represent 28 electric and 28 magnetic charges. The number of supersymmetries preserved by the extremal black hole solutions of the $N = 8$ theory depends on whether or not I_4 , $\partial_J I_4$ and $\partial_J \partial_K I_4$ vanish [8].

GENERALIZED SPACE-TIMES DEFINED BY JORDAN ALGEBRAS

Generalized rotation, Lorentz and conformal groups

In the previous sections we saw how Jordan algebras arise in a fundamental way within the framework of supergravity theories. In this section I will review how Jordan algebras appear naturally within the framework of attempts to generalize four dimensional Minkowski spacetime and its symmetry groups. The first proposal to use Jordan algebras to define generalized spacetimes was made in the early days of spacetime supersymmetry while searching for the super analogs of the exceptional Lie algebras [11].

[3] Split exceptional Jordan algebra $J_3^{\mathbb{O}_s}$ is realized in terms of hermitian 3×3 matrices over the split octonions $\mathbb{O}_s$, which is not a division algebra. As a consequence $J_3^{\mathbb{O}_s}$ is not a Euclidean (formally real) Jordan algebra in contrast to the real exceptional Jordan algebra $J_3^{\mathbb{O}}$ of the exceptional $N = 2$ theory, which is defined over the division algebra of real octonions $\mathbb{O}$. A Jordan algebra J is called Euclidean if for any pair of elements $X, Y \in J$ the equation $X^2 + Y^2 = 0$ implies that $X = 0$ and $Y = 0$.

Now the twistor formalism in four-dimensional space-time ($d = 4$) leads naturally to the representation of four vectors in terms of 2×2 Hermitian matrices over the field of complex numbers $\mathbb{C}$. In particular, the coordinate four vectors x_μ can be represented as :

$$x = x_\mu \sigma^\mu . \tag{21}$$

Since the Hermitian matrices over the field of complex numbers close under the symmetric anti-commutator product one can regard the coordinate vectors as elements of a Jordan algebra denoted as $J_2^{\mathbb{C}}$ [11, 12]. Then the rotation, Lorentz and conformal groups in $d = 4$ can be identified with the automorphism , reduced structure and Möbius (linear fractional) groups of the Jordan algebra of 2×2 complex Hermitian matrices $J_2^{\mathbb{C}}$ [11, 12]. The reduced structure group $Str_0(J)$ of a Jordan algebra J is simply the invariance group of its norm form $N(J)$. (The structure group $Str(J) = Str_0(J) \times SO(1,1)$,on the other hand, is simply the invariance group of $N(J)$ up to an overall constant scale factor.) Furthermore, this interpretation leads one naturally to define generalized space-times whose coordinates are parameterized by the elements of Jordan algebras [11]. The rotation $Rot(J)$, Lorentz $Lor(J)$ and conformal $Con(J)$ groups of these generalized space-times are then identified with the automorphism $Aut(J)$, reduced structure $Str_0(J)$ and Möbius Mö(J) groups of the corresponding Jordan algebra [11, 12, 13, 14]. Denoting as $J_n^{\mathbb{A}}$ the Jordan algebra of $n \times n$ Hermitian matrices over the *division* algebra $\mathbb{A}$ and the Jordan algebra of Dirac gamma matrices in d (Euclidean) dimensions as $\Gamma(d)$ one finds the following symmetry groups of generalized space-times defined by simple Euclidean (formally real) Jordan algebras:

J	$Rotation(J)$	$Lorentz(J)$	$Conformal(J)$
$J_n^{\mathbb{R}}$	$SO(n)$	$SL(n,\mathbb{R})$	$Sp(2n,\mathbb{R})$
$J_n^{\mathbb{C}}$	$SU(n)$	$SL(n,\mathbb{C})$	$SU(n,n)$
$J_n^{\mathbb{H}}$	$USp(2n)$	$SU^*(2n)$	$SO^*(4n)$
$J_3^{\mathbb{O}}$	F_4	$E_{6(-26)}$	$E_{7(-25)}$
$\Gamma(d)$	$SO(d)$	$SO(d,1)$	$SO(d,2)$

The symbols $\mathbb{R}$, $\mathbb{C}$, $\mathbb{H}$, $\mathbb{O}$ represent the four division algebras. For the Jordan algebras $J_n^{\mathbb{A}}$ the norm form is the determinental form (or its generalization to the quaternionic and octonionic matrices). For the Jordan algebra $\Gamma(d)$ generated by Dirac gamma matrices Γ_i $(i = 1,2,...d)$

$$\{\Gamma_i, \Gamma_j\} = \delta_{ij}\mathbf{1}; \quad i,j,\ldots = 1,2,\ldots,d \tag{22}$$

the norm of a general element $x = x_0\mathbf{1} + x_i\Gamma_i$ of $\Gamma(d)$ is quadratic and given by

$$N(x) = x\bar{x} = x_0^2 - x_i x_i \tag{23}$$

where $\bar{x} = x_0\mathbf{1} - x_i\Gamma_i$. Its automorphism, reduced structure and Möbius groups are simply the rotation, Lorentz and conformal groups of $(d+1)$-dimensional Minkowski spacetime. One finds the following special isomorphisms between the Jordan algebras of 2×2 Hermitian matrices over the four division algebras and the Jordan algebras of gamma matrices:

$$J_2^{\mathbb{R}} \simeq \Gamma(2) \quad ; \quad J_2^{\mathbb{C}} \simeq \Gamma(3) \quad ; \quad J_2^{\mathbb{H}} \simeq \Gamma(5) \quad ; \quad J_2^{\mathbb{O}} \simeq \Gamma(9) \; . \tag{24}$$

The Minkowski spacetimes they correspond to are precisely the critical dimensions for the existence of super Yang-Mills theories as well as of the classical Green-Schwarz superstrings. These Jordan algebras are all quadratic and their norm forms are precisely the quadratic invariants constructed using the Minkowski metric.

We should note two remarkable facts about the above table. First , the conformal groups of generalized space-times defined by Euclidean (formally real) Jordan algebras all admit positive energy unitary representations [4]. Hence they can be given a causal structure with a unitary time evolution as in four dimensional Minkowski space-time. Second is the fact that the maximal compact subgroups of the generalized conformal groups of formally real Jordan algebras are simply the compact forms of their structure groups (which are the products of their generalized Lorentz groups with dilatations), whose importance will be explained in the next subsection.

Covariant quantum fields over generalized spacetimes and the positive energy unitary representations of their conformal groups

The Lie algebra g of a non-compact group G that admits unitary lowest weight representations (ULWR) (positive energy) has a 3-grading with respect to the Lie algebra h of its maximal compact subgroup H i.e

$$g = g^{-1} \oplus g^0 \oplus g^{+1} , \tag{25}$$

where $g^0 = h$ and we have the formal commutation relations

$$[g^{(m)}, g^{(n)}] \subseteq g^{(m+n)} , \qquad m,n = \mp 1, 0$$

and $g^{(m)} \equiv 0$ for $|m| > 1$.

In [15] the general oscillator construction of unitary lowest weight representations of non-compact groups was given. To construct the ULWR's of a noncompact group G one realizes its generators in terms of bilinears of bosonic oscillators transforming

[4] Similarly, the generalized conformal groups defined by Hermitian Jordan triple systems all admit positive energy unitary representations [14]. In fact the conformal groups of simple Hermitian Jordan triple systems exhaust the list of simple noncompact groups that admit positive energy unitary representations. They include the conformal groups of simple Euclidean Jordan algebra since the latter form an hermitian Jordan triple system under the Jordan triple product [14].

in a certain representation of H. Then in the corresponding Fock space $\mathscr{F}$ of all the oscillators one chooses a set of states $|\Omega>$, referred to as the "lowest weight vector" (lwv), which transforms irreducibly under H and which are annihilated by the generators belonging to the g^{-1} space. Then by acting on $|\Omega>$ repeatedly with the generators belonging to the g^{+1} space one obtains an infinite set of states

$$|\Omega>, \; g^{+1}|\Omega>, \; g^{+1}g^{+1}|\Omega>, \ldots \tag{26}$$

that form the basis of an irreducible unitary lowest weight representation of g. (The irreducibility of the representation of g follows from the irreducibility of the lwv $|\Omega>$ under h. We shall refer to this basis as the compact "particle" basis.

For example, the conformal group of the Jordan algebra $J_2^{\mathbb{C}}$ corresponding to the four dimensional Minkowski space is $SU(2,2)$ with a maximal compact subgroup $SU(2) \times SU(2) \times U(1)$ which is simply the compact form of the structure group $SL(2,\mathbb{C}) \times SO(1,1)$. In [17] it was explicitly shown how to go from the compact $SU(2) \times SU(2) \times U(1)$ basis of the ULWR's of $SU(2,2)$ to the manifestly covariant $SL(2,\mathbb{C}) \times SO(1,1)$ basis. The transition from the compact to the covariant basis corresponds simply to going from a compact "particle" basis to a noncompact coherent state basis of the ULWR. The noncompact coherent states are labelled by the elements of $J_2^{\mathbb{C}}$ i.e by the coordinates of four dimensional Minkowski space. One thereby establishes a one-to-one correspondence between irreducible ULWR's of $SU(2,2)$ and the fields transforming irreducibly under the Lorentz group $SL(2,\mathbb{C})$ with a definite conformal dimension. Thus one can associate with irreducible ULWR's of $SU(2,2)$ conformal fields transforming covariantly under the Lorentz group with a definite conformal dimension.

Similarly, the conformal group $SO^*(8)$ of the Jordan algebra $J_2^{\mathbb{H}}$ parameterizing the six dimensional Minkowski space has a maximal compact subgroup $U(4)$ which is the compact form of the structure group $SU^*(4) \times SO(1,1)$. In [18] it was shown explicitly how to go from the compact $U(4)$ particle basis of the ULWR's of $SO^*(8)$ to the non-compact $SU^*(4) \times SO(1,1)$ coherent state basis. ($SU^*(4) \times SO(1,1)$ is simply the covering group of the Lorentz group in six dimensions times dilatations) . The coherent states of the non-compact basis are again labelled by the elements of $J_2^{\mathbb{H}}$, i.e the coordinates of 6d Minkowski space. Thus each irreducible ULWR of $SO^*(8)$ can be identified with a field transforming covariantly under the Lorentz group $SU^*(4)$ with a definite conformal dimension.

The results obtained explicitly for the conformal groups of $J_2^{\mathbb{C}}$ and $J_2^{\mathbb{H}}$ extend to the conformal groups of all formally real Jordan algebras and of Hermitian Jordan triple systems [16, 19]. The general theory can be summarized as follows: Let g be the Lie algebra of the conformal group of a formally real Jordan algebra and g^0 the Lie algebra of its maximal compact subgroup. Then g has a three-graded decomposition with respect to g^0:

$$g = g^- + g^0 + g^+ \ , \tag{27}$$

where the grading is determined by the "conformal energy operator". Now let n^0 be the Lie algebra of the structure group (Lorentz group times dilatations) of the Euclidean Jordan algebra (or of a Hermitian Jordan triple system). As stated above the Lie algebra g has a 3-graded decomposition with respect to n^0 as well:

$$g = n^- + n^0 + n^+ \ , \tag{28}$$

where the grading is defined by the generator of scale transformations. In the compact basis an irreducible ULWR of $Conf(J)$ is uniquely determined by a lowest weight vector $|\Omega\rangle$ transforming irreducibly under the maximal compact subgroup K that is annihilated by the operators belonging to g^-

$$g^-|\Omega\rangle = 0. \tag{29}$$

As was done explicitly for the conformal groups in 4 and 6 dimensions [17, 18] one can show that there exists a rotation operator W in the representation space with the property that the states $W|\Omega\rangle$ are annihilated by all the generators belonging to n^-

$$n^- W|\Omega\rangle = 0 \tag{30}$$

and transform in a finite dimensional non-unitary representation of the non-compact structure group. Remarkably the transformation properties of $W|\Omega\rangle$ under the structure group coincide with the transformation properties of $|\Omega\rangle$ under the maximal compact subgroup K. In particular, the conformal dimension of the vector $W|\Omega\rangle$ is simply the negative of the conformal energy of $|\Omega\rangle$. If one chooses a basis e_μ for the Jordan algebra J and denote the generators of generalized translations in the space n^+ corresponding to e_μ as P_μ, then the noncompact coherent states defined by the action of generalized translations on $W|\Omega\rangle$

$$|\Phi(x_\mu\rangle := e^{ix^\mu P_\mu} W|\Omega\rangle \tag{31}$$

form the covariant basis of the ULWR of the generalized conformal group $Con(J)$ [5] . The coherent states $|\Phi(x_\mu\rangle$ labelled by the coordinates correspond to conformal fields transforming covariantly under the Lorentz group with a definite conformal dimension. Since the state $W|\Omega\rangle$ is annihilated by the generators of special conformal transformations K_μ belonging to the space n^- this proves that the irreducible ULWR's are equivalent to representations induced by finite dimensional irreps of the Lorentz group with a definite conformal dimension and trivial special conformal transformation properties. This generalizes the well-known construction of the positive energy representations of the four dimensional conformal group $SU(2,2)$ [21] to all generalized conformal groups of formally real Jordan algebras and Hermitian Jordan triple systems. They are simply induced representations with respect to the maximal parabolic subgroup $Str(J) \odot S_J$ where $\odot$ denotes semi-direct product and S_J is the Abelian subgroup generated by generalized special conformal transformations.

We should perhaps note that the generalized Poincaré groups associated with the spacetimes defined by Jordan algebras are of the form

$$\mathscr{PG}(J) := Lor(J) \odot T_J \tag{32}$$

where T_J is the Abelian subgroup generated by generalized translations P_μ. For quadratic Jordan algebras, $\Gamma(d)$, $\mathscr{PG}(\Gamma(d))$ is simply the Poincaré group in d dimensional Minkowski space. The group $\mathscr{PG}(\Gamma(d))$ has a quadratic Casimir operator $M^2 = P_\mu P^\mu$

---

[5] We should note that the (super) coherent states associated with ULWR's of non-compact (super) groups introduced in [20] are labelled by complex (super) "coordinates" in the compact basis. These (super) coordinates parametrize the (super) hermitian symmetric space G/H.

which is simply the mass operator. For Jordan algebras J of degree n the generalized Poincaré group $\mathscr{PG}(J)$ has a Casimir invariant of order n constructed out of the generalized translation generators P_μ. For example for the real exceptional Jordan algebra $J_3^{\mathbb{O}}$ the corresponding Casimir invariant is cubic and has the form

$$M^3 = C_{\mu\nu\rho} P^\mu P^\nu P^\rho \tag{33}$$

where $C_{\mu\nu\rho}$ is the symmetric invariant tensor of the generalized Lorentz group $E_{6(-26)}$ of $J_3^{\mathbb{O}}$ ($\mu,\nu,\rho,.. = 0,1,...26$)[6].

CONFORMAL AND QUASI-CONFORMAL EXTENSIONS OF U-DUALITY GROUPS IN FIVE AND FOUR DIMENSIONS

As discussed above we can associate an element $J = \sum_{I=0}^{26} e_I q^I$ of the real (split) exceptional Jordan algebra $J_3^{\mathbb{O}}(J_3^{\mathbb{O}_s})$ with a BPS black hole solution of the exceptional $N=2(N=8)$ supergravity theory with charges q^I in five dimensions. Its entropy S is given by the square root of the cubic norm $\mathscr{N}(J)$ of J, whose invariance group is $E_{6(-26)}$ ($E_{6(6)}$).

Acting on an element $J = \sum_{I=0}^{26} e_I q^I$ of $J_3^{\mathbb{O}}(J_3^{\mathbb{O}_s})$ by its conformal group $E_{7(-25)}(E_{7(7)})$ changes its norm and hence the corresponding entropy. Thus one can regard $E_{7(-25)}(E_{7(7)})$ as a spectrum generating symmetry in the charge space of BPS black hole solutions of the exceptional $N=2$ ($N=8$) supergravity in five dimensions [22, 23, 26]. If one defines a distance function between any two solutions with charges q^I and q'^I as the cubic norm of their difference

$$d(q,q') \equiv \mathscr{N}_3(J-J') \tag{34}$$

one finds that the light like separations are preserved under the conformal action of $E_{7(-25)}(E_{7(7)})$ [22, 26]. The explicit action of $E_{7(7)}$ and $E_{7(-25)}$ on the corresponding 27 dimensional spaces are given in [22] and [26] , respectively. Let us review briefly the conformal action of $E_{7(7)}$ given in [22]. Lie algebra of $E_{7(7)}$ has a 3-graded decomposition

$$\mathbf{133} = \mathbf{27} \oplus (\mathbf{78} \oplus \mathbf{1}) \oplus \overline{\mathbf{27}} \tag{35}$$

under its $E_{6(6)} \times \mathscr{D}$ subgroup, where $\mathscr{D}$ represents the dilatation group $SO(1,1)$. Under its maximal compact subalgebra USp(8) Lie algebra $E_{6(6)}$ decomposes as a symmetric tensor $\tilde{G}^{ij}$ in the adjoint **36** of USp(8) and a fully antisymmetric symplectic traceless tensor $\tilde{G}^{ijkl}$ transforming as the **42** of USp(8) (indices $1 \leq i,j,\ldots \leq 8$ are USp(8) indices). $\tilde{G}^{ijkl}$ is traceless with respect to the real symplectic metric $\Omega_{ij} = -\Omega_{ji} = -\Omega^{ij}$

---

[6] We should note that we are relabelling the indices $I,J,..$ with Greek characters $\mu,\nu,...$ so as to emphasize the fact that ,in this section, we are looking at Jordan algebras from a spacetime point of view.

(thus $\Omega_{ik}\Omega^{kj} = \delta_i^j$). The symplectic metric is used to raise and lower indices, with the convention that this is always to be done from the left. The other generators of conformal $E_{7(7)}$ consist of a dilatation generator $\tilde{H}$, translation generators $\tilde{E}^{ij}$ and the nonlinearly realized "special conformal" generators $\tilde{F}^{ij}$, transforming as **27** and $\overline{\mathbf{27}}$, respectively.

The fundamental **27** of $E_{6(6)}$ on which $E_{7(7)}$) acts nonlinearly can be represented as the symplectic traceless antisymmetric tensor $\tilde{Z}^{ij}$ transforming as [7]

$$\begin{aligned} \tilde{G}^i{}_j(\tilde{Z}^{kl}) &= 2\,\delta^k_j \tilde{Z}^{il}\,, \\ \tilde{G}^{ijkl}(\tilde{Z}^{mn}) &= \tfrac{1}{24}\varepsilon^{ijklmnpq}\tilde{Z}_{pq}\,, \end{aligned} \tag{36}$$

where $\tilde{Z}_{ij} := \Omega_{ik}\Omega_{jl}\tilde{Z}^{kl} = (\tilde{Z}^{ij})^*$ and $\Omega_{ij}\tilde{Z}^{ij} = 0$. The conjugate $\overline{\mathbf{27}}$ representation transforms as

$$\begin{aligned} \tilde{G}^i{}_j(\tilde{Z}^{kl}) &= 2\,\delta^k_j \tilde{Z}^{il}\,, \\ \tilde{G}^{ijkl}(\tilde{Z}^{mn}) &= -\tfrac{1}{24}\varepsilon^{ijklmnpq}\tilde{Z}_{pq}\,. \end{aligned} \tag{37}$$

The cubic invariant of $E_{6(6)}$ in the **27** is given by

$$\mathcal{N}_3(\tilde{Z}) := \tilde{Z}^{ij}\tilde{Z}_{jk}\tilde{Z}^{kl}\Omega_{il}\,. \tag{38}$$

The generators $\tilde{E}^{ij}$ act as translations on the space with coordinates $\tilde{Z}^{ij}$ as :

$$\tilde{E}^{ij}(\tilde{Z}^{kl}) = -\Omega^{i[k}\Omega^{l]j} - \tfrac{1}{8}\Omega^{ij}\Omega^{kl} \tag{39}$$

and $\tilde{H}$ by dilatations

$$\tilde{H}(\tilde{Z}^{ij}) = \tilde{Z}^{ij}\,. \tag{40}$$

The " special conformal generators" $\tilde{F}^{ij}$ in the $\overline{\mathbf{27}}$ are realized nonlinearly:

$$\begin{aligned} \tilde{F}^{ij}(\tilde{Z}^{kl}) :=\; & -2\tilde{Z}^{ij}(\tilde{Z}^{kl}) + \Omega^{i[k}\Omega^{l]j}(\tilde{Z}^{mn}\tilde{Z}_{mn}) + \tfrac{1}{8}\Omega^{ij}\Omega^{kl}(\tilde{Z}^{mn}\tilde{Z}_{mn}) \\ & + 8\tilde{Z}^{km}\tilde{Z}_{mn}\Omega^{n[i}\Omega^{j]l} - \Omega^{kl}(\tilde{Z}^{im}\Omega_{mn}\tilde{Z}^{nj})\,. \end{aligned} \tag{41}$$

The norm form needed to define the $E_{7(7)}$ invariant “light cones” is constructed from the cubic invariant of $E_{6(6)}$. If we define the "distance" between $\tilde{X}$ and $\tilde{Y}$ as $\mathcal{N}_3(\tilde{X} - \tilde{Y})$ then it is manifestly invariant under $E_{6(6)}$ and under the translations $\tilde{E}^{ij}$. Under $\tilde{H}$ it transforms by a constant factor, whereas under the action of $\tilde{F}^{ij}$ one finds

$$\tilde{F}^{ij}\Big(\mathcal{N}_3(\tilde{X} - \tilde{Y})\Big) = (\tilde{X}^{ij} + \tilde{Y}^{ij})\mathcal{N}(\tilde{X} - \tilde{Y})\,, \tag{42}$$

[7] Throughout we use the convention that indices connected by a bracket are antisymmetrized with weight one.

which proves that the light cone in $\mathbb{R}^{27}$ with base point $\tilde{Y}$ defined by

$$\mathcal{N}_3(\tilde{X}-\tilde{Y})=0 \tag{43}$$

is indeed invariant under $E_{7(7)}$.

The above formulas carry over in a straightforward manner to the conformal realization of $E_{7(-25)}$ on a 27 dimensional space coordinatized by the real exceptional Jordan algebra $J_3^{\mathbb{O}}$. In this case the cubic form is invariant under $E_{6(-26)}$ which has $USp(6,2)$ as a subgroup. The $USp(8)$ covariant formulas above for $E_{7(7)}$ are then replaced by $USp(6,2)$ covariant formulas [26, 27]. For the generic Jordan family of $N=2$ MESGT's in five dimensions the conformal extensions of U-duality groups $SO(1,1)\times SO(n-1,1)$ are the groups

$$SO(2,1)\times SO(n,2)\,, \tag{44}$$

which act as spectrum generating symmetry groups in $d=5$. These conformal groups of generic $d=5$ MESGT's as well as the conformal groups $E_{7(7)}$ and $E_{7(-25)}$ acting on the 27 dimensional charge spaces of the $N=8$ and the exceptional $N=2$ supergravity in five dimensions are isomorphic to the U-duality groups of the corresponding four dimensional theories obtained by dimensional reduction. We should stress the obvious fact that the conformal extensions of U-duality groups act nonlinearly on the charge space of five dimensional theories , whereas the isomorphic four dimensional U-duality groups act linearly on the charge space spanned by both electric and magnetic charges.

One may wander whether there exist "conformal extensions " of the four dimensional U-duality groups that act nonlinearly on the 4d charge spaces as spectrum generating symmetry groups and are isomorphic to the U-duality groups of the corresponding three dimensional theories obtained by dimensional reduction. This question was investigated in [22] and it was shown there that in the case of maximal supergravity, even though there is no conformal action of $E_{8(8)}$, which is the corresponding U-duality group in $d=3$, it has a quasi-conformal group action on a 57 dimensional space which is an extension of the 56 dimensional charge space by an extra coordinate. For BPS black hole solutions in $d=4$ this extra coordinate can be taken to be the entropy [22].

The realization of quasi-conformal action of $E_{8(8)}$ uses the 5-graded decomposition of its Lie algebra with respect to the Lie algebra of its $E_{7(7)}\times\mathcal{D}$ subgroup

$$\begin{array}{ccccc} \mathfrak{g}^{-2}\oplus\mathfrak{g}^{-1}\oplus & \mathfrak{g}^{0} & \oplus\,\mathfrak{g}^{+1}\oplus\mathfrak{g}^{+2} \\ \mathbf{1}\;\oplus\;\mathbf{56}\;\oplus & (\mathbf{133}\oplus\mathbf{1}) & \oplus\;\mathbf{56}\;\oplus\;\mathbf{1} \end{array} \tag{45}$$

with $\mathcal{D}$ representing dilatations, whose generator together with grade ± 2 elements generate an $SL(2,\mathbb{R})$ subgroup. It turns out to be very convenient to work in a basis covariant with respect to the $\mathrm{SL}(8,\mathbb{R})$ subgroup of $E_{7(7)}$ [22]. Let us denote the $\mathrm{SL}(8,\mathbb{R})$ covariant generators belonging to the grade $-2,-1,0,1$ and 2 subspaces in the above decomposition as follows:

$$E\oplus\{E^{ij},E_{ij}\}\oplus\{G^{ijkl},\,G^{i}{}_{j};H\}\oplus\{F^{ij},F_{ij}\}\oplus F\;, \tag{46}$$

where $i,j,..=1,2...,8$ are now $\mathrm{SL}(8,\mathbb{R})$ indices.

Consider now a 57-dimensional real vector space with coordinates

$$\mathscr{X} := (X^{ij}, X_{ij}, x)\,, \tag{47}$$

where X^{ij} and X_{ij} transform in the 28 and $\tilde{28}$ of SL$(8,\mathbb{R})$ and x is a singlet. The generators of $E_{7(7)}$ subalgebra act linearly on this space

$$\begin{aligned}
G^i{}_j(X^{kl}) &= 2\,\delta^{k}_{j}X^{il} - \tfrac{1}{4}\delta^i_j X^{kl}\,, & G^{ijkl}(X^{mn}) &= \tfrac{1}{24}\varepsilon^{ijklmnpq}X_{pq}\,,\\
G^i{}_j(X_{kl}) &= -2\,\delta^i_k X_{jl} + \tfrac{1}{4}\delta^i_j X_{kl}\,, & G^{ijkl}(X_{mn}) &= \delta^{[ij}_{mn}X^{kl]}\,,\\
G^i{}_j(x) &= 0\,, & G^{ijkl}(x) &= 0\,.
\end{aligned} \tag{48}$$

The generator H of dilatations acts as

$$H(X^{ij}) = X^{ij}\,, \qquad H(X_{ij}) = X_{ij}\,, \qquad H(x) = 2x\,, \tag{49}$$

and the generator E acts as translations on x:

$$E(X^{ij}) = 0\,, \qquad E(X_{ij}) = 0\,, \qquad E(x) = 1\,. \tag{50}$$

The grade ± 1 generators act as

$$\begin{aligned}
E^{ij}(X^{kl}) &= 0\,, & E^{ij}(X_{kl}) &= \delta^{ij}_{kl}\,, & E^{ij}(x) &= -X^{ij}\,,\\
E_{ij}(X^{kl}) &= \delta^{kl}_{ij}\,, & E_{ij}(X_{kl}) &= 0\,, & E_{ij}(x) &= X_{ij}\,.
\end{aligned} \tag{51}$$

The positive grade generators are realized nonlinearly. The generator F acts as

$$\begin{aligned}
F(X^{ij}) &= 4X^{ik}X_{kl}X^{lj} + X^{ij}X^{kl}X_{kl}\\
&\quad - \tfrac{1}{12}\varepsilon^{ijklmnpq}X_{kl}X_{mn}X_{pq} + X^{ij}x\,,\\
F(X_{ij}) &= -4X_{ik}X^{kl}X_{lj} - X_{ij}X^{kl}X_{kl}\\
&\quad + \tfrac{1}{12}\varepsilon_{ijklmnpq}X^{kl}X^{mn}X^{pq} + X_{ij}x\,,\\
F(x) &= 4\,\mathscr{I}_4(X^{ij}, X_{ij}) + x^2\,,
\end{aligned} \tag{52}$$

where $\mathscr{I}_4$ is the quartic invariant of $E_{7(7)}$

$$\begin{aligned}
\mathscr{I}_4 \equiv{}& X^{ij}X_{jk}X^{kl}X_{li} - \tfrac{1}{4}X^{ij}X_{ij}X^{kl}X_{kl} + \tfrac{1}{96}\,\varepsilon^{ijklmnpq}X_{ij}X_{kl}X_{mn}X_{pq}\\
&+ \tfrac{1}{96}\,\varepsilon_{ijklmnpq}X^{ij}X^{kl}X^{mn}X^{pq}\,.
\end{aligned} \tag{53}$$

The action of the remaining generators of $E_{8(8)}$ are as follows:

$$F^{ij}(X^{kl}) = -4X^{i[k}X^{l]j} + \tfrac{1}{4}\,\varepsilon^{ijklmnpq}X_{mn}X_{pq}\,,$$

$$
\begin{aligned}
F^{ij}(X_{kl}) &= +8\,\delta_k^{[i}X^{j]m}X_{ml}+\delta_{kl}^{ij}X^{mn}X_{mn}+2X^{ij}X_{kl}-\delta_{kl}^{ij}x\,,\\
F_{ij}(X^{kl}) &= -8\,\delta_{[i}^{k}X_{j]m}X^{ml}+\delta_{ij}^{kl}X^{mn}X_{mn}-2X_{ij}X^{kl}-\delta_{ij}^{kl}x\,,\\
F_{ij}(X_{kl}) &= 4X_{ki}X_{jl}-\tfrac{1}{4}\,\varepsilon_{ijklmnpq}X^{mn}X^{pq}\,,\\
F^{ij}(x) &= 4X^{ik}X_{kl}X^{lj}+X^{ij}X^{kl}X_{kl}\\
&\quad -\tfrac{1}{12}\,\varepsilon^{ijklmnpq}X_{kl}X_{mn}X_{pq}+X^{ij}x\,,\\
F_{ij}(x) &= 4X_{ik}X^{kl}X_{lj}+X_{ij}X^{kl}X_{kl}\\
&\quad -\tfrac{1}{12}\,\varepsilon_{ijklmnpq}X^{kl}X^{mn}X^{pq}-X_{ij}x\,.
\end{aligned}
\tag{54}
$$

The above action of $E_{8(8)}$ was called quasiconformal in [22] since it leaves a certain norm invariant up to an overall factor. Since the standard difference $(\mathscr{X}-\mathscr{Y})$ of two vectors in the 57 dimensional space is not invariant under "translations" generated by (E^{ij},E_{ij}), one defines a nonlinear difference that is invariant under these translations as:

$$
\delta(\mathscr{X},\mathscr{Y}) := (X^{ij}-Y^{ij},X_{ij}-Y_{ij}\,;\,x-y+\langle X,Y\rangle) = -\delta(\mathscr{Y},\mathscr{X})\,, \tag{55}
$$

where $\langle X,Y\rangle := X^{ij}Y_{ij}-X_{ij}Y^{ij}$. One defines the norm of a vector $\mathscr{X}$ in the 57 dimensional space as

$$
\mathscr{N}_4(\mathscr{X}) \equiv \mathscr{N}_4(X^{ij},X_{ij};x) := \mathscr{I}_4(X)-x^2\,. \tag{56}
$$

Then the "distance" between any two vectors $\mathscr{X}$ and $\mathscr{Y}$ defined as $\mathscr{N}_4(\delta(\mathscr{X},\mathscr{Y}))$ is invariant under $E_{7(7)}$ and translations generated by E^{ij},E_{ij} and E. Under the action of the remaining generators of $E_{8(8)}$ one finds that

$$
\begin{aligned}
F\Big(\mathscr{N}_4(\delta(\mathscr{X},\mathscr{Y}))\Big) &= 2(x+y)\,\mathscr{N}_4(\delta(\mathscr{X},\mathscr{Y}))\,,\\
F^{ij}\Big(\mathscr{N}_4(\delta(\mathscr{X},\mathscr{Y}))\Big) &= 2(X^{ij}+Y^{ij})\,\mathscr{N}_4(\delta(\mathscr{X},\mathscr{Y}))\,,\\
H\Big(\mathscr{N}_4(\delta(\mathscr{X},\mathscr{Y}))\Big) &= 4\,\mathscr{N}_4(\delta(\mathscr{X},\mathscr{Y}))\,.
\end{aligned}
$$

Therefore, for every $\mathscr{Y}\in\mathbb{R}^{57}$ the “light cone” with base point $\mathscr{Y}$, defined by the set of $\mathscr{X}\in\mathbb{R}^{57}$ satisfying

$$
\mathscr{N}_4(\delta(\mathscr{X},\mathscr{Y}))=0 \tag{57}
$$

is preserved by the full $E_{8(8)}$ group.

The quasiconformal realization of the other real noncompact form $E_{8(-24)}$ with the maximal compact subgroup $E_7\times SU(2)$ is given in [26, 27]. In going to $E_{8(-24)}$ the role played by the subgroup $SL(8,\mathbb{R})$ of $E_{7(7)}$ is played by the subgroup $SU^*(8)$ of $E_{7(-25)}$. The quasiconformal groups $E_{8(8)}$ and $E_{8(-24)}$ are isomorphic to the U-duality groups of the maximal $N = 16$ supergravity and the $N = 4$ exceptional supergravity in three dimensions. Since their action changes the "norm" in the charge-entropy space of the corresponding four dimensional theories they can be interpreted as spectrum generating symmetry groups. The quasiconformal realizations of $E_{8(8)}$ and $E_{8(-24)}$ can be consistently truncated to quasiconformal realizations of other exceptional subgroups.

For the generic Jordan family of $N = 2$ MESGT's the quasi-conformal extensions of their U-duality groups $SO(2,1) \times SO(n,2)$ in $d = 4$ are $SO(n+4,4)$, which are isomorphic to their U-duality groups in $d = 3$.

THE MINIMAL UNITARY REPRESENTATIONS OF U-DUALITY GROUPS AS QUASI-CONFORMAL GROUPS

As we saw above the noncompact U-duality groups of four dimensional $N = 2$ MESGT's defined by Jordan algebras of degree three can also arise as spectrum generating conformal symmetry groups in five dimensions. These groups all admit positive energy unitary representations which can be constructed using the oscillator method outlined in previous sections. Here I would like to discuss the unitary representations of exceptional groups that one obtains by quantization of their quasiconformal realizations [8]. For $E_{8(8)}$ this was done in [23] and for $E_{8(-24)}$ in [26]. Remarkably, the quantization of the quasi-conformal realizations of $E_{8(8)}$ and $E_{8(-24)}$ yield their minimal unitary representations. The concept of a minimal unitary representation of a non-compact group G was first introduced by A. Joseph [28] and is defined as a unitary representation on a Hilbert space of functions depending on the minimal number of coordinates for a given non-compact group. Here we shall summarize the results mainly for $E_{8(8)}$ and indicate how they extend to $E_{8(-24)}$. By truncation one can obtain the minimal unitary realizations of smaller exceptional groups as quasiconformal groups as well as unitary realizations of $E_{7(7)}$ and other exceptional subgroups as conformal groups [23, 26, 27].

Since the positive graded generators form an Heisenberg algebra one introduces 28 coordinates X^{ij} and 28 momenta $P_{ij} \equiv X_{ij}$, and one extra real coordinate y to represent the central term. By quantizing

$$[X^{ij}, P_{kl}] = \mathrm{i} \tag{58}$$

we can realize the positive grade generators of $E_{8(8)}$ as

$$E^{ij} := yX^{ij}, \quad E_{ij} := yP_{ij}, \quad E := \tfrac{1}{2}y^2. \tag{59}$$

To realize the other generators of $E_{8(8)}$ one introduces a momentum conjugate to the coordinate y representing the central charge of the Heisenberg algebra:

$$[y, p] = \mathrm{i}. \tag{60}$$

---

[8] We should note that both the quasiconformal extensions of four dimensional U-duality groups and the isomorphic three dimensional U-duality groups act nonlinearly in their respective dimensions.

Then the remaining generators are given by

$$
\begin{aligned}
H &:= \tfrac{1}{2}(yp+py)\,, \\
F^{ij} &:= -pX^{ij} + 2\mathrm{i}y^{-1}\left[X^{ij}, I_4(X,P)\right] \\
&= -4y^{-1}\overbrace{X^{ik}P_{kl}X^{lj}} - \tfrac{1}{2}y^{-1}(X^{ij}P_{kl}X^{kl} + X^{kl}P_{kl}X^{ij}) \\
&\quad + \tfrac{1}{12}y^{-1}\varepsilon^{ijklmnpq}P_{kl}P_{mn}P_{pq} - pX^{ij}\,, \\
F_{ij} &:= -pP_{ij} + 2\mathrm{i}y^{-1}\left[P_{ij}, I_4(X,P)\right] \\
&= 4y^{-1}\underbrace{P_{ik}X^{kl}P_{lj}} + \tfrac{1}{2}y^{-1}(P_{ij}X^{kl}P_{kl} + P_{kl}X^{kl}P_{ij}) \\
&\quad - \tfrac{1}{12}y^{-1}\varepsilon_{ijklmnpq}X^{kl}X^{mn}X^{pq} - pP_{ij}\,, \\
F &:= \tfrac{1}{2}p^2 + 2y^{-2}I_4(X,P) \\
G^i{}_j &:= 2X^{ik}P_{kj} + \tfrac{1}{4}X^{kl}P_{kl}\,\delta^i_j\,, \\
G^{ijkl} &:= -\tfrac{1}{2}X^{[ij}X^{kl]} + \tfrac{1}{48}\varepsilon^{ijklmnpq}P_{mn}P_{pq}\,.
\end{aligned}
\tag{61}
$$

The hermiticity of all generators is manifest. Here $I_4(X,P)$ is the fourth order differential operator

$$
\begin{aligned}
I_4(X,P) &:= -\tfrac{1}{2}(X^{ij}P_{jk}X^{kl}P_{li} + P_{ij}X^{jk}P_{kl}X^{li}) \\
&\quad + \tfrac{1}{8}(X^{ij}P_{ij}X^{kl}P_{kl} + P_{ij}X^{ij}P_{kl}X^{kl}) \\
&\quad - \tfrac{1}{96}\,\varepsilon^{ijklmnpq}P_{ij}P_{kl}P_{mn}P_{pq} \\
&\quad - \tfrac{1}{96}\,\varepsilon_{ijklmnpq}X^{ij}X^{kl}X^{mn}X^{pq} + \tfrac{547}{16}\,,
\end{aligned}
\tag{62}
$$

and represents the quartic invariant of $E_{7(7)}$ because

$$
[G^i{}_j, I_4(X,P)] = [G^{ijkl}, I_4(X,P)] = 0\,. \tag{63}
$$

The above unitary realization of $E_{8(8)}$ in terms of position and momentum operators (Schröedinger picture) can be reformulated in terms of annihilation and creation operators (oscillator realization) (Bargman-Fock picture) [23]. The transition from the Schrodinger picture to the Bargmann-Fock picture corresponds to going from the $SL(8,\mathbb{R})$ basis to the $SU(8)$ basis of $E_{7(7)}$.

The quadratic Casimir operator of $E_{8(8)}$ reduces to a number for the above realization and one can show that all the higher Casimir operators must also reduce to numbers as required for an irreducible unitary representation. Thus by exponentiating the above generators we obtain the minimal unitary irreducible representation of $E_{8(8)}$ over the Hilbert space of square integrable complex functions in 29 variables.

The $SL(2,\mathbb{R})$ subgroup generated by the grade ± 2 elements E,F and the dilatation generator H are precisely of the form that arises in conformal quantum mechanics [24]. The quadratic Casimir of this $SL(2,\mathbb{R})$ subgroup is simply

$$C_2(SL(2,\mathbb{R}) = I_4 - \tfrac{3}{16} = \frac{g}{4} - \frac{3}{16}, \tag{64}$$

showing that the role played by the coupling constant g in conformal quantum mechanics is played by the quartic invariant I_4 of $E_{7(7)}$, which is an $SL(2,\mathbb{R})$ singlet. The role of conformal quantum mechanics in the description of black hole solutions of supergravity was discussed in [25].

In the minimal unitary realization of the other noncompact real form $E_{8(-24)}$ with the maximal compact subgroup $E_7 \times SU(2)$ given explicitly in [26] the relevant 5-graded decomposition of its Lie algebra $\mathfrak{e}_{8(-24)}$ is with respect to its subalgebra $\mathfrak{e}_{7(-25)} \oplus \mathfrak{so}(1,1)$

$$\mathfrak{e}_{8(-24)} = \mathbf{1} \oplus \mathbf{56} \oplus \left(\mathfrak{e}_{7(-25)} \oplus \mathfrak{so}(1,1)\right) \oplus \mathbf{56} \oplus \mathbf{1}\,. \tag{65}$$

The Schrödinger picture for the minimal unitary representation of $E_{8(-24)}$ corresponds to working in the $SU^*(8)$ basis of the $E_{7(-25)}$ subgroup. The position and momentum operators transform in the $\mathbf{28}$ and $\tilde{\mathbf{28}}$ of this $SU^*(8)$ subgroup and the above formulas for $E_{8(8)}$ carry over to those of $E_{8(-24)}$ with some subtle differences[26]. The Bargmann-Fock picture for the minimal unitary realization of $E_{8(-24)}$ in terms of annihilation and creation operators is obtained by going from the $SU^*(8)$ basis to the $SU(6,2)$ basis of the $E_{7(-25)}$ subgroup of $E_{8(-24)}$.

One can obtain the minimal unitary realizations of certain subgroups of $E_{8(8)}$ and $E_{8(-24)}$ by truncating their minimal realizations. However, we should stress that since the minimal realizations of $E_{8(8)}$ [23] and $E_{8(-24)}$ [26] are nonlinear consistent truncations exist for only certain subgroups. It turns out that the minimal unitary realizations of all lower rank noncompact exceptional groups can be obtained from those of $E_{8(8)}$ and $E_{8(-24)}$ [22, 26]. The relevant subalgebras of $\mathfrak{e}_{8(-24)}$ and $\mathfrak{e}_{8(8)}$ are those that are realized as quasi-conformal algebras , i.e those that have a 5-grading

$$\mathfrak{g} = \mathfrak{g}^{-2} \oplus \mathfrak{g}^{-1} \oplus \mathfrak{g}^{0} \oplus \mathfrak{g}^{+1} \oplus \mathfrak{g}^{+2},$$

such that $\mathfrak{g}^{\pm 2}$ subspaces are one-dimensional and $\mathfrak{g}^0 = \mathfrak{h} \oplus \Delta$ where Δ is the generator that determines the 5-grading.. Hence they all have an $\mathfrak{sl}(2,\mathbb{R})$ subalgebra generated by elements of $\mathfrak{g}^{\pm 2}$ and the generator Δ. For the truncated subalgebra, the quartic invariant $\mathscr{I}_4$ will now be that of a subalgebra $\mathfrak{h}$ of $\mathfrak{e}_{7(-25)}$ or of $\mathfrak{e}_{7(7)}$. Furthermore, this subalgebra must act on the grade ± 1 spaces via symplectic representation. Below we give the main chain of such subalgebras [26]

$$\mathfrak{h} = \mathfrak{e}_{7(-25)} \supset \mathfrak{so}^*(12) \supset \mathfrak{su}(3,3) \supset \mathfrak{sp}(6,\mathbb{R}) \supset \oplus_1^3 \mathfrak{sp}(2,\mathbb{R}) \supset \mathfrak{sp}(2,\mathbb{R}) \supset \mathfrak{u}(1). \tag{66}$$

Corresponding quasi-conformal subalgebras read as follows

$$\mathfrak{g} = \mathfrak{e}_{8(-24)} \supset \mathfrak{e}_{7(-5)} \supset \mathfrak{e}_{6(2)} \supset \mathfrak{f}_{4(4)} \supset \mathfrak{so}(4,4) \supset \mathfrak{g}_{2(2)} \supset \mathfrak{su}(2,1). \tag{67}$$

The corresponding chains for the other real form $\mathfrak{e}_{8(8)}$ are

$$\mathfrak{h} - \mathfrak{e}_{7(7)} \supset \mathfrak{so}(6,6) \supset \mathfrak{sl}(6,\mathbb{R} \supset \mathfrak{sp}(6,\mathbb{R}) \supset \oplus_1^3 \mathfrak{sp}(2,\mathbb{R}) \supset \mathfrak{sp}(2,\mathbb{R}) \supset \mathfrak{u}(1)\,, \quad (68)$$

$$\mathfrak{g} = \mathfrak{e}_{8(8)} \supset \mathfrak{e}_{7(7)} \supset \mathfrak{e}_{6(6)} \supset \mathfrak{f}_{4(4)} \supset \mathfrak{so}(4,4) \supset \mathfrak{g}_{2(2)} \supset \mathfrak{su}(2,1)\,. \quad (69)$$

The minimal unitary realizations of $\mathfrak{e}_{8(8)}$ and of $\mathfrak{e}_{8(-24)}$ can also be consistently truncated to unitary realizations of certain subalgebras that act as regular conformal algebras with a 3-grading. For $\mathfrak{e}_{8(8)}$ we have the following chain of consistent truncations to conformal subalgebras $\mathfrak{conf}$:

$$\mathfrak{conf} = \mathfrak{e}_{7(7)} \supset \mathfrak{so}(6,6) \supset \mathfrak{sl}(6,\mathbb{R} \supset \mathfrak{sp}(6,\mathbb{R}) \supset \oplus_1^3 \mathfrak{sp}(2,\mathbb{R}) \supset \mathfrak{sp}(2,\mathbb{R})\,. \quad (70)$$

The corresponding chain of consistent truncations to conformal subalgebras for $\mathfrak{e}_{8(-24)}$ is

$$\mathfrak{conf} = \mathfrak{e}_{7(-25)} \supset \mathfrak{so}^*(12) \supset \mathfrak{su}(3,3) \supset \mathfrak{sp}(6,\mathbb{R}) \supset \oplus_1^3 \mathfrak{sp}(2,\mathbb{R}) \supset \mathfrak{sp}(2,\mathbb{R})\,. \quad (71)$$

CONCLUDING REMARKS

The minimal unitary realizations of $E_{8(8)}$ and $E_{8(-24)}$ and their subgroups given in [23, 26] can be extended to all noncompact groups and supergroups [27] [9] and formulated in a unified manner. This unified construction follows closely the formalism of unified construction of nonlinear superconformal and quasi-superconformal algebras in two dimensions [29]. The realization of the distinguished $SL(2,\mathbb{R})$ subgroup generated by the grade ± 2 elements in the unified construction is always of the form that arises in conformal or superconformal quantum mechanics. The coherent state formulation of the minimal unitary representations of noncompact groups and supergroups analogous to those of conformal groups is currently under investigation.

As I stated at the beginning of my talk, the discrete subgroups $E_{(11-d)(11-d)}(\mathbb{Z})$ of the U-duality groups of M-theory toroidally compactified to d dimensions are expected to be non-perturbative symmetries of M-theory [4]. One expects similarly that certain discrete subgroups of the U-duality groups of matter coupled supergravity theories obtained by other compactifications of M/superstring theory to be non-perturbative symmetries in the respective dimensions. The major goal of the work summarized in this talk is to find the unitary realizations of these discrete subgroups and understand how the spectra of compactified M/Superstring theory fit into these representations.

ACKNOWLEDGMENTS

I would like to thank the organizers of the XIXth Max Born Symposium in Wroclaw for their kind hospitality. The work presented here was done mostly in collaboration

---

[9] For $SL(2,\mathbb{R})$ the minimal realization reduces to the conformal realization.

with Kilian Koepsell, Hermann Nicolai and Olexander Pavlyk, which I would like to acknowledge with pleasure.

REFERENCES

1. E. Cremmer, B. Julia and J. Scherk, Phys. Lett. , **B76**, 409 (1978).
2. E. Witten, Nucl. Phys. **B443** , 85 (1995).
3. B Julia, *"Group Disintegrations"*, in: *Superspace and Supergravity*, Eds. S.W.Hawking and M. Rocek, Cambridge University Press, 1981.
4. C. Hull and P.K. Townsend, Nucl. Phys. **B438**, 109 (1995).
5. M. Günaydin, G. Sierra and P.K. Townsend, Nucl. Phys. **B242**, 244 (1984); Phys. Lett. **133B**, 72 (1983).
6. M. Günaydin and M. Zagermann, JHEP **0307**, 023 (2003).
7. S. Ferrara and R. Kallosh, Phys. Rev. **D54**, 5344 (1996); hep-th/9603090. For further references on the subject see [8, 9].
8. See S. Ferrara and J. Maldacena, Class. Quant. Grav. **15**, 749 (1998) and the references therein.
9. S. Ferrara and M. Gunaydin, Int. J. Mod. Phys. **A13**, 2075 (1998).
10. R. Kallosh and B. Kol, *Phys. Rev.* **D53**, 1525 (1996), hep-th/9602014. For further references on the subject see [8, 9].
11. M. Günaydin, Nuovo Cimento **29A**, 467 (1975).
12. M. Günaydin, Ann. Israel Physical Society **3**, 279 (1980).
13. M. Günaydin, in *"Elementary Particles and the Universe: Essays in Honor of Murray Gell-Mann"*, Ed. by J.H. Schwarz (Cambridge University Press, 1991);
    M. Günaydin, J. Math. Phys. **31**, 1776 (1990).
14. M. Günaydin, Mod. Phys. Lett. **A8**, 1407 (1993), hep-th/9301050.
15. M. Günaydin and C. Saclioglu, Comm. Math. Phys., **87**, 159 (1982); Phys. Lett. **B108**, 180 (1982).
16. M. Günaydin, *"AdS/CFT Dualities and the Unitary Representations of Non-compact Groups and Supergroups: Wigner versus Dirac"*, Invited talk in the proceedings of the VIth International Wigner Symposium (Istanbul, 1999), hep-th/0005168, Ed. by M. Arik, Bogazici University Press, 2002, pp. 55-69.
17. M. Günaydin, D. Minic and M. Zagermann, hep-th/9810226, Nucl. Phys. **B544**, 737–758 (1999), hep-th/9806042, Nucl. Phys. **B534**, 96–120 (1998).
18. M. Günaydin and S. Takemae, *"Unitary Supermultiplets of $OSp(8^*|4)$ and the AdS_7/CFT_6 Duality"*, hep-th/9910110;
    S. Fernando, M. Günaydin and S. Takemae, Nucl. Phys. **B628**, 79 (2002), hep-th/0106161.
19. M. Günaydin, in preparation.
20. I. Bars and M. Günaydin, Comm. Math. Phys. **91**, 21 (1983).
21. G. Mack and A. Salam, Ann. Phys. **53**, 174 (1969); G. Mack, Comm. Math. Phys. **55**, 1 (1977).
22. M. Gunaydin, K. Koepsell and H. Nicolai, Commun. Math. Phys. **221**, 57 (2001).
23. M. Gunaydin, K. Koepsell and H. Nicolai, Adv. Theor. Math. Phys. **5**, 923 (2002).
24. V. de Alfaro, S. Fubini, G. Furlan. Nuovo Cimm. **A34**, 569 (1976).
25. P. Claus, M. Derix, R. Kallosh, J. Kumar, P. Townsend and A. Van Proeyen, Phys. Rev. Lett. **81**, 4553 (1998).
26. M. Gunaydin and O. Pavlyk, JHEP **01**, 019 (2005), hep-th/0409272.
27. M. Gunaydin and O. Pavlyk, in preparation.
28. A. Joseph, Commun. Math. Phys. **36**, 325 (1974); Ann. Sci. Ec. Norm. Super., IV. Ser. **9**, 1 (1976). For further references on the minimal unitary representations of noncompact exceptional groups see [23].
29. B. Bina and M. Günaydin. Nucl. Phys. **B502**, 713 (1997).

Non–Anticommutative Deformations of Gauge Fields and Hypermultiplets in N=(1,1) Superspace

E. Ivanov

Bogoliubov Laboratory of Theoretical Physics, JINR,
141980, Dubna, Moscow Region, Russia
e-mail: eivanov@thsun1.jinr.ru

Abstract. We overview the SO(4)×SU(2) invariant and $N{=}(1,0)$ supersymmetry-preserving non-anticommutative deformations of the Euclidean $N{=}(1,1)$ supersymmetric gauge theories and hypermultiplets (neutral and charged) interacting with an abelian gauge multiplet, starting from their off-shell formulation in $N{=}(1,1)$ harmonic superspace. The corresponding component actions are presented and the Seiberg-Witten-type transformations to the undeformed component fields are explicitly given. Mass terms and scalar potentials for the hypermultiplets can be generated via the Scherk-Schwarz mechanism and Fayet-Iliopoulos term in analogy to the undeformed case. The neutral hypermultiplet action is invariant under $N{=}(2,0)$ supersymmetry and describes a deformed $N{=}(2,2)$ gauge theory. The string theory origin of the considered singlet deformation is exhibited.

1. INTRODUCTION

Some special backgrounds in string theory are known to induce various types of space-time noncommutativity in the low-energy limit. For instance, a constant Neveu-Schwarz B-field background in type IIB string theory implies that the relevant low-energy dynamics is described by a gauge theory defined on non-commutative flat space, with

$$[x^i, x^j] = \mathrm{i}\theta^{ij},$$

where θ^{ij} is a constant skew-symmetric matrix [1]. Recently it was found that certain string backgrounds give rise to supersymmetric field theories living on superspaces with non-anticommuting Grassmann coordinates [2]–[5]. In particular, a specific four-dimensional compactification of the type IIB string in the presence of a constant self-dual graviphoton background $F^{\alpha\beta}$ yields a superspace whose odd coordinates obey the Clifford algebra

$$\{\theta^\alpha, \theta^\beta\} = \alpha'^2 F^{\alpha\beta}, \tag{1}$$

instead of the standard Grassmann algebra [4]. This superspace and supersymmetry realized in it must be of Euclidean signature since a real field strength can be self-dual only in Euclidean (or Kleinian) but not in Minkowski space. The classical and quantum properties of theories defined on such nilpotently-deformed (or non-anticommutative) Euclidean $N{=}(\frac{1}{2},\frac{1}{2})$ superspace were analyzed in [4, 5] and in subsequent works (see

CP767, *Fundamental Interactions and Twistor-Like Methods,* edited by J. Lukierski and D. Sorokin

[6] for a review). This deformation breaks the original $N=(\frac{1}{2},\frac{1}{2})$ supersymmetry down to $N=(\frac{1}{2},0)$, but preserves the notion of chirality. The basic technical device of constructing the corresponding superfield theories is the Moyal-Weyl star product generalized to Grassmann coordinates [7, 8, 9].

A natural next step to the analysis of non-anticommutative $N=(\frac{1}{2},\frac{1}{2})$ superspace is the study of analogous nilpotent chiral deformations of Euclidean $N=(1,1)$ superfield theories. This study was initiated in [10, 11] and then continued in [12]–[15]. Both the D-type and Q-type deformations were considered, with either spinor covariant derivatives D^i_α or supersymmetry generators Q^i_α as the building blocks of the bi-differential Poisson operator specifying the relevant star products [7, 8, 9].

The Q-deformations generically break the $N=(1,1)$ supersymmetry by half, i.e. down to $N=(1,0)$, but preserve both chirality and anti-chirality. The simplest $N=(1,1)$ Q-deformation is the singlet one ('QS-deformation'), based on the Poisson operator

$$P_s = -I\overleftarrow{Q}^i_\alpha \overrightarrow{Q}^\alpha_i \,, \qquad \text{with} \quad (P_s)^5 = 0 \,, \tag{2}$$

where I is a real parameter. While breaking half the supersymmetry, it preserves the internal $SU(2)_R \times Spin(4)$ symmetry. It can be given a stringy interpretation [12] along the line of [4, 2, 3, 16]. Such a non-anticommutative $N=(1,1)$ superspace naturally arises for the $N=4$ superstring coupled to a complex axion background. The Q-deformations and their QS-subclass also preserve Grassmann harmonic analyticity [10, 11] which is the fundamental notion in theories with manifest extended supersymmetry [17, 18]. In view of these motivations, a complete understanding of the geometric and quantum structure of Q-deformed $N=(1,1)$ theories (and of their higher-N counterparts) is of the same importance as the analogous exploration of non-anticommuting $N=(\frac{1}{2},\frac{1}{2})$ theories.

Until now, the main attention was concentrated on the Q-deformation of $N=(1,1)$ pure gauge theories. In [12], the detailed superfield and component structure of the QS-deformed $N=(1,1)$ U(1) and U(n) gauge theories was explored.[1] In particular, an analog of the Seiberg-Witten (SW) map to quantities with undeformed gauge and supersymmetry transformation laws was explicitly worked out.

An important class of $N=(1,1)$ theories are those including matter hypermultiplets interacting with themselves and with gauge multiplets. The first type of theories, i.e. self-interacting hypermultiplets, yields, in the bosonic sector, Euclidean versions of hyper-Kähler sigma models. The second type, i.e. hypermultiplets coupled to $N=(1,1)$ gauge multiplets, could be of relevance from the phenomenological point of view. The system of a gauge superfield minimally coupled to a hypermultiplet in the adjoint representation of the gauge group provides an off-shell $N=(1,1)$ superfield formulation of $N=(2,2)$ supersymmetric gauge theory which is the Euclidean analog of the renowned $N=4$ super Yang-Mills theory. With these motivations in mind, it is of obvious interest to elaborate on the structure of Q-deformed $N=(1,1)$ hypermultiplet theories. Note that, in contrast, the singlet D-deformation (which is the unique D-type deformation preserving Grassmann harmonic analyticity [10, 11]) does not at all affect $N=(1,1)$ analytic superfield Lagrangians, including those for hypermultiplets. So only the Q-deformations are of

[1] The QS-deformed U(1) theory was independently considered in [14].

interest in the hypermultiplet context. The QS-deformations of the $N=(1,1)$ supersymmetric coupled systems of hypermultiplets and gauge superfields were a subject of recent paper [19].

The goal of the present contribution is to review, both in the superfield and component-field formulations, the QS-deformation of $N=(1,1)$ gauge theories, as well as of two simple models with hypermultiplets coupled to the U(1) gauge superfield. Our exposition is based on the papers [10, 12, 19].

We start by sketching the basics of the Euclidean $N=(1,1)$ superspace (sect 2). Then we describe the superfield and component formulations of the QS-deformed abelian and non-abelian $N=(1,1)$ gauge theories (sect. 3 and 4). In sect. 5 - 7 we present the QS-deformed actions of two $N=(1,1)$ models with hypermultiplets, respectively neutral and charged ones, coupled to U(1) gauge superfield. The peculiarities of the superfield realizations of the QS-deformed U(1) gauge group on hypermultiplets are discussed and a few mechanisms of generating hypermultiplet masses and scalar potentials in the deformed case are described. The action with neutral hypermultiplet is shown to respect an extra hidden $N=(1,1)$ supersymmetry and so corresponds to the QS-deformed $N=(2,2)$ gauge theory with the residual $N=(2,0)$ supersymmetry.[2] In section 8 the stringy origin of the QS-deformation is explained. The last section is an outline of some open problems for further study.

2. Q-DEFORMATIONS OF $N=(1,1)$ SUPERSPACE

We start with a short account of the Q-deformed Euclidean harmonic $N=(1,1)$ superspace as a necessary background for considering the QS-deformations of $N=(1,1)$ gauge theories with and without hypermultiplets.

The basic concepts of the $N=(1,1), D=4$ Euclidean harmonic superspace, which is an extension of the standard $N=(1,1)$ superspace by the $SU(2)_R/U(1)_R$ harmonics $u^\pm_i$, are collected in [10, 11, 12] (see also [18]). The standard (central) coordinates of the $N=(1,1)$ harmonic superspace are $(Z, u^\pm_i) = (x^m, \theta^\alpha_k, \bar\theta^{\dot\alpha k}, u^\pm_i)$. We shall also use the chiral coordinates

$$Z_L = (x^m_L, \theta^\alpha_k, \bar\theta^{\dot\alpha k}), \quad x^m_L = x^m + i(\sigma^m)_{\alpha\dot\alpha}\theta^\alpha_k\bar\theta^{\dot\alpha k}, \tag{3}$$

the chiral-analytic coordinates

$$Z_C = (x^m_L, \theta^{\pm\alpha}, \bar\theta^{\pm\dot\alpha}), \tag{4}$$

and the analytic coordinates

$$Z_A = (x^m_A, \theta^{\pm\alpha}, \bar\theta^{\pm\dot\alpha}), \quad x^m_A = x^m_L - 2i(\sigma^m)_{\alpha\dot\alpha}\theta^{-\alpha}\bar\theta^{+\dot\alpha}, \tag{5}$$

where $\theta^{\pm\alpha} = u^\pm_k\theta^{\alpha k}$ and $\bar\theta^{\pm\dot\alpha} = u^\pm_k\bar\theta^{\dot\alpha k}$. It is worth mentioning that all these coordinates are (pseudo)real with respect to the basic conjugation $\widetilde{\ }$ [10]. For instance, the condition

[2] Nilpotent Q-deformations of the on-shell superfield constraints of the Euclidean $N=(2,2)$ gauge theory were studied in [20].

of reality can be consistently imposed on the Euclidean chiral superfields. An important invariant pseudoreal subspace of the harmonic superspace $(Z_A, u_i^{\pm})$ is the analytic harmonic superspace

$$(\zeta, u_i^{\pm}),$$

where

$$\zeta = (x_A^m, \theta^{+\alpha}, \bar{\theta}^{+\dot{\alpha}}).$$

The supersymmetry-preserving spinor derivatives $D_{\alpha}^{\pm}, \bar{D}_{\dot{\alpha}}^{\pm}$ and the harmonic derivatives $D^{\pm\pm}$ in different coordinate bases are given in [10, 12]. An analytic superfield is defined by the conditions

$$(D_{\alpha}^{+}, \bar{D}_{\dot{\alpha}}^{+})\Phi(Z, u_i^{\pm}) = 0 \quad \leftrightarrow \quad \Phi = \Phi_A(\zeta, u_i^{\pm}). \tag{6}$$

In what follows we shall use harmonic projections of the supersymmetry generators

$$Q_{\alpha}^{k} = u^{+k} Q_{\alpha}^{-} - u^{-k} Q_{\alpha}^{+}, \quad \bar{Q}_{\dot{\alpha}k} = u_k^{+} \bar{Q}^{-} - u_k^{-} \bar{Q}^{+}. \tag{7}$$

For instance, in the analytic coordinates

$$Q_{\alpha}^{+} = \partial_{-\alpha} - 2i\bar{\theta}^{+\dot{\alpha}} \partial_{\alpha\dot{\alpha}}, \quad Q_{\alpha}^{-} = -\partial_{+\alpha}, \tag{8}$$

where $\partial_{\pm\alpha} = \partial/\partial\theta^{\pm\alpha}$ and $\partial_{\alpha\dot{\alpha}} = (\sigma_m)_{\alpha\dot{\alpha}}\partial_m$.

In this contribution we shall deal with the Q-singlet (or QS-) deformation which is associated with the SO(4)×SU(2) invariant Poisson bracket

$$AP_s B = -I(-1)^{p(A)} Q_{\alpha}^{k} A Q_k^{\alpha} B = -I(-1)^{p(A)} (Q_{\alpha}^{+} A Q^{-\alpha} B + Q^{-\alpha} A Q_{\alpha}^{+} B). \tag{9}$$

The noncommutative QS-star product of two analytic superfields has the following simple form:

$$\Lambda \star \Phi = \Lambda\Phi + \Lambda P_s \Phi + \tfrac{1}{2}\Lambda P_s^2 \Phi. \tag{10}$$

Since $\partial_{-}^{\alpha} A = \partial_{-}^{\alpha} B = 0$ in the analytic basis, we can omit $\partial_{-\alpha}$ in Q_{α}^{+} in this basis. In particular,

$$AP_s B = 2iI(-1)^{p(A)} \bar{\theta}^{+\dot{\alpha}} (\partial_{+}^{\alpha} A \partial_{\alpha\dot{\alpha}} B - \partial_{\alpha\dot{\alpha}} A \partial_{+}^{\alpha} B). \tag{11}$$

In the $\star$-commutator of two bosonic analytic superfields $[A, B]_{\star}$ with $[A, B] = 0$ only the term $\sim P_s$ survives:

$$[A, B]_{\star} = 2AP_s B, \tag{12}$$

while the $\star$-anticommutator reads

$$\{A, B\}_{\star} = 2AB + AP_s^2 B. \tag{13}$$

We shall make use of these general formulas in the sections 6 and 7.

3. QS-DEFORMATION OF $N{=}(1,1)$ U(1) GAUGE THEORY

The basic superfield of the $N{=}(1,1)$ gauge theory is the analytic gauge potential V^{++}. The QS-deformed gauge transformation of the U(1) potential V^{++} in the analytic basis reads

$$\begin{aligned}\delta_\Lambda V^{++} &= D^{++}\Lambda + [V^{++},\Lambda]_\star = D^{++}\Lambda + 2V^{++}P_s\Lambda \\ &= D^{++}\Lambda + 4iI\bar\theta^{+\dot\alpha}(\partial^\alpha_+ V^{++}\partial_{\alpha\dot\alpha}\Lambda - \partial_{\alpha\dot\alpha}V^{++}\partial^\alpha_+\Lambda)\,. \end{aligned} \tag{14}$$

Here Λ is an analytic gauge parameter satisfying $\tilde\Lambda = -\Lambda$. The gauge potential in the Wess-Zumino gauge has the following $\theta^{+\alpha}, \bar\theta^{+\dot\alpha}$-expansion

$$\begin{aligned}V^{++}{}_{WZ} &= (\theta^+)^2\bar\phi + (\bar\theta^+)^2\phi + 2\theta^+\sigma_m\bar\theta^+A_m + 4(\theta^+)^2\bar\theta^+_{\dot\alpha}u^-_k\bar\Psi^{\dot\alpha k} \\ &\quad + 4(\bar\theta^+)^2\theta^{+\alpha}u^-_k\Psi^k_\alpha + 3(\theta^+)^2(\bar\theta^+)^2u^-_k u^-_l\mathcal{D}^{kl}\,, \end{aligned} \tag{15}$$

where all the component fields are functions of x^m_A. The residual U(1) gauge transformations with the parameter $\Lambda_r = ia(x_A)$ act on the component fields in (15) as

$$\begin{aligned}\delta_r\bar\phi &= 0, \qquad \delta_r\phi = -8IA_m\partial_m a\,, \qquad \delta_r A_m = (1+4I\bar\phi)\partial_m a\,, \\ \delta_r\Psi^k_\alpha &= -4I\bar\Psi^{\dot\alpha k}\partial_{\alpha\dot\alpha}a \qquad \delta_r\bar\Psi^k_{\dot\alpha} = 0, \quad \delta_r\mathcal{D}^{kl} = 0. \end{aligned} \tag{16}$$

The deformed $N = (1,0)$ supersymmetry transformations of the U(1) gauge multiplet component fields defined in (15) are as follows [12]:

$$\begin{aligned}\delta_\varepsilon\phi &= 2\varepsilon^{\alpha k}\Psi_{\alpha k}, \quad \delta_\varepsilon\bar\phi = 0, \quad \delta_\varepsilon A_m = \varepsilon^{\alpha k}(\sigma_m)_{\alpha\dot\alpha}\bar\Psi^{\dot\alpha}_k\,, \\ \delta_\varepsilon\Psi^k_\alpha &= -\varepsilon_{\alpha l}\mathcal{D}^{kl} + \tfrac{1}{2}(1+4I\bar\phi)(\sigma_{mn}\varepsilon^k)_\alpha F_{mn} - 4iI\varepsilon^k_\alpha A_m\partial_m\bar\phi\,, \\ \delta_\varepsilon\bar\Psi^k_{\dot\alpha} &= -i\varepsilon^{\alpha k}(1+4I\bar\phi)\partial_{\alpha\dot\alpha}\bar\phi\,, \\ \delta_\varepsilon\mathcal{D}^{kl} &= i\partial_m[(\varepsilon^k\sigma_m\bar\Psi^l + \varepsilon^l\sigma_m\bar\Psi^k)(1+4I\bar\phi)]\,, \end{aligned} \tag{17}$$

where $F_{mn} = \partial_m A_n - \partial_n A_m$. They are produced by the transformation of $V^{++}{}_{WZ}$ which is a sum of the standard supertranslation and the compensating gauge transformation with the parameter

$$\begin{aligned}\Lambda_\varepsilon &= 2(\varepsilon^-\theta^+)\bar\phi - 2\bar\theta^+_{\dot\alpha}\varepsilon^-_\alpha(A^{\alpha\dot\alpha} + 2\theta^{+\alpha}\bar\Psi^{-\dot\alpha}) \\ &\quad + 2(\bar\theta^+)^2[(\varepsilon^-\Psi^-) + (\varepsilon^-\theta^+)\mathcal{D}^{--}]\,. \end{aligned} \tag{18}$$

The non-polynomial superfield action of the Q-deformed gauge theory has been given in [10] as an integral over the full superspace in the chiral coordinates, by analogy with the undeformed $N{=}2$ superfield action [21]. It was shown in [12] that the QS-deformed U(1) gauge action can be conveniently rewritten as the integral over the chiral superspace

$$S = \frac{1}{4}\int d^4x_L d^4\theta\,\mathcal{A}^2 \tag{19}$$

where $\mathcal{A}(x_L, \theta^+, \theta^-, u)$ is the deformed chiral superfield strength. The latter is the lowest component in the $\bar\theta^{+\dot\alpha}$ expansion of the covariantly chiral superfield strength $\mathcal{W}$:

$$\mathcal{W} \equiv -\tfrac{1}{4}(\bar D^+)^2 V^{--} = \mathcal{A} + \bar\theta^+_{\dot\alpha}\tau^{-\dot\alpha} + (\bar\theta^+)^2\tau^{-2}\,. \tag{20}$$

The composite harmonic connection V^{--} is related to the basic potential V^{++} via the deformed harmonic zero curvature equation [10]

$$D^{++}V^{--} - \Delta^{--}V^{++} + [V^{++}, V^{--}]_\star = 0. \tag{21}$$

Here, in the chiral-analytic basis,

$$D^{--} = \partial^{--} + \theta^{-\alpha}\partial_{+\alpha} + \bar\theta^{-\dot\alpha}\partial_{+\dot\alpha}\,, \quad \partial^{--} = u^{-i}\frac{\partial}{\partial u^{+i}}\,.$$

The action (19) can be written through $\mathscr{W}$ as

$$S = \frac{1}{4}\int d^4x_L d^4\theta\, \mathscr{W}^2\,. \tag{22}$$

It can be shown that the remaining two components in (20) do not contribute to (22).

The component action can be straightforwardly obtained from (19) or (22) by performing there Berezin integration. The most readable form of the action is achieved in terms of the fields having standard transformation properties with respect to both residual U(1) gauge and $N = (1,0)$ supersymmetry transformations (these standard transformations are obtained by setting I=0 in (16), (17)). The Seiberg-Witten (SW) transformation to the undeformed U(1) gauge supermultiplet $\varphi, \bar\phi, a_m, \psi^\alpha_k, \bar\psi^{\dot\alpha}_k, d^{kl}$ is defined as

$$\begin{aligned}
\phi &= (1+4I\bar\phi)^2\varphi - 4I(1+4I\bar\phi)^{-1}[A_m^2 + 4I^2(\partial_m\bar\phi)^2]\,,\\
A_m &= (1+4I\bar\phi)a_m\,, \qquad\qquad \bar\Psi^k_{\dot\alpha} = (1+4I\bar\phi)\bar\psi^k_{\dot\alpha}\,,\\
\Psi^k_\alpha &= (1+4I\bar\phi)^2\psi^k_\alpha - 4I(1+4I\bar\phi)a_{\alpha\dot\alpha}\bar\psi^{\dot\alpha k}\,,\\
\mathscr{D}^{kl} &= (1+4I\bar\phi)^2 d^{kl} - 8I(1+4I\bar\phi)\bar\psi^k_{\dot\alpha}\bar\psi^{\dot\alpha l}\,.
\end{aligned} \tag{23}$$

Using this SW-map, one can express the component Lagrangian of the deformed U(1) gauge theory through the standard undeformed free N=$(1,1)$ gauge theory Lagrangian [12, 14]

$$L_g = (1+4I\bar\phi)^2[-\tfrac{1}{2}\varphi\Box\bar\phi + \tfrac{1}{4}f^2_{mn} + \tfrac{1}{8}\varepsilon_{mnrs}f_{mn}f_{rs} - \mathrm{i}\psi^\alpha_k\partial_{\alpha\dot\alpha}\bar\psi^{\dot\alpha k} + \tfrac{1}{4}(d^{kl})^2]\,, \tag{24}$$

where $f_{mn} = \partial_m a_n - \partial_n a_m$. After redefining the involved fields as

$$\hat\varphi = (1+4I\bar\phi)^2\varphi\,, \;\; \hat\psi^\alpha_k = (1+4I\bar\phi)^2\psi^\alpha_k\,, \;\; \hat d^{kl} = (1+4I\bar\phi)d^{kl}\,, \tag{25}$$

the Lagrangian (24) acquires the form in which it differs from the free Lagrangian only by a simple interaction term:

$$L_g = -\tfrac{1}{2}\hat\varphi\Box\bar\phi - \mathrm{i}\hat\psi^\alpha_k\partial_{\alpha\dot\alpha}\bar\psi^{\dot\alpha k} + \tfrac{1}{4}(1+4I\bar\phi)^2\left(f^2_{mn} + \tfrac{1}{2}\varepsilon_{mnrs}f_{mn}f_{rs}\right) + \tfrac{1}{4}(\hat d^{kl})^2. \tag{26}$$

4. GENERALIZATION TO THE NON-ABELIAN CASE

The above straightforwardly generalizes to the non-abelian U(n) case (n $\geq$ 2). We use the WZ-gauge for the U(n) potential (15), and the corresponding deformed component gauge transformations are

$$\begin{aligned} \delta_a\bar{\phi} &= -i[a,\bar{\phi}], \qquad \delta_r\bar{\Psi}^k_{\dot{\alpha}} = -i[a,\bar{\Psi}^k_{\dot{\alpha}}], \qquad \delta_r\mathscr{D}^{kl} = -i[a,\mathscr{D}^{kl}], \\ \delta_a A_m &= \partial_m a + i[A_m,a] + 2I\{\bar{\phi},\partial_m a\}, \\ \delta_a\phi &= -i[a,\phi] - 4I\{A_m,\partial_m a\} - 4iI^2[\Box a,\bar{\phi}], \\ \delta_a\Psi^k_\alpha &= -i[a,\Psi^k_\alpha] - 2I(\sigma_m)_{\alpha\dot{\alpha}}\{\bar{\Psi}^{\dot{\alpha}k},\partial_m a\}. \end{aligned} \tag{27}$$

The undeformed harmonic chiral U(n) superfield A has the following component expansion

$$\begin{aligned} A &= \varphi + 2\theta^+\psi^- - 2\theta^-\psi^+ + (\theta^+)^2 d^{--} + (\theta^+\theta^-)([\varphi,\bar{\phi}] - 2d^{+-}) + (\theta^-)^2 d^{++} \\ &\quad + (\theta^+\sigma_{mn}\theta^-) f_{mn} + 2(\theta^-)^2\theta^+\left(i\xi^+ - [\bar{\phi},\psi^+]\right) + 2i(\theta^+)^2\theta^-\xi^- \\ &\quad - (\theta^+)^2(\theta^-)^2\left(p + [\bar{\phi},d^{+-}]\right), \end{aligned} \tag{28}$$

where all the component fields are $n \times n$ matrices and the following short-hand notation is used:

$$\begin{aligned} &\nabla_m = \partial_m + i[a_m,\ ], \qquad f_{mn} = \partial_m a_n - \partial_n a_m + i[a_m,a_n], \\ &\xi^k_\alpha = (\sigma_m)_{\alpha\dot{\alpha}}\nabla_m\bar{\psi}^{\dot{\alpha}k}, \qquad p = \nabla^2_m\bar{\phi} + \{\bar{\psi}^{\dot{\alpha}k},\bar{\psi}_{\dot{\alpha}k}\} + \frac{1}{2}[\bar{\phi},[\bar{\phi},\varphi]]. \end{aligned} \tag{29}$$

The deformed chiral U(n) superfield can be written as a sum of two $N{=}(1,0)$ covariant objects

$$\mathscr{A}(x_L,\theta^+,\theta^-,u) = L^2 R_\theta A(x_L,\theta^+,\theta^-,u) - 4I^2\hat{A}(x_L,\theta^+,u), \tag{30}$$

$$R_\theta A = [1 + (L^{-1}-1)(\theta^-\partial_-) - \frac{1}{4}(L^{-1}-1)^2(\theta^-)^2(\partial_-)^2]A \tag{31}$$

where A is the undeformed U(n) superfield (28) and L^{-1} is the inverse of the matrix operator L defined as

$$L = 1 + 2I\{\bar{\phi},\ \} \tag{32}$$

($\bar{\phi}$-dependent matrix operators commute with $\theta^{\pm\alpha}$ and $\partial_{-\alpha}$ and act on all matrix quantities standing to the right). The second part $\hat{A}$ is a traceless chiral-analytic $N{=}(1,0)$ superfield

$$\begin{aligned} \hat{A}(x_L,\theta^+,u) &= \hat{p} - [\bar{\phi},d^{+-}] + 2\theta^{+\alpha}(i[\bar{\phi},\xi^-_\alpha] - [\bar{\phi},[\bar{\phi},\psi^-_\alpha]]) \\ &\quad + (\theta^+)^2[\bar{\phi},[\bar{\phi},d^{--}]], \qquad \hat{p} = p - \tfrac{1}{n}\mathrm{Tr}\,p. \end{aligned} \tag{33}$$

Both parts of $\mathscr{A}$ are thus expressed in terms of the undeformed field components of the superfield A (28). The $N{=}(1,0)$ supersymmetry transformation of $\mathscr{A}$ has the following form:

$$\delta_\varepsilon\mathscr{A} = 2(\varepsilon^-\theta^+)[\bar{\phi},\mathscr{A}] + L\varepsilon^{-\alpha}\partial_{-\alpha}\mathscr{A} + \varepsilon^{+\alpha}\partial_{+\alpha}\mathscr{A}. \tag{34}$$

The QS-deformed U(n) gauge theory component action can be directly obtained from the superfield chiral action

$$\mathscr{S}_n = \frac{1}{4}\int d^4x_L d^8\theta\,\mathrm{Tr}\,\mathscr{A}^2 = \int d^4x_L d^8\theta\,\mathrm{Tr}\{\frac{1}{4}(LA)^2 - 2I^2\hat{A}A\}\,, \tag{35}$$

using relations (28) and (33). In the limit $I \to 0$ the first term yields the action of the undeformed U(n) gauge theory. The non-standard second term contains higher derivative terms, in particular $I^2(\Box\bar{\phi})^2$, which can hopefully be removed by a redefinition of the scalar field φ (for the time being we have checked this only for the free part of the total action).

5. REPRESENTATIONS OF THE QS-DEFORMED GAUGE GROUP ON HYPERMULTIPLETS

The free q^+ hypermultiplet superfield action [18] is not deformed in the non-anticommutative superspace [10, 11]:

$$S_0(q^+) = -\int du\,d\zeta^{-4}\,\tilde{q}^+D^{++}q^+ = \tfrac{1}{2}\int du\,d\zeta^{-4}\,q^+_a D^{++}q^{+a}\,. \tag{36}$$

Here $d\zeta^{-4} = d^4x_A(D^-)^4$ and the additional $SU(2)_{PG}$ indices $a, b = 1, 2$ are introduced, $q^{+a} = \varepsilon^{ab}q^+_b = (\tilde{q}^+, q^+) = \widetilde{q^+_a}$. After passing to the component fields, integrating over $\theta^+_\alpha, \bar{\theta}^+_{\dot{\alpha}}$ and eliminating the infinite tower of the auxiliary fields by their algebraic equations of motion, the superfield free action (36) yields the free action for (4+8) physical fields of the hypermultiplet, viz. scalars $f^{ak}(x_A)$ and fermions $\rho^a_\alpha(x_A)$ and $\chi^a_{\dot{\alpha}}(x_A)$. We do not quote it here; it can be easily reproduced as the $I \to 0$ limit of the QS-deformed actions to be given below.

Let us study how the QS-deformed U(1) gauge transformations can be implemented on the superfield q^{+a}. Obviously, these transformations should have the same Lie bracket structure as the QS-deformed U(1) gauge transformations (14) of V^{++}. It is easy to see that the Lie bracket of two such transformations is again of the form (14) with $\Lambda_{br} = [\Lambda_2, \Lambda_1]_\star$. Then a simple analysis shows that only two non-equivalent realizations of the deformed U(1) gauge group with such Lie bracket structure are possible for the hypermultiplet doublet q^{+a} (besides the trivial realization $\delta_\Lambda q^{+a} = 0$)

$$1.\qquad \delta_\Lambda q^{+a} = [q^{+a}, \Lambda]_\star\,, \tag{37}$$

$$2.\qquad \delta_\Lambda q^{+a} = \tfrac{1}{2}[q^{+a}, \Lambda]_\star + \tfrac{1}{2}(\tau_3)^a_b\{q^{+b}, \Lambda\}_\star\,, \tag{38}$$

where τ_3 is the diagonal Pauli matrix. These representations of the deformed U(1) group can be called 'adjoint' and 'fundamental', respectively. Indeed, (37) has the same form as the non-abelian part of the deformed U(1) transformation (14) of the gauge superfield, while (38) can be equivalently written as

$$\delta_\Lambda q^+ = -\Lambda \star q^+\,,\quad \delta_\Lambda \tilde{q}^+ = \tilde{q}^+ \star \Lambda\,. \tag{39}$$

One can wonder whether the charged hypermultiplets with an arbitrary value of the charge e can be properly QS-deformed. Surprisingly, for this one should modify the QS-deformed U(1) gauge transformation law (14) in the following way

$$\delta_\Lambda V^{++} = D^{++}\Lambda + e\,[V^{++},\Lambda]_\star \tag{40}$$

and the hypermultiplet transformation laws (37) and (38) as

$$1.\quad \delta_\Lambda q^{+a} = e\,[q^{+a},\Lambda]_\star\,, \tag{41}$$

$$2.\quad \delta_\Lambda q^{+a} = \frac{1}{2}e\,[q^{+a},\Lambda]_\star + \frac{1}{2}e\,(\tau_3)^a_b\{q^{+b},\Lambda\}_\star\,. \tag{42}$$

Eq. (42) can be rewritten as

$$\delta_\Lambda q^+ = -e\,\Lambda\star q^+\,,\quad \delta_\Lambda \tilde{q}^+ = e\,\tilde{q}^+\star\Lambda \tag{43}$$

and it goes into the standard U(1) transformation law of the hypermultiplet of the charge $|e|$ in the commutative limit. Any gauge transformation from the one-parameter family (40) in this limit goes into the standard abelian transformation of V^{++}. The modified transformation laws have the same Lie bracket structure for all superfields, now with the bracket parameter $e[\Lambda_2,\Lambda_1]_\star$. For the deformed U(1) gauge system, as well as in the coupled system of the U(1) gauge superfield and 'adjoint' hypermultiplet, this modification is in fact unessential: the parameter e can be removed from the superfield transformation laws and the relevant invariant actions by rescaling the deformation parameter as $eI \to I'$. In the system of V^{++} and the 'fundamental' hypermultiplet the charge e cannot be removed since it is present there already in the undeformed limit. Thus e can be treated as an additional deformation parameter of the analytical superfield gauge group, and gauge transformations for different e are not equivalent. Note that both in the undeformed and deformed cases one can introduce the gauge U(1) group in such a way that it will commute with full $SU(2)_{PG}$. This can be achieved by introducing *two* independent hypermultiplets, which amounts to making q^{+a} complex, $\widetilde{q^+_a} \neq q^{+a}$. Then one can gauge U(1) which multiplies the whole q^{+a} by phase and so commutes with $SU(2)_{PG}$ acting on the doublet indices. The corresponding analog of the QS-deformed transformation rules (43) is

$$\delta_\Lambda q^{+a} = -e\,\Lambda\star q^{+a}\,,\quad \delta_\Lambda \tilde{q}^+_a = e\,\tilde{q}^+_a\star\Lambda\,. \tag{44}$$

In what follows we concentrate mainly on the simplest case of one hypermultiplet.

The QS-deformed superfield hypermultiplet actions corresponding to the two realizations of U(1) gauge group presented above (for arbitrary e) are constructed by the general rule of ref. [10], viz. via the replacement

$$D^{++}q^{+a} \Rightarrow \nabla^{++}q^{+a} \tag{45}$$

in the action (36), where, for two choices (41), (42),

$$1.\quad \nabla^{++}q^{+a} = D^{++}q^{+a} + e\,[V^{++},q^{+a}]_\star = D^{++}q^{+a} + 2eV^{++}P_s q^{+a}\,, \tag{46}$$

$$2.\quad \nabla^{++}q^{+a} = D^{++}q^{+a} + \frac{e}{2}[V^{++},q^{+a}]_\star - \frac{e}{2}(\tau_3)^a_b\{V^{++},q^{+b}\}_\star\,. \tag{47}$$

The total QS-deformed action in both cases is the sum of the superfield action of the QS-deformed U(1) gauge multiplet given in [12, 14] and the gauge invariant hypermultiplet action

$$S(V,q) = \tfrac{1}{2}\int du\, d\zeta^{-4}\, q^+_a \nabla^{++} q^{+a}. \tag{48}$$

In what follows we shall consider the component structure of both types of the gauge multiplet-hypermultiplet action.

6. QS-DEFORMED NEUTRAL HYPERMULTIPLET

The invariant action (48) for this case reads

$$\begin{aligned} S_n(V,q) &= \tfrac{1}{2}\int du d\zeta^{-4} q^+_a \Big[D^{++}q^{+a} + 4\mathrm{i}I\,\bar\theta^{+\dot\alpha} \\ &\quad (\partial^\alpha_+ V^{++}\partial_{\alpha\dot\alpha}q^{+a} - \partial_{\alpha\dot\alpha}V^{++}\partial^\alpha_+ q^{+a})\Big], \end{aligned} \tag{49}$$

where we made use of Eqs. (12), (11) and (46) and put $e=1$ since the dependence on e can be absorbed into a redefinition of I. To obtain the component action, we should substitute into (49) the WZ form of V^{++}, Eq. (15), and the θ-expansion of q^{+a}

$$\begin{aligned} q^{+a} &= f^{+a} + \theta^{+\alpha}\pi^a_\alpha + \bar\theta^+_{\dot\alpha}\kappa^{\dot\alpha a} + \theta^+\sigma_m\bar\theta^+ r_m^{-a} + (\theta^+)^2 g^{-a} + (\bar\theta^+)^2 h^{-a} \\ &\quad + (\bar\theta^+)^2\theta^{+\alpha}\Sigma_\alpha^{--a} + (\theta^+)^2\bar\theta^{+\dot\alpha}\bar\Sigma_{\dot\alpha}^{--a} + (\theta^+)^2(\bar\theta^+)^2\omega^{-3a}, \end{aligned} \tag{50}$$

where all component fields are functions of x_A and u. Then we should integrate in (49) over $\theta^+, \bar\theta^+$, and eliminate the infinite tower of the auxiliary fields contained in (50) using the non-dynamical equations of motion.

Skipping details, we find the following solution for the components in (50) in terms of the remaining 4 physical bosonic fields $f^{ia}(x_A)$ and 8 physical fermions $\rho^a_\alpha(x_A)$, $\chi^{\dot\alpha a}(x_A)$

$$\begin{aligned} f^{+a} &= f^{ak}u^+_k, \qquad \pi^a_\alpha = \rho^a_\alpha, \quad \kappa^{\dot\alpha a} = \chi^{\dot\alpha a}, \qquad r_m^{-a} = r_m^{ak}u^-_k, \quad g^{-a} = 0, \\ h^{-a} &= h^{ak}u^-_k, \qquad\qquad \Sigma_\alpha^{--a} = \Sigma_\alpha^{kla}u^-_k u^-_l, \qquad \bar\Sigma_{\dot\alpha}^{--a} = 0, \qquad \omega^{-3a} = 0, \\ r^{ak}_m &= 2\mathrm{i}(1+4I\bar\phi)\partial_m f^{ak}, \quad h^{ak} = -8\mathrm{i}IA_m\partial_m f^{ak}, \\ \Sigma^{kla}_\alpha &= -4\mathrm{i}I(\bar\Psi^{\dot\alpha k}\partial_{\alpha\dot\alpha}f^{al} + \bar\Psi^{\dot\alpha l}\partial_{\alpha\dot\alpha}f^{ak}). \end{aligned} \tag{51}$$

Actually, for deducing the component results below it is of no need to know the explicit form of the solution for h^{-a} and Σ^{--a}; it was presented for completeness.

The final Lagrangian in the x-space is written in terms of the physical fields as

$$\begin{aligned} L_1 &= \tfrac{1}{2}(1+4I\bar\phi)^2\partial_m f^{ak}\partial_m f_{ak} + \tfrac{1}{2}\mathrm{i}(1+4I\bar\phi)\rho^{\alpha a}\partial_{\alpha\dot\alpha}\chi^{\dot\alpha}_a + 4\mathrm{i}I\bar\Psi^{\dot\alpha}_k\rho^\alpha_a\partial_{\alpha\dot\alpha}f^{ak} \\ &\quad + 2\mathrm{i}I\rho^{\alpha a}A_m\partial_m\rho_{\alpha a} + \mathrm{i}I\rho^{\beta a}\rho^\alpha_a\partial_{(\alpha\dot\alpha}A^{\dot\alpha}_{\beta)}. \end{aligned} \tag{52}$$

Note that only the components $\bar\phi, A_m$ and $\bar\Psi^{\dot\alpha}_k$ of the gauge multiplet interact with the hypermultiplet fields in this model. We shall see that in the sum of L_1 and the U(1) gauge multiplet Lagrangian L_g, Eq. (24), most of the interaction terms can be removed by making a proper redefinition of fields.

Let us look at symmetries of (52). The adjoint hypermultiplet gauge transformation (37) in the unfolded form reads

$$\delta_\Lambda q^{+a} = 4iI\bar{\theta}^+_{\dot{\alpha}}(\partial^{\alpha\dot{\alpha}}\Lambda\partial_{+\alpha}q^{+a} - \partial_{+\alpha}\Lambda\partial^{\alpha\dot{\alpha}}q^{+a}), \tag{53}$$

while the unbroken $N{=}(1,0)$ supersymmetry transformation is simply

$$\delta_\varepsilon q^{+a} = (\varepsilon^{-\alpha}Q^+_\alpha - \varepsilon^{+\alpha}Q^-_\alpha)\,q^{+a} = \left(\varepsilon^{+\alpha}\partial_{+\alpha} - 2i\varepsilon^{-\alpha}\bar{\theta}^{+\dot{\alpha}}\partial_{\alpha\dot{\alpha}}\right)q^{+a}, \tag{54}$$

where $\varepsilon^{\pm\alpha} = \varepsilon^{ai}u^\pm_i$. In the WZ-gauge for the U(1) potential (15) the corresponding residual gauge and $N{=}(1,0)$ supersymmetry transformations of the hypermultiplet are

$$\delta_r q^{+a} = -4I\bar{\theta}^+_{\dot{\alpha}}\partial^{\alpha\dot{\alpha}}a(x_A)\partial_{+\alpha}q^{+a}, \tag{55}$$

$$\begin{aligned}\delta_\varepsilon q^{+a} &= (\varepsilon^{+\alpha}\partial_{+\alpha} + 2i\bar{\theta}^+_{\dot{\alpha}}\varepsilon^-_\alpha\partial^{\alpha\dot{\alpha}})q^{+a}\\ &\quad +4iI\bar{\theta}^+_{\dot{\alpha}}(\partial_{+\alpha}\Lambda_\varepsilon\partial^{\alpha\dot{\alpha}}q^{+a} - \partial^{\alpha\dot{\alpha}}\Lambda_\varepsilon\partial_{+\alpha}q^{+a}),\end{aligned} \tag{56}$$

where the field-dependent compensating gauge parameter Λ_ε was defined in (18). After making use of Eqs. (51) for the component fields in these formulas we obtain the residual gauge transformations of the physical fields of q^{+a} as

$$\delta_r f^{ak} = 0, \quad \delta_r\rho^a_\alpha = 0, \quad \delta_r\chi^{\dot{\alpha}a} = -4I\partial^{\alpha\dot{\alpha}}a\rho^a_\alpha \tag{57}$$

and the corresponding $N = (1,0)$ transformations as

$$\delta_\varepsilon f^{ak} = \varepsilon^{\alpha k}\rho^a_\alpha, \quad \delta_\varepsilon\rho^a_\alpha = 0, \quad \delta_\varepsilon\chi^a_{\dot{\alpha}} = 2i\varepsilon^{\alpha k}(1+4I\bar{\phi})\partial_{\alpha\dot{\alpha}}f^a_k\,. \tag{58}$$

It is straightforward to construct an analog of the SW-type transformation for the physical fields of the deformed hypermultiplet

$$\begin{aligned} f^{ak}_0 &= (1+4I\bar{\phi})f^{ak}, \qquad \rho^{\alpha a}_0 = (1+4I\bar{\phi})\rho^{\alpha a},\\ \chi^{\dot{\alpha}a}_0 &= \chi^{\dot{\alpha}a} + 4I(1+4I\bar{\phi})^{-1}A^{\alpha\dot{\alpha}}\rho^a_\alpha - 8I(1+4I\bar{\phi})^{-1}\bar{\Psi}^{\dot{\alpha}k}f^a_k\,.\end{aligned} \tag{59}$$

The redefined hypermultiplet fields $f^{ak}_0, \rho^{\alpha a}_0, \chi^{\dot{\alpha}a}_0$ possess the standard undeformed transformation properties: they are neutral with respect to the gauge group U(1), i.e. their $a(x)$ variations are equal to zero, and their $N = (1,0)$ supersymmetry transformations look as the $I = 0$ case of (58).

Let us pass to the 'undeformed' component fields in the Lagrangian (52). The straightforward computation shows that, up to a total derivative, it acquires the following simple form

$$\begin{aligned} L_1 &= \tfrac{1}{2}\partial_m f^{ak}_0\partial_m f_{0ak} + \tfrac{1}{2}i\rho^{\alpha a}_0\partial_{\alpha\dot{\alpha}}\chi^{\dot{\alpha}}_{0a} + 2iI(1+4I\bar{\phi})^{-1}\rho^{\beta a}_0\rho^\alpha_{0a}\partial_{(\alpha\dot{\alpha}}a^{\dot{\alpha}}_{\beta)}\\ &\quad + 2I(1+4I\bar{\phi})^{-1}(f^{ak}_0 f_{0ak})\Box\bar{\phi} + 4iI(1+4I\bar{\phi})^{-1}(\rho^{\alpha a}_0 f_{0ak})\partial_{\alpha\dot{\alpha}}\bar{\psi}^{\dot{\alpha}k}\,. \end{aligned} \tag{60}$$

Now it is easy to observe that in the total gauge multiplet-hypermultiplet Lagrangian $L = L_g + L_1$ the last two terms in (60) can be removed by the appropriate redefinition of the fields φ and ψ_k^α of the U(1) gauge multiplet

$$\begin{aligned} \hat{\varphi} &= (1+4I\bar{\phi})^2\varphi - 4I(1+4I\bar{\phi})^{-1}(f_0^{ak} f_{0ak}), \\ \hat{\psi}_k^\alpha &= (1+4I\bar{\phi})^2\psi_k^\alpha - 4I(1+4I\bar{\phi})^{-1}(\rho_0^{\alpha a} f_{0ak}). \end{aligned} \tag{61}$$

In terms of the newly introduced fields the total on-shell Lagrangian can be rewritten as the sum of the free gauge multiplet-hypermultiplet action and the simple interaction term

$$L = L_g + L_1 = L_0 + L_{int}, \tag{62}$$

$$\begin{aligned} L_0 = &-\tfrac{1}{2}\hat{\varphi}\Box\bar{\phi} + \tfrac{1}{2}\partial_m f_0^{ak}\partial_m f_{0ak} - \tfrac{1}{16}f^{\alpha\beta}f_{\alpha\beta} - \mathrm{i}\hat{\psi}_k^\alpha\partial_{\alpha\dot{\alpha}}\bar{\psi}^{\dot{\alpha}k} \\ &+ \tfrac{1}{2}\mathrm{i}\rho_0^{\alpha a}\partial_{\alpha\dot{\alpha}}\chi_{0a}^{\dot{\alpha}} + \tfrac{1}{4}(\hat{d}_{kl})^2, \end{aligned} \tag{63}$$

$$L_{int} = -\tfrac{1}{2}I\bar{\phi}(1+2I\bar{\phi})f^{\alpha\beta}f_{\alpha\beta} + I(1+4I\bar{\phi})^{-1}\rho_0^{\beta a}\rho_{0a}^{\alpha}f_{\alpha\beta}, \tag{64}$$

where $f_{\alpha\beta} = \mathrm{i}\partial_{\alpha\dot{\alpha}}a_\beta^{\dot{\alpha}} + \mathrm{i}\partial_{\beta\dot{\alpha}}a_\alpha^{\dot{\alpha}} = (\sigma_{mn})_{\alpha\beta}f_{mn}$ and $\hat{d}_{kl} = (1+4I\bar{\phi})d_{kl}$.

In the limit $I = 0$ the Lagrangian (62) becomes a sum of free Lagrangians of the vector gauge multiplet and hypermultiplet, and so it represents the Euclidean $N{=}(2,2)$ supersymmetric abelian gauge theory. Hence in this limit it should reveal a hidden on-shell $N{=}(1,1)$ supersymmetry which forms $N{=}(2,2)$ supersymmetry together with the manifest $N{=}(1,1)$. At $I \neq 0$ the Lagrangian (62) can be regarded as a QS-deformed version of the $N{=}(2,2)$ gauge theory Lagrangian, and it is expected to enjoy half of the original $N{=}(2,2)$ supersymmetry. Indeed, it can be checked that (62) is invariant, up to a total derivative, under the following extra $N{=}(1,0)$ supersymmetry:

$$\begin{aligned} &\delta_\eta\hat{\varphi} = 0,\ \delta_\eta\bar{\phi} = -2\eta^{\alpha a}\rho_{0\alpha a},\ \delta_\eta a_{\alpha\dot{\alpha}} = 2\eta_\alpha^a\chi_{0a\dot{\alpha}},\ \delta_\eta\hat{\psi}_k^\alpha = 0,\ \delta_\eta\hat{d}_{kl} = 0, \\ &\delta_\eta\bar{\psi}^{\dot{\alpha}k} = 2\mathrm{i}\eta_\alpha^a\partial^{\alpha\dot{\alpha}}f_{0a}^k,\quad \delta_\eta f_0^{ak} = -2\eta^{\alpha a}\hat{\psi}_\alpha^k,\quad \delta_\eta\chi_{0a}^{\dot{\alpha}} = -2\mathrm{i}\eta_{\alpha a}\partial^{\alpha\dot{\alpha}}\hat{\varphi}, \\ &\delta_\eta\rho_0^{\alpha a} = (1+4I\bar{\phi})^2\eta_\beta^a f^{\alpha\beta} - 8I(1+4I\bar{\phi})^{-1}\eta_\beta^a\rho_0^{\beta b}\rho_{0b}^\alpha, \end{aligned} \tag{65}$$

where $\eta^{\alpha a}$ are the corresponding Grassmann parameters. In terms of the redefined fields the manifest $N{=}(1,0)$ supersymmetry is realized by the transformations

$$\begin{aligned} \delta_\varepsilon\hat{\varphi} &= 2\varepsilon^{\alpha k}\hat{\psi}_{\alpha k}, \qquad \delta\bar{\phi} = 0, \qquad \delta_\varepsilon a_{\alpha\dot{\alpha}} = 2\varepsilon_\alpha^k\bar{\psi}_{\dot{\alpha}k}, \\ \delta_\varepsilon\hat{\psi}_k^\alpha &= (1+4I\bar{\phi})\varepsilon^{\alpha l}\hat{d}_{kl} + \tfrac{1}{2}(1+4I\bar{\phi})^2\varepsilon_{\beta k}f^{\alpha\beta} - 4I(1+4I\bar{\phi})^{-1}\varepsilon_{\beta k}(\rho_0^{\alpha a}\rho_{0a}^\beta), \\ \delta_\varepsilon\bar{\psi}^{\dot{\alpha}k} &= \mathrm{i}\varepsilon_\alpha^k\partial^{\alpha\dot{\alpha}}\bar{\phi}, \qquad \delta_\varepsilon\hat{d}_{kl} = \mathrm{i}(1+4I\bar{\phi})(\varepsilon_k^\alpha\partial_{\alpha\dot{\alpha}}\bar{\psi}_l^{\dot{\alpha}} + \varepsilon_l^\alpha\partial_{\alpha\dot{\alpha}}\bar{\psi}_k^{\dot{\alpha}}), \\ \delta_\varepsilon f_0^{ak} &= \varepsilon^{\alpha k}\rho_{0\alpha}^a, \qquad \delta_\varepsilon\chi_{0a}^{\dot{\alpha}} = -2\mathrm{i}\varepsilon_\alpha^k\partial^{\alpha\dot{\alpha}}f_{0ak}, \quad \delta_\varepsilon\rho_0^{\alpha a} = 0. \end{aligned} \tag{66}$$

The additional η-transformations can be checked to commute on-shell with themselves and also with the $N{=}(1,0)$ ε-transformations.

Presumably, the transformations (65) can be deduced from a superfield transformation which is analogous to the one used in the $N{=}2$ superfield formulation of $N{=}4$ gauge theory in the Minkowski space [18].

7. QS-DEFORMED CHARGED HYPERMULTIPLET

According to Eqs. (48), (47), the superfield action for this case reads

$$S_e(V,q) = \tfrac{1}{2}\int du\, d\zeta^{-4}\, q^+_a \left(D^{++}q^{+a} + \frac{1}{2}e[V^{++},q^{+a}]_\star - \frac{1}{2}e(\tau_3)^a_b\{V^{++},q^{+b}\}_\star\right). \quad (67)$$

The detailed component structure of this action can be found like in the previous case, applying the general formulas (12), (13), inserting the $\theta^+, \bar\theta^+$ expansions (15), (50) and performing integration over the Grassmann and (at the final step) harmonic variables. Solving the harmonic equations for the auxiliary fields, we finally express q^{+a} in terms of off-shell fields of the gauged multiplet and the physical hypermultiplet fields:

$$\begin{aligned} q^{+a}_e \;&=\; u^+_k f^{ak} + \theta^{+\alpha}\rho^a_\alpha + (\theta^+)^2 u^-_k g^{ak} \\ &+\bar\theta^+_{\dot\alpha}[\chi^{\dot\alpha a} + (\theta^+)^2 u^-_k u^-_l \sigma^{\dot\alpha akl}] + \theta^+\sigma_m\bar\theta^+ r^{ak}_m u^-_k \\ &+(\bar\theta^+)^2[u^-_k h^{ak} + \theta^{+\alpha}u^-_k u^-_l \Sigma^{akl}_\alpha + (\theta^+)^2 u^-_k u^-_l u^-_j X^{aklj}], \end{aligned} \quad (68)$$

where

$$\begin{aligned} g^{ak} &= e(\tau_3)^a_b\bar\phi f^{bk}, \quad r^{ak}_m = 2\mathrm{i}(1+2eI\bar\phi)\partial_m f^{ak} + 2e(\tau_3)^a_b A_m f^{bk}, \\ h^{ak} &= -4\mathrm{i}eIA_m\partial_m f^{ak} + e(\tau_3)^a_b\left(\phi f^{bk} + 2eI^2\bar\phi\Box f^{bk}\right), \\ \sigma^{\dot\alpha akl} &= 2e(\tau_3)^a_b\bar\Psi^{\dot\alpha(k}f^{bl)}, \quad \Sigma^{akl}_\alpha = -4\mathrm{i}eI\bar\Psi^{\dot\alpha(k}\partial_{\alpha\dot\alpha}f^{al)} + 2e(\tau_3)^a_b\Psi^{(k}_\alpha f^{bl)}, \\ X^{aklj} &= e(\tau_3)^a_b\mathscr{D}^{(kl}f^{bj)}. \end{aligned} \quad (69)$$

The deformed residual U(1) gauge transformations of the charged hypermultiplet components are

$$\begin{aligned} &\delta_r f^{ak} = \mathrm{i}a\,e(\tau_3)^a_b f^{bk}, \quad \delta_r\rho^a_\alpha = \mathrm{i}a\,e(\tau_3)^a_b\rho^b_\alpha, \\ &\delta_r\chi^{\dot\alpha a} = \mathrm{i}a\,e(\tau_3)^a_b\chi^{\dot\alpha b} - 2eI\partial^{\alpha\dot\alpha}a\,\rho^a_\alpha. \end{aligned} \quad (70)$$

The corresponding component unbroken $N{=}(1,0)$ supersymmetry transformations read

$$\begin{aligned} &\delta_\varepsilon f^{ak} = \varepsilon^{\alpha k}\rho^a_\alpha, \quad \delta_r\rho^a_\alpha = 2\varepsilon^k_\alpha g^a_k = 2\varepsilon^k_\alpha e(\tau_3)^a_b\bar\phi f^b_k, \\ &\delta_\varepsilon\chi^a_{\dot\alpha} = -\varepsilon^\alpha_k r^{ak}_{\alpha\dot\alpha} = -2\varepsilon^\alpha_k[\mathrm{i}(1+2eI\bar\phi)\partial_{\alpha\dot\alpha}f^{ak} + e(\tau_3)^a_b A_{\alpha\dot\alpha}f^{bk}]. \end{aligned} \quad (71)$$

The SW-map for the charged hypermultiplet fields is given by the relations which are similar to those for the 'adjoint' hypermultiplet, Eqs. (59), though slightly differ in some coefficients

$$\begin{aligned} &f^{ak}_0 = (1+2eI\bar\phi)f^{ak}, \qquad \rho^{\alpha a}_0 = (1+2eI\bar\phi)\rho^{\alpha a}, \\ &\chi^a_{\dot\alpha 0} = \chi^a_{\dot\alpha} - 2eI(1+4eI\bar\phi)^{-1}A_{\alpha\dot\alpha}\rho^{\alpha a} + 4eI(1+4eI\bar\phi)^{-1}\bar\Psi_{\dot\alpha k}f^{ak}. \end{aligned} \quad (72)$$

The gauge and supersymmetry transformations of fields f^{ak}_0, $\rho^{\alpha a}_0$ and $\chi^a_{\dot\alpha 0}$ look just as the $I=0$ limit of (70), (71). While checking this, one should take into account that in the considered case of $e \neq 1$ the gauge and supersymmetry transformations of the gauge multiplet fields are obtained by the replacement $I \to eI$ in Eqs. (16) and (17).

The deformed charged hypermultiplet Lagrangian for the physical fields is given by

$$\begin{aligned} L_e &= \tfrac{1}{2}(1+4eI\bar{\phi})\partial_m f_{ak}\partial_m f^{ak} + \mathrm{i}e(\tau_3)^b_a A_m f_{bk}\partial_m f^{ak} + \tfrac{1}{2}e^2(A_m)^2(f^{ak})^2 \\ &+ \tfrac{1}{2}e^2\phi\bar{\phi}(f^{ak})^2 + I^2e^2(f^{ak})^2\Box(\bar{\phi}^2) - \tfrac{1}{2}e(\tau_3)^a_b f^k_a f^{bl}\mathscr{D}_{kl} + 2\mathrm{i}eI\bar{\Psi}^{\dot\alpha}_k\rho^\alpha_a\partial_{\alpha\dot\alpha}f^{ak} \\ &+ e(\tau_3)^a_b\Psi^\alpha_k\rho_{\alpha a}f^{bk} + e(\tau_3)^a_b f^k_a\bar{\Psi}_{\dot\alpha k}\chi^{\dot\alpha b} + \tfrac{1}{2}\mathrm{i}(1+2eI\bar{\phi})\rho^{\alpha a}\partial_{\alpha\dot\alpha}\chi^{\dot\alpha}_a \\ &- \tfrac{1}{2}e(\tau_3)^a_b\rho^\alpha_a A_{\alpha\dot\alpha}\chi^{\dot\alpha b} + \tfrac{1}{4}e(\tau_3)^a_b(\bar{\phi}\chi_{\dot\alpha a}\chi^{\dot\alpha b} + \phi\rho^\alpha_a\rho^b_\alpha) + \mathrm{i}eI\rho^{\alpha a}A_m\partial_m\rho_{\alpha a} \\ &+ \tfrac{1}{2}\mathrm{i}eI\rho^{\beta a}\rho^\alpha_a\partial_{(\alpha\dot\alpha}A^{\dot\alpha}_{\beta)} + I^2e(\tau_3)^a_b\bar{\phi}\partial_{\alpha\dot\alpha}\rho_{\beta a}\partial^{\beta\dot\alpha}\rho^{\alpha b}. \end{aligned} \tag{73}$$

It has to be combined with the $e \neq 1$ modification of the U(1) gauge theory Lagrangian (24) rewritten in terms of the original deformed fields. As distinct from the case of neutral hypermultiplet, passing to the undeformed fields in the total action with the help of the transformation (72) and the $e \neq 1$ version of (23) does not give rise to radical simplifications. Here we present the scalar potential of the model in terms of the deformed fields. It arises as the result of integrating out the gauge multiplet auxiliary field $\mathscr{D}^{kl}$ from the corresponding piece of the total action

$$\tfrac{1}{4}(1+4eI\bar{\phi})^{-2}\mathscr{D}^{kl}\mathscr{D}_{kl} - \tfrac{1}{2}e(\tau_3)^a_b f^k_a f^{bl}\mathscr{D}_{kl} + \tfrac{1}{2}e^2\phi\bar{\phi}(f^{ak})^2, \tag{74}$$

and is given by the following positively-definite expression

$$V = \tfrac{1}{8}e^2(f^{ak})^2[(1+4eI\bar{\phi})^2(f^{ak})^2 + 4\phi\bar{\phi}]. \tag{75}$$

It is worth noting that there is one more mechanism of generating scalar potentials in the considered deformed hypermultiplet-gauge multiplet systems. Namely, one can add to the superfield actions (49) or (67) the analytic Fayet-Iliopoulos (FI) superfield term

$$S_{FI} = \int du\, d\zeta^{-4}\left[3\mathrm{i}c^{++} + c_0(\bar{\theta}^+)^2\right]V^{++}, \tag{76}$$

where $c^{++} = c^{(ik)}u^+_i u^+_k$, $\widetilde{c^{++}} = c^{++}$, and $c^{(ik)}$ and c_0 are some harmonic-independent constants. This term is manifestly invariant under $N=(1,0)$ supersymmetry and, up to a total harmonic derivative, under gauge transformations (14) or (40). In WZ gauge for V^{++} FI-term gives rise to the following contribution to the component Lagrangian

$$\mathrm{i}c^{kl}\mathscr{D}_{kl} + c_0\bar{\phi}. \tag{77}$$

Being combined with (74), this term results in the $SU(2)_R$-breaking addition to (75)

$$\Delta V = \left[e(\tau_3)^a_b(f^k_a f^{bl}c_{kl}) + c^2\right](1+4eI\bar{\phi})^2 + c_0\bar{\phi}. \tag{78}$$

It contains the tadpole $\sim \bar{\phi}$ which could destabilize the theory in the quantum case, giving rise to vacuum transitions. Such an unwanted term vanishes under the choice

$$c_0 = -8\,eI\,c^2\,. \tag{79}$$

The same mechanism applies to the case of the adjoint hypermultiplet considered in the previous section. The relevant potential can be obtained from (78) via the replacement $eI \to I$ with setting $e = 0$ afterwards. Note that such a potential breaks the second (hidden) $N{=}(1,0)$ supersymmetry of this model.

Finally, let us briefly outline how mass terms for the hypermultiplets can be introduced in the QS-deformed case. The mechanism of such a generation is of the Scherk-Schwarz type and it is similar to the one known in the $N{=}2$ Minkowski case.

Let us introduce the constant U(1) analytic potential

$$B^{++} = \bar{m}(\theta^+)^2 + m(\bar{\theta}^+)^2\,, \tag{80}$$

where m and $\bar{m}$ are independent real constants (they are mutually conjugated in the Minkowski case). Like in the undeformed case [21, 22], B^{++} is a background solution of the deformed U(1) gauge theory equations. The generation of mass terms can be interpreted as a result of interaction with these constant background 'scalar fields' m and $\bar{m}$ in (80) [22, 23, 24]. It is convenient to define new shifted scalar fields of the gauge U(1) supermultiplet

$$\phi = m + \phi'\,, \qquad \bar{\phi} = \bar{m} + \bar{\phi}'\,. \tag{81}$$

The $N{=}(1,0)$ supersymmetry of both gauge field-hypermultiplet models survives after this shift, although its realization on the component fields slightly changes, as can be explicitly seen by substituting (81) into (58) and (71). Performing this shift in the Lagrangians (52), (73), we see that the genuine mass terms appear only for the charged hypermultiplet and they survive in the commutative limit. Masses of the left- and right-handed fermions ρ^α_a and $\chi^{\dot{\alpha}}_a$ are proportional to m and $\bar{m}$ and so are independent. Also, as a specific feature of the non-anticommutative case, background $\bar{m}$ triggers proper renormalizations of the kinetic terms. The only impact of this background on the Lagrangian of the neutral hypermultiplet is such renormalizations of the kinetic terms.

The new 'free' parts of the superfield actions (49) and (67) are obtained by substituting there B^{++} for V^{++}, and the shift (81) corresponds to decomposing

$$V^{++} = \hat{V}^{++} + B^{++}\,. \tag{82}$$

While the free part of the massless hypermultiplet actions, i.e. (36), is invariant under the standard realization of $N{=}(1,0)$ supersymmetry, the free actions with mass terms are invariant under modified supersymmetry transformations which are combinations of the standard $N = (1,0)$ supertranslations and the particular case of U(1) gauge transformations (14) (or (40) for $e \neq 1$) and (37), (38) (or (41), (42)), with

$$\hat{\Lambda} = -2\bar{m}\varepsilon^-\theta^+\,.$$

The modified transformations close on the constant-parameter subgroup of the corresponding QS-deformed U(1) group, and the appropriate generator is identified with the central charge of the $N=(1,0)$ superalgebra. Since $\star$-commutators of a constant parameter Λ with both V^{++} and q^{+a} are vanishing, the modified $N=(1,0)$ transformations have a non-zero closure only on the charged hypermultiplet and the anticommutator of the corresponding supercharges is proportional to

$$\sim (\tau_3)^a_b q^{+b} ,$$

precisely as in the undeformed case. The free massive hypermultiplet actions are invariant just under the modified $N=(1,0)$ transformations and by no means under the original ones (which mix the free actions with the interaction terms $\sim \hat{V}^{++}$).

As mentioned in sect. 5, in the case of the complex hypermultiplet there are two mutually commuting rigid U(1) groups. In addition to the transformation (44) of the complex hypermultiplet, one can consider the following SU(2)$_{PG}$-breaking gauge transformation of the same complex superfield:

$$\delta_{\Lambda'} q^{+a} = \tfrac{e}{2} q^{+b} \star \Lambda' [\delta^a_b + (\tau_3)^a_b] , \tag{83}$$

where Λ' is an independent analytic superfield parameter. This 'right' transformation commutes with the 'left' transformation (44). So we can introduce two independent deformed U(1) gauge potentials, interacting with the complex hypermultiplet superfield, or e.g. one potential V^{++} for the first gauge group and the constant background potential for the second one. In the second case we shall obtain SU(2)$_{PG}$ breaking mass terms which cannot be generated by shifts of the scalar fields of the 'left' gauge multiplet. Of course we could introduce background fields for the first U(1) group and gauge the second one, then SU(2)$_{PG}$ invariant mass terms and 'right' gauge interactions could be generated.

Finally we note that the potential (78) induced by the FI-term provides another independent $N=(1,0)$ supersymmetric source of generating masses for the scalar physical fields of the hypermultiplet (and simultaneously for the field $\bar{\phi}$). The superfield FI-term (76) is invariant under both the original and central-charge modified $N=(1,0)$ supersymmetries.

8. STRING THEORY ORIGIN OF QS-DEFORMATION

Here, following ref. [12], we present a possible interpretation of the QS-deformation as originating from string theory with a specially chosen constant background, along the lines of [2, 4].

From the point of view of the compactification of the type IIB background one may say that the self-dual tensor $F^{\alpha\beta}$ appearing in (1) originates from the Ramond-Ramond sector, more precisely, from the five-form in ten dimensions wrapped around the (3,0)-form of the Calabi-Yau manifold viewed as an orbifold $\mathbb{C}^3/Z_2 \times Z_2$. If we want to find the origin of the singlet scalar deformation parameter I of our QS-deformation, we should look in a different part of the Ramond-Ramond sector, namely at the one-form. This time we should choose a compactification on $\mathbb{C} \times \mathbb{C}^2/Z_2$. The one-form is the derivative of the axion field, $F_\mu = \partial_\mu C$. After the compactification we restrict it to just the two dimensions of the torus, $F_a = \partial_a C$, $a = 4,5$, or in the complex notation, $F = \partial_\tau C$, $\tau = x^4 + ix^5$. Further, the kinetic term of the axion is of the form

$$\partial_a C \partial_a C = \partial_\tau C \partial_{\bar{\tau}} C\,. \tag{84}$$

Our (Euclidean) choice will be

$$C = \tau F_0 \tag{85}$$

where F_0 is a real constant, i.e. C will be a real analytic function. Then, since $\partial_{\bar{\tau}} C = 0$, we see that the kinetic term (84), and with it the contribution to the stress tensor, vanishes. We conclude that this is a consistent choice of the background.

The rest closely follows the N=1 case [2, 4, 16]. We couple this background to an N=4 superstring and obtain the effective Lagrangian

$$\mathscr{L}_{\text{eff}} = \left(\frac{1}{\alpha'^2 F_0}\right) \varepsilon_{\alpha\beta}\varepsilon_{ij}\,\partial\tilde{\theta}^{i\alpha}\bar{\partial}\theta^{j\beta}\,. \tag{86}$$

The boundary conditions cut the number of θs by half, and the resulting propagators give rise to the deformed superspace anticommutator

$$\{\theta^{i\alpha}, \theta^{j\beta}\} = \alpha'^2 F_0 \varepsilon^{\alpha\beta}\varepsilon^{ij}\,, \tag{87}$$

where the product of constants $\alpha'^2 F_0$ plays the rôle of the deformation parameter I.

9. OUTLOOK

The results reported in this talk can be extended in several directions. In particular, it would be tempting to find out possible phenomenological uses of the considered QS-deformations as well as of their non-abelian generalizations. Deformations of this sort may possibly provide a geometrical way of generating soft supersymmetry breaking. A closely related quantum issue is the renormalization and finiteness properties of the coupled gauge multiplet-hypermultiplet systems given above, which may be studied like in the $N=(\frac{1}{2},\frac{1}{2})$ case [6]. The $N=(2,2)$ gauge theory is the Euclidean analog of N=4 super Yang-Mills which supplied the first example of an ultraviolet-finite quantum field theory and displays many other remarkable properties (e.g. in the context of the AdS/CFT correspondence). It is of obvious interest to study the $N=(2,2)$ gauge theory from this point of view and to inspect whether its nilpotent deformation as presented in section 6 and its non-abelian generalization preserve the basic quantum and geometric properties of the undeformed theory. Nilpotent deformations do not induce noncommutativity for the bosonic space-time coordinates; hence, such deformed quantum theories are expected not to suffer from such problems as UV-IR mixing. The issues of their vacuum structure and classical solutions are also interesting tasks for the future study.

As another conceivable application of the deformed minimal coupling of a U(1) gauge multiplet with hypermultiplets it is worthwhile to mention a generalization of the quotient approach to constructing hyper-Kähler metrics in N=2 supersymmetric sigma models (see e.g. [25]). Its basic ingredient is a non-propagating N=2 U(1) gauge multiplet coupled to hypermultiplets. Using the relations given in sect. 6 and 7 it is straightforward to generalize this approach to QS-deformed $N=(1,1)$ theories and to investigate the possible deformation of the hyper-Kähler target space geometry [26]. Finally, we point out that it is also desirable to elaborate on the implications of more general (non-singlet) Q-deformations in $N=(1,1)$ (and perhaps $N=(2,2)$) supersymmetric systems [27].

ACKNOWLEDGMENTS

This work was partially supported by the INTAS grant 00-00254, by the DFG grant 436 RUS 113/669-2, by the RFBR grants 03-02-17440 and 04-02-04002, by the NATO grant PST.GLG.980302 and by a grant of the Heisenberg-Landau program. The author should like to thank the Organizers of XIXth Max Born Symposium and, personally, J. Lukierski for inviting him to give this talk. He is grateful to S. Ferrara, O. Lechtenfeld, E. Sokatchev and B. Zupnik in collaboration with whom most of the results overviewed here were obtained.

REFERENCES

1. N. Seiberg, E. Witten, JHEP **9909**, 032 (1999) [hep-th/9908142].
2. H. Ooguri, C. Vafa, Adv. Theor. Math. Phys. **7**, 53 (2003) [hep-th/0302109].
3. J. de Boer, P.A. Grassi, P. van Nieuwenhuizen, Phys. Lett. **B574**, 98 (2003) [hep-th/0302078].
4. N. Seiberg, JHEP **0306**, 010 (2003) [hep-th/0305248].
5. N. Berkovits, N. Seiberg, JHEP **0307**, 010 (2003) [hep-th/0306226].
6. M.T. Grisaru, S. Penati, A. Romagnoni, Class. Quant. Grav. **21**, S1391 (2004) [hep-th/0401174].
7. S. Ferrara, M.A. Lledó, JHEP **05**, 008 (2000) [hep-th/0002084].
8. D. Klemm, S. Penati, L. Tamassia, Class. Quant. Grav. **20**, 2905 (2003) [hep-th/0104190].
9. S. Ferrara, M.A. Lledó, O. Maciá, JHEP **09**, 068 (2003) [hep-th/0307039].
10. E. Ivanov, O. Lechtenfeld, B. Zupnik, JHEP **0402**, 012 (2004) [hep-th/0308012].
11. S. Ferrara, E. Sokatchev, Phys. Lett. **B579**, 226 (2004) [hep-th/0308021].
12. S. Ferrara, E. Ivanov, O. Lechtenfeld, E. Sokatchev, B. Zupnik, Nucl. Phys. **B704**, 154 (2005) [hep-th/0405049].
13. T. Araki, K. Ito, A. Ohtsuka, JHEP **0401**, 046 (2004) [hep-th/0401012].
14. T. Araki, K. Ito, Phys. Lett. **B595**, 513 (2004) [hep-th/0404250].
15. S.V. Ketov, S. Sasaki, Phys. Lett. **B595**, 530 (2004) [hep-th/0404119];
    Phys. Lett. **B597**, 105 (2004) [hep-th/0405278];
    SU(2)×U(1) non-anticommutative N=2 supersymmetric gauge theory, [hep-th/0407211].
16. M. Billo, M. Frau, I. Pesando, A. Lerda, JHEP **0405**, 023 (2004) [hep-th/0402160].
17. A. Galperin, E. Ivanov, V. Ogievetsky, E. Sokatchev, JETP Lett. **40**, 912 (1984) [Pis'ma ZhETF **40** (1984) 155];
    A. Galperin, E. Ivanov, S. Kalitzin, V. Ogievetsky, E. Sokatchev, Class. Quant. Grav. **1**, 469 (1984).
18. A. Galperin, E. Ivanov, V. Ogievetsky, E. Sokatchev, *Harmonic superspace*, Cambridge University Press, 2001, 306 p.
19. E. Ivanov, O. Lechtenfeld, B. Zupnik, *Non-anticommutative deformation of N=(1,1) hypermultiplets*, [hep-th/0408146] (Nucl. Phys B, in press).
20. C. Sämann, M. Wolf, JHEP **0403**, 048 (2004) [hep-th/0401147].
21. B.M. Zupnik, Sov. J. Nucl. Phys. **44**, 512 (1986) [Yad. Fiz. **44**, 794 (1986)].
22. E. Ivanov, S. Ketov, B. Zupnik, Nucl. Phys. **B509**, 53 (1998) [hep-th/9706078].
23. I.L. Buchbinder, S.M. Kuzenko, Class. Quant. Grav. **14**, L157 (1997) [hep-th/9704002].
24. E.I. Buchbinder, I.L. Buchbinder, E.A. Ivanov, S.M. Kuzenko, Mod. Phys. Lett. **A13**, 1071 (1998) [hep-th/9803176].
25. N.J. Hitchin, A. Karlhede, U. Lindström, M. Roček, Commun. Math. Phys. **108**, 535 (1987);
    A. Galperin, E. Ivanov, V. Ogievetsky, P.K. Townsend, Class. Quant. Grav. **3**, 625 (1986).
26. E. Ivanov, A. Shcherbakov, B. Zupnik, work in preparation.
27. A. De Castro, E. Ivanov, O. Lechtenfeld, L. Quevedo, work in preparation.

Canonical Quantization and Black Hole Perturbations

Zoltán Perjés [1] * and Árpád Lukács*

*KFKI Research Institute for Particle and Nuclear Physics, Budapest 114, P.O.Box 49, H-1525 Hungary

Abstract. We examine the possibility of a constraint-free quantization of linearized gravity, based on the Teukolsky equation for black hole perturbations.

We exhibit a simple quadratic (but complex) Lagrangian for the Teukolsky equation, leading to the interpretation that the elementary excitations (gravitons bound to the Kerr black hole) are unstable.

INTRODUCTION

Quite a number of attempts have been made and are being made to quantize Einstein's gravitation theory. One of the earliest attempts to use canonical quantization of linearized gravity goes back to Gupta [1]. The canonical approach to quantum gravity using the 3+1 decomposition has been worked out by Arnowitt, Deser and Misner [2]. A different idea has been put forward by Penrose [3] introducing twistors. Ashtekar [4] introduced spinor variables to handle the constraints. Nowadays loop quantum gravity is being actively pursued, see the review by Rovelli [5]. In spite of the remarkable progress achieved using these beautiful constructions, there remains a lot to be done.

Undoubtedly most of the difficulties in all canonical approaches to quantum gravity come from the constraints. Another problem, closely related to the treatment of constraints is the choice of the canonical coordinates. A good choice of coordinates leads to less constraint equations, and thus it makes quantization easier.

In general relativity, the physical state of space-time is described by a symmetric, second rank tensor that can be parametrized by ten components. Due to the constraints, we should introduce less than ten independent canonical coordinates, and their conjugate momenta.

The well known Teukolsky equation [6], which is a linear PDE, describes the perturbations of a Kerr space-time with a single scalar quantity, ψ, one of the Newman–Penrose spin coefficients.

Chrzanowski, Misner [7] and Ori [8] have shown the way to reconstruct the metric perturbations from the solution of the Teukolsky equation by acting on it with a differential operator.

[1] Talk presented by Zoltán Perjés

CP767, *Fundamental Interactions and Twistor-Like Methods*, edited by J. Lukierski and D. Sorokin

This makes the quantity ψ a promising candidate as a canonical coordinate in a linearized quantum gravity, provided a Hamiltonian formulation of the Teukolsky equation (1) can be given. This problem will be examined in the next section.

One would think, that the easiest way to obtain a canonical formalism based on the Teukolsky equation is to reconstruct the perturbations of the metric tensor, and subsequently the quadratic part of the corresponding perturbed Lagrangian. This could then be taken as a starting point for a canonical quantization. However, the formulae for the metric perturbations are very complicated, and contain fourth order derivatives of ψ.

One could just as well try to find a Lagrangian directly for Teukolsky's equation in itself, and use that for quantization. Unfortunately, as we will see later, there is no quadratic, first order real function, which gives (1) as an Euler–Lagrange equation.

We have identified a complex Lagrangian for the Teukolsky equation. A possible solution to quantize the Teukolsky equation could be considering the complex part of the Lagrangian as an interaction term. The interaction-free part can be then simply quantized, and it's quanta may be identified as gravitons. As a result of the perturbations (the complex part of the Lagrangian), these excitations decay.

IS A LAGRANGIAN FORMULATION OF TEUKOLSKY'S EQUATION POSSIBLE?

Some basic features of the Teukolsky equation can be found in the Appendix. Here we recall that this equation governs the dynamics of one of the Riemann tensor's tetrad components in a linearized approximation over a Kerr background.

Using the notations of the Appendix, Teukolsky's equation [6] can be written as

$$\begin{aligned}&\left[\frac{(r^2+a^2)^2}{\Delta}-a^2\sin^2\vartheta\right]\frac{\partial^2\psi}{\partial t^2}+\frac{4Mar}{\Delta}\frac{\partial^2\psi}{\partial t\partial\varphi}+\left[\frac{a^2}{\Delta}-\frac{1}{\sin^2\vartheta}\right]\frac{\partial^2\psi}{\partial\varphi^2}\\&-\Delta^{-s}\frac{\partial}{\partial r}\left(\Delta^{s+1}\frac{\partial\psi}{\partial r}\right)-\frac{1}{\sin\vartheta}\frac{\partial}{\partial\vartheta}\left(\sin\vartheta\frac{\partial\psi}{\partial\vartheta}\right)-2s\left[\frac{a(r-M)}{\Delta}+\frac{i\cos\vartheta}{\sin^2\vartheta}\right]\frac{\partial\psi}{\partial\varphi}\\&-2s\left[\frac{M(r^2-a^2)}{\Delta}-r-ia\cos\vartheta\right]\frac{\partial\psi}{\partial t}+(s^2\mathrm{ctg}^2\vartheta-s)\psi=4\pi\Sigma T,\end{aligned}\tag{1}$$

where T is a source term, calculated from the energy–momentum tensor of other fields. In the following, we concentrate on the free gravity case ($T=0$).

Over a Kerr geometry, this equation describes massless scalar fields ($s=0$), electromagnetic perturbations ($s=\pm1$) and gravity ($s=\pm2$).

In this section, we examine, whether it is possible to obtain equation (1) as an Euler–Lagrange equation from a *real* Lagrangian of the form

$$\mathscr{L}=ag^{ik}\overline{\partial_i(b\psi)}\partial_k(b\psi)+c\overline{\psi}\psi,\tag{2}$$

which is the most general first order quadratic local Lagrangian of a scalar field. The motivation to try to find such a Lagrangian (eg. local, not containing higher derivatives) is the possibility to quantize it easily.

In order for $\mathscr{L}$ be real, a and c have to be real, and b complex arbitrary functions of the coordinates. From the above function, the Euler–Lagrangian equation is given by the formula

$$\partial_i \frac{\partial \sqrt{-q}\mathscr{L}}{\partial \partial_i \overline{\psi}} - \frac{\partial \sqrt{-g}\mathscr{L}}{\partial \overline{\psi}} = 0, \tag{3}$$

where

$$\sqrt{-g} = (r^2 + a^2 \cos^2 \vartheta) \sin \vartheta,$$

is the determinant of the metric tensor.

Calculating the above equation, and taking the coefficients of ϕ's derivatives in (3) and (1) yields equations for a, b, c and their derivatives. When trying to solve these equations, one finds, that the integrability conditions (eg. the r derivative of $\partial a/\partial t$ should be equal to the t derivative of $\partial a/\partial r$) cannot be fulfilled. This clearly shows that no real Lagrangian of the form (2) can lead to Teukolsky's equation.

While this work was in progress, we have learnt from the summary of Stephen Anco's talk at the Montreal Workshop on the Interaction of Gravity with External Fields [9], that Teukolsky's equation has no Lagrange function. According to Anco [9], this is because the non self-adjointness of (1), which means that

$$\int \overline{\phi} \mathrm{D} \psi \neq \overline{\int \overline{\psi} \mathrm{D} \phi},$$

where D denotes the differential operator in (1).

To show that the equation is indeed not self-adjoint, one should take the terms in $\int \overline{\phi} \mathrm{D} \psi$ and do the partial integrations to calculate its adjoint, and compare it with the complex conjugate of $\int \overline{\psi} \mathrm{D} \phi$.

In the first term, self-adjointness is obvious: the coefficient of $\partial^2/\partial t^2$ does not depend on t. Similar is the case with the second and the third term.

However, the non-self-adjointness can be seen on the fourth term:

$$\int \overline{\phi} \Delta^{-s} \frac{\partial}{\partial r} \left(\Delta^{s+1} \frac{\partial \psi}{\partial r} \right), \tag{4}$$

where substituting Δ, and expanding the derivatives yields in the $s = 2$ case

$$\int \phi \frac{3(r^2 - 2Mr + a^2)^2 (2r - 2M) \frac{\partial \psi}{\partial r} + (r^2 - 2Mr + a^2)^3 \frac{\partial^2 \psi}{\partial r^2}}{(r^2 - 2Mr + a^2)^2},$$

which takes the form

$$\int \psi \left[(r^2 - 2Mr + a^2) \frac{\partial^2 \phi}{\partial r^2} - (2r - 2M) \frac{\partial \phi}{\partial r} - 4\phi \right] \tag{5}$$

when doing the partial integrations. One can see, that exchanging the roles of ψ and ϕ in the original integral, and conjugating does not give such terms, and no other term can cancel the r-derivatives in the above.

Bini, Cherubini and Jantzen [10] gave another form of the above equation, closely resembling an ordinary Klein–Gordon equation. They have proven, that the Teukolsky equation is a generalization of the Klein–Gordon equation for arbitrary spin weight s. Examining the following form of the equation, it is easy to point out the cause of the non self-adjointness:

$$\left[(\nabla^i + s\Gamma^i)(\nabla_i + s\Gamma_i) - 4s^2\Psi_2^A\right]\psi = 4\pi T, \tag{6}$$

where Γ is a four-vector with components

$$\begin{aligned}
\Gamma^t &= -\frac{1}{\Sigma}\left[\frac{M(r^2-a^2)}{\Delta} - (r + ia\cos\vartheta)\right], \\
\Gamma^r &= -\frac{1}{\Sigma}(r-M), \\
\Gamma^\vartheta &= 0, \\
\Gamma^\varphi &= -\frac{1}{\Sigma}\left[\frac{a(r-M)}{\Delta} + i\frac{\cos\vartheta}{\sin^2\vartheta}\right].
\end{aligned} \tag{7}$$

It is easy to show that

$$\nabla_i\Gamma^i = -\frac{1}{\Sigma} \quad \text{és} \quad \Gamma_i\Gamma^i = \frac{1}{\Sigma}\text{ctg}^2\vartheta + 4\Psi_2^A .$$

Here Ψ_2^A denotes the value of the Ψ_2 spin coefficient in an unperturbed Kerr space-time,

$$\Psi_2^A = -\frac{M}{(r - ia\cos\vartheta)^3} .$$

Now we can see, that the equation is non self-adjoint, because the Γ_i-components are not purely imaginary, unlike the Klein–Gordon case, where $\Gamma_i = iA_i$, where A_i is the electromagnetic four-potential. These complex terms make the equation absorptive.

PERTURBATIVE QUANTIZATION OF THE TEUKOLSKY EQUATION

In the previous section we have seen that a direct canonical quantization of the Teukolsky equation is not possible because of the absorptive terms in the Lagrangian. On the other hand, the Bini–form (6) of the Teukolsky equation (1) is in close resemblance to the Klein–Gordon equation. This suggests the following solution: we separate the absorptive part of the equation to treat it as a perturbation, and this way we can construct a Lagrangian for the self-adjoint part. This will be real, and so the Fock space construction is well defined, a particle interpretation can be given.

Introducing the notation of

$$\begin{aligned}
A_i &= \Im(\Gamma_i), \\
B_i &= \Re(\Gamma_i),
\end{aligned} \tag{8}$$

the Bini form of the equation can be written as

$$\left[(\nabla^i + s(B^i + iA^i))(\nabla_i + s(B_i + iA_i)) - 4s^2\Psi_2^A\right]\psi = 4\pi T, \tag{9}$$

and if we keep only the A_is, we get an equation which is formally a Klein–Gordon equation in an external electromagnetic field. The (complex) Lagrangian for this equation is:

$$\mathscr{L} = \left\{\nabla^i - s(B^i + iA^i)\right\}\overline{\psi}\left\{\nabla_i + s(B_i + iA_i)\right\}\psi + 4s^2\Psi_2^A\overline{\psi}\psi, \tag{10}$$

which can be written as a sum of an unperturbed part, and a perturbation term:

$$\mathscr{L} = \mathscr{L}_{KG} + \mathscr{L}_1,$$

where

$$\mathscr{L}_{KG} = \overline{(\nabla^i + isA^i)\psi}(\nabla_i + isA_i)\psi \tag{11}$$

is the Klein-Gordon Lagrangian, and

$$\mathscr{L}_1 = -s(B^i\overline{\psi}\nabla_i\psi - B^i\psi\nabla_i\overline{\psi}) + 2isB^iA_i\overline{\psi}\psi + 4is^2\Psi_2^A\overline{\psi}\psi \tag{12}$$

is the perturbation term.

The canonical momenta are given then by the well-known formula

$$\begin{aligned}\Pi &= \frac{\partial L}{\partial\partial_0\overline{\psi}} = (\partial^0 + isA^0)\psi,\\ \overline{\Pi} &= \frac{\partial L}{\partial\partial_0\psi} = (\partial^0 - isA^0)\overline{\psi},\end{aligned} \tag{13}$$

and the Hamiltonian can be expressed as

$$\mathscr{H}_0 = \Pi\partial_0\psi + \overline{\Pi}\partial_0\overline{\psi} - \mathscr{L}_0 = \overline{\Pi}\Pi + \overline{(\nabla_\alpha + iA_\alpha)\psi}(\nabla^\alpha + iA^\alpha)\psi. \tag{14}$$

In the following discussion $\mathscr{H}_0$ will be taken as the "unperturbed" Hamiltonian density for gravitons. In the next section, we examine the "unperturbed" equation obtained by taking the variation of the Lagrangian $\mathscr{L}_{KG}$.

THE "UNPERTURBED" EQUATION. SEPARATION OF VARIABLES

The Euler-Lagrange equations can be obtained by using either (3) or the well known tensorial - and thus simpler - formula

$$\nabla_i\frac{\partial\mathscr{L}_0}{\partial\nabla_i\overline{\psi}} - \frac{\partial\mathscr{L}_0}{\partial\overline{\psi}} = 0\,.$$

The resulting equation, as mentioned above, is formally a Klein–Gordon equation with an external electromagnetic four-potential:

$$(\nabla_i - isA_i)(\nabla^i - isA_i)\psi = 0. \tag{15}$$

In the following paragraphs, we examine this equation in detail. An explicit computation yields

$$\left(\left((2M-r)r-a^2\right)\psi s^2+\frac{\partial^2\psi}{\partial\varphi^2}-4\frac{\partial^2\psi}{\partial\varphi\partial t}aMr-(a^2-2Mr+r^2)\frac{\partial^2\psi}{\partial r^2}\right.$$
$$-(a^4-4a^2Mr-r^4)\frac{\partial^2\psi}{\partial t^2}-\left((2M-r)r-a^2\right)\left(2(M-r)\frac{\partial\psi}{\partial r}-\frac{\partial^2\psi}{\partial\vartheta^2}\right)\Bigg)\cos^2\vartheta$$
$$-\left(\left((2M-r)\frac{\partial^2\psi}{\partial\varphi^2}-4\frac{\partial^2\psi}{\partial\varphi\partial t}aM\right)r-(a^2-2Mr+r^2)^2\frac{\partial^2\psi}{\partial r^2}\right.$$
$$\left(a^2\cos^4\vartheta\frac{\partial^2}{\partial t^2}+ias\cos^3\vartheta\frac{\partial\psi}{\partial t}+\frac{\partial^2\psi}{\partial\vartheta^2}-2(m-r)\frac{\partial\psi}{\partial r}\right)\left((2m-r)r-a^2)\right)$$
$$+\left((2m+r)a^2+r^2\right)\frac{\partial\psi}{\partial t^2}-\left(ias\frac{\partial\psi}{\partial t}-\sin\vartheta\frac{\partial\psi}{\partial\vartheta}+is\frac{\partial\psi}{\partial\varphi}\right)\left((2m-r)r-a^2\right)\cos\vartheta\Bigg)=0. \tag{16}$$

It can be shown that the variables in the above equation are separable. Let us look for solutions in the form

$$\psi(t,r,\vartheta,\varphi)=e^{-i\omega t}R(r)T(\vartheta)e^{im\varphi}. \tag{17}$$

The Euler–Lagrange equations can be normalized so, that the coefficient of $\frac{\partial^2\psi}{\partial\vartheta^2}$ will be 1, similarly to the original Teukolsky equation (1).

After substituting the ansatz (17) into the above equation, it can be seen, that none of the terms contains mixed derivatives, and the angle dependence can be factorized in the coefficients of the radial derivatives, and similarly the r-dependence can be factorized in the coefficients of the ϑ-derivatives.

The above properties of the equation make the separation of the variables possible. Straightforward but tedious calculation yields for the radial equation

$$\begin{aligned}&(a^4-4a^2Mr+2a^2r^2+4M^2r^2-4Mr^3+r^4)R''(r)\\&\quad+2(-a^2M+a^2r+2M^2r-3Mr^2+r^3))R'(r)=\lambda_rR(r),\end{aligned} \tag{18}$$

where

$$\lambda_r=a^4\omega^2-4a^2M\omega^2r-a^2m^2-a^2s^2-4aMm\omega r+2Mrs^2-\omega^2r^4-r^2s^2-k,$$

in which k is the separation constant.

Similarly the angular equation is

$$(\cos^2\vartheta-1)T''(\vartheta)-\cos\vartheta\sin\vartheta T'(\vartheta)=\lambda_\vartheta T(\vartheta), \tag{19}$$

where

$$\lambda_\vartheta=((\cos\vartheta a\omega-s)\cos^2\vartheta a\omega+(a\omega+m)s)\cos\vartheta+m^2+s^2-a^2\omega^2+k.$$

Introducing $x=\cos\vartheta$ as a new variable in the angular equation we get

$$(-x^4+2x^2-1)T''(x)+2x(1-x^2)T'(x)=\lambda_\vartheta T(t)\,. \tag{20}$$

Both equations are boundary value problems for second order linear differential operators, with regular boundary conditions at the horizon and at $r = \infty$. Unfortunately we have not succeeded to express the solutions of equations (18) and (20) using elementary functions. Let ψ_k denote the kth eigenfunction of (16), then the general solution can be written as

$$\psi = \sum_k a_k \psi_k \,. \tag{21}$$

Now the canonical quantization of the unperturbed problem is completely straightforward. The modes (21) of equation (16) can be interpreted as gravitons bound to the black hole. Defining the corresponding a_k and $a_k^\dagger$ absorption and emission operators for each mode, the Hamiltonian can be expressed as

$$\mathscr{H} = \sum_k E_k a_k^\dagger a_k \,. \tag{22}$$

The E_k energy levels can be determined numerically, as well as the eigenmodes.

Taking into account the complex part of the Hamiltonian (12) of the full Teukolsky equation, the "gravitons" of (22) become unstable. This may not come as a complete surprise, as we know that the Kerr black hole admits only quasi normal modes.

The decay constants for these excitations can be calculated using the matrix elements of the perturbation part of the Lagrangian, similarly to the way decaying states are described with an imaginary energy contribution in nuclear physics.

APPENDIX

In this paper, we quantize the perturbations of a Kerr black hole. We use the Boyer–Lindquist coordinate system: the coordinates are the time, t, the radius, r, and the usual spherical angles, ϑ and φ. Using these coordinates the line element can be expressed as

$$\begin{aligned} ds^2 = \left(1 - \frac{2Mr}{\Sigma}\right) dt^2 + \frac{4Mar\sin^2\vartheta}{\Sigma} dt\,d\varphi - \frac{\Sigma}{\Delta} dr^2 - \Sigma d\vartheta^2 \\ - \sin^2\vartheta \frac{r^2 + a^2 + 2Ma^2 r \sin^2\vartheta}{\Sigma} d\varphi^2, \end{aligned} \tag{23}$$

where we used

$$\Sigma = r^2 + a^2\cos^2\vartheta$$

and

$$\Delta = r^2 - 2Mr + a^2$$

for simplicity. Here M denotes the mass of the black hole, and

$$a = \frac{J}{M}$$

is the fraction of its angular momentum (J) and mass. It's an important property of these coordinates when deriving the Teukolsky equation, that these handle the two double principal directions of the space-time symmetrically.

To derive the Teukolsky equation, the Newman–Penrose null tetrad method is utilized. When using Boyer–Lindquist coordinates, a usual choice for the null tetrad is the Kinnersley tetrad:

$$
\begin{aligned}
(l^i) &= (\frac{r^2+a^2}{\Delta}, 1, 0, \frac{a}{\Delta}), \\
(n^i) &= \frac{1}{2\Sigma}(r^2+a^2, -\Delta, 0, a), \\
(m^i) &= \frac{1}{\sqrt{2}(r+ia\cos\vartheta)}(ia\sin\vartheta, 0, 1, \frac{i}{\sin\vartheta}).
\end{aligned}
\tag{24}
$$

Using these, the non-vanishing spin coefficients are

$$
\begin{aligned}
\rho &= -\frac{1}{r-ia\cos\vartheta}, \\
\beta &= \bar{\rho}\frac{\mathrm{ctg}\vartheta}{2\sqrt{2}}, \\
\pi &= ia\rho^2\frac{\sin\vartheta}{\sqrt{2}}, \\
\tau &= -ia\bar{\rho}\rho\frac{\sin\vartheta}{\sqrt{2}}, \\
\mu &= \rho^2\bar{\rho}\frac{\Delta}{2}, \\
\gamma &= \mu+\bar{\rho}\rho\frac{r-2M}{2}, \\
\alpha &= \pi-\bar{\beta}.
\end{aligned}
\tag{25}
$$

Here we recall the basic properties of the Teukolsky equation [6]. This equation is obtained using the Newman–Penrose tetrad method. This calculation can in fact be done for any space-time in the D Petrov class (ie. two pairs of degenerate principal directions).

The perturbations of a Kerr space-time are described by the Teukolsky equation

$$
\begin{aligned}
&\left[\frac{(r^2+a^2)^2}{\Delta}-a^2\sin^2\vartheta\right]\frac{\partial^2\psi}{\partial t^2}+\frac{4Mar}{\Delta}\frac{\partial^2\psi}{\partial t\partial\varphi}+\left[\frac{a^2}{\Delta}-\frac{1}{\sin^2\vartheta}\right]\frac{\partial^2\psi}{\partial\varphi^2} \\
&-\Delta^{-s}\frac{\partial}{\partial r}\left(\Delta^{s+1}\frac{\partial\psi}{\partial r}\right)-\frac{1}{\sin\vartheta}\frac{\partial}{\partial\vartheta}\left(\sin\vartheta\frac{\partial\psi}{\partial\vartheta}\right)-2s\left[\frac{a(r-M)}{\Delta}+\frac{i\cos\vartheta}{\sin^2\vartheta}\right]\frac{\partial\psi}{\partial\varphi} \\
&-2s\left[\frac{M(r^2-a^2)}{\Delta}-r-ia\cos\vartheta\right]\frac{\partial\psi}{\partial t}+(s^2\mathrm{ctg}^2\vartheta-s)\psi=4\pi\Sigma T
\end{aligned}
\tag{26}
$$

then the calculation of T from the energy-momentum tensor of other fields is not included here, but it can be found in Teukolsky's original paper [6].

In this equation, the variables can be separated, using an ansatz

$$
\psi = e^{-i\omega t}e^{im\varphi}S(\theta)R(r)
$$

for the dependent variable. The radial equation is

$$\Delta^{-s}\frac{d}{dr}\left(\Delta^{s+1}\frac{dR}{r}\right)+\left(\frac{K^2-2is(r-M)K}{\Delta}+4is\omega r-\lambda\right)R=0, \tag{27}$$

and the angular equation is

$$\frac{1}{\sin\vartheta}\frac{d}{d\vartheta}\left(\sin\vartheta\frac{dS}{d\vartheta}\right) + \left(a^2\omega^2\cos^2\vartheta-\frac{m^2}{\sin^2\vartheta}-2a\omega s\cos\vartheta-\frac{2ms\cos\vartheta}{\sin^2\vartheta}-s^2\mathrm{ctg}^2\vartheta+s+A\right)S=0, \tag{28}$$

then, where

$$K=(r^2+a^2)\omega-am$$

and

$$\lambda=A+a^2\omega^2-2am\omega\,.$$

The boundary conditions demand ψ being regular at $\vartheta=0,\pi$. The solutions are examined in a paper by Mano, Suzuki and Takasugi [11].

According to Bini, Cherubini, Jantzen and Ruffini [10], the above equation is the direct generalization of the Klein–Gordon equation for arbitrary s spin weight. In their paper, they have brought the equation into a form closely resembling the Klein–Gordon equation

$$\left[(\nabla^i+s\Gamma^i)(\nabla_i+s\Gamma_i)-4s^2\Psi_2^A\right]\psi=4\pi T, \tag{29}$$

where Γ is the following four-vector:

$$\begin{aligned}
\Gamma^t &= -\frac{1}{\Sigma}\left[\frac{M(r^2-a^2)}{\Delta}-(r+ia\cos\vartheta)\right],\\
\Gamma^r &= -\frac{1}{\Sigma}(r-M),\\
\Gamma^\vartheta &= 0,\\
\Gamma^\varphi &= -\frac{1}{\Sigma}\left[\frac{a(r-M)}{\Delta}+i\frac{\cos\vartheta}{\sin^2\vartheta}\right].
\end{aligned} \tag{30}$$

ACKNOWLEDGMENTS

Á. L. would like to thank his supervisor, Zoltán Perjés for all what he has learnt from him, and for suggesting this problem. The tragic loss of Zoltán is a great blow for all Hungarian physics.

Á. L. also would like to thank György Pócsik and Péter Gnädig for discussions, and Péter Forgács for his help in preparing this manuscript.

This work has been supported by OTKA grant no. TS044665.

REFERENCES

1. S. N. Gupta, *Proc. Phys. Soc. A*, **65**, 161–169 (1952).
   S. N. Gupta, *Proc. Phys. Soc. A*, **65**, 608–619 (1952).
2. R. L. Arnowitt, S. Deser, and C. W. Misner, "Canonical Analysis of General Relativity," in *Recent Developments in General Relativity*, 1962.
   R. L. Arnowitt, S. Deser, and C. W. Misner, "The Dynamics of General Relativity," in *Gravitation: An Introduction to Current Research*, edited by L. Witten, 1962.
3. R. Penrose, *Phys. Reports*, **6**, 241–316 (1972).
   R. Penrose, "Twistor theory: it's aims and achievements," in *Quantum Gravity: An Oxford Symposium* , eds. C. J. Isham, R. Penrose, D. W. Sciama, Calrendon Press, Oxford, 1975.
4. A. Ashtekar, *Phys. Rev.*, **D36**, 1587–1602 (1987).
   A. Ashtekar, *Lectures on non-perturbative canonical gravity*, World Scientific, 1991.
5. C. Rovelli, *Loop Quantum Gravity*, Living Rev. Relativity, **1**, 1998.
6. S. A. Teukolsky, *ApJ*, **185**, 635–647 (1973).
   S. A. Teukolsky, and W. H. Press, *ApJ*, **185**, 649–672 (1973).
7. P. L. Chrzanowski, *Phys. Rev.*, **D11**, 2042–2062 (1975).
   P. L. Chrzanowski, and C. W. Misner, *Phys. Rev.*, **D10**, 1701–1721 (1974).
8. A. Ori, *Phys. Rev.*, **D67**, 124010 (2003).
9. S. Anco, Conservation Laws of the Teukolsky equation for massless spin *s* fields in Kerr spacetime (2003).
10. D. Bini, C. Cherubini, R. Jantzen, and R. Ruffini, *Prog.Theor.Phys.*, **107**, 967–992 (2002).
11. S. Mano, H. Suzuki, and E. Takasugi, *Prog. Theor. Phys.*, **95**, 1079–1096 (1996).

Quaternionic and Octonionic Spinors

F. Toppan

CCP/CBPF, Rua Dr. Xavier Sigaud 150, cep 22290-180 Rio de Janeiro (RJ), Brazil

Abstract. Quaternionic and octonionic spinors are introduced and their fundamental properties (such as the space-times supporting them) are reviewed. The conditions for the existence of their associated Dirac equations are analyzed. Quaternionic and octonionic supersymmetric algebras defined in terms of such spinors are constructed. Specializing to the $D = 11$-dimensional case, the relation of both the quaternionic and the octonionic supersymmetries with the ordinary M-algebra are discussed.

INTRODUCTION

The division algebras are responsible for many important mathematical structures of interest for physicists (as an example one can cite the Hopf fibrations).
In this talk we review at first the well-known, see e.g. [1], [2] connection between division algebras and Clifford algebras, explaining also in which sense we can extend the (associative) notion of Clifford algebra in order to accommodate an alternative (i.e. non-associative) structure as the one given by the octonions. This motivates us to introduce quaternionic and octonionic spinors following [3] for later studying their free dynamics (namely, their associated Dirac-type of equations) for any space-time supporting quaternionic or octonionic spinors.
The potentially most interesting physical applications of the formalism here discussed concern supersymmetry. There are several reasons for that. Division algebras, including quaternions and octonions, naturally enter the classification of supersymmetry, [4]. Octonions (for a review on them one can consult [5]), which on mathematical side are the most interesting structure (taking into account that they are the maximal division algebra and are associated with the existence of the exceptional Lie algebras, [6]), so far have not found any concrete application in physics, despite many attempts to introduce them in several different contexts (e.g. in order to explain the strong interactions and the confinement of quarks, [7]). On the other hand, see [8] and references therein, one very appealing possibility could be found at the very heart of the unification program of the interactions which goes under the name of M-Theory. Indeed in the past [9] octonionic-valued superstrings have been described. More recently, [10] and [11], it was shown that an octonionic-valued version of the 11-dimensional M-algebra can be constructed and admits very peculiar properties. This opens the way to a possible octonionic formulation of the M-theory, which could correspond to some suggestions put forward in [8] and [12].
For what concerns the quaternionic supersymmetry, its mathematical aspects have been clarified and classified (see [13]). It must be said that no concrete physical implemen-

CP767, *Fundamental Interactions and Twistor-Like Methods,* edited by J. Lukierski and D. Sorokin

tation has yet been investigated. The most closely related relevant application so far concerns the analytic continuation of the M-algebra to the Euclidean [14]. It is based on complex spinors living on a quaternionic spacetime.

ON CLIFFORD ALGEBRAS AND DIVISION ALGEBRAS

The basic relation defining a Clifford algebra is given by

$$\Gamma^{\mu}\Gamma^{\nu}+\Gamma^{\mu}\Gamma^{\nu} = 2\eta^{\mu\nu}, \tag{1}$$

with $\eta^{\mu\nu}$ being a diagonal matrix of (p,q) signature (i.e. p positive, $+1$, and q negative, -1, diagonal entries).
On the other hand the four division algebra of real (**R**) and complex (**C**) numbers, quaternions (**H**) and octonions (**O**) possess respectively 0, 1, 3 and 7 imaginary elements e_i satisfying the relations

$$e_i \cdot e_j = -\delta_{ij}+C_{ijk}e_k, \tag{2}$$

(i,j,k are restricted to take the value 1 in the complex case, $1,2,3$ in the quaternionic case and $1,2,\ldots,7$ in the octonionic case; furthermore, the sum over repeated indices is understood), with C_{ijk} the totally antisymmetric division-algebra structure constants. The octonionic division algebra is the maximal, since quaternions, complex and real numbers can be obtained as its restriction. Its totally antisymmetric octonionic structure constants can be expressed as

$$C_{123}=C_{147}=C_{165}=C_{246}=C_{257}=C_{354}=C_{367}=1 \tag{3}$$

(and vanishing otherwise).
The octonions are the only non-associative, however alternative (see [5]), division algebra. It is therefore clear, due to the antisymmetry of C_{ijk}, that (1) can be realized, for the $(0,3)$ and the $(0,7)$ signatures, in terms of, respectively, the imaginary quaternions and the imaginary octonions.
With an abuse of language (due to their non-associativity) when in the following we will speak about "octonionic Clifford algebra" we will always have in mind the above connection.
For our later purposes it is of particular importance the notion of division-algebra principal conjugation. Any element X in the given division algebra can be expressed through the sum

$$X = x_0+x_ie_i, \tag{4}$$

where x_0 and x_i are real, the summation over repeated indices is understood and the positive integral i are restricted up to 1, 3 and 7 in the **C**, **H** and **O** cases respectively. The principal conjugate X^* of X is defined to be

$$X^* = x_0-x_ie_i. \tag{5}$$

It allows introducing the division-algebra norm through the product X^*X. The normed-one restrictions

$$X^*X = 1 \tag{6}$$

select the three parallelizable spheres S^1, S^3 and S^7 in association with **C**, **H** and **O** respectively.
Further comments on the division algebras and their relations with Clifford algebras can be found in [3].
The connection between division algebras and Clifford algebras can be extended to other signature spacetimes as well. The two following algorithms can be used to lift d-dimensional Gamma matrices (denoted as γ_i) of a $D = p+q$ spacetime with (p,q) signature into $2d$-dimensional $D+2$ Gamma matrices (denoted as Γ_j) of a $D+2$ spacetime, produced according to either

$$\begin{aligned}\Gamma_j &\equiv \begin{pmatrix} 0 & \gamma_i \\ \gamma_i & 0 \end{pmatrix}, \begin{pmatrix} 0 & \mathbf{1}_d \\ -\mathbf{1}_d & 0 \end{pmatrix}, \begin{pmatrix} \mathbf{1}_d & 0 \\ 0 & -\mathbf{1}_d \end{pmatrix} \\ (p,q) &\mapsto (p+1,q+1)\end{aligned} \tag{7}$$

or

$$\begin{aligned}\Gamma_j &\equiv \begin{pmatrix} 0 & \gamma_i \\ -\gamma_i & 0 \end{pmatrix}, \begin{pmatrix} 0 & \mathbf{1}_d \\ \mathbf{1}_d & 0 \end{pmatrix}, \begin{pmatrix} \mathbf{1}_d & 0 \\ 0 & -\mathbf{1}_d \end{pmatrix} \\ (p,q) &\mapsto (q+2,p)\,.\end{aligned} \tag{8}$$

The relation (1) can therefore be realized, for specific spacetimes, in terms of quadratic matrices with either quaternionic or octonionic entries. The spacetimes supporting such Clifford realizations can be easily computed. In the octonionic case, up to $D = 13$, we obtain the following list of octonionic spacetimes

TABLE 1.

D=7	(0,7), (7,0)
D=8	(0,8), (8,0)
D=9	(0,9), (9,0), (1,8), (8,1)
D=10	(1,9), (9,1)
D=11	(1,10), (10,1),(2,9),(9,2)
D=12	(2,10), (10,2)
D=13	(3,10),(10,3),(2,11), (11,2)

An analogous list can be produced in the quaternionic case too.

A comment on the octonionic realization

One should be aware of the properties of the non-associative realizations of the relation (1), in terms of Gamma-matrices with octonionic-valued entries. In the octonionic case the commutators

$$\Sigma_{\mu\nu} = [\Gamma_\mu, \Gamma_\nu] \tag{9}$$

no longer correspond, as in the associative case, to the generators of the Lorentz group. They correspond instead to the generators of the coset $SO(p,q)/G_2$, being G_2 the 14-dimensional exceptional Lie algebra of automorphisms of the octonions. This point can be easily illustrated with the basic example of the Euclidean 7-dimensional case expressed by the imaginary octonions. Their commutators give rise to $7 = 21 - 14$ generators, isomorphic to the imaginary octonions. Indeed

$$[e_i, e_j] = 2C_{ijk}e_k \,. \tag{10}$$

The alternativity property satisfied by the octonions implies that the seven-dimensional commutator algebra among imaginary octonions is not a Lie algebra, the Jacobi identity being replaced by a weaker condition that endorses (10) with the structure of a Malcev algebra (see [5]).
Such an algebra admits a nice geometrical interpretation [15, 3]. Indeed, the normed 1 unitary octonions $X = x_0 + x_i e_i$ satisfying the (6) condition describe the seven-sphere S^7. The latter is a parallelizable manifold with a quasi (due to the lack of associativity) group structure.
On the seven sphere, infinitesimal homogeneous transformations which play the role of the Lorentz algebra can be introduced through

$$\delta X = a \cdot X \,, \tag{11}$$

with a an infinitesimal constant octonion. The requirement of preserving the unitary norm (6) implies the vanishing of the a_0 component, so that $a \equiv a_i e_i$. Therefore, the above commutator algebra (10), generated by the seven e_i, can be interpreted as the algebra of "quasi" Lorentz transformations acting on the seven sphere S^7. At least in this specific example we discovered a nice geometrical setting underlining the use of the octonionic realization of the Clifford relation (1) with $(0,7)$ signature. Indeed, while the associative representation (realized by 8×8 real matrices) of the seven dimensional Clifford algebra is required for describing the Euclidean 7-dimensional flat space, the non-associative realization describes the geometry of S^7.

The Weyl condition

Spinors can simply be introduced as column-vectors with entries valued in the given division algebra and carrying a representation of the Lorentz-algebra generators $\Sigma_{\mu\nu}$ introduced in (9) (Octonionic spinors, as discussed in the previous subsection, carry a representation of the G_2 coset).
A particular case arises for those space-time whose associated Gamma-matrices can be chosen to be block-antidiagonal, i.e. of the form

$$\Gamma^\mu = \begin{pmatrix} 0 & \sigma^\mu \\ \tilde{\sigma}^\mu & 0 \end{pmatrix} . \tag{12}$$

The corresponding signatures can be easily recovered from the introduced algorithmic constructions (7) and (8). We will call the Gamma-matrices satisfying (12) as "generalized Weyl matrices". The generators $\Sigma_{\mu\nu}$ of (9) in this case carry fundamental spinor-representations, realized by either upper or lower Weyl spinors, whose number of components is only half of their size.
In the Weyl case two projectors $P_\pm$ can be introduced through

$$\begin{aligned} P_\pm &= \frac{1}{2}(\mathbf{1}_{2d} \pm \overline{\Gamma}) , \\ \overline{\Gamma} &= \begin{pmatrix} \mathbf{1}_d & 0 \\ 0 & -\mathbf{1}_d \end{pmatrix} \end{aligned} \tag{13}$$

and the corresponding chiral (upper components) and antichiral (lower components) Weyl spinors, whose number of components is half of the size of the corresponding Γ-matrices, can be defined as satisfying

$$\Psi_\pm = P_\pm \Psi . \tag{14}$$

QUATERNIONIC AND OCTONIONIC DIRAC EQUATIONS

In the previous section we have introduced all the necessary ingredients (division-algebra valued Clifford algebras and the associated spinors) to define the quaternionic and octonionic versions of the Dirac equation.
In this section we introduce such equations and provide the full classification [3] of the spacetimes supporting them (and under which condition, i.e. the possible presence of massive or pseudomassive terms, Weyl spinors, etc.). The results will be presented in a series of tables.
Let us introduce at first the needed conventions. A matrix A, given by the product of all time-like Gamma matrices and generalizing the role of Γ^0 in the Minkowskian case is used to define barred spinors ($\overline{\psi} = \psi^\dagger A$). (Pseudo)-kinetic and (pseudo)-massive terms can be introduced for full (NW) and Weyl (W) spinors according to the following prescriptions (in order to avoid unnecessary repetitions, it is sufficient to list here the octonionic case, the quaternionic one being easily recovered from the given formulas).
The (pseudo)-kinetic terms are given according to

$$\begin{aligned} K_X &= \frac{1}{2}tr[(\Psi^\dagger A\Gamma^\mu X)\partial_\mu\Psi] + \frac{1}{2}tr[\Psi^\dagger(A\Gamma^\mu X\partial_\mu\Psi)] , \\ K_{//X} &= \frac{1}{2}tr[(\Psi_+^\dagger A\Gamma^\mu X)\partial_\mu\Psi_+] + \frac{1}{2}tr[\Psi_+^\dagger(A\Gamma^\mu X\partial_\mu\Psi_+)] , \\ K_{\perp X} &= \frac{1}{2}tr[(\Psi_+^\dagger A\Gamma^\mu X)\partial_\mu\Psi_-] + \frac{1}{2}tr[\Psi_+^\dagger(A\Gamma^\mu X\partial_\mu\Psi_-)] + \\ & \frac{1}{2}tr[(\Psi_-^\dagger A\Gamma^\mu X)\partial_\mu\Psi_+] + \frac{1}{2}tr[\Psi_-^\dagger(A\Gamma^\mu X\partial_\mu\Psi_+)] . \end{aligned}$$

Some remarks are in order. The first line refers to full spinors, while the suffices "$//$" and "$\perp$" are used to denote bilinear terms constructed with Weyl spinors of, respectively,

same and opposite chiralities. Please notice that, due to the non-associativity of the octonions, we need to specify the correct order in which the operations are taken (this is not necessary in the quaternionic case). The symbol "X" denotes the possibility of introducing, depending on the given space-time, external, extra-type of Gamma matrices, which could be either of time-like, or of space-like nature. More specifically, in the tables below, X will be denoted as T, S, J or F whether it will be associated with external time-like Gamma matrices (T), space-like (S) ones, the product of two of them (J) or, finally, of three of them (F). The presence of an extra-number specifies how many inequivalent choices for the introduction of such matrices can be given.
It should also be noticed that, in the octonionic case, the symbol "tr" introduced on the r.h.s. denotes the projection over the octonionic identity (while in the quaternionic case it coincides with the usual definition of the trace).
In full analogy with the (pseudo)-kinetic terms, (pseudo)-massive terms in a lagrangian action can be introduced through

$$\begin{aligned} M_X &= tr(\Psi^\dagger AX\Psi), \\ M_{//X} &= tr(\Psi_+^\dagger AX\Psi_+), \\ M_{\perp X} &= tr(\Psi_+^\dagger AX\Psi_- + \Psi_-^\dagger AX\Psi_+)\,. \end{aligned}$$

Due to the anticommuting character of the spinors and of their basic components, the non-vanishing (pseudo)kinetic and (pseudo)massive terms are only allowed in given spacetimes. In the next two subsection we report the full classification of the allowed Dirac equations in, respectively, the quaternionic and octonionic cases.

The quaternionic Dirac equations

We present here the tables of the allowed free (pseudo)-kinetic and (pseudo)-massive terms for quaternionic spinors. The columns are labeled by $t\ mod\ 4$ and the rows by $t-s\ mod\ 8$ (t,s denoting the number of time-like and space-like directions of the given space-time), while the symbols used in the entries have been explained above.
For full spinors (NW) case we have the Table 2

TABLE 2.

	0	1	2	3
0	$K_{J_j}, K_F, M_{S_j}, M_{J_j}$	K, K_F, M_{J_j}	K, K_{S_j}, M, M_F	$K_{S_j}, K_{J_j}, M, M_{S_j}, M_F$
5		K	K, M	M
6	M_S	K	K, K_S, M	K_S, M, M_S
7	K_J, M_{S_i}, M_J	K, M_J	K, K_{S_i}, M	K_{S_i}, K_J, M, M_{S_i}

For Weyl spinors (W case) we get the Table 3
Please notice that in the Tables 2 and 3 the suffix "j" denotes the existence of three

TABLE 3.

	0	1	2	3
1	$K_{//T_j}, K_{\perp J_j}$, $M_{//J_j}$	$K_{//}, K_{\perp T_j}$, $M_{//F}, M_{//J_j}, M_{\perp J_j}$	$K_{//F}, K_{\perp}$, $M_{//}, M_{\perp F}, M_{\perp T_j}$	$K_{\perp F}, K_{//J_j}$, $M_{\perp}$
2	$K_{//T_i}, K_{\perp J}$, $M_{//J}$	$K_{//}, K_{\perp T_i}$, $M_{//T_i}, M_{\perp J}$	$K_{\perp}$, $M_{//}, M_{\perp T_i}$	$K_{//J}$, $M_{\perp}$
3	$K_{//T}$	$K_{//}, K_{\perp T}$, $M_{//T}$	$K_{\perp}$, $M_{//}, M_{\perp T}$	$M_{\perp}$
4		$K_{//}$	$K_{\perp}$, $M_{//}$	$M_{\perp}$

inequivalent choices for the corresponding matrices (e.g., the three distinct space-like matrices S_j), while the suffix "i" denotes the existence of two inequivalent choices.

The octonionic Dirac equations

In full analogy with the previous case, we are able to produce the tables corresponding to the allowed (pseudo)-kinetic and (pseudo)-massive terms in octonionic Dirac equations. We get, in the (NW) case

TABLE 4.

	0	1	2	3
1		K	K, M	M
2	M_S	K	K, K_S, M	K_S, M, M_S
3	K_J, M_{S_i}, M_J	K, M_J	K, K_{S_i}, M	K_{S_i}, K_J, M, M_{S_i}
4	$K_{J_j}, K_F, M_{S_j}, M_{J_j}$	K, K_F, M_{J_j}, M_F	K, K_{S_j}, M, M_F	$K_{S_j}, K_{J_j}, M, M_{S_j}$

while in the W (Weyl) case we obtain (Table 5)

TABLE 5.

	0	1	2	3
0		$K_{//}$	$K_{\perp}$, $M_{//}$	$M_{\perp}$
5	$K_{//T_j}, K_{\perp J_j}$, $M_{//J_j}, M_{\perp F}$	$K_{//}, K_{\perp T_j}$, $M_{//T_j}, M_{\perp J_j}$	$K_{\perp}, K_{//F}$, $M_{//}, M_{\perp T_j}$	$K_{//J_j}, K_{\perp F}$, $M_{\perp}, M_{//F}$
6	$K_{//T_i}, K_{\perp J}$, $M_{//J}$,	$K_{//}, K_{\perp T_i}$, $M_{//T_i}, M_{\perp J}$	$K_{\perp}$, $M_{//}, M_{\perp T_i}$	$K_{//J}$, $M_{\perp}$
7	$K_{//T}$	$K_{//}, K_{\perp T}$, $M_{//T}$	$K_{\perp}$, $M_{//}, M_{\perp T}$	$M_{\perp}$

The introduced symbols have the same meaning as before.

QUATERNIONIC AND OCTONIONIC SUPERSYMMETRIES

It comes to no surprise that the most potentially interesting physical applications of the formalism and results previously introduced concern supersymmetry. After all, supersymmetry is nowadays (in the superstring/M-theory scenario) a necessary ingredient for our present understanding of fundamental interactions.
If octonions should play any role at all in physics, this is quite likely being in relation with the unification of all interactions realized by supersymmetry in higher-dimensional spacetimes. The mere fact that an octonionic formulation of the M-algebra is available [10] gives us some hopes that this program could one day be carried out.
In this section we will briefly review how generalized supersymmetries in higher-dimensional space-times (which are regarded as the scenarios for the unification of interactions and must be supplemented by some dimensional-reduction mechanism such as the Kaluza-Klein compactifications) can be constructed in terms of quaternionic and octonionic spinors (depending on the spacetime, either within or without a Weyl projection). The scheme of this section is therefore the following, at first the necessary ingredients to define (division-algebra valued) generalized supersymmetries are introduced. Next, the two cases of quaternionic and octonionic supersymmetries are more closely analyzed. We especially focus on results concerning the $D = 11$-dimensional spacetimes, mostly because this is the space-time dimensionality where the supposed M-theory should live.
In terms of n-component real spinors Q_a, the most general real supersymmetry algebra is represented by

$$\{Q_a, Q_b\} = \mathscr{Z}_{ab}, \tag{15}$$

where the matrix $\mathscr{Z}$ appearing in the r.h.s. is the most general $n \times n$ symmetric matrix with total number of $\frac{n(n+1)}{2}$ components. For any given space-time we can easily compute its associated decomposition of $\mathscr{Z}$ in terms of the antisymmetrized products of k-Gamma matrices, namely

$$\mathscr{Z}_{ab} = \sum_k (A\Gamma_{[\mu_1\ldots\mu_k]})_{ab} Z^{[\mu_1\ldots\mu_k]}, \tag{16}$$

where the values k entering the sum in the r.h.s. are restricted by the symmetry requirement for the $a \leftrightarrow b$ exchange and are specific for the given spacetime. The coefficients $Z^{[\mu_1\ldots\mu_k]}$ are the rank-k abelian tensorial central charges.
In the Minkowskian $(10,1)$ space-time, supporting 32-component real spinors, the bosonic r.h.s. is split into the $11+55+462 = 528$ bosonic components sectors $M_1 + M_2 + M_5$, where the k in M_k, for $k = 1,2,5$, specifies the level of the rank-k antisymmetric tensors.
When the the fundamental spinors entering the supersymmetry algebra belong to a division algebra other than the real one (this is evidently true in the quaternionic and octonionic cases), an extra possibility is available. The most general supersymmetry algebra can be expressed in terms of anticommutators among the fundamental spinors Q_a and their conjugate ${Q^*}_{\dot a}$, where the conjugation refers to the principal conjugation in the given division algebra. One should remember that the principal conjugation, restricted

to real spinors, acts as the identity, see (5). In the quaternionic and octonionic (as well as complex) cases we have

$$\{Q_a, Q_b\} = \mathscr{Z}_{ab}, \qquad \{Q^*{}_{\dot a}, Q^*{}_{\dot b}\} = \mathscr{Z}^*{}_{\dot a\dot b}, \tag{17}$$

together with

$$\{Q_a, Q^*{}_{\dot b}\} = \mathscr{W}_{a\dot b}, \tag{18}$$

where the matrix $\mathscr{Z}_{ab}$ ($\mathscr{Z}^*{}_{\dot a\dot b}$ is its conjugate and does not contain new degrees of freedom) is symmetric, while $\mathscr{W}_{a\dot b}$ is hermitian.
Two big classes of subalgebras, respecting the Lorentz-covariance, can be obtained from (17) and (18) by imposing division-algebra constraints, obtained by setting identically equal to zero either $\mathscr{Z}$ or $\mathscr{W}$, namely $\mathscr{Z}_{ab} \equiv \mathscr{Z}^*{}_{\dot a\dot b} \equiv 0$, so that the only bosonic degrees of freedom enter the hermitian matrix $\mathscr{W}_{a\dot b}$ or, conversely, $\mathscr{W}_{a\dot b} \equiv 0$, so that the only bosonic degrees of freedom enter $\mathscr{Z}_{ab}$ and its conjugate matrix $\mathscr{Z}^*{}_{\dot a\dot b}$. The first type of constraint will be referred as the one giving rise to the "hermitian" generalized supersymmetries, while the generalized supersymmetries satisfying the second constraint will be referred to as "holomorphic" generalized supersymmetries.
Several other constraints can be imposed, for instance one can consistently set, for complex spinors, the matrix Z entering (17) to be real. However, for our purposes, it is enough to concentrate on hermitian and holomorphic supersymmetries.

Quaternionic supersymmetries

Both the hermitian and holomorphic quaternionic supersymmetries can be classified with the help of tables specifying the number and type of bosonic elements (abelian tensorial central charges of rank k) entering the r.h.s. It is worth noticing that the

TABLE 6.

spacetime	bosonic sectors	bosonic components
$D=3$	M_0+M_1	$1+3=4$
$D=4$	M_1	4
$D=5$	M_1	5
$D=6$	−	−
$D=7$	−	−
$D=8$	−	−
$D=9$	M_0	1
$D=10$	M_0+M_1	$1+10=11$
$D=11$	M_0+M_1	$1+11=12$
$D=12$	M_1	12
$D=13$	M_1	13

results do not depend on the signature of the associated space-time, but only on its

dimensionality D, provided of course that the associated spacetime is actually carrying quaternionic spinors.
For what concerns the quaternionic hermitian supersymmetry we get the Table 6. The last column in the Table 6 denotes the number of bosonic components enetring the rank-k decomposition. It is worth noticing that the hermitian quaternionic supersymmetry saturates the bosonic sector.
This property is not hold by the holomorphic quaternionic supersymmetry. The reason can be traced to the fact that if we try implementing transposition on imaginary quaternions we are in conflict with their product since, e.g., $(e_1 \cdot e_2)^T = {e_2}^T \cdot {e_1}^T = -e_3 \neq {e_3}^T$. Indeed the only consistent operation respecting the composition law would correspond to setting ${e_i}^T = -e_i$, but this in fact coincides with the principal conjugation employed in the construction of quaternionic hermitian matrices and quaternionic hermitian supersymmetries. The holomorphic analog of the previous Table 6 is given by the Table 7

TABLE 7.

spacetime	bosonic sectors	bosonic components
$D=3$	M_0	1
$D=4$	M_0	1
$D=5$	M_0+M_1	$1+5=6$
$D=6$	M_1	6
$D=7$	M_1+M_2	$7+21=28$
$D=8$	M_2	28
$D=9$	M_2+M_3	$36+84=120$
$D=10$	M_3	120
$D=11$	$M_0+M_3+M_4$	$1+165+330=496$
$D=12$	M_0+M_4	$1+495=496$
$D=13$	$M_0+M_1+M_4+M_5$	$1+13+715+1287=2016$

The results in the Table 7 can be interpreted as follows: quaternionic holomorphic supersymmetry cannot admit bosonic tensorial central charges of rank $k \geq 2$. At most a single bosonic central charge (M_0), depending on the dimensionality of the space-time, can exist. In some dimensions, no consistent quaternionic holomorphic supersymmetry can be defined.

From physical point of view, so far, the most interesting application of supersymmetries of quaternionic spacetimes does not directly concern the quaternionic supersymmetry, but the complex holomorphic supersymmetry which can be realized with the quaternionic spinors entering the 11-dimensional quaternionic spacetime $(0,11)$. The bosonic components correspond to the 528 bosonic components of the real M algebra and the $11D$ complex Euclidean holomorphic supersymmetry can be regarded as the Euclideanized version of the M algebra, see [14].

Octonionic supersymmetries

Let us discuss now the peculiar features of the octonionic supersymmetries which are consequences of the non-associativity of octonions. The octonionic supersymmetries exist for the spacetimes entering the Table 1. It is worth mentioning that here we limit ourselves to consider only "hermitian" octonionic supersymmetries.
In a D-dimensional spacetime described in terms of the octonions, $D-7$ Clifford Gamma matrices are purely real, while the remaining 7 of them are given by the imaginary octonions e_i, $i = 1, 2, \ldots, 7$, multiplying a common real matrix. In describing the antisymmetric product of $k > 2$ octonionic Γ-matrices a correct prescription must be specified to take into account the non-associativity of the octonions. As a matter of fact, the correct prescription can be induced by assuming a given prescription for the antisymmetrized product of $k > 2$ imaginary octonions e_i. The correct prescription can be uniquely specified by assuming the validity of the Hodge duality and an irreducibility requirement, namely that the rank-k antisymmetric product of k imaginary octonions are either proportional to the octonionic identity or to the imaginary octonions. In full generality, this prescription corresponds at taking the following antisymmetrized product of k octonionic Gamma matrices

$$[\Gamma_1 \cdot \Gamma_2 \cdot \ldots \cdot \Gamma_k] \equiv \frac{1}{k!} \sum_{perm.} (-1)^{\varepsilon_{i_1 \ldots i_k}} (\Gamma_{i_1} \cdot \Gamma_{i_2} \cdot \ldots \cdot \Gamma_{i_k}), \tag{19}$$

where $(\Gamma_1 \cdot \Gamma_2 \ldots \cdot \Gamma_k)$ denotes the symmetric product

$$(\Gamma_1 \cdot \Gamma_2 \cdot \ldots \cdot \Gamma_k) \equiv \frac{1}{2}(.((\Gamma_1 \Gamma_2)\Gamma_3 \ldots)\Gamma_k) + \frac{1}{2}(\Gamma_1(\Gamma_2(\ldots \Gamma_k)).) . \tag{20}$$

The usefulness of this prescription is due to the fact that the product

$$A[\Gamma_1 \cdot \Gamma_2 \cdot \ldots \cdot \Gamma_k] , \tag{21}$$

(where A is the matrix, product of the time-like Gamma matrices, already introduced at the beginning of this section) has a definite (anti)-hermiticity property. The different (21) tensors, for different choices of the Gamma's, are all hermitian or antihermitian, depending only on the value of k and not of the Γ's themselves. In odd-dimensions D we get the table, whose columns are labeled by the antisymmetric tensors rank k, specifying the number of independent bosonic components in each rank-k antisymmetric product (19).

TABLE 8.

	0	1	2	3	4	5	6	7	8	9	10	11	12	13
$D=7$	$\underline{1}$	7	7	$\underline{1}$	$\underline{1}$	7	7	$\underline{1}$						
$D=9$	$\underline{1}$	$\underline{9}$	22	22	$\underline{10}$	$\underline{10}$	22	22	$\underline{9}$	$\underline{1}$				
$D=11$	1	$\underline{11}$	$\underline{41}$	75	76	$\underline{52}$	$\underline{52}$	76	75	$\underline{41}$	$\underline{11}$	1		
$D=13$	1	13	$\underline{64}$	$\underline{168}$	267	279	$\underline{232}$	$\underline{232}$	279	267	$\underline{168}$	$\underline{64}$	13	1

An analogous table can be produced in even-dimensional spacetimes as well.
In the above Table 8 the k sectors corresponding to hermitian matrices (and therefore

entering the r.h.s. of a generalized supersymmetry) are underlined.
The table 8 shows the existence of identities relating higher-rank antisymmetric octonionic tensors. Let us discuss the $D = 11$ example. The 52 independent components of an octonionic hermitian (4×4) matrix can be expressed either as a rank-5 antisymmetric tensors (simbolically denoted as "$M5$"), or as the combination of the 11 rank-1 ($M1$) and the 41 rank-2 ($M2$) tensors. The relation between $M1 + M2$ and $M5$ can be made explicit as follows. The 11 vectorial indices μ are split into 4 real indices, labeled by $a, b, c, \ldots$ and 7 octonionic indices labeled by $i, j, k, \ldots$. We get, on one side,

$$\begin{array}{cc} 4 & M1_a \\ 7 & M1_i \\ 6 & M2_{[ab]} \\ 4 \times 7 = 28 & M2_{[ai]} \\ 7 & M2_{[ij]} \equiv M2_i \end{array}$$

while, on the other side,

$$\begin{array}{cl} 7 & M5_{[abcdi]} \equiv M5_i \\ 4 \times 7 = 28 & M5_{[abcij]} \equiv M5_{[ai]} \\ 6 & M5_{[abijk]} \equiv M5_{[ab]} \\ 4 & M5_{[aijkl]} \equiv M5_a \\ 7 & M5_{[ijklm]} \equiv \widetilde{M5}_i \end{array}$$

which shows the equivalence of the two sectors, as far as the tensorial properties are concerned. Please notice that the correct total number of 52 independent components is recovered

$$52 = 2 \times 7 + 28 + 6 + 4 . \tag{22}$$

The octonionic equivalence of different antisymmetric tensors can be symbolically expressed, in odd space-time dimensions, through

TABLE 9.

$D = 7$	$M0 \equiv M3$
$D = 9$	$M0 + M1 \equiv M4$
$D = 11$	$M1 + M2 \equiv M5$
$D = 13$	$M2 + M3 \equiv M6$
$D = 15$	$M3 + M4 \equiv M0 + M7$

The octonionic M-algebra

We are now in the position to introduce the octonionic M algebra [10].
It corresponds to replace the real supersymmetry algebra in the $(10, 1)$ spacetime, given by

$$\{Q_a, Q_b\} = (A\Gamma_\mu)_{ab} P^\mu + (A\Gamma_{[\mu\nu]})_{ab} Z^{[\mu\nu]} + (A\Gamma_{[\mu_1 \ldots \mu_5]})_{ab} Z^{[\mu_1 \ldots \mu_5]} \tag{23}$$

with its two octonionic-valued variants, given by 4-component octonionic spinors Q_a (together with their conjugate spinors $Q^*{}_b$) and the 52 octonionic-valued 4×4 hermitian matrices which can be expressed, either as the $11+41$ bosonic generators entering

$$\{Q_a, Q^*{}_b\} = P^\mu (A\Gamma_\mu)_{ab} + Z^{\mu\nu}_{\mathbf{O}} (A\Gamma_{\mu\nu})_{ab} , \tag{24}$$

or as the 52 bosonic generators entering

$$\{Q_a, Q^*{}_b\} = Z^{[\mu_1 \ldots \mu_5]}_{\mathbf{O}} (A\Gamma_{\mu_1 \ldots \mu_5})_{ab} . \tag{25}$$

Associated to the above octonionic M algebra, its superconformal extension, given by the $Osp(1,8|\mathbf{O})$ superalgebra, can be constructed [11].

CONCLUSIONS

In this paper we have quickly reviewed the fundamental issues concerning the employment of quaternionic and octonionic spinors. In particular we have described a general construction which allows us to specify the spacetimes supporting quaternionic and octonionic spinors. The specific problems raised by the non-associativity of the octonions have been discussed and clarified. It was shown, in particular, that octonionic spinors naturally encode the geometry of the Euclidean seven-sphere S^7.
With our tools we were able to construct and classify all free Dirac-type equations involving quaternionic and octonionic spinors. The concepts of Weyl spinors, the presence of (pseudo)-kinetic and (pseudo)-massive terms in association with different space-times have been fully investigated and the complete list of results has been reported.
In the last part of this talk we took a further step. In view of studying the possible physical consequences of the formalism here introduced we applied the above investigation to the construction and the classification of the generalized supersymmetries supported by quaternionic and octonionic spinors. Some recent results on this subject have been reported, like the notion of division-algebra constrained (in the quaternionic case) hermitian and holomorphic suprsymmetries.
By far the most intriguing possible application of the ideas related with the octonionic spinors concern the M-theory investigations, as suggested in [10] and [11]. Indeed, it is quite remarkable that an octonionic structure can be introduced, instead of the standard real structure, in defining the (octonionic version of) the M-algebra. Peculiar identities, relating different rank-k antisymmetric tensors of the bosonic sectors, are an absolute novel feature of the octonionic formulation, finding no counterparts in the standard formulation. It is worth noticing that, in a somewhat different context, octonions have been suggested [8] to be linked to a possible exceptional formulation [12] for a single unifying theory of all interactions.
The investigations concerning the dynamics of octonionic spinors are a necessary preliminary step to unveil this challenging and fascinating present area of research.

ACKNOWLEDGMENTS

The present paper reports results based on a series of works done in collaboration with J. Lukierski and with H.L. Carrion and M. Rojas, who I am pleased to acknoweldge.

REFERENCES

1. I.R. Porteous, *Clifford Algebras and the Classical Groups*, Cambridge Un. Press, 1995.
2. S. Okubo, J. Math. Phys. **32**, 1657 (1991); *ibid.* **32**, 1669 (1991).
3. H.L. Carrion, M. Rojas F. and Toppan, JHEP **04**, 040 (2003).
4. T. Kugo and P. Townsend, Nucl. Phys. **B221**, 357 (1983).
5. J. Baez, *The Octonions*, math.RA/0105155.
6. C.A. Barton and T. Sudbery, math.RA/0203010.
7. M. Günaydin and F. Gürsey, Lett. Nuovo Cim. **6**, 401 (1973).
8. L. Boya, *Octonions and M-theory*, hep-th/0301037.
9. D.B. Fairlie and A.C. Manogue, Phys. Rev. **D34**, 1832 (1986).
10. J. Lukierski and F. Toppan, Phys. Lett. **B539**, 266 (2002).
11. J. Lukierski and F. Toppan, Phys. Lett. **B567**, 125 (2003).
12. P. Ramond, *Algebraic Dreams*, hep-th/0112261.
13. F. Toppan, JHEP **09**, 016 (2004).
14. J. Lukierski and F. Toppan, Phys. Lett. **B584**, 315 (2004).
15. J. Lukierski and P. Minnaert, Phys. Lett. **B129**, 392 (1983).

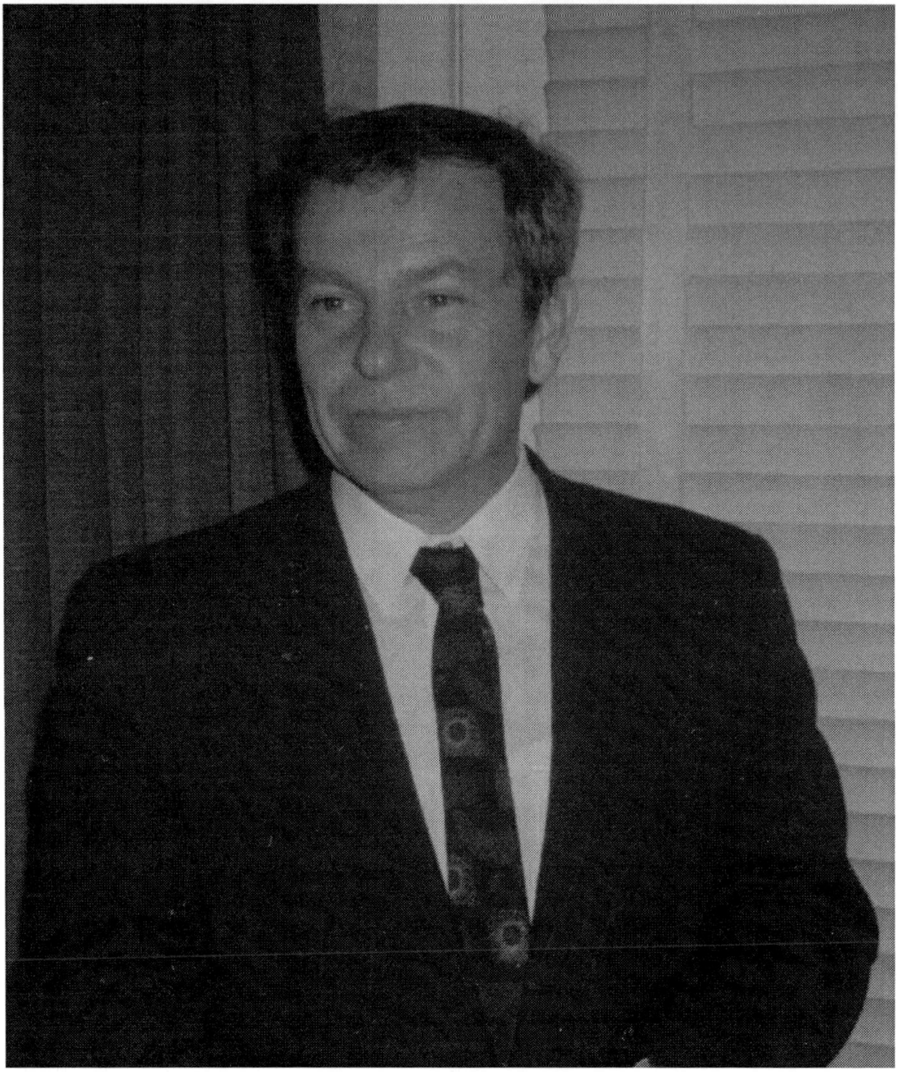

Perjés Zoltán

(1943-2004)

Obituary

Perjés Zoltán (1943-2004)

Perjés Zoltán, the internationally well known and widely appreciated physicist left us totally unexpectedly on the 27th of October 2004. He has succumbed in a hospital in Budapest to unforseen consequences of a diabetes he has been living with since long years. He has left two daughters and his wife, Ildikó.

The bulk of Zoltán's abundant scientific activities belongs mostly to the field of General Relativity where he excelled in several domains. It is simply not possible for me to even try to present here an appreciation of Zoltán's many important contributions to General Relativity in detail, suffice it to mention some of the large fields where has made very important discoveries or substantial contributions; exact solutions, twistor theory, black hole perturbations, problems of gravitational radiation. It is easy to recall some of the most important exact solutions which now bear his name: the Perjés-Israel-Wilson [7] and the Kóta-Perjés [9] solutions. But many of his works on exact solutions should be mentioned, the early ones such as [5] and [6] and the series of papers with the Pittsburg group [37,38,39,42,43,44] mostly on the problem of stationary vacuum metrics with conformally flat spatial sections. In the field of exact solutions one of Zoltán's favorite subjects has been the stationary, axisymmetric problem to which he has often returned during his career. He has always been interested to apply his skills developed to a very high degree together with a special and deep intuition about the structure of the Ernst equations. While developing his very original "parametric manifold" approach to General Relativity [63,64, 68, 69,70 75,77,80, 82,87,93] (with G. Fodor) he continued to work on exact solutions. He has given the complete solution to the problem of generating vacuum solutions of the Einstein equations from other vacuum solutions by applying the Kerr-Schild map (together with L.Á. Gergely). They have also shown a uniqueness theorem by proving that the only members of the so-called shearing class are the Kóta-Perjés and the Kasner solutions, ([73,74,81,84,85,88]).

More recently he has done some fine work on rotating perfect fluids mostly with M. Bradley, G. Fodor and L.Á. Gergely [110,111,115,116,117,118,121,129] The above list is far from being exhaustive on Zoltán's important works just in the domain of exact solutions of Einstein's equations.

Many of Zoltán's works are on twistor theory, which has been one of his early darlings and his hope to quantize gravity [15,18,20,22, 23,24, 29, 30, 33, 34,100,105], but he has also applied in a very original way twistor methods to particle classification.

Zoltán has started the study of the problem of gravitational radiation together with L.Á. Gergely and M. Vasúth. They have investigated in a post-Newtonian framework the radiation reaction on compact spinning binaries in eccentric motion. They have derived the secular contribution to the radiation losses due to the spin-orbit coupling by

CP767, *Fundamental Interactions and Twistor-Like Methods,* edited by J. Lukierski and D. Sorokin

developing a new and efficient averaging method in the perturbed two-body problem based on the residue theorem ([103,104,107,108,109,114,120] and [126]). Zoltán has equally worked on perturbations of charged black holes. In a particular choice of gauge he obtained a pair of wave equations for the electromagnetic and gravitational perturbations.

Zoltán has kept on working with a never decreasing intensity in spite of some gradually aggravating sight troubles (linked to his diabetes). The problems of cosmological perturbations in the presence of cosmological constant and the Sachs-Wolfe effect were the latest fields in which he was working on, with V.Czinner who is just about completing his PhD on this subject. Together with M. Vasúth, V. Czinner, D. Eriksson and Á. Lukács, he has given the complete analytic solution to the problem of the first order perturbations in a flat Friedman-Robertson- Walker spacetime with pressureless ideal fluid matter source in the presence of cosmological constant.

Undoubtedly Zoltán has singlehandedly founded and managed the Hungarian Relativity school, and he has an important number of pupils, many of them known researchers by now. Without trying to give a complete list one should mention some of his (ex)students who became his longer term Hungarian collaborators such as Gy. Fodor, L.Á. Gergely and M. Vasúth, all of whom are by now well established researchers in GR. He has been organizing the series of Hungarian Relativity workshops and has particular merits that the Hungarian physics community has got well recognized in the international arena.

Let me finally mention that Physics has only been a part (although the most important one) of Zoltán's activity. Zoltán has been known for the unusual ability, that his brain has always been very actively working (in contrast to more ordinary mortals, who get tired after some time not like him) and he always could produce some problem he has been thinking about. He has always been interested in the functioning of the human brain, and in various combinatorial mathematical problems and games, for example MacMahon cubes, Rubik's cube, colouring of the torus, etc.

Zoltán's loss is severe blow not only to his Family, Friends, to the Hungarian physics community, to the Institute of Nuclear and Particle Physics (RMKI, Budapest) where he has been always working, but also to the whole scientific community.

Peter Forgacs

LIST OF SCIENTIFIC PUBLICATIONS OF ZOLTÁN PERJÉS

1. B. Kardon, D. Kiss, Z. Perjés, Z. Seress and Z. Zámori: Measurement of the g-factor of excited nuclei by the method of perturbed angular correlations, KFKI Közl. **15**, 63 (1967).
2. Z. Perjés: Anwendung der Hypermatrizen für die Untersuchung eines Widerstandnetzes, Studia Sci. Math. **2**, 275 (1967).
3. Z. Perjés: Some properties of cylindrical electrovac fields, Acta Physica Sci. Hung. **25**, 393 (1968).
4. Z. Perjés: A method for constructing certain axially symmetric Einstein-Maxwell fields, Nuovo Cimento **55B**, 600 (1968).
5. Z. Perjés: 3-dimensional 'relativity' for axisymmetric stationary space-times, Commun. Math. Phys. **12**, 275 (1969).
6. Z. Perjés: Spinor treatment of stationary space-times, J. Math. Phys. **11**, 3383 (1970).

7. Z. Perjés: Solutions of the coupled Einstein-Maxwell equations representing the fields of spinning sources, Phys. Rev. Letters **27**, 1668 (1971).
8. Perjés Z.: Előadások a mai gravitáció elméletről I-II. Fizikai Szemle, **XXI/2**, 48 (1971), Fizikai Szemle, **XXI/3**, 1 (1971).
9. J. Kóta and Z. Perjés: All stationary vacuum metrics with shearing geodesic eigenrays, J. Math. Phys. **13** , 1695 (1972),
   Erratum: J. Math. Phys. **39**, 6242 (1998).
10. Z. Perjés: Applications of twistor theory in weak interactions, in *Proceedings or the Neutrino'72 Conference*, Budapest, 1972.
11. Z. Perjés: SU(l,l) spin coefficients, Acta Phys. Sci. Hung. **32**, 207 (1972).
12. Z. Perjés: Twistor kvantálás, Magyar Fizikai Folyóirat **XXI**, 407 (1973).
13. B. Lukács and Z. Perjés: Electrovac fields with geodesic eigenrays, Gen. Rel. Gravitation, **4**, 161 (1973).
14. Z. Perjés: Classification of stationary space-times, Gravity Research Foundation essay, Int. J. Theor. Physics **10**, 217 (1974).
15. Z. Perjés: Twistor variables of relativistic mechanics, Phys. Rev. **D11**, 2031 (1975).
16. Z. Perjés: Stacionárius gravitációs terek tulajdonságai, Magyar Fizikai Folyóirat **XXIV**, 173 (1976).
17. B. Lukács and Z. Perjés: Time-dependent Maxwell fields in a stationary geometry, in *Proc. of the First Marcel Grossmann Meeting*, Ed.: R. Ruffini, North-Holland, 1976, p. 281.
18. K. P. Tod and Z. Perjés: Two examples of massive scattering using twistor Hamiltonians, Gen. Rel. Gravitation, **7**, 903 (1976).
19. Z. Perjés: Introduction to spinors and Petrov types in General Relativity, Acta Phys. Sci. Hung. **41**, 173 (1976).
20. Z. Perjés: Perspectives of Penrose theory in particle physics, Reps. Math. Phys., **12**, 193 (1977).
21. Z. Perjés: An irradiated Schwarzschild object, Gen. Rel. Gravitation **8**, 689 (l977).
22. Z. Perjés and G. A. J. Sparling: The twistor structure of hadrons, KFKI-76-62 preprint, in *Advances in Twistor Theory*, Eds.: L. P. Hughston and R. S. Ward, Pitman, 1979.
23. Z. Perjés: Unitary space of particle internal states, Phys. Rev. **D20**, 1857 (1979).
24. Z. Perjés: Picturing intrinsic properties of particles, Twistor Newsletter **8**, 10 (1979).
25. B. Lukács, Z. Perjés and A. Sebestyén: Null Killing vectors, J. Math. Phys. **22**, 1249 (1981)
26. Z. Perjés: Tori of colored cubes, KFKI-1981-57 preprint.
27. Z. Perjés: Twistors and unitary space, *85th Summer Meeting of the American Mathematical Society*, 457 (1981).
28. I. Bialynicki-Birula, E. T. Newman, J. Porter, J. Winicour, B. Lukács, Z. Perjés and A. Sebestyén: A note on helicity, J. Math. Phys. **22**, 2530 (1981).
29. Z. Perjés: Twistor Theory - A particle physicist attitude, *Proc. 2nd Marcel Grossman Conference*, Ed. R. Ruffini, North-Holland, 1982.
30. B. Lukács, Z. Perjés, A. Sebestyén, E. T. Newman and J. Porter: Structure of three-twistor particles, J. Math. Phys. **23**, 2108. (1982)
31. B. Lukács and Z. Perjés: Note on conformastat vacuum space-times, Phys. Letters **A88**, 267 (1982).
32. J. Kóta, B. Lukács and Z. Perjés: Solutions of the spin coefficient equations with nongeodesic eigenrays, *Proc. of the 2nd Marcel Grossmann Meeting*, Ed. R.Ruffini, North-Holland, 1982, p. 203.
33. Z. Perjés: Internal symmetries in twistor theory, Czech. J. Phys. **B32**, 540 (1982).
34. Z. Perjés and G. A. J. Sparling: An ISU(3) hadron mass formula, Surveys in High-Energy Phys. **3** , 27 (1982).
35. Z. Perjés: Introduction to twistor particle theory, in *Twistor Geometry and Non-Linear Systems*, Eds.: H.D. Doebner and T. D. Palev, Lecture Notes in Mathematics, Vol. 970, Springer, 1982.
36. Z. Perjés: Chaotic behavior in the evolution of the Universe, in *Káosz*, Eds.: P. Szépfalusy and T. T'l, Akadémiai Kiadó, Budapest, 1982 (in Hungarian).
37. B. Lukács, Z. Perjés, A. Sebestyén and A. Valentini: Stationary vacuum fields with a conformally flat three-space II. Proof of axial symmetry, KFKI-1982-19 preprint.
38. B. Lukács, Z. Perjés, A. Sebestyén and G. A. J. Sparling: Stationary vacuum fields with a conformally flat three- space I. General theory, Gen. Rel. Grav. **15**, 511 (1983).
39. B. Lukács, Z. Perjés, A. Sebestyén and G. A. J. Sparling: Stationary vacuum fields with a conformally flat three-space III. The conformal condition, KFKI-1983-3l preprint.

40. Z. Perjés: Twistor internal symmetry groups, in *Group Theoretical Methods in Physics*, Ed.: M.A. Markov, Nauka and Harwood Acad. Publishing Co., 1983, p. 631.
41. Z. Perjés: Twistors and unitary space, in *Global Analysis - Analysis on Manifolds*, Ed.: T. Rassias, Teubner, 1983, p. 252.
42. B. Lukács, Z. Perjés, J. Porter and A. Sebestyén: Lyapunov functional approach to radiative metrics, Gen. Rel. Gravitation **16**, 691 (1984).
43. Z. Perjés, B. Lukács, A. Sebestyén, A. Valentini and G. A. J. Sparling: Solution of the stationary vacuum equations of relativity for conformally flat three-spaces, Phys. Letters **100A**, 405 (1984).
44. B. Lukács, Z. Perjés, A. Sebestyén, and G. A. J. Sparling: Conformally flat 3-spaces with axial symmetry, in *Topics in Differential Geometry , in Colloquia Mathematica Societatis János Bolyai*, **46**, 1984, p.743.
45. Z. Perjés: Twistor Theory, in *Quantum Gravity*, Eds: M. A. Markov and P. C. West, Plenum, 1984, p.631.
46. Z. Perjés: Improved characterization of the Kerr metric, KFKI-1984-115 preprint, in the volume *Quantum Gravity* 3, Ed. M.A. Markov, World Scientific Publishing Co., Singapore, 1985.
47. Z. Perjés: An almost conformal approach to axial symmetry, in *Galaxies, Axisymmetric Systems and Relativity*, Ed. M. A. H. MacCallum, Cambridge Univ. Press, 1985, p.166.
48. Z. Perjés: Stationary vacuum fields with a conformally flat three-space II. Proof of axial symmetry, Gen. Rel. Gravitation **18**, 511 (1986).
49. Z. Perjés: Stationary vacuum fields with a conformally flat three-space III. Complete solution, INS preprint (Tokyo Univ., 1984), Gen. Rel. Grav. **18**, 531 (1986).
50. Z. Perjés: The conformal potential of the stationary axisymmetric vacuum, Astron. Nachr. **5**, 321 (1986).
51. Z. Perjés: Stationary vacuum space-times in Ernst coordinates, *Proceedings of the 4th Marcel Grossmann Meeting*, Ed.: R. Ruffini, North-Holland, 1986, pp.1003.
52. G. S. Hall, T. Morgan and Z. Perjés: Three-dimensional space-times, Gen. Rel. Gravitation **19**, 1137 (1987).
53. Z. Perjés: Dynamics of Robinson-Trautman space-times, KFKI-1986-98/B preprint, in: *Non-Perturbative Methods in Quantum Field Theory*, Eds.: Z.Horváth, L.Palla and A.Patkós, World Scientific, Singapore, 1987, p. 53.
54. J. Bicak and Z. Perjés: Asymptotic behaviour of Robinson-Trautman pure radiation solutions, Class. Quantum Grav. **4**, 595 , (1987).
55. Z. Perjés: Ernst coordinates, KFKI-1986-33/B preprint Acta Physica Hung. **63**, 89 (1988).
56. Z. Perjés: Approaches to axisymmetry by man and machine, in: *Relativity Today*, Ed.: Z. Perjés, World Scientific, Singapore, 1988, p. 125.
57. Z. Perjés: Robinson-Trautman space-times, in: *Proceedings of the 5th Marcel Grossmann Meeting*, Eds. D. G. Blair, M. J. Buckingham and R. Ruffini, World Scientific, Singapore, 1989, p. 493.
58. Z. Perjés: Parametric manifolds, in: *Proceedings of the 5th Marcel Grossmann Meeting*, Eds. D. G. Blair, M. J. Buckingham and R. Ruffini, World Scientific, Singapore, 1989, p. 429.
59. Z. Perjés: Twistor string models, in: *Quantum Gravity 4*, Ed. M. A. Markov, V. Berezin and V. P. Frolov, World Scientific Publishing Co., Singapore, 1988, p. 477.
60. Z. Perjés: Unitary spinor methods in general relativity, in: *Conference on Mathematical Relativity*, Ed.: R. Bartnik, Proceedings of the CMA, Australian National University, Vol. 19, Canberra, 1988, p. 207.
61. Z. Perjés: Factor structure of the Tomimatsu-Sato metrics, J. Math. Phys. **30**, 2197 (1989).
62. G. Fodor, C. Hoenselaers and Z. Perjés: Multipole moments of axisymmetric systems in relativity, J. Math. Phys. **30**, 2252 (1989).
63. Z. Perjés: Parametric spinor approach to gravity, Univ. Cape Town preprint, (1989).
64. Z. Perjés: The parametric manifold picture of space-time, in: *Physics and Geometry*, Eds. L-L. Chau and W. Nahm, Plenum, New York, (1990).
65. Z. Perjés: Tomimatsu-Sato structures, in: *Relativity Today*, Ed.: Z. Perjés, Nova, New York, 1990.
66. C. Hoenselaers and Z. Perjés: Multipole moments of axisymmetric electrovacuum space-times, Class. Quantum Grav., **7**, 1819, (1990).
67. C. Hoenselaers and Z. Perjés: Factor structure of rational vacuum metrics, Class. Quantum Grav., **7**, 2215, (1990).

68. G. Fodor and Z. Perjés: Ashtekar variables without hypersurfaces, in: *Quantum Gravity 5*, Ed. M. A. Markov, V. Berezin and V. P. Frolov, World Scientific Publishing Co., Singapore, 1991, p. 183.
69. Z. Perjés: The parametric manifold picture of space-time, KFKI-1991-42 preprint.
70. G. Fodor and Z. Perjés: Canonical gravity in the parametric manifold picture, KFKI-1991-42 preprint.
71. R. P. Kerr, Z. Perjés and C. Hoenselaers, 13th Int. Conf. on General Relativity and Gravitation, Cordoba, Argentina, Abstracts of Contributed Papers, p. 50 (1992).
72. C. Hoenselaers and Z. Perjés: Remarks on the Robinson-Trautman solutions, Class. Quantum Grav., **10**, 375, (1993).
73. L. A. Gergely and Z. Perjés: Kerr-Schild metrics revisited I. The ground state KFKI-1993-07/B preprint.
74. L. A. Gergely and Z. Perjés: Kerr-Schild metrics revisited II. The homogeneous integrals, KFKI-1993-09/B preprint.
75. Z. Perjés: The parametric manifold picture of space-time, Nucl. Phys. **B403**, 809, (1993).
76. Z. Perjés: Parametric manifolds and canonical gravity, in:*Proceedings of the ICGC-91 Conference*, Ahmedabad , Ed. C. Vishveshwara, Wiley Eastern Ltd., New Delhi, 1993, pp. 92-100.
77. Z. Perjés: Classical Aspects of Gravitation - Report on the workshop, in:*Proceedings of the ICGC-91 Conference*, Ahmedabad , Ed. C. Vishveshwara, Wiley Eastern Ltd., New Delhi, 1993, pp. 101-106.
78. Z. Perjés: Einstein-Maxwell fields with no vacuum counterpart, Class. Quantum Grav, **10**, 1649-1652 (1993).
79. Z. Perjés: Post-Everett kvantummechanika, in:*Paradox kvantummechanika*, Szerk. L. Szabó, 1993, pp. 85-96.
80. Z. Perjés: Parametric spinor approach to gravity, Mod. Phys. Letters, **8**, 1969-1973 (1993).
81. L. A. Gergely and Z. Perjés: Solution of the vacuum Kerr-Schild problem Phys. Letters **A181**, 345 (1993).
82. G. Fodor and Z. Perjés: Canonical structure of parametric manifolds, in: *Relativity Today*, Ed.: R. Kerr and Z. Perjés, Akadémiai Kiadó (1994).
83. R. P. Kerr, Z. Perjés and C. Hoenselaers: Generalised Kerr-Schild spaces, in: *Relativity Today*, Ed.: R. Kerr and Z. Perjés, Akadémiai Kiadó (1994), pp. 31-36.
84. L. A. Gergely and Z. Perjés: Kerr-Schild metrics revisited I. The ground state J. Math. Phys. **35**, 2438 (1994).
85. L. A. Gergely and Z. Perjés: Kerr-Schild metrics revisited II. The complete vacuum solution, J. Math. Phys. **35**, 2448 (1994).
86. R. Arianrhod, A. W. C. Lun, C. B. G. McIntosh and Z. Perjés: Magnetic Curvatures Class. Quantum Grav. **11**, 2331 (1994).
87. G. Fodor and Z. Perjés: Canonical gravity in the parametric manifold picture, Gen. Rel. Gravitation, **26** 759-779. (1994)
88. L. A. Gergely and Z. Perjés: Vacuum Kerr-Schild metrics generated by nontwisting congruences, Ann. der Phys. **3**, 609-619 (1994).
89. Z. Perjés: Einstein-Maxwell fields with no vacuum counterpart I, Int. Conf. on General Relativity and Gravitation, Florence, Italy, Abstracts of Contributed Papers, p. A59 (1995).
90. M. Melvin and Z. Perjés: Einstein-Maxwell fields with no vacuum counterpart II, Int. Conf. on General Relativity and Gravitation, Florence, Italy, Abstracts of Contributed Papers, p. A59 (1995).
91. A. Komárik and Z. Perjés: Perturbations of a universe filled with dust and radiation, Int. J. Theor. Phys. **34**, 2275 (1995).
92. J. B. Hartle and Z. Perjés: Solutions of the Regge equations on some triangulations of CP2 UCSBTH-95-28 preprint (1995).
93. G. Fodor and Z. Perjés: The parametric-manifold approach to canonical gravity, in New Developments in Differential Geometry, Eds. J. Szenthe and L. Tamássy, Kluwer, l996.
94. Z. Perjés: A guide to basic exact solutions, in General Relativity, Eds. G. S. Hall and J. Pulham, Scottish Universities Summer School in Physics 34, pp. 83-106, IOP, Bristol, 1996.
95. M. Melvin and Z. Perjés: Einstein-Maxwell fields with no vacuum counterpart II, UCSBTH-96-18 preprint (1996).
96. Z. Perjés and D. Kramer: An electrovacuum space-time satisfying the oval equation of general relativity Class. Quantum Grav, **13**, 3241-3244 (1996).
97. L. Gergely and Z. Perjés: The evolution of radiating debris trapped by a black hole, in Proceedings

of the International Conference on Gravitational Waves, Eds. I. Ciufolini and F. Fidecaro, World Scientific (1997), pp. 41-43.
98. J. B. Hartle and Z. Perjés: Solutions of the Regge equations on some triangulations of CP2 J. Math. Phys. **38**, 2577-2586 (1997).
99. M. Marklund and Z. Perjés: Stationary rotating matter in general relativity J. Math. Phys. **38**, 5280-5292 (1997).
100. Z. Perjés: Four-twistor particles, Twistor Newsletter **43**, 7 (1997).
101. J. B. Hartle and Z. Perjés: Triangulations of CP2, in: *Relativity Today*, Ed.: C. Hoenselaers and Z. Perjés, Akadémiai Kiadó (1997).
102. M. Melvin and Z. Perjés: Einstein-Maxwell fields satisfying the oval equation, in: *Relativity Today*, Ed.: C. Hoenselaers and Z. Perjés, Akadémiai Kiadó (1997).
103. L. Gergely, Z. Perjés and M. Vasúth: Spin effects in gravitational radiation backreaction I. The Lense-Thirring approximation Phys. Rev. **D57**, 876-884 (1998).
104. L. Gergely, Z. Perjés and M. Vasúth: Spin effects in gravitational radiation backreaction II. Finite mass effects Phys. Rev. **D57**, 3423-3432 (1998).
105. Z. Perjés and G. Sparling: The abstract twistor space of the Schwarzschild space-time ESI-520 preprint (1998), Twistor Newsletter **44**, 12 (1998), Nucl. Phys. (Proc. Suppl) **B80**, (2000), Texas Symposium on Relativistic Astrophysics, Eds. E. Aubourg, T. Montmerle, J. Paul and P. Peter, CD ROM 12/21.
106. Z. Perjés, G. Fodor, L. Gergely and M. Marklund: A rotating incompressible perfect fluid space-time gr-qc/9806095.
107. L. Gergely, Z. Perjés and M. Vasúth: Spin effects in gravitational radiation backreaction III. Compact binaries with two spinning components gr-qc/9808063, Phys. Rev. **D58**, 124001 (1998).
108. L. Gergely, Z. Perjés and M. Vasúth: Spin effects in radiating compact binaries gr-qc/9811055, Relativity and Gravitation in General, Proceedings of the ERES98 Conference, Salamanca, Eds. J. Martin, E. Ruiz, F. Atrio and A. Molina, World Scientific (1999), pp. 259-262.
109. L. A. Gergely, Z. Perjés and G. Fodor: Rotating incompressible perfect fluid source of the NUT metric gr-qc/9811056, Relativity and Gravitation in General, Proceedings of the ERES98 Conference, Salamanca, Eds. J. Martin, E. Ruiz, F. Atrio and A. Molina, World Scientific (1999), pp. 255-258.
110. G. Fodor, M. Marklund and Z. Perjés: Axistationary perfect fluids - a tetrad approach gr-qc/9805017, Class. Quantum Grav, **16**, 453-463 (1999).
111. M. Bradley, G. Fodor, L. Gergely, M. Marklund and Z. Perjés: Rotating perfect fluid sources of the NUT metric, gr-qc/9807058, Class. Quantum Grav, **16**, 1667-1675 (1999).
112. G. Fodor, M. Marklund and Z. Perjés: Nonholonomic approach to rotating matter in general relativity Proc. 8th Marcel Grossman Meeting, Ed. Tsvi Piran, World Scientific (1999), pp. 342-344.
113. L. Gergely, Z. Perjés and M. Vasúth: Radiation backreaction in spinning binaries Proc. 8th Marcel Grossman Meeting, Ed. Tsvi Piran, World Scientific (1999), pp. 1098-1100, gr-qc/9801002.
114. L. Gergely, Z. Perjés and M. Vasúth: The true and eccentric anomaly parametrizations of the perturbed Kepler motion gr-qc/9908015, Astrophys. J. Supplement **126**, 79-84 (2000).
115. M. Bradley, G. Fodor, M. Marklund and Z. Perjés: The Wahlquist metric cannot describe an isolated rotating body gr-qc/9910001, Class. Quantum Grav, **17**, 351-359 (2000).
116. G. Fodor and Z. Perjés: Petrov types of slowly rotating fluid balls Gen. Rel. Gravitation **32**, 2319-2332 (2000), gr-qc/9911068.
117. Z. Perjés: Rotating perfect fluid models in general relativity gr-qc/9911113, Ann. Physik, **9**, 368-377 (2000).
118. M. Bradley, G. Fodor, and Z. Perjés: A slowly rotating perfect fluid body in an ambient vacuum gr-qc/0002014, Class. Quantum Grav, **17**, 2635-2640 (2000).
119. Z. Perjés: A class of Einstein-Maxwell fields generalizing the equilibrium solutions gr-qc/0003102.
120. L. Gergely, Z. Perjés and M. Vasúth: Spin effects in gravitational radiation back reaction Nucl. Phys. (Proc. Suppl) **B80**, (2000), Texas Symposium on Relativistic Astrophysics, Eds. E. Aubourg, T. Montmerle, J. Paul and P. Peter, CD ROM 07/07.
121. L. A. Gergely, Z. Perjés, G. Fodor and M. Marklund: A rotating incompressible perfect fluid space-time as source for the NUT metric Nucl. Phys. (Proc. Suppl) **B80**, (2000), Texas Symposium on Relativistic Astrophysics, Eds. E. Aubourg, T. Montmerle, J. Paul and P. Peter, CD ROM 12/11.
122. M. Bradley, G. Fodor, M. Marklund and Z. Perjés: Some results on the integrability of Einstein's field equations for stationary perfect fluids gr-qc/0101072.

123. Z. Perjés: Perturbed Friedmann cosmologies filled with dust and radiation astro-ph/0102187.
124. M. Bradley, G. Fodor, M. Marklund and Z. Perjés: The Wahlquist metric cannot be matched to an asymptotically flat vacuum region Proc. 9th Marcel Grossman Meeting, Eds. V.G.Gurzadyan, R.T. Jantzen and R.Ruffini, World Scientific,2002, pp. 869-8709 gr-qc/0101072.
125. M. Bradley, G. Fodor, M. Marklund and Z. Perjés: Some results on the integrability of Einstein's field equations for axisymmetric perfect fluids Proc. 9th Marcel Grossman Meeting, Eds. V.G.Gurzadyan, R.T. Jantzen and R.Ruffini, World Scientific,2002, pp. 777-778 gr-qc/0101072.
126. L. Gergely, Z. Perjés and M. Vasúth : Generalized true- and eccentric anomaly parametrization for the perturbed Kepler motion Proc. 9th Marcel Grossman Meeting, Eds. V.G.Gurzadyan, R.T. Jantzen and R.Ruffini, World Scientific,2002, pp. 1751-1752.
127. Z. Perjés: A class of Einstein-Maxwell fields generalizing the equilibrium solutions Proc. 9th Marcel Grossman Meeting, Eds. V.G.Gurzadyan, R.T. Jantzen and R.Ruffini, World Scientific,2002, pp. 939-941 gr-qc/0003102.
128. Z. Perjés: The slow-rotation approximation as a tool for spotting and evading troubles with perfect fluid models Proc. 9th Marcel Grossman Meeting, Eds. V.G.Gurzadyan, R.T. Jantzen and R.Ruffini, World Scientific,2002, pp. 461-467.
129. G. Fodor, Z. Perjés and M. Bradley : Slowly rotating charged fluid balls and their matching to an exterior domain Phys.Rev. D66, 084012 (2002) gr-qc/0207099.
130. Z. Perjés: Fundamental equations for the gravitational and electromagnetic perturbations of a charged black hole gr-qc/0206088.
131. Z. Perjés: Perturbations of rotating cosmological black holes gr-qc/0211059.
132. Z. Perjés: Wave equations for the perturbations of a charged black hole Gen. Rel. Gravitation 35, 1291-1297 (2003). gr-qc/0205069
133. Z. Perjés and M. Vasúth: Charged black holes: Wave equations for the gravitational and electromagnetic perturbations ApJ., 582, 342-346 (2003) gr-qc/0211036.
134. Z. Perjés: Black hole perturbations Bolyai Memorial Volume , Ed. A. Prékopa Kluwer, 2003.
135. Z. Perjés and M. Vasúth: Principal null directions of perturbed black holes Class. Quantum Grav. **20**, 5241-5252 (2003) gr-qc/0310022.
136. C. Cherubini, D. Bini, M. Bruni and Z. Perjés: Petrov classification of perturbed space-times: the Kasner example Class.Quant.Grav. **21** (2004) 4833-4844, gr-qc/0404075.
137. C. Cherubini, D. Bini, M. Bruni and Z. Perjés: The speciality index as invariant indicator in the BKL Mixmaster dynamics gr-qc/0408040.
138. Z. Perjés, M. Vasúth, V.Czinner and D. Eriksson: C^∞ perturbations of FRW models with a cosmological constant astro-ph/0402069.
139. Viktor Czinner, Mátyás Vasúth, Árpád Lukács, Zoltán Perjés: Covariant Linear Perturbations in a Concordance Model, gr-qc/0501009.

Author Index